Operational Amplifiers and Linear Integrated Circuits:

Theory and Applications

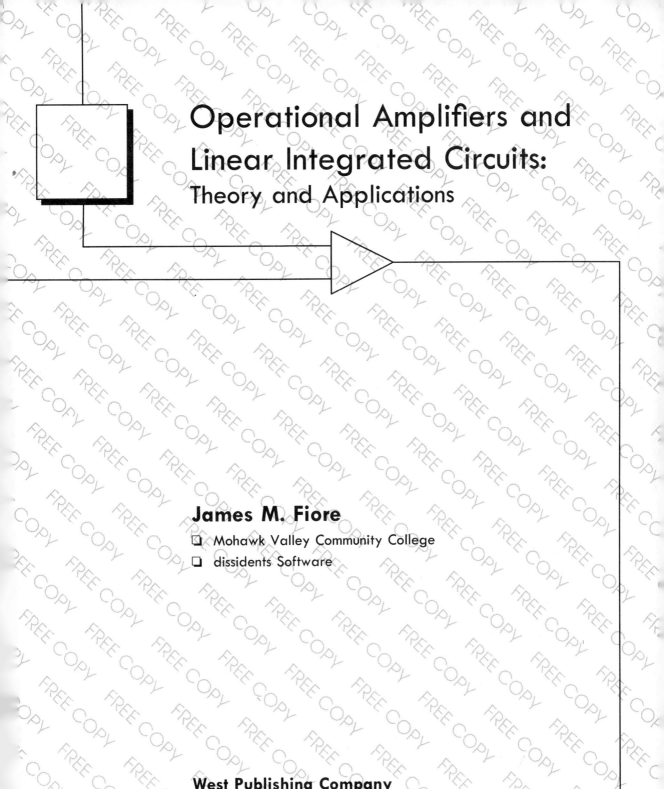

Operational Amplifiers and Linear Integrated Circuits:
Theory and Applications

James M. Fiore
❑ Mohawk Valley Community College
❑ dissidents Software

West Publishing Company

St. Paul   New York   Los Angeles   San Francisco

❏ Copyediting   Caroline E. Jumper
❏ Proofreading   Amy Mayfield and Linda J. McPhee
❏ Interior Design   Brian Betsill, TECH*arts*
❏ Interior Artwork   Kevin Tucker and Caroline E. Jumper
❏ Composition   Syntax International

Printed in the United States of America

99 98 97 96 95 94 93 92   8 7 6 5 4 3 2 1 0

Library of Congress Cataloging-in-Publication Data

Fiore, James M.
  Operational amplifiers and linear integrated circuits : theory and
applications / James M. Fiore.
    p.   cm.
  ISBN 0-314-90893-5
    1. Operational amplifiers.   2. Linear integrated circuits.
I. Title.
TK7871.58.O6F56   1992
621.39′5—dc20   91-37254
                                              ∞   CIP

Contents

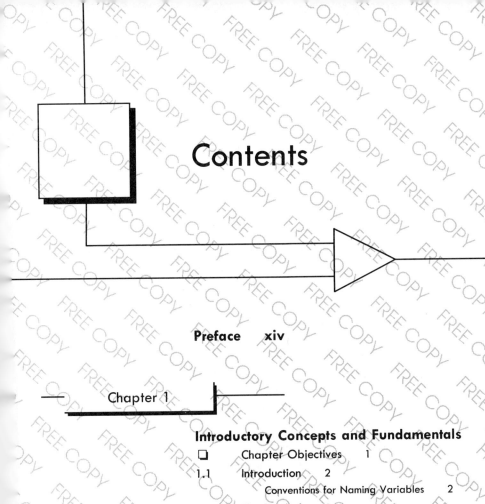

Chapter 2

Operational Amplifier Internals   61

Chapter 3

Negative Feedback   87

Chapter 4

Basic Op Amp Circuits   126

Chapter 5

Practical Limitations of Op Amp Circuits   168

Chapter 6

Specialized Op Amps   218

— Chapter 7 ⌐

Nonlinear circuits 271

— Chapter 8 ⌐

Voltage Regulation 333

Chapter 9

Oscillators and Frequency Generators   385

Chapter 10

Integrators and Differentiators   441

Chapter 11

---   Chapter 12

---   Appendixes

Appendix B: Answers to Selected Problems   B1

Index   C1

Preface

The goal of this text, as its name implies, is to encourage the reader to become proficient in the analysis and design of circuits utilizing modern linear integrated circuits. It progresses from the fundamental circuit building blocks through to analog/digital conversion systems. The text is designed to be used in a second-year operational amplifiers course at the associate level, or for a junior-level course at the baccalaureate level. To make effective use of this text, students already should have taken a course in basic discrete transistor circuits, and have a solid background in algebra and trigonometry, along with exposure to phasors. Calculus is used in certain sections of the text, but for the most part, its use is kept to a minimum. For students without a calculus background, these sections may be skipped without a loss of continuity. (The sole exception is Chapter 10, Integrators and Differentiators, which hinges upon knowledge of calculus.)

In writing this text, I have tried to make it ideal for both the teacher and the student. Instead of inundating the student with page after page of isolated formulas and collections of disjunct facts and figures, this text first builds a sound foundation. While it may take just a little bit longer to "get into" the operational amplifier than a more traditional approach, the initial outlay of time is rewarded with a deeper understanding and better retention of the later material. I tried to avoid presenting formulas out of thin air, as is often done for the sake of expediency in technical texts. Instead, I strived to provide sufficient background material and proofs so that the student is never left wondering where particular formulas came from, or worse, coming to the conclusion that they are either too difficult to understand completely or are somehow "magic."

The text can be broken into two major sections. The first section, comprised of Chapters 1 through 6, can be seen as the foundation of the operational amplifier. Therein, a methodical step-by-step presentation introduces the basic idealized operational amplifier, and eventually examines practical limitations in great detail. These chapters should be presented in order. The remaining six chapters comprise a selection of popular applications, including voltage regulation, oscillators, and active filters, to name a few. While it is not imperative that these chapters be presented in the order given (or for that matter, that they all be covered), the present arrangement will probably result in the most natural progression.

There are a few points worth noting about certain chapters. Chapter 1 presents material on decibels, Bode plots, and the differential amplifier, which is the heart of most operational amplifiers. This chapter may be skipped if your curriculum covers these topics in a discrete semiconductor or circuit analysis course. In this case, it is recommended that students scan the chapter in order to familiarize themselves with the nomenclature used later in the text. As part of the focus on fundamentals, a separate chapter on negative feedback (Chapter 3) is presented. This chapter examines the four basic negative feedback connections that can be used with an operational amplifier, and details the action of negative feedback in general. Chapter 6 presents a variety of modern special-purpose devices for specific tasks, such as high load current, programmable operation, and very high speed. An extensive treatment of active filters is given in Chapter 11. Unlike many texts that use a "memorize this" approach to this topic, Chapter 11 strives to explain the underlying operation of the circuits, yet it does so without the more advanced math requirement of the classical engineering treatment. The final chapter, Chapter 12, serves as a bridge between the analog and digital worlds, covering analog-to-digital and digital-to-analog conversion schemes.

At the beginning of each chapter is a set of chapter objectives. These point out the major items that will be discussed. Following the chapter objectives is an introduction. The introduction sets the stage for the upcoming discussion, and puts the chapter into perspective within the text as a whole. At the end of each chapter is a summary and a set of self test questions. The self test questions are of a general nature, and are designed to test for retention of circuit and system concepts. Finally, each chapter has a problem set. For most chapters, the problems are broken into four categories: analysis problems, design problems, challenge problems, and SPICE problems. An important point here was to include problems of both sufficient variety and number. An optional extended topic is found at the end of many chapters. An extended topic is designed to present extra discussion and detail about a certain portion of the chapter. This allows the chapter presentation to be customized to a certain degree. Not everyone will want to cover every extended topic. Even if they aren't used as part of the normal run of the course, these sections do allow for interesting side reading, possible outside assignments, and launching points for more involved discussions.

Integrated with the text are examples using the popular SPICE circuit simulation package. If you are not using SPICE in your courses, these items can be skipped with little loss of continuity. However, if you are using SPICE, I think that you will find that this integrated approach can be very worthwhile. SPICE should never be used in place of a traditional analysis, and with this in mind, many of the SPICE examples are used to verify the results of manual calculations, or to investigate second-order effects which may be too time consuming otherwise. SPICE is also very valuable in the classroom and laboratory as a means of posing "what if" questions. All of the SPICE input files in this text are generic, and should work with virtually any modern version of SPICE. These files have been tested using an MS-DOS® "clone" computer running MicroSim's PSpice®, and with a Commodore Amiga® personal computer running the public domain AmigaSPICE program. A portion of the output graphs were generated using MicroSim Corporation's Probe graphics post-processor.

Finally, there are two elements included in this text that I always longed for when I was a student. First, besides the traditional circuit examples used to explain circuit operation, a number of schematics of commercial products have been included to show how a given type of circuit might be used in the real world. It is one thing to read how a precision rectifier works, it is quite another thing to see how one is used in an audio power amplifier as part of an overload protection scheme.

The second element is the writing style of this text. I have tried to be direct and conversational, without being overly cute or "chatty." The body of this text is not written in a passive formal voice. It is not meant to be impersonal and cold. Instead, it is meant to sound as if someone were explaining the topics to you over your shoulder. It is intended to draw you into the topic, and to hold your attention. Over the years, I have noticed that some people feel that, in order to be taken seriously, a topic must be addressed in a detached, almost antiseptic manner. While I will agree that this mindset is crucial in order to perform a good experiment, it does not translate particularly well to textbooks, especially at the undergraduate level. The result, unfortunately, is a thorough but thoroughly unreadable book. If teachers find a text to be uninteresting, Heaven help the students. I hope that you will find this text to be complete and engaging.

Acknowledgements

In a project of this size, there are many people who have done their part to shape the result. Some have done so directly, others indirectly, and still others without knowing that they did anything at all. First, I'd like to thank my family and friends for being such "good eggs" about the project, even if it did mean that I disappeared for hours on end in order to work on it. For their encouragement and assistance, my fellow faculty members and students. For turning my ideas into reality, the folks at West Publishing and Tom Tucker. For copyediting and perspective, Caroline Jumper. For comments that helped shape this text's final form, the reviewers listed on page xviii. Finally, I'd like to thank the various IC and equipment manufacturers for providing pertinent data sheets, schematics, and photographs.

For authors and those interested in the mechanics, the original manuscript was created using the following tools: a Commodore Amiga® personal computer, with disED text editor (thanks to Jeff Glatt), Professional Page DTP™, GoldSpell™, and Amiga T_EX™, along with a few utilities I created myself. The filter graphs found in Chapter 11 were created using Tony Richardson's public domain port of PLPLOT and the SAS/Lattice 5.0 C compiler.

Finally, it is no lie to say that I couldn't have completed this project without my personal computer, a stack of Frank Zappa, Kate Bush, and King Crimson CDs, and far more peanut butter and banana sandwiches than any normal person should consume.

Jim Fiore

Reviewers

Walter Banzhaf
University of Hartford

H. M. Franzman
Red River Community College

Leonard J. Geis
DeVry Institute of Technology,
Lombard

Al Girth
Corning Community
College

Rick Hardman
State University of New York, Alfred

Ronald L. Harris
Weber State University

Gary Hecht
Portland Community College

Larry G. Keating
Metropolitan State University

Leo Kent
Southern College of Technology

Alex Kisha
DeVry Institute of Technology,
Columbus

Joe Kopec
Niagara College of Applied Arts and
Technology

Jonathan R. Lambert
Black Hawk College

Robert W. Lehman
Clackamas Community College

Vincent J. Loizzo
DeVry Institute of Technology,
Chicago

Edwin Nave
Delta College

Ellis Nuckolls
Oklahoma State University

Mark E. Oliver
Monroe Community College

Albert E. Pistilli
Hudson Valley Community College

Clay Rawlins
Eastfield College

William D. Sheehan
State University of New York, Alfred

Charles F. Solomon
Texas State Technical Institute

Charles Swain
Rochester Institute of Technology

Joseph J. Tashetta
Purdue University

Mel Turner
Oregon Institute of Technology

Neal Voke
Triton College

Neal Willison
Oklahoma State University

Ethan Wilson
State University of New York,
Morrisville

1

Introductory Concepts and Fundamentals

After completing this chapter, you should be able to:

- ❏ Convert between ordinary and decibel-based power and voltage gains.
- ❏ Utilize decibel-based voltage and power measurements during circuit analysis.
- ❏ Define and graph a general Bode plot.
- ❏ Detail the differences between lead and lag networks, and graph Bode plots for each.
- ❏ Combine the effects of several lead and lag networks in order to determine a system Bode plot.
- ❏ Describe a Nyquist plot and note its similarities to and differences from a Bode plot.
- ❏ Describe the use of digital computers in circuit simulation.
- ❏ Analyze differential amplifiers for a variety of AC characteristics, including single-ended and differential voltage gains.
- ❏ Define *common-mode gain* and *common-mode rejection*.
- ❏ Describe a current mirror, and note typical uses for it.

1.1 Introduction

Before we can begin our study of the operational amplifier, it is very important that certain background elements be in place. The purpose of this chapter is to present the very useful analysis concepts and tools associated with the decibel measurement scheme and the frequency domain. We will also be examining the differential amplifier, which serves as the heart of most operational amplifiers. With a thorough working knowledge of these items, you will find that circuit design and analysis will proceed at a much faster and more efficient pace. Consider this chapter as an investment in time, and treat it appropriately.

The decibel measurement scheme is in very wide use, particularly in the field of communications. We shall be examining its advantages over the ordinary system of measurement, and discussing how to convert values of one form into the other. One of the more important parameters of a circuit is its frequency response. To this end, we shall be looking at the general frequency domain representations of a circuit's gain and phase. This will include both manual and computer-generated analysis and graphing techniques. While our main emphasis will eventually be on application with operational amplifiers, the techniques explored can be applied equally to discrete circuits. Indeed, our initial examples will use simple discrete or black box circuits exclusively.

Finally, we will examine the DC and AC operation of the differential amplifier. This is an amplifier which utilizes two active devices and offers dual inputs. It offers certain features which make it suitable as the first section of most operational amplifiers.

Conventions for Naming Variables

One item which often confuses students of almost any subject is nomenclature. It is important, then, that we decide upon a consistent naming convention at the outset. Throughout this text, we will be examining numerous circuits containing several passive and active components. We will be interested in a variety of parameters and signals. While we will utilize the standard conventions, such as f_c for critical frequency and X_c for capacitive reactance, a great number of other possibilities exist. In order to keep confusion to a minimum, we will use the following conventions in our equations for naming devices and signals which haven't been standardized:

R	Resistor (DC, or actual circuit component)
r	Resistor (AC equivalent, where phase is 0 or ignored)
C	Capacitor
L	Inductor
Q	Transistor (bipolar or FET)

D	Diode
V	Voltage (DC)
v	Voltage (AC, where spectrum may include 0 Hz, i.e., DC)
A	Voltage or current gain
G	Power gain
I	Current (DC)
i	Current (AC, where spectrum may include 0 Hz, i.e., DC)

A resistor, capacitor, or inductor is differentiated via a subscript, which usually refers to the active device it is connected to. For example, R_E is a DC bias resistor connected to the emitter of a transistor, while r_C refers to the AC equivalent resistance seen at a transistor's collector. C_E refers to a capacitor connected to a transistor's emitter lead (most likely a bypass or coupling capacitor). Note that the device-related subscripts are always shown in upper case, with one exception: If the resistance or capacitance is part of the device model, the subscript will be shown in lower case to distinguish it from the external circuit components. For example, the AC dynamic resistance of a diode would be called r_d. If no active devices are present, or if several items exist in the circuit, a simple numbering scheme is used, such as R_1. In very complex circuits, a specific name will be given to particularly important components, as in R_{source}.

A voltage is normally given a two-letter subscript indicating the nodes at which it is measured. V_{CE} is the DC potential from the collector to the emitter of a transistor, while v_{BE} indicates the AC signal appearing across a transistor's base emitter junction. A single-letter subscript, as in V_B, indicates a potential relative to ground (in this case, base-to-ground potential). The exceptions to this rule are power supplies, which are given a double-letter subscript indicating the connection point (V_{CC} is the collector power supply), and particularly important potentials which are directly named, as in v_{in} (AC input voltage) and V_{R_1} (DC voltage appearing across R_1). Currents are named in a similar way, but generally use a single subscript referring to the measurement node (I_C is the DC collector current). All other items are directly named. By using this scheme, you will always be able to determine whether the item expressed in an equation is a DC or AC equivalent, its approximate circuit location, and other relevant information.

1.2 The Decibel

Most people are familiar with the term "decibel" in reference to sound pressure. It's not uncommon to hear someone say something like "It was 110 decibels at the concert last night, and my ears are still ringing." This popular use is somewhat inaccurate, but does show that decibels indicate some sort of quantity—in this case, sound pressure level.

Decibel Representation of Power and Voltage Gains

In its simplest form, the decibel is used to measure some sort of gain, such as power or voltage gain. Unlike the ordinary gain measurements that you may be familiar with, the decibel form is logarithmic. Because of this, it can be very useful for showing ratios of change, as well as absolute change. The base unit is the *bel*. To convert an ordinary gain to its bel counterpart, just take the common log (base 10) of the gain. In equation form:

$$\text{bel gain} = \log_{10}(\text{ordinary gain})$$

Note that on most hand calculators common log is denoted as `log`, while the natural log is given as `ln`. Unfortunately, many programming languages use `log` to indicate natural log and `log10` for common log. More than one student has been bitten by this bug, so be forewarned! As an example, if a circuit produces an output power of 200 mW for an input of 10 mW, we would normally say that its power gain (*G*) is found as follows:

$$G = \frac{p_{out}}{p_{in}}$$

$$G = \frac{200 \text{ mW}}{10 \text{ mW}}$$

$$G = 20$$

For the bel version, just take the log of this result:

$$G' = \log_{10} G$$

$$G' = \log_{10} 20$$

$$G' = 1.301$$

The bel gain is 1.3 bels. The term "bels" is not a unit in the strict sense of the word (as in "watts"), but is simply used to indicate that this is not an ordinary gain. In contrast, ordinary power and voltage gains are sometimes given units of W/W and V/V to distinguish them from bel gains. Also, note that the symbol for bel power gain is G' and not G. All bel gains are denoted with the prime (') notation to avoid confusion.

Since bels tend to be rather large, we typically use one-tenth of a bel as the norm. The resulting unit is the decibel (one-tenth bel). To convert to decibels, simply multiply the number of bels by 10. Our gain of 1.3 bels is equivalent to 13 decibels. The unit is commonly shortened to dB. Consequently, we may say

$$G' = 10 \log_{10} G \qquad \text{where the result is in dB} \qquad (1.1)$$

At this point, you may be wondering what the big advantage of the decibel system is. To answer this, recall a few log identities: Normal multiplication becomes addition in the log system, and division becomes subtraction. Likewise, powers and roots become multiplication and division. Because of this, two important facts arise. First, ratios of change become constant offsets in the decibel system, and second, the entire range of values diminishes in size. The result is that a very wide range of gains can be represented within a fairly small scope of values, and the corresponding calculations can become faster.

With the aid of your hand calculator, it is very easy to show the following:

Factor	dB Value (using $G' = 10 \log_{10} G$)
1	0 dB
2	3.01 dB
4	6.02 dB
8	9.03 dB
10	10 dB

We can also look at fractional factors (i.e., losses instead of gains):

Factor	dB Value
.5	−3.01 dB
.25	−6.02 dB
.125	−9.03 dB
.1	−10 dB

In addition to those available from your calculator, there are two values which are useful to commit to memory. If you look carefully at the tables of factors, you will notice that a doubling is represented by an increase of approximately 3 dB. A factor of 4 is, in essence, two doublings. Therefore, it is equivalent to 3 dB + 3 dB, or 6 dB. Remember, since we are using logs, multiplication turns into simple addition. In a similar manner, a halving is represented by approximately −3 dB. The negative sign indicates a reduction. To simplify things a bit, think of factors of 2 as ±3 dB, the sign indicating whether you are increasing (multiplying), or decreasing (dividing). As you can see, factors of 10 work out to a very convenient 10 dB. By remembering these two factors, you can often estimate a decibel conversion without the use of your calculator. For example, we could rework our initial conversion problem:

The amplifier has a gain of 20

20 can be written as 2 × 10

The factor of 2 is 3 dB, the factor of 10 is 10 dB

The answer must be 3 dB + 10 dB, or 13 dB

This verifies our earlier result.

Example 1.1   An amplifier has a power gain of 800. What is the decibel power gain?

$$G' = 10 \log_{10} G$$

$$G' = 10 \log_{10} 800$$

$$G' = 10 \times 2.903$$

$$G' = 29.03 \text{ dB}$$

We could also use our estimation technique:

$G = 800 = 8 \times 10^2$

8 is equivalent to 3 factors of 2, or 2 × 2 × 2, which can be expressed as 3 dB + 3 dB + 3 dB, which is, of course, 9 dB

10^2 is equivalent to 2 factors of 10, or 10 dB + 10 dB = 20 dB

The result is 9 dB + 20 dB, or 29 dB.

Note that if the leading digit is not a power of 2, the estimation will not be exact. For example, if the gain is 850, you know that the decibel gain is just a bit over 29 dB. You also know that it must be less than 30 dB ($1000 = 10^3$, which is 3 factors of 10, or 30 dB). As you can see, by using the decibel form, you tend to concentrate on the magnitude of gain, and not so much on trailing digits.

Example 1.2   An attenuator reduces signal power by a factor of 10,000. What is this loss expressed in decibels?

$$G' = 10 \log_{10} \frac{1}{10,000}$$

$$G' = 10 \times (-4)$$

$$G' = -40 \text{ dB}$$

By using the approximation, we can say,

$$\frac{1}{10,000} = 10^{-4}$$

The negative exponent tells us we have a loss (negative decibel value), and 4 factors of 10.

$$G' = -10 \text{ dB} - 10 \text{ dB} - 10 \text{ dB} - 10 \text{ dB} = -40 \text{ dB}$$

Remember, if an increase in signal is produced, the result will be a positive decibel value. A decrease in signal will always result in a negative decibel value. A signal which is unchanged indicates a gain of unity, or 0 dB.

To convert from decibels to ordinary form, just invert the steps.

$$G = \log_{10}^{-1} \frac{G'}{10}$$

On most hand calculators, base 10 antilog is denoted as 10^x. In most computer languages, you just raise 10 to the appropriate power, as in G = 10.0^(Gprime/10.0) (**BASIC**), or use an exponent function, as in pow(10.0, Gprime/10.0) (**C**).

Example 1.3 An amplifier has a power gain of 23 dB. If the input is 1 mW, what is the output?

In order to find the output power, we need to find the ordinary power gain, G.

$$G = \log_{10}^{-1} \frac{G'}{10}$$

$$G = \log_{10}^{-1} \frac{23}{10}$$

$$G = 199.5$$

Therefore, $P_{out} = 199.5 \times 1$ mW or 199.5 mW.

You can also use the approximation technique in reverse. To do this, break up the decibel gain into chunks of 10 dB and 3 dB:

$$23 \text{ dB} = 3 \text{ dB} + 10 \text{ dB} + 10 \text{ dB}$$

Now replace each chunk with the appropriate factor, and multiply them together. (Remember, when going from log to ordinary form, addition turns into multiplication.)

$$3 \text{ dB} = 2X \qquad \text{and} \qquad 10 \text{ dB} = 10X$$

so

$$G = 2 \times 10 \times 10$$
$$G = 200$$

While the approximation technique appears to be slower than the calculator, practice will show otherwise. Being able to quickly estimate decibel values can prove to be a very handy skill in the electronics field. This is particularly true in larger, multistage designs.

Example 1.4   A three-stage amplifier has gains of 10 dB, 16 dB, and 14 dB per section. What is the total decibel gain?

Since decibel gains are a log form, just add the individual stage gains to arrive at the system gain:

$$G'_{total} = G'_1 + G'_2 + G'_3$$
$$G'_{total} = 10 \text{ dB} + 16 \text{ dB} + 14 \text{ dB}$$
$$G'_{total} = 40 \text{ dB}$$

As you may have noticed, all of the examples up to this point have used power gain and not voltage gain. You may be tempted to use the same equations for voltage gain. In a word, DON'T. If you think back for a moment, you will recall that power varies as the square of voltage. In other words, a doubling of voltage will produce a quadrupling of power. If you were to use the same decibel conversions, a doubling of voltage would be 3 dB, yet, since the power has quadrupled, this would indicate a 6 dB rise. Consequently, voltage gain (and current gain as well) are treated in a slightly different fashion. We would rather have our doubling of voltage work out to 6 dB, so that it matches the power calculation. The correction factor is very simple: Since power varies as the square of voltage, the decibel form should be twice as large for voltage. (Remember, exponentiation turns into multiplication when using logs.) Applying this factor to Equation 1.1 yields

$$A_v' = 20 \log_{10} A_v \qquad\qquad (1.2)$$

Be careful, though—the bel voltage gain equals the bel power gain only if the input and output impedances of the system are matched. (You may recall from your earlier work that it is quite possible to design a circuit with vastly different voltage and power gains. A voltage follower, for example, exhibits moderate power gain with a voltage gain of unity. It is quite likely that the follower will not exhibit matched impedances.) If we were to recalculate our earlier tables of common factors, we would find that a doubling of voltage gain is equivalent to a 6 dB rise, and a tenfold increase is equivalent to a 20 dB rise, twice the size of their power gain counterparts.

Note that current gain may be treated in the same manner as voltage gain (although this is less commonly done in practice).

Example 1.5 An amplifier has an output signal of 2 V for an input of 50 mV. What is A_v'? First find the ordinary gain:

$$A_v = \frac{2}{.05} = 40$$

Now convert to decibel form.

$$A_v' = 20 \log_{10} 40$$
$$A_v' = 20 \times 1.602$$
$$A_v' = 32.04 \text{ dB}$$

The approximation technique yields $40 = 2 \times 2 \times 10$, or $6\text{ dB} + 6\text{ dB} + 20\text{ dB} = 32\text{ dB}$.

To convert A_v' to A_v, reverse the process:

$$A_v = \log_{10}^{-1} \frac{A_v'}{20}$$

Example 1.6   An amplifier has a gain of 26 dB. If the input signal is 10 mV, what is the output?

$$A_v = \log_{10}^{-1} \frac{A_v'}{20}$$

$$A_v = \log_{10}^{-1} \frac{26}{20}$$

$$A_v = 19.95$$

$$v_{out} = A_v v_{in}$$

$$v_{out} = 19.95 \times 10\ \text{mV}$$

$$v_{out} = .1995\ \text{V}$$

The final point to note in this section is that, as in the case of power gain, a negative decibel value indicates a loss. Therefore, a 2:1 voltage divider would have a gain of −6 dB.

Signal Representation in dBW and dBV

As you can see from the preceding section, it is possible to spend considerable time converting between decibel gains and ordinary voltages and powers. Since the decibel form does offer advantages for gain measurement, it would make sense to use a decibel form for power and voltage levels as well. This is a relatively straightforward process. There is no reason why we can't express a power or voltage in a logarithmic form. Since a decibel value simply indicates a ratio, all we need to do is decide on a reference (i.e., a comparative base for the ratio). For power measurements, a likely choice would be 1 watt. In other words, we can describe a power as being X decibels above or below 1 watt. Positive values will indicate powers greater than 1 watt, while negative values will indicate powers less than 1 watt. In general equation form,

$$p' = 10 \log_{10} \frac{p}{\text{reference}} \tag{1.3}$$

The answer will have units of dBW, that is, decibels relative to 1 watt.

Example 1.7 A power amplifier has a maximum output of 120 W. What is this power in dBW?

$$p' = 10 \log_{10} \frac{p}{1 \text{ W}}$$

$$p' = 10 \log_{10} \frac{120 \text{ W}}{1 \text{ W}}$$

$$p' = 20.8 \text{ dBW}$$

There is nothing sacred about the 1-watt reference, short of its convenience. We could just as easily choose a different reference. Other common reference points are 1 milliwatt (for units of dBm) and 1 femtowatt (for units of dBf). Obviously, dBf is used for very low signal levels, such as those coming from an antenna. dBm is in very wide use in the communications industry. To use these other references, just divide the given power by the new reference.

Example 1.8 A small personal audio tape player delivers 200 mW to its headphones. What is this output power in dBW? In dBm?

For an answer in units of dBW, use the 1-watt reference:

$$p' = 10 \log_{10} \frac{p}{1 \text{ W}}$$

$$p' = 10 \log_{10} \frac{200 \text{ mW}}{1 \text{ W}}$$

$$p' = -7 \text{ dBW}$$

For units of dBm, use a 1-milliwatt reference:

$$p' = 10 \log_{10} \frac{p}{1 \text{ mW}}$$

$$p' = 10 \log_{10} \frac{200 \text{ mW}}{1 \text{ mW}}$$

$$p' = 23 \text{ dBm}$$

200 mW, -7 dBW, and 23 dBm are three ways of saying the same thing. Note that the dBW and dBm values are 30 dB apart. This will always be true, since the references are a factor of 1000 (30 dB) apart.

In order to transfer a value in dBW (or similar units) into watts, reverse the process:

$$p = \log_{10}^{-1} \frac{p'}{10} \times \text{reference}$$

Example 1.9 A studio microphone produces a 12 dBm signal while recording normal speech. What is the output power in watts?

$$p = \log_{10}^{-1} \frac{p'}{10} \times \text{reference}$$

$$p = \log_{10}^{-1} \frac{12 \text{ dBm}}{10} \times 1 \text{ mW}$$

$$p = 15.8 \text{ mW} \quad \text{or} \quad .0158 \text{ W}$$

For voltages, we can use a similar system. A logical reference is 1 volt, with the resulting units being dBV. As before, these voltage measurements will require a multiplier of 20 instead of 10:

$$v' = 20 \log_{10} \frac{v}{\text{reference}} \tag{1.4}$$

Example 1.10 A test oscillator produces a 2 volt signal. What is this value in dBV?

$$v' = 20 \log_{10} \frac{v}{\text{reference}}$$

$$v' = 20 \log_{10} \frac{2 \text{ V}}{1 \text{ V}}$$

$$v' = 6.02 \text{ dBV}$$

When both circuit gains and signal levels are specified in decibel form, analysis can be very quick. Given an input level, simply add the gain to it in order to find the output level. Given input and output levels, subtract them in order to find the gain.

Example 1.11   A floppy disk read/write amplifier exhibits a gain of 35 dB. If the input signal is −42 dBV, what is the output signal?

$$v'_{out} = v'_{in} + A'_v$$

$$v'_{out} = -42 \text{ dBV} + 35 \text{ dB}$$

$$v'_{out} = -7 \text{ dBV}$$

Note that the final units are dBV and not dB, thus indicating a voltage and not merely a gain.

Example 1.12   A guitar power amp needs an input of 20 dBm to achieve an output of 25 dBW. What is the gain of the amplifier in decibels?
First, it is necessary to convert the power readings so that they share the same reference unit. Since dBm represents a reference 30 dB smaller than the dBW reference, just subtract 30 dB to compensate.

$$20 \text{ dBm} = -10 \text{ dBW}$$

$$G' = p'_{out} - p'_{in}$$

$$G' = 25 \text{ dBW} - (-10 \text{ dBW})$$

$$G' = 35 \text{ dB}$$

Note that the units are dB and not dBW. This is very important! Saying that the gain is "so many" dBW is the same as saying the gain is "so many" watts. Obviously, gains are "pure" numbers and do not carry units such as watts or volts.

The usage of a decibel-based system is shown graphically in Figure 1.1. Note how the stage gains are added to the input signal to form the output. Even large circuits can be quickly analyzed in this form. To make life in the

Figure 1.1
Multistage

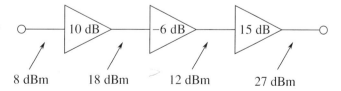

8 dBm   18 dBm   12 dBm   27 dBm

lab even easier, it is possible to take measurements directly in decibel form. By doing this, you need never convert while troubleshooting a design. For general-purpose work, voltage measurements are the norm, and therefore a dBV scale is often used.

Items of Interest in the Laboratory

When using a digital meter on a dBV scale it is possible to "underflow" the meter if the signal is too weak. This will happen if you try to measure 0 volts, for example. If you attempt to calculate the corresponding dBV value, your calculator will probably show "error." The effective value is negative infinite dBV. The meter will certainly have a hard time showing this value! Another item of interest revolves around the use of dBm measurements. It is common to use a voltmeter to make dBm measurements, in lieu of a wattmeter. While the connections are considerably simpler, a voltmeter cannot measure power. How is this accomplished then? Well, as long as the circuit impedance is known, power can be derived from a voltage measurement. A common impedance in communication systems (such as recording studios) is 600 ohms, so a meter can be calibrated to give correct dBm readings by using the Power Law. If this meter is used on a non-600 ohm circuit, the readings will no longer reflect accurate dBm values (but will still properly reflect relative changes in dB).

1.3 Bode Plots

The Bode plot is a graphical response-prediction technique which is useful for both circuit design and analysis. A Bode plot is, in actuality, a pair of plots: one graphs the gain of a system versus frequency, while the other details the circuit phase versus frequency. Both of these items are very important in the design of well-behaved, optimal operational amplifier circuits.

Generally, Bode plots are drawn with logarithmic frequency axes, a decibel gain axis, and a phase axis in degrees. First, let's take a look at the gain plot. A typical gain plot is shown in Figure 1.2.

Note how the plot is relatively flat in the middle, or midband, region. The gain value in this region is known as the midband gain. At either extreme of the midband region, the gain begins to decrease. The gain plot shows two important frequencies, f_1 and f_2. f_1 is the lower break frequency while f_2 is the upper break frequency. The gain at the break frequencies is 3 dB less than the midband gain. These frequencies are also known as the half-power points, or corner frequencies. Normally, amplifiers are only used for signals between f_1 and f_2. The exact shape of the rolloff regions will depend on the design of the

Figure 1.2
Gain plot

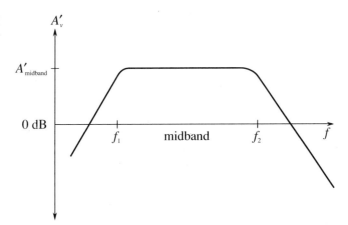

circuit. It is possible to design amplifiers with no lower break frequency (i.e., a DC amplifier); however, all amplifiers will exhibit an upper break. The break points are caused by the presence of circuit reactances, typically coupling and stray capacitances. The gain plot is a summation of the midband response with the upper and lower frequency limiting networks. Let's take a look at the lower break, f_1.

Lead Network Gain Response

Reduction in low-frequency gain is caused by lead networks. A generic lead network is shown in Figure 1.3. It gets its name from the fact that the output voltage developed across R leads the input. At very high frequencies the circuit will be essentially resistive. Conceptually, think of this as a simple voltage divider. The divider ratio depends on the reactance of C. As the input frequency drops, X_c increases. This makes v_{out} decrease. At very high frequencies, where $X_c \ll R$, v_{out} is approximately equal to v_{in}. This can be seen graphically in

Figure 1.3
Lead network

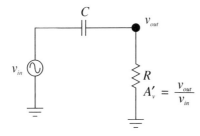

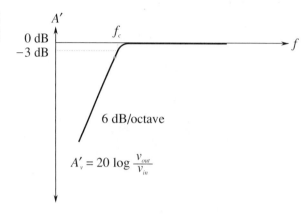

Figure 1.4. The break frequency is defined by the standard equation

$$f_c = \frac{1}{2\pi RC}$$

The response below f_c will be a straight line if a decibel gain axis and a logarithmic frequency axis are used. This makes for very quick and convenient sketching of circuit response. The slope of this line is 6 dB per octave. (An octave is a doubling or halving of frequency, e.g., 800 Hz is 3 octaves above 100 Hz.) This slope may also be expressed as 20 dB per decade, where a decade is a factor of 10 in frequency. With reasonable accuracy, this curve may be approximated as two line segments, called *asymptotes*, as shown in Figure 1.5. The shape of this curve is the same for any lead network. Because of this, it is very easy to find the approximate gain at any given frequency as long as f_c is known. It is not necessary to go through reactance and phasor calculations.

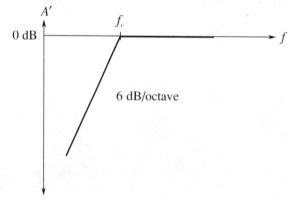

To create a general response equation, start with the voltage divider rule to find the gain:

$$A_v = \frac{v_{out}}{v_{in}} = \frac{R}{R - jX_c}$$

$$A_v = \frac{R \angle 0}{\sqrt{R^2 + X_c^2} \angle -\arctan \dfrac{X_c}{R}}$$

The magnitude of this is

$$|A_v| = \frac{R}{\sqrt{R^2 + X_c^2}}$$

$$|A_v| = \frac{1}{\sqrt{1 + \dfrac{X_c^2}{R^2}}} \qquad (1.5)$$

Recalling that

$$f_c = \frac{1}{2\pi RC}$$

we may say

$$R = \frac{1}{2\pi f_c C}$$

For any frequency of interest, f,

$$X_c = \frac{1}{2\pi f C}$$

Equating the two preceding equations yields

$$\frac{f_c}{f} = \frac{X_c}{R} \qquad (1.6)$$

Substituting Equation 1.6 in Equation 1.5 gives

$$A_v = \frac{1}{\sqrt{1 + \dfrac{f_c^2}{f^2}}} \qquad (1.7)$$

To express A_v in decibels, substitute Equation 1.7 into Equation 1.2:

$$A_v' = 20 \log_{10} \frac{1}{\sqrt{1 + \dfrac{f_c^2}{f^2}}}$$

After simplification, the final result is

$$A_v' = -10 \log_{10}\left(1 + \frac{f_c^2}{f^2}\right) \tag{1.8}$$

where

f_c is the critical frequency

f is the frequency of interest

A_v' is the decibel gain at the frequency of interest

Example 1.13   An amplifier has a lower break frequency of 40 Hz. How much gain is lost at 10 Hz?

$$A_v' = -10 \log_{10}\left(1 + \frac{f_c^2}{f^2}\right)$$

$$A_v' = -10 \log_{10}\left(1 + \frac{40^2}{10^2}\right)$$

$$A_v' = -12.3 \text{ dB}$$

In other words, the gain is 12.3 dB lower than it is in the midband. Note that 10 Hz is 2 octaves below the break frequency. Since the cutoff slope is 6 dB per octave, each octave loses 6 dB. Therefore, the approximate result is -12 dB, which double-checks the exact result. Without the lead network, the gain would stay at 0 dB all the way down to DC (0 Hz).

Lead Network Phase Response

At very low frequencies, the circuit of Figure 1.3 is largely capacitive. Because of this, the output voltage developed across R leads by 90°. At very high frequencies the circuit will be largely resistive. At this point v_{out} will be in phase with v_{in}. At the critical frequency, v_{out} will lead by 45°. A general phase graph is shown in Figure 1.6. As with the gain plot, the phase plot shape is the same

Figure 1.6
Lead phase (exact)

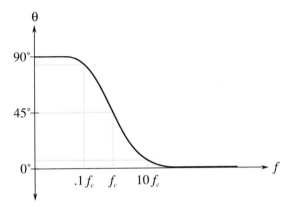

for any lead network. The general phase equation may be obtained from the voltage divider:

$$\frac{v_{out}}{v_{in}} = \frac{R}{R - jX_c}$$

$$\frac{v_{out}}{v_{in}} = \frac{R\angle 0}{\sqrt{R^2 + X_c^2}\angle - \arctan\dfrac{X_c}{R}}$$

The phase portion of this is

$$\theta = \arctan\frac{X_c}{R}$$

By using Equation 1.6, this simplifies to

$$\theta = \arctan\frac{f_c}{f} \qquad\qquad (1.9)$$

where

f_c is the critical frequency

f is the frequency of interest

θ is the phase angle at the frequency of interest

Often, an approximation such as Figure 1.7 (page 20) is used in place of the graph shown in Figure 1.6. By using Equation 1.9, you can show that the approximation is off by no more than 6° at the corners.

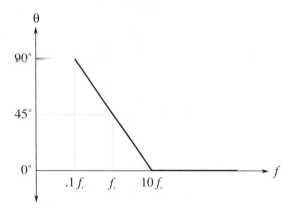

Figure 1.7

Lead phase (approximate)

Example 1.14   A telephone amplifier has a lower break frequency of 120 Hz. What is the phase response one decade below and one decade above?

One decade below 120 Hz is 12 Hz, while one decade above is 1.2 kHz.

$$\theta = \arctan \frac{f_c}{f}$$

$$\theta = \arctan \frac{120\ \text{Hz}}{12\ \text{Hz}}$$

$\theta = 84.3°$ one decade below f_c (i.e., approaching 90°)

$$\theta = \arctan \frac{120\ \text{Hz}}{1.2\ \text{kHz}}$$

$\theta = 5.71°$ one decade above f_c (i.e., approaching 0°)

Remember, if an amplifier is direct-coupled, and has no lead networks, the phase will remain at 0° right back to 0 Hz (DC).

Lag Network Response

Unlike its lead network counterpart, all amplifiers contain lag networks. In essence, a lag network is little more than an inverted lead network. As you can see from Figure 1.8, it simply transposes the R and C locations. Because of this, the response tends to be inverted as well. In terms of gain, X_c is very large at low frequencies, and thus v_{out} equals v_{in}. At high frequencies, X_c decreases, and v_{out} falls. The break point occurs when X_c equals R. The general gain plot is shown in Figure 1.9. Like the lead network response, the slope of

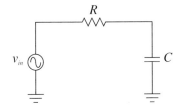

Figure 1.8
Lag network

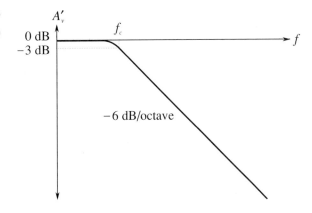

Figure 1.9
Lag gain (exact)

this curve is −6 dB per octave (or −20 dB per decade). Note that the slope is negative instead of positive. A straight line approximation is shown in Figure 1.10 (page 22). We can derive a general gain equation for this circuit in virtually the same manner as we did for the lead network. The derivation is left as an exercise.

$$A'_v = -10 \log_{10}\left(1 + \frac{f^2}{f_c^2}\right) \tag{1.10}$$

where

f_c is the critical frequency

f is the frequency of interest

A'_v is the decibel gain at the frequency of interest

Note that this equation is almost the same as Equation 1.8. The only difference is that f and f_c have been transposed.

In a similar vein, we can examine the phase response. At very low frequencies, the circuit is basically capacitive. Since the output is taken across C, v_{out} will be in phase with v_{in}. At very high frequencies, the circuit is essentially resistive. Consequently, the output voltage across C will lag by 90°. At the break frequency the phase will be −45°. A general phase plot is shown in Figure 1.11, with the approximate response detailed in Figure 1.12. As with the lead network,

Figure 1.10
Lag gain (approximate)

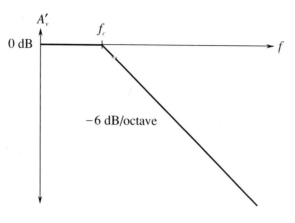

Figure 1.11
Lag phase (exact)

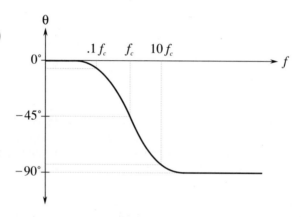

Figure 1.12
Lag phase (approximate)

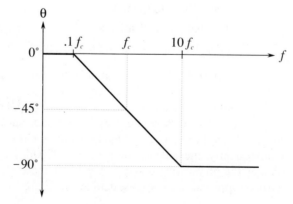

we can derive a phase equation. Again, the exact steps are very similar, and left as an exercise.

$$\theta = -90° + \arctan\frac{f_c}{f}$$ (1.11)

where

    f_c is the critical frequency
    f is the frequency of interest
    θ is the phase angle at the frequency of interest

Example 1.15   A medical ultrasound transducer feeds a lag network with an upper break frequency of 150 kHz. What are the gain and phase values at 1.6 MHz?

    Since this represents a little more than a 1-decade increase, the approximate values are -20 dB and $-90°$, from Figures 1.10 and 1.12, respectively. The exact values are

$$A'_v = -10\log_{10}\left(1 + \frac{f^2}{f_c^2}\right)$$

$$A'_v = -10\log_{10}\left(1 + \frac{1.6\text{ MHz}^2}{150\text{ kHz}^2}\right)$$

$$A'_v = -20.6\text{ dB}$$

$$\theta = -90° + \arctan\frac{f_c}{f}$$

$$\theta = -90° + \arctan\frac{150\text{ kHz}}{1.6\text{ MHz}}$$

$$\theta = -84.6°$$

    The complete Bode plot for the network in Example 1.15 is shown in Figure 1.13 (page 24). It is very useful to examine both plots simultaneously. In this manner you can find the exact phase change for a given gain quite easily. This information is very important when the application of negative feedback is considered (Chapter 3). For example, if you look carefully at the plots of Figure 1.13, you will note that at the critical frequency of 150 kHz, the total phase

——— **Figure 1.13**
Bode plot for 150 kHz lag

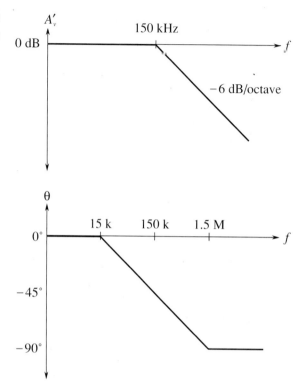

change is $-45°$. Since this circuit involved the use of a single lag network, this is exactly what you would expect.

Risetime versus Bandwidth

For pulse-type signals, the speed of an amplifier is often expressed in terms of its *risetime*. If a square pulse such as Figure 1.14a is passed into a simple lag network, the capacitor charging effect will produce a rounded variation, as seen in Figure 1.14b. This effect places an upper limit on the duration of pulses which a given amplifier can handle without producing excessive distortion. By definition, risetime is the amount of time it takes for the signal to traverse from 10% to 90% of the peak value of the pulse. The shape of this pulse is defined by the standard capacitor charge equation examined in earlier course work, and is valid for any system with a single clearly dominant lag network.

$$v_{out} = V_{peak}(1 - \varepsilon^{-t/RC}) \tag{1.12}$$

Figure 1.14
Pulse-risetime effect

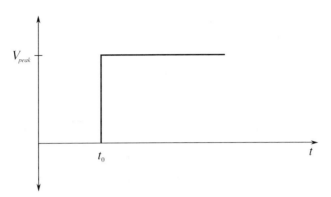

a. Input to lag network

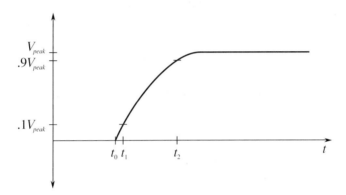

b. Output of lag network

In order to find the time interval from the initial starting point to the 10% point, set v_{out} to $.1V_{peak}$ in Equation 1.12 and solve for t_1:

$$.1V_{peak} = V_{peak}(1 - \varepsilon^{-t_1/RC})$$

$$.1V_{peak} = V_{peak} - V_{peak}\varepsilon^{-t_1/RC}$$

$$.9V_{peak} = V_{peak}\varepsilon^{-t_1/RC}$$

$$.9 = \varepsilon^{-t_1/RC}$$

$$\log .9 = \frac{-t_1}{RC}$$

$$t_1 = .105RC \tag{1.13}$$

To find the interval up to the 90% point, follow the same technique using $.9V_{peak}$. Doing so yields

$$t_2 = 2.303RC \tag{1.14}$$

The risetime, T_r, is the difference between t_1 and t_2:

$$T_r = t_2 - t_1$$
$$T_r = 2.303RC - .105RC$$
$$T_r \approx 2.2RC \tag{1.15}$$

Equation 1.15 ties the risetime to the lag network's R and C values. These same values also set the critical frequency f_2. By combining Equation 1.15 with the basic critical frequency relationship, we can derive an equation relating f_2 to T_r:

$$f_2 = \frac{1}{2\pi RC}$$

Solving Equation 1.15 in terms of RC, and substituting, yields

$$f_2 = \frac{2.2}{2\pi T_r}$$

$$f_2 = \frac{.35}{T_r} \tag{1.16}$$

where

f_2 is the upper critical frequency

T_r is the risetime of the output pulse

Example 1.16 Determine the risetime for a lag network critical at 100 kHz.

$$f_2 = \frac{.35}{T_r}$$

$$T_r = \frac{.35}{f_2}$$

$$T_r = \frac{.35}{100\ \text{kHz}}$$

$$T_r = 3.5\ \mu\text{S}$$

The Nyquist plot combines the two parts of the Bode plot into a single plot. It is a plot of the network's complex gain. In this manner, it is similar in appearance to a complex phasor impedance plot. A typical Nyquist plot is shown in Figure 1.15. A given gain is represented as a vector comprised of real and imaginary components. A gain with a phase of 0° would lie directly on the horizontal axis to the right of the origin. A positive phase value would produce a counterclockwise displacement, while a negative phase response would produce clockwise displacement. An inverting amplifier ($-180°$ shift) would produce a vector lying on the horizontal axis, but extending to the left of the origin. Normally, gain values are plotted in a nondecibel form. The complete response for a given amplifier normally follows some form of closed curve. Note that frequency information is not discernible from the Nyquist plot. Sometimes, specific points along the curve are labeled with their corresponding frequencies, which can prove quite helpful.

Figure 1.15
Nyquist plot of complex gain

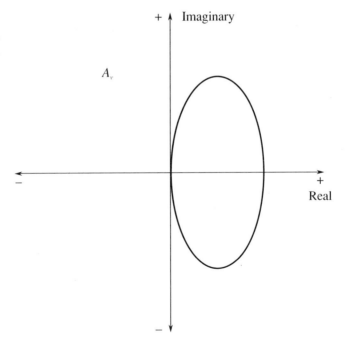

─────   **Figure 1.16**
Nyquist plot over the
range 1 kHz to 20 kHz

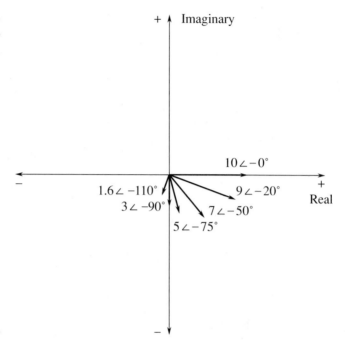

a. Individual gain vectors

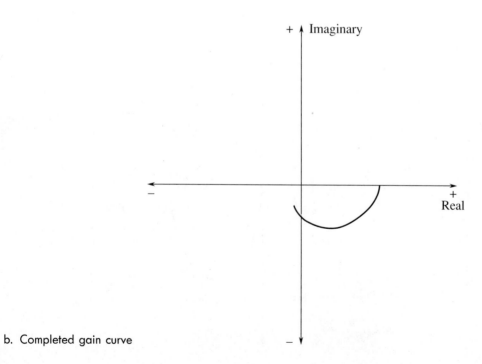

b. Completed gain curve

As an example, let's produce a Nyquist plot for an amplifier. Assume that the following amplitude and phase components have been measured in the laboratory.

Frequency (Hz)	Gain (V/V)	Phase (degrees)
1000	10	0
2000	9	−20
3000	7	−50
5000	5	−75
10,000	3	−90
20,000	1.6	−110

First, draw the real and imaginary axes. The first point to be plotted is at 1 kHz. Draw a vector 10 units long from the origin, horizontally and to the right. This direction indicates a 0° phase angle. For the second vector, a length of 9 units is required. It must be rotated 20° counterclockwise from the horizontal. The third unit is rotated 50° from the horizontal, and is 7 units in length. If you keep adding vectors in this fashion, you should wind up with a plot similar to the one in Figure 1.16a. In Figure 1.16b, the individual data points have been joined with a smooth curve. This curve is the Nyquist plot of the amplifier over the 1-kHz to 20 kHz range.

For the curve to be accurate, it is very important to take a sufficient number of measurements at close intervals. If this is not done, then certain response aberrations or resonances may go unnoticed. For example, if there had been a sudden gain increase in a narrow region around 15 kHz, this plot would not show it.

While the Nyquist plot can be very useful, we shall concentrate on the Bode plot since it tends to present a bit more information.

1.5 Combining the Elements: Multistage Effects

A complete gain or phase plot combines three elements: 1) the midband response, 2) the lead response, and 3) the lag response. Normally, a particular design will contain multiple lead and lag networks. The complete response is the summation of the individual responses. For this reason, it is useful to find the dominant lead and lag networks. These are the networks which affect the midband response first. For lead networks, the dominant one will be the one with the highest f_c. Conversely, the dominant lag network will be the one with the lowest f_c. It is very common to approximate the complete system response by drawing straight line segments such as those given in Figures 1.5 and 1.10. The process goes something like this:

1. Locate all f_c's on the frequency axis.

2. Draw a straight line between the dominant lag and lead f_c's at the midband gain. If the system does not contain any lead networks, continue the midband gain line down to DC.

3. Draw a slope of 6 dB per octave between the dominant lead and the next lower lead network.

4. Since the effects of the networks are cumulative, draw a slope of 12 dB per octave between the second lead f_c and the third f_c. After the third f_c, the slope should be 18 dB per octave, after the fourth, 24 dB per octave, and so on.

5. Draw a slope of -6 dB per octave between the dominant lag f_c and the next highest f_c. Again, the effects are cumulative, so increase the slope by -6 dB at every new f_c.

Example 1.17 Draw the Bode gain plot for the following amplifier: A'_v midband $= 26$ dB, one lead network critical at 200 Hz, one lag network critical at 10 kHz, and another lag network critical at 30 kHz.

The dominant lag network is 10 kHz. There is only one lead network, so it's dominant by default.

1. Draw a straight line between 200 Hz and 10 kHz at an amplitude of 26 dB.

2. Draw a slope of 6 dB per octave below 200 Hz: Drop down one octave (100 Hz) and subtract 6 dB from the present gain (26 dB $-$ 6 dB $=$ 20 dB). The line will start at the point 200 Hz/26 dB, and pass through the point 100 Hz/20 dB. Since there are no other lead networks, this line may be extended to the left edge of the graph.

3. Draw a slope of -6 dB per octave between 10 kHz and 30 kHz. The construction point will be 20 kHz/20 dB. Continue this line to 30 kHz. The gain at the 30 kHz intersection should be around 16 dB. The slope above this second f_c will be -12 dB per octave. Therefore, the second construction point should be at 60 kHz/4 dB (one octave above 30 kHz, and 12 dB down from the 30 kHz gain.) Since this is the final lag network, this line may be extended to the right edge of the graph.

Figure 1.17   Gain plot for complete amplifier

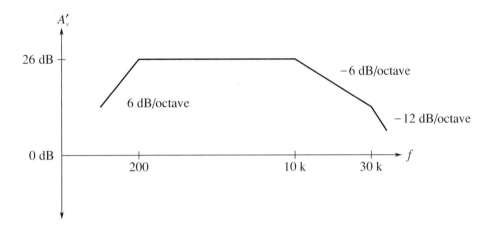

A completed graph is shown in Figure 1.17. There is one item that should be noted before we leave this section, and that is the concept of narrowing. Narrowing occurs when two or more networks share similar critical frequencies, and one of them is a dominant network. The result is that the true -3 dB breakpoints may be altered. Here is an extreme example: Assume that a circuit has two lag networks, both critical at 1 MHz. A Bode plot would indicate that the breakpoint is 1 MHz. This is not really true. Remember, the effects of lead and lag networks are cumulative. Since each network produces a 3 dB loss at 1 MHz, the net loss at this frequency is actually 6 dB. The true -3 dB point will have been shifted. The Bode plot only gives you the approximate shape of the response.

1.6 Computer-Generated Output

With the advent of low-cost personal computers, there are many alternatives to sketching plots by hand. One method involves the use of commercial or public domain software packages designed for circuit analysis. One common package is the public domain program SPICE. SPICE is an acronym that stands for **S**imulation **P**rogram with **I**ntegrated **C**ircuit **E**mphasis, and was originally written in the mid-1970s by Dr. Laurence Nagel of the University of California. This program is available for many different computing platforms. SPICE also serves as the core for a number of commercial packages. The commercial versions generally add features like graphical input and output of data, and large device libraries. All of the SPICE examples in this text were tested using the public domain AmigaSPICE program on a Commodore Amiga® personal computer, as well as the commercial PSpice package, produced by MicroSim, on a standard MS DOS "clone" computer.[1] In order to use SPICE, a circuit is described with a special data file. The data file is created using an ordinary text editor. This data file is then used by the SPICE program to estimate circuit response. Outputs can include items such as Bode gain and phase plots.

An example schematic with SPICE file and output are shown in Figure 1.18.[2] This circuit is a simple lag network. Before the data file is created, each node of the circuit must be labelled. By default, ground is node 0. The form of the SPICE data file can be broken into three broad categories: 1) component descriptions, 2) desired types of analysis, and 3) comments. Each of these elements generally takes up a single line. Comments are easy to spot since they always start with an asterisk (*). As with any programming language, comments are ignored by the computer program and are intended for documentation only. Component descriptions begin with a name for the item, where the first letter indicates the type of component, such as C for capacitor and R for resistor. After the component name come the nodes to which the item is connected, followed by the value of the component. For example, the line

C1   2   0   1U

indicates that we have a capacitor called C1 connected between nodes 2 and 0, and that its value is 1 μF. (SPICE includes many convenient shortcuts, such as K for Kilo, U for Micro, etc.)[3] The action elements start with a period. These items describe signal sources and desired analysis types. The SPICE output file includes the original input data file and the output tables and graphs requested. These are also shown in Figure 1.18.

SPICE is by no means a small or trivial program. It has many features and options, not to mention the variations produced by the many commercial versions. This text does not attempt to teach all of the intricacies of SPICE. For that, you should consult your SPICE user's manual, or one of the books devoted to SPICE. The simulation examples in this book assume that you already have some familiarity with SPICE.

---

[1] Amiga is a registered trademark of Commodore-Amiga, Inc.

[2] The representation of the PSpice/Probe screens used by permission of MicroSim Corporation. Copyright 1990 MicroSim Corporation. This material is proprietary to MicroSim Corporation. Unauthorized use, copying, or duplication is strictly prohibited.

PSpice is a registered trademark of MicroSim Corporation.

Probe is a trademark of MicroSim Corporation.

[3] Plotters, analysis packages, and many printers traditionally recognize standard ANSI characters only. Check your manuals for the ANSI substitutions for the Greek and math symbols used in this text (such as the use of the letter "U" for the non-ANSI character "μ" for "micro", "*" instead of "×" for multiplication, etc.).

Figure 1.18 SPICE example for a simple lag network

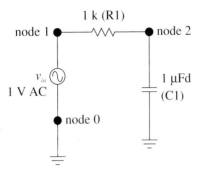

1 k (R1)

node 1 ●——/\/\/——● node 2

v_{in} ⊙
1 V AC

1 μFd
(C1)

● node 0

a. SPICE schematic

```
SIMPLE LAG OUTPUT
*Rl from node 1 to node 2, 1 kΩ
Rl 1 2 lk
*Cl from node 2 to node 0, 1 μFd
Cl 2 0 1U
*Vin from node 1 to node 0, 1 V AC
VIN 1 0 AC 1
*Calculate AC response, 8 points per
*decade from 10 Hz to 2 kHz.
.AC DEC 8 10 2k
*Print,the output voltage in dB, and
*the phase in degrees. output is node 2
.PRINT AC VDB(2) VP(2)
*Plot the Bode gain and phase graphs
.PLOT AC VDB(2)
.PLOT AC VP(2)
.END
```

b. SPICE data file (comments are preceded by asterisks)

```
*********************************AMIGASPICE 2.3/2G.6 GWS********************************
SIMPLE LAG OUTPUT
****        INPUT LISTING                        TEMPERATURE=27.000 DEG C
**************************************************************************************
*Rl from node 1 to node 2, 1 kΩ
Rl 1 2 lk
*Cl from node 2 to node 0, 1 μFd
Cl 2 0 1U
*Vin from node 1 to node 0, 1 V AC
VIN 1 0 AC 1
*Calculate AC response, 8 points per
*decade from 10 Hz to 2 kHz.
.AC DEC 8 10 2k
*Print the output voltage in dB, and
*the phase in degrees. output is node 2
.PRINT AC VDB(2) VP(2)
*Plot the Bode gain and phase graphs
.PLOT AC VDB(2)
.PLOT AC VP(2)
.END
*********************************AMIGASPICE 2.3/2G.6 GWS********************************
SIMPLE LAG OUTPUT
****        SMALL SIGNAL BIAS SOLUTION           TEMPERATURE=27.000 DEG C
**************************************************************************************
NODE    VOLTAGE NODE VOLTAGE
(1)     .0000   (2)   .0000
VOLTAGE SOURCE CURRENTS
NAME    CURRENT
VIN     0.000D-01
TOTAL POWER DISSIPATION 0.00D-01 WATTS
```

c. SPICE output: header and DC bias

(Continued)

———   **Figure 1.18**   *(Continued)*

```
**********************************AMICAOTION S.5/SG.6 GWS********************************

SIMPLE LAG OUTPUT
****          AC ANALYSIS                     TEMPERATURE=27.000 DEG C

********************************************************************************************

FREQ      VDB(2)   VP(2)
X

1.000E+01  -1.711E-02  -3.595E+00
1.334E+01  -3.038E-02  -4.789E+00
1.778E+01  -5.388E-02  -6.375E+00
2.371E+01  -9.536E-02  -8.475E+00
3.162E+01  -1.682E-01  -1.124E+01
4.217E+01  -2.947E-01  -1.484E+01
5.623E+01  -5.109E-01  -1.946E+01
7.499E+01  -8.707E-01  -2.523E+01
1.000E+02  -1.445E+00  -3.214E+01
1.334E+02  -2.310E+00  -3.996E+01
1.778E+02  -3.519E+00  -4.817E+01
2.371E+02  -5.079E+00  -5.613E+01
3.162E+02  -6.944E+00  -6.328E+01
4.217E+02  -9.042E+00  -6.932E+01
5.623E+02  -1.130E+01  -7.420E+01
7.499E+02  -1.365E+01  -7.802E+01
1.000E+03  -1.607E+01  -8.096E+01
1.334E+03  -1.853E+01  -8.319E+01
1.778E+03  -2.100E+01  -8.489E+01
2.371E+03  -2.348E+01  -8.616E+01
Y
```

d. SPICE output: gain and phase table

------- **Figure 1.18** *(Continued)*

```
*********************************AMIGASPICE 2.3/2G.6 GWS*********************************

SIMPLE LAG OUTPUT
****          AC ANALYSIS                    TEMPERATURE=27.000 DEG C

***************************************************************************************

X

FREQ     VDB(2)

                   -3.000D+01    -2.000D+01   -1.000D+01    0.000D-01    1.000D+01
                   ----------------------------------------------------------------
1.000D+01  -1.711D-02  .              .            .            *            .
1.334D+01  -3.038D-02  .              .            .            *            .
1.778D+01  -5.388D-02  .              .            .            *            .
2.371D+01  -9.536D-02  .              .            .            *            .
3.162D+01  -1.682D-01  .              .            .            *            .
4.217D+01  -2.947D-01  .              .            .            *            .
5.623D+01  -5.109D-01  .              .            .           *             .
7.499D+01  -8.707D-01  .              .            .          *.             .
1.000D+02  -1.445D+00  .              .            .         *               .
1.334D+02  -2.310D+00  .              .            .       *                 .
1.778D+02  -3.519D+00  .              .            .     *                   .
2.371D+02  -5.079D+00  .              .            .   *                     .
3.162D+02  -6.944D+00  .              .            . *                       .
4.217D+02  -9.042D+00  .              .           .*                         .
5.623D+02  -1.130D+01  .              .         *.                           .
7.499D+02  -1.365D+01  .              .       *                              .
1.000D+03  -1.607D+01  .              .    *                                 .
1.334D+03  -1.853D+01  .              .  *                                   .
1.778D+03  -2.100D+01  .           *.                                        .
2.371D+03  -2.348D+01  .         *    .                                      .
                   ----------------------------------------------------------------

Y
```

e. SPICE output: gain plot

(Continued)

———————   **Figure 1.18**   *(Continued)*

```
**********************AMIGASPICE 2.3/2G.6 GWS*********************************

SIMPLE LAG OUTPUT
****        AC ANALYSIS              TEMPERATURE=27.000 DEG C

***************************************************************************************

X

FREQ      VP(2)

              -1.500D+02    -1.000D+02   -5.000D+01    0.000D-01    5.000D+01
          ------------------------------------------------------------------
1.000D+01  -3.595D+00  .            .            .            * .            .
1.334D+01  -4.789D+00  .            .            .            * .            .
1.778D+01  -6.375D+00  .            .            .           * .             .
2.371D+01  -8.475D+00  .            .            .          * .              .
3.162D+01  -1.124D+01  .            .            .         * .               .
4.217D+01  -1.484D+01  .            .            .        * .                .
5.623D+01  -1.946D+01  .            .            .      * .                  .
7.499D+01  -2.523D+01  .            .            .    * .                    .
1.000D+02  -3.214D+01  .            .            .   * .                     .
1.334D+02  -3.996D+01  .            .            . * .                       .
1.778D+02  -4.817D+01  .            .           .* .                        .
2.371D+02  -5.613D+01  .            .        * . .                          .
3.162D+02  -6.328D+01  .            .      * .   .                          .
4.217D+02  -6.932D+01  .            .    * .     .                          .
5.623D+02  -7.420D+01  .            .   * .      .                          .
7.499D+02  -7.802D+01  .            .  * .       .                          .
1.000D+03  -8.096D+01  .          * .   .        .                          .
1.334D+03  -8.319D+01  .         * .    .        .                          .
1.778D+03  -8.489D+01  .        * .     .        .                          .
2.371D+03  -8.616D+01  .       * .      .        .                          .
          ------------------------------------------------------------------

Y

JOB CONCLUDED
```

f. SPICE output: phase plot

Figure 1.18   (*Continued*)

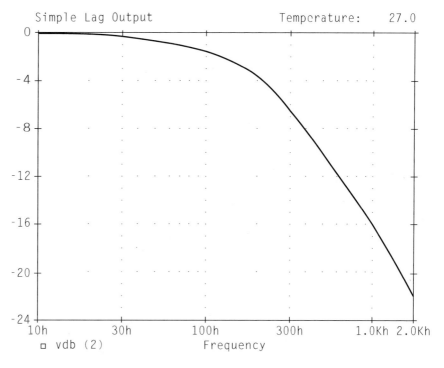

g. SPICE output: gain plot
using Probe

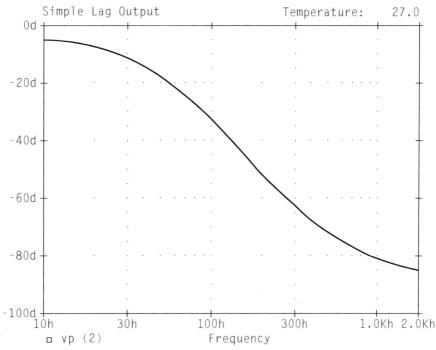

h. SPICE output: phase
plot using Probe

We shall be using SPICE in the following chapters for various purposes. One thing that you should always bear in mind is that simulation tools should not be used in place of a normal "human" analysis. Doing so can cause no end of grief. Simulations are only as good as the models used with them. It is easy to see that if the description of the circuit or the components within the circuit is not accurate, the simulation will not be accurate. Simulation tools are best used as a form of double-checking a design, not as a substitute for proper analysis.

1.7 The Differential Amplifier

Most modern operational amplifiers utilize a differential amplifier front end. In other words, the first stage of the operational amplifier is a differential amplifier. This circuit is commonly referred to as a diff amp or as a long-tailed pair. A diff amp utilizes a minimum of two active devices, although four or more may be used in more complex designs. Our purpose here is to examine the basics of the diff amp so that we can understand how it relates to the larger operational amplifier. Therefore, we shall not be investigating the more esoteric designs. To approach this in an orderly fashion, we shall examine the DC analysis first, and then follow with the AC small-signal analysis.

DC Analysis

A simplified diff amp is shown in Figure 1.19. This circuit utilizes a pair of NPN bipolar transistors, although the circuit could just as easily be built with

Figure 1.19
Simplified diff amp

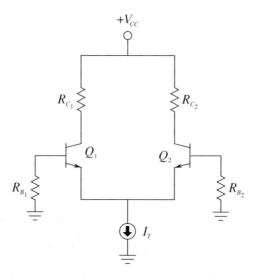

Figure 1.20
Diff amp analysis of
Figure 1.19

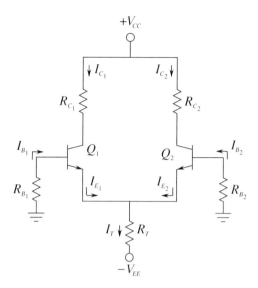

PNPs or FETs. Note the inherent symmetry of the circuit. If you were to slice
the circuit in half vertically, all of the components on the left half would have
a corresponding component on the right half. Indeed, for optimal performance,
we shall see that these component pairs should have identical values. For
critical applications, a matched pair of transistors would be used. In this case,
the transistor parameters, such as β, would be very closely matched for the
two devices.

In Figure 1.20, the circuit currents are noted, and the generalized current
source has been replaced with a resistor/negative-power-supply combination.
This is, in essence, an emitter bias technique. Assuming that the base voltages
are negligible and that V_{BE} is equal to .7 V, we can see that the emitter of each
device is at approximately $-.7$ V. Kirchoff's Voltage Law indicates that the
bulk of the negative supply potential must drop across R_T.

$$V_{R_T} = |V_{EE}| - .7 \text{ V}$$

Knowing this, we may find the current through R_T, which is known as the
tail current, I_T.

$$I_T = \frac{|V_{EE}| - .7 \text{ V}}{R_T}$$

If the two halves of the circuit are well matched, the tail current will split
equally into two portions, I_{E_1} and I_{E_2}. Given identical emitter currents, it
follows that the remaining currents and voltages in the two halves must be

identical as well. These potentials and currents are found through the application of Kirchoff's Voltage and Current Laws just as in any other transistor bias analysis.

Example 1.18   Find the tail current, the two emitter currents, and the two collector-to-ground voltages in the circuit of Figure 1.21. You may assume that the two transistors are very closely matched.

The first step is to find the tail current:

$$I_T = \frac{|V_{EE}| - .7\ \text{V}}{R_T}$$

$$I_T = \frac{10\ \text{V} - .7\ \text{V}}{2\ \text{k}\Omega}$$

$$I_T = \frac{9.3\ \text{V}}{2\ \text{k}\Omega}$$

$$I_T = 4.65\ \text{mA}$$

The tail current is the combination of the two equal emitter currents, so

$$I_{E_1} = I_{E_2} = \frac{I_T}{2}$$

$$I_{E_1} = I_{E_2} = \frac{4.65\ \text{mA}}{2}$$

$$I_{E_1} = I_{E_2} = 2.325\ \text{mA}$$

Figure 1.21
Diff amp for Example 1.18

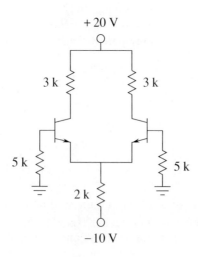

If we make the approximation that collector and emitter currents are equal, we can find the collector voltage by calculating the voltage drop across the collector resistor, and subtracting the result from the positive power supply:

$$V_C = V_{CC} - I_C R_C$$

$$V_C = 20 \text{ V} - 2.325 \text{ mA} \times 3 \text{ k}\Omega$$

$$V_C = 20 \text{ V} - 6.975 \text{ V}$$

$$V_C = 13.025 \text{ V}$$

Again, since we have identical values for both halves of the circuit, $V_{C_1} = V_{C_2}$. If we continue with this and assume a typical β of 100, we find that the two base currents are identical as well.

$$I_B = \frac{I_C}{\beta}$$

$$I_B = \frac{2.325 \text{ mA}}{100}$$

$$I_B = 23.25 \text{ }\mu\text{A}$$

Noting that the base currents flow through the 5 k base resistors, we can find the base voltages. Note that this is a negative potential since the base current is flowing from ground into the transistor's base.

$$V_B = -I_B R_B$$

$$V_B = -23.25 \text{ }\mu\text{A} \times 5 \text{ k}\Omega$$

$$V_B = -116.25 \text{ mV}$$

(This result indicates that the actual emitter voltage is closer to $-.8$ V than $-.7$ V and thus, the tail current is actually a little less than our approximation of 4.65 mA. This error is probably within the error we can expect by using the .7 V junction potential approximation.)

Input Offset Current and Voltage

As you have no doubt guessed, it is impossible to make both halves of the circuit identical, and thus, the currents and voltages will never be exactly the same. Even a small resistor tolerance variation will cause an upset. If the base resistors are mismatched, this will cause a direct change in the two base potentials. A variation in collector resistance will cause a mismatch in the collector potentials. A simple β or V_{BE} mismatch can cause variations in the base currents and base voltages, as well as smaller changes in emitter currents and collector potentials. It is desirable then to quantify the circuit's performance so that we can see just how well balanced it is. We can judge a diff amp's DC performance by measuring its input offset current and its input and output offset voltages. In simple terms, the difference between the two base currents is the input offset current. The difference between the two collector voltages is the output offset voltage. The DC potential required at one of the bases to counteract the output offset voltage is called the input offset voltage (this is little more than the output offset voltage divided by the DC gain of the amplifier). In an ideal diff amp, all three of these factors are equal to 0. We will take a much closer look at these parameters and how they relate to operational amplifiers in later chapters. For now, it is only important that you understand that these inaccuracies exist, and what can cause them.

AC Analysis

Figure 1.22 shows a typical circuit with input and output connections. In order to minimize confusion with the DC circuit, AC equivalent values will be shown in lower case. Small emitter degeneration resistors, r_{E_1} and r_{E_2}, have been added to this diff amp. This circuit has two signal inputs and two signal outputs. It is possible to configure a diff amp so that only a single input and/or output is used. This means that there are four variations on the theme:

1. differential (also called dual- or double-ended) input, differential output

2. differential input, single-ended output

Figure 1.22
A typical diff amp with input and output connections

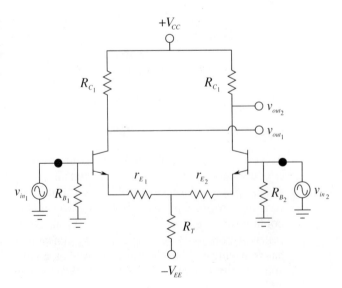

3. single-ended input, differential output

4. single-ended input, single-ended output

These variations are shown in Figure 1.23. For use in operational amplifiers, the differential input/single-ended output variation is the most common. We will examine the most general case, the differential-input/differential-output version.

Figure 1.23
The four different diff amp input/output configurations

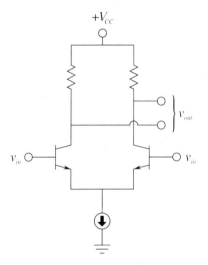

a. Differential input and output

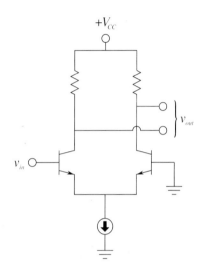

b. Single-ended input and differential output

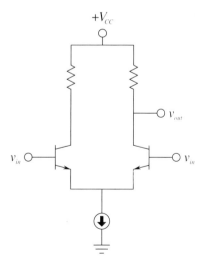

c. Differential input and single-ended output

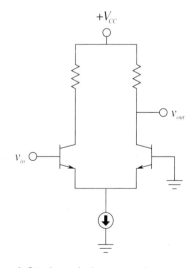

d. Single-ended input and output

Since the diff amp is a linear circuit, we can use the principle of superposition to independently determine the output contribution from each of the inputs. Utilizing the circuit of Figure 1.22, we will first determine the gain equation from v_{in_1} to either output. To do this, we replace v_{in_2} with a short circuit. The AC equivalent circuit is shown in Figure 1.24. For the output on collector number 1, transistor number 1 forms the basis of a common emitter amplifier. The voltage across r_{C1} is found via Ohm's Law.

$$v_{rc_1} = -i_{c_1} r_{c_1}$$

The negative sign comes from the fact that AC ground is used as our reference (i.e., for a positive input, current flows from AC ground down through r_{c_1}, and into the collector). To a reasonable approximation, we can say that the collector and emitter currents are identical.

$$v_{rc_1} = -i_{E_1} r_{c_1}$$

We must now determine the AC emitter current in relation to v_{in_1}. In order to better visualize the process, the circuit of Figure 1.24 is altered to include simplified transistor models, as shown in Figure 1.25. The value r'_e is the dynamic resistance of the base-emitter junctions and is inversely proportional to the DC emitter current. You may recall the following equation from your prior course work:

$$r'_e = \frac{26 \text{ mV}}{I_E}$$

where

r'_e is the dynamic base-emitter junction resistance

I_E is the DC emitter current

For typical circuits, the values of r'_e and r_E are much smaller than the tail current biasing resistor, R_T. Because of its large size, we can ignore the parallel

Figure 1.24
The circuit of Figure 1.22
redrawn for AC analysis

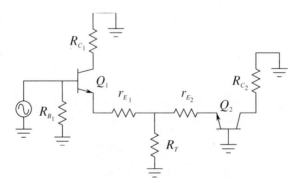

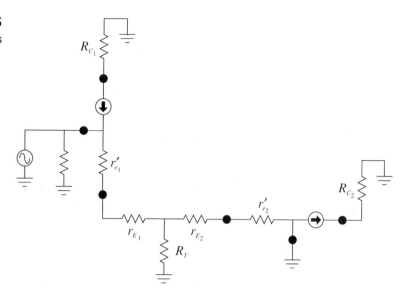

Figure 1.25
AC analysis

effect of R_T. By definition, the AC emitter current must equal the AC emitter potential divided by the AC resistance in the emitter section. If you trace the signal flow from the base of transistor number 1 to ground, you find that it passes through r'_{e_1}, r_{E_1}, r'_{e_2}, and r_{E_2}. You will also notice that the magnitude of i_{E_1} is the same as that of i_{E_2}, although they are out of phase.

$$i_E = \frac{v_{in_1}}{r'_{e_1} + r_{E_1} + r'_{e_2} + r_{E_2}}$$

Since the circuit values should be symmetrical for best performance, this equation may be simplified to

$$i_E = \frac{v_{in_1}}{2(r'_e + r_E)}$$

If we now solve for voltage gain,

$$A_v = \frac{v_{out}}{v_{in}}$$

$$A_v = \frac{-i_E r_C}{v_{in}}$$

$$A_v = -\frac{\dfrac{v_{in}}{2(r'_e + r_E)} r_C}{v_{in}}$$

$$A_v = -\frac{r_C}{2(r'_e + r_E)}$$

where

A_v is the voltage gain

r_C is the AC equivalent collector resistance

r_E is the AC equivalent emitter resistance

r'_e is the dynamic base-emitter junction resistance

The final negative sign indicates that the collector voltage at transistor number 1 is 180° out of phase with the input signal. Earlier, we noted that i_{E_2} is the same magnitude as i_{E_1}, the only difference being that it is out of phase. Because of this, the magnitude of the collector voltage at transistor number 2 will be the same as that on the first transistor. Since the second current is out of phase with the first, it follows that the second collector voltage must be out of phase with the first. This means that the voltage at the second collector is in phase with the first input signal. Its gain equation is

$$A_v = \frac{r_C}{2(r'_e + r_E)}$$

The various waveforms are depicted in Figure 1.26. The equation above is often referred to as the single-ended-input/single-ended-output gain equation since it describes the single change from one input to one output. The output signal will be in phase if we are examining the opposite transistor, and out of

Figure 1.26
Waveforms for a single
input

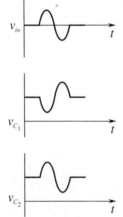

phase if we are looking at the input transistor. Since the circuit is symmetrical, we will get similar results when we examine the second input. The voltage between the two collectors is 180° apart. If we were to use a differential output, that is, derive the output from collector to collector rather than from one collector to ground, we would see an effective doubling of the output signal. If the reason for this is not clear to you, consider the following: Assume that each collector has a 1 V peak sine wave riding on it. When collector number 1 is at +1 V, collector number 2 is at −1 V, making +2 V total. Likewise, when collector number 1 is at its negative peak, collector number 2 is at its positive peak, producing a total of −2 V. The single-ended-input/differential-output gain therefore is

$$A_v = \frac{r_C}{r'_e + r_E}$$

Example 1.19 Using the circuit of Figure 1.22, determine the single-ended-input/differential-output and single-ended-input/single-ended-output voltage gains. Use the following component values: $V_{CC} = 15$ V, $V_{EE} = -8$ V, $R_T = 10$ kΩ, $R_C = 8$ kΩ, $r_E = 30$.

In order to find r'_e we must find the DC current.

$$I_T = \frac{|V_{EE}| - .7 \text{ V}}{R_T}$$

$$I_T = \frac{7.3 \text{ V}}{10 \text{ k}\Omega}$$

$$I_T = .73 \text{ mA}$$

$$I_E = \frac{I_T}{2}$$

$$I_E = \frac{.73 \text{ mA}}{2}$$

$$I_E = .365 \text{ mA}$$

$$r'_e = \frac{26 \text{ mV}}{I_E}$$

$$r'_e = \frac{26 \text{ mV}}{.365 \text{ mA}}$$

$$r'_e = 71.2$$

For the single-ended output gain,

$$A_v = \frac{r_L}{2(r_e' + r_E)}$$

$$A_v = \frac{8 \text{ k}\Omega}{2(71.2 \text{ }\Omega + 30 \text{ }\Omega)}$$

$$A_v = \frac{8 \text{ k}\Omega}{202.4 \text{ }\Omega}$$

$$A_v = 39.5$$

The differential output gain is twice this value, or 79.

Since it is possible to drive a diff amp with two distinct inputs, a wide variety of outputs may be obtained. It is useful to investigate two specific cases:

1. Two identical inputs in both phase and magnitude

2. Two inputs with identical magnitude, but 180° out of phase

Let's consider the collector potentials for the first case. Assume that a diff amp has a single-ended-input/single-ended-output gain of 100 and a 10 mV signal is applied to both bases. Using superposition, we find that the outputs due to each input are 100 × 10 mV, or 1 V in magnitude. For the first input, the voltages are sketched in Figure 1.27a. For the second input, the voltages are sketched in Figure 1.27b. Note that each collector sees both a sine wave and an inverted sine wave, both of equal amplitude. When these two signals are added, the result is zero, as seen in Figure 1.27c. In equation form,

$$v_{C_1} = v_{in_1}(-A_v) + v_{in_2}A_v$$

$$v_{C_1} = A_v(v_{in_2} - v_{in_1})$$

Since v_{in_1} and v_{in_2} are identical, the output is ideally zero given a perfectly matched and biased diff amp. The exact same effect is seen on the opposite collector.

This last equation is very important. It says that the output voltage is equal to the gain times the difference between the two inputs. This is how the differential amplifier got its name. In this case, the two inputs are identical, and thus their difference is zero.

Figure 1.27   Input-output waveforms for common mode

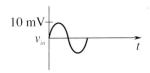

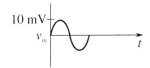

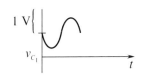

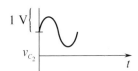

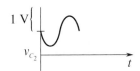

a. Waveforms for first input

b. Waveforms for second input

c. Combinations of a and b

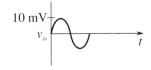

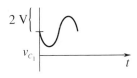

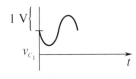

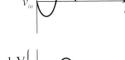

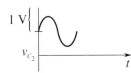

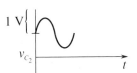

d. Waveforms for identical second input

e. Waveforms for inverted second input

f. Combinations of d and e

On the other hand, if we were to invert one of the input signals (case 2), we find a completely different result:

$$v_{in_1} = -v_{in_2}$$

$$v_{C_1} = A_v(v_{in_2} - v_{in_1})$$

$$v_{C_1} = A_v(v_{in_2} - (-v_{in_1}))$$

$$v_{C_1} = 2A_v v_{in_2}$$

Thus, if one input is inverted, the net result is a doubling of gain. This effect is shown graphically in Figures 1.27d through 1.27f. In short, a differential amplifier suppresses in-phase signals while simultaneously boosting out-of-phase signals. This can be a very useful attribute, particularly in the area of noise reduction.

Common-Mode Rejection

By convention, in-phase signals are known as common-mode signals. An ideal differential amplifier will perfectly suppress these common-mode signals, and thus, its common-mode gain is said to be zero. In the real world, a diff amp will never exhibit perfect common-mode rejection. The common-mode gain may be made very small, but it is never zero. For a common-mode gain of zero, the two halves of the circuit have to be perfectly matched, and all circuit elements must be ideal. This is impossible to achieve as errors may arise from several sources. The most obvious error sources are resistor tolerance variations and transistor parameter spreads. The basic design of the circuit will also affect the common-mode gain. With some circuit rearrangements, it is possible to determine a common-mode gain for the circuits we have been using. The circuit of Figure 1.22 has been redrawn in Figure 1.28 in order to emphasize its parallel symmetry. Since the DC potentials are identical in both halves, and identical signals drive both inputs, we can combine resistors in parallel in order to arrive at the circuit of Figure 1.29. Although it is not shown explicitly on the diagram,

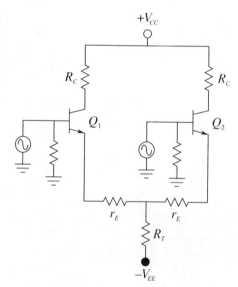

Figure 1.28
The circuit of Figure 1.22 redrawn for common mode rejection ratio (CMRR) analysis

Figure 1.29
Common mode gain
analysis

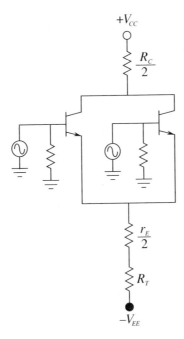

Figure 1.29
Common mode gain
analysis

the internal dynamic resistances (r'_e) may also be combined ($r'_e/2$). This circuit has been effectively reduced to a simple common emitter stage. Based on our earlier work, the gain for this circuit is

$$A_{v_{cm}} = \frac{\dfrac{r_C}{2}}{R_T + \dfrac{r'_e}{2} + \dfrac{r_E}{2}}$$

This is the common-mode voltage gain. If R_T is considerably larger than r_C, then this circuit will exhibit good common-mode rejection (assuming that the other parts are matched, naturally). R_T is the effective resistance of the tail current source. A very high internal resistance (i.e., an ideal current source) is desirable.

There are many ways of creating a more nearly ideal current source. One way is to use a third bipolar transistor as shown in Figure 1.30 (page 52). The tail current is found by determining the potential across R_2 and subtracting the .7 V V_{BE} drop. The remaining potential appears across R_3. Given the voltage and resistance, Ohm's Law will let you find the tail current. In this circuit, R_2 is sometimes replaced with a zener diode. This can help to reduce temperature-induced current fluctuations. In any case, the effective resistance of this current

—————   Figure 1.30
Improved current source

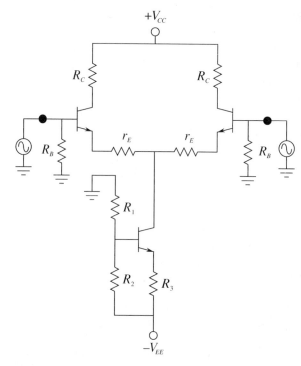

source is considerably larger than the simple tail-resistor variation. It is largely dependent on the characteristics of the tail current transistor, and can easily be in the megohm region.

Current Mirror

A very popular biasing technique in integrated circuits involves the current mirror. Current mirrors are also employed as active loads in order to optimize a circuit's gain. A simple current mirror is shown in Figure 1.31. This circuit requires that the transconductance curves of the diode and the transistor be very closely matched. One way to guarantee this is to use two transistors, and form one of them into a diode by shorting its collector to its base. If we use an approximate forward bias potential of .7 V and ignore the small base current, the current through the diode is

$$I_D = \frac{V_{CC} - .7\,\text{V}}{R}$$

Figure 1.31
Current mirror

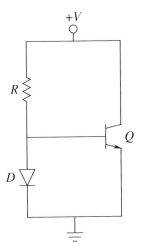

In reality, the diode potential will probably not be exactly .7 V. This will have little effect on I_D, though. Since the diode is in parallel with the transistor's base-emitter junction, we know that $V_D = V_{BE}$. If the two devices have identical transconductance curves, the transistor's emitter current will equal the diode current. You can think of the transistor as mirroring the diode's current, hence the circuit's name. If the two device curves are slightly askew, then the two currents will not be identical. This is shown graphically in Figure 1.32.

Figure 1.32 Transfer curve mismatch

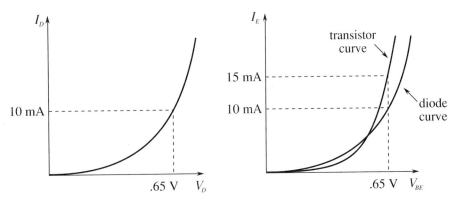

Assume actual $V_D = .65$ V

Figure 1.33
Current mirror bias

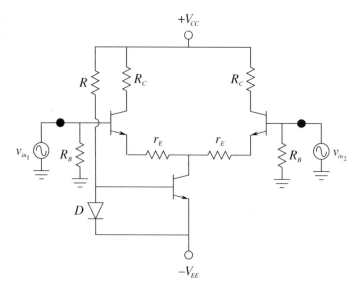

Figure 1.33
Current mirror bias

A current mirror could be used in the circuit of Figure 1.30. The result is shown in Figure 1.33. If the positive power supply is 15 V, the negative supply is − 10 V, and R is 10 kΩ, the tail current will be

$$I_D = \frac{V_{CC} - V_{EE} - V_D}{R}$$

$$I_D = \frac{15 \text{ V} - (-10 \text{ V}) - .7 \text{ V}}{10 \text{ k}\Omega}$$

$$I_D = 2.43 \text{ mA}$$

Since the tail current is the mirror current,

$$I_T = I_D$$

$$I_T = 2.43 \text{ mA}$$

Biasing of this type is very popular in operational amplifiers. Another use for current mirrors is in the application of active loads. Instead of using simple resistors for the collector loads, a current mirror may be used instead. A PNP-based current mirror suitable for use as an active load in our previous circuits is shown in Figure 1.34. To use this, we simply remove the two collector resistors from a circuit such as the one in Figure 1.33, and drop in the current mirror. The result of this operation is shown in Figure 1.35. The current-mirror active load produces a very high internal impedance, thus contributing to a very high differential gain. In effect, by using a constant current source in the

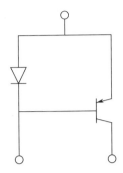

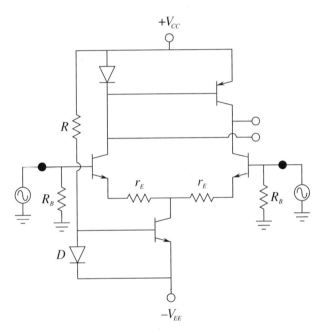

collectors, all AC current is forced into the following stage. You should also note that the number of resistors used in the circuit has decreased considerably.

Summary

We have seen how to convert gains and signals into a decibel form for both powers and voltages. This is convenient because what would require multiplication and division under the ordinary scheme requires only simple addition and subtraction in the decibel scheme. Also, decibel measurement is used almost exclusively for Bode gain plots. A Bode plot details a system's gain magnitude and phase response. For gain, the amplitude is measured in decibels, while the frequency is normally presented in log form. For a phase plot, phase is measured in degrees, and again, the frequency axis is logarithmic. The changes in gain and phase at the frequency extremes are caused by lead and lag networks. Lead networks cause the low-frequency gain to roll off. The rolloff rate is 6 dB per octave per network. The phase will change from $+90°$ to $0°$ per network. Lag networks cause the high-frequency gain to roll off at a rate of -6 dB per octave per network. The phase change per lag network is from $0°$ to $-90°$. A Nyquist plot combines the Bode gain and phase plots into a single plot of complex phasor gain. It was noted that computers can be used to quickly tabulate the response of complex networks such as these. Among the popular simulation programs is SPICE.

Differential amplifiers are symmetrical circuits, employing a minimum of two active devices. They may be configured with single or dual inputs, and single or dual outputs. Diff amps are commonly used as the first stage of an operational amplifier. They tend to amplify the difference in the input signals while simultaneously supressing in-phase (common-mode) signals. Current mirrors are widely used for biasing purposes and as active loads. Active loads offer the advantage of producing higher gains than ordinary resistive loads.

─── Self Test Questions ├──────

1. What are the advantages of using the decibel scheme over the ordinary scheme?

2. How do decibel power and voltage gain calculations differ?

3. What does the third letter in a decibel-based signal measurement indicate (as in dBV or dBm)?

4. What is a Bode plot?

5. What is a lead network? What general response does it yield?

6. What is a lag network? What general response does it yield?

7. What do the terms f_1 and f_2 indicate about a system's response?

8. What are the rolloff slopes for lead and lag networks?

9. What are the phase changes produced by individual lead and lag networks?

10. How is risetime related to upper break frequency?

11. How do multiple lead or lag networks interact to form an overall system response?

12. What is a Nyquist plot?

13. What is SPICE?

14. What is common-mode rejection?

15. What is a current mirror?

16. What is the advantage of using an active load?

Problem Set

▷ **Analysis Problems**

1. Convert the following power gains into decibel form:

a. 10   b. 80   c. 500

d. 1   e. .2   f. .03

2. Convert the following decibel power gains into ordinary form:

a. 0 dB   b. 12 dB   c. 33.1 dB

d. .2 dB   e. −5.4 dB   f. −20 dB

3. An amplifier has an input signal of 1 mW, and produces a 2 W output. What is the power gain in decibels?

4. A hi-fi power amplifier has a maximum output of 50 W and a power gain of 19 dB. What is the maximum input signal power?

5. An amplifier with a power gain of 27 dB is driven by a 25 mW source. Assuming the amplifier doesn't clip, what is the output signal in watts?

6. Convert the following voltage gains into decibel form:

a. 10   b. 40   c. 250

d. 1   e. .5   f. .004

7. Convert the following decibel voltage gains into ordinary form:

a. .5 dB   b. 0 dB   c. 46 dB

d. 10.7 dB   e. −8 dB   f. −14.5 dB

8. A guitar pre-amp has a gain of 44 dB. If the input signal is 12 mV, what is the output signal?

9. A video amplifier has a 140 mV input and a 1.2 V output. What is the voltage gain in decibels?

10. The pre-amp in a particular tape deck can produce a maximum signal of 4 V. If this amplifier has a gain of 18 dB, what is the maximum input signal?

11. Convert the following powers into dBW:

    a. 1 W   b. 23 W   c. 6.5 W

    d. .2 W   e. 2.3 mW   f. 1.2 kW

    g. .045 mW   h. .3 μW   i. 5.6×10^{-18} W

12. Repeat Problem 11 for units of dBm.

13. Repeat Problem 11 for units of dBf.

14. Convert the following voltages into dBV:

    a. 12.4 V   b. 1 V   c. .25 V   d. 1.414 V

    e. .1 V   f. 10.6 kV   g. 13 mV   h. 2.78 μV

15. A two-stage power amplifier has power gains of 12 dB and 16 dB. What is the total gain in decibel and in ordinary form?

16. If the amplifier of Problem 15 has an input of -18 dBW, what is the final output in dBW? In dBm? In watts?

17. Referring to Figure 1.1, what are the various stages' outputs if the input is changed to -4 dBm? To -34 dBW?

18. Which amplifier has the greatest power output?

    a. 50 W   b. 18 dBW   c. 50 dBm

19. Which amplifier has the greatest power output?

    a. 200 mW   b. -10 dBW   c. 22 dBm

20. A three-stage amplifier has voltage gains of 20 dB, 5 dB, and 12 dB, respectively. What is the total voltage gain in decibels and in ordinary form?

21. If the circuit of Problem 20 has input of voltage of -16 dBV, what are the outputs of the various stages in dBV? In volts?

22. Repeat Problem 21 for an output of 12 mV.

23. Which amplifier produces the largest output voltage?

    a. 15 V   b. 16 dBV

24. Given a lead network critical at 3 kHz, what are the gain and phase values at 100 Hz, 3 kHz, and 40 kHz?

25. Given a lag network tuned to 700 kHz, what are the gain and phase values at 50 kHz, 700 kHz, and 10 MHz? What is the risetime?

26. A noninverting amplifier has a midband voltage gain of 18 dB and a single lag network at 200 kHz. What are the gain and phase values at 30 kHz, 200 kHz, and 1 MHz? What is the risetime?

27. Repeat Problem 26 for an inverting ($-180°$) amplifier.

28. Draw the Bode plot for the circuit of Problem 26.

29. Draw the Bode plot for the circuit of Problem 27.

30. An inverting ($-180°$) amplifier has a midband gain of 32 dB and a single lead network critical at 20 Hz. (Assume the lag network f_c is high enough to ignore for low frequency calculations.) What are the gain and phase values at 4 Hz, 20 Hz, and 100 Hz?

31. Repeat Problem 30 with a noninverting amplifier.

32. Draw the Bode plot for the circuit of Problem 30.

33. Draw the Bode plot for the circuit of Problem 31.

34. A noninverting amplifier used for ultrasonic applications has a midband gain of 41 dB, a lag network critical at 250 kHz, and a lead network critical at 30 kHz. Draw its gain Bode plot.

35. Find the gain and phase at 20 kHz, 100 kHz, and 800 kHz for the circuit of Problem 34.

36. If the circuit of Problem 34 has a second lag network added at 300 kHz, what are the new gain and phase values at 20 kHz, 100 kHz, and 800 kHz?

37. Draw the gain Bode plot for the circuit of Problem 36.

38. What are the maximum and minimum phase shifts across the entire frequency spectrum for the circuit of Problem 36?

39. A noninverting DC amplifier has a midband gain of 36 dB, and lag networks at 100 kHz, 750 kHz, and 1.2 MHz. Draw its gain Bode plot.

40. What are the maximum and minimum phase shifts across the entire frequency spectrum for the circuit of Problem 39?

41. What is the maximum rate of high-frequency attenuation for the circuit of Problem 39 in dB/decade?

42. If an amplifier has two lead networks, what is the maximum rate of low-frequency attenuation in dB/octave?

43. Given the circuit of Figure 1.22, determine the single-ended-input/single-ended-output gain for the following values: $R_B = 5$ k, $R_T = 7.5$ k, $R_C = 12$ k, $V_{CC} = 25$ V, $V_{EE} = -9$ V, $r_E = 50$ Ω.

44. Determine the differential voltage gain in the circuit of Figure 1.30 if $R_B = 15$ k, $R_1 = 5$ k, $R_2 = 7$ k, $R_3 = 10$ k, $R_C = 20$ k, $V_{CC} = 22$ V, $V_{EE} = -12$ V, $r_E = 75$ Ω.

45. For the circuit of Problem 44, determine the output at collector number 2 if $v_{in_1}(t) = .001 \sin 2\pi 1000t$ and $v_{in_2}(t) = -.001 \sin 2\pi 1000t$.

46. Determine the differential voltage gain in the circuit of Figure 1.33 if $R_B = 8$ k, $R_{mirror} = 22$ k, $R_C = 10$ k, $V_{CC} = 18$ V, $V_{EE} = -15$ V, $r_E = 25$ Ω.

47. For the circuit of Problem 46, determine the output at collector number 1 if $v_{in_1}(t) = -.005 \sin 2\pi 2000t$ and $v_{in_2}(t) = .005 \sin 2\pi 2000t$.

48. Determine the tail and emitter currents in the circuit of Figure 1.35 if $R_B = 6$ k, $R_{mirror} = 50$ k, $V_{CC} = 15$ V, $V_{EE} = -15$ V, $r_E = 0$ Ω.

▷ Challenge Problems

49. You would like to use a voltmeter to take dBm readings in a 600-ohm system. What voltage should produce 0 dBm?

50. Assuming that it takes about an 8 dB increase in sound pressure level in order to produce a sound that is subjectively twice as loud to the human ear, can a hi-fi using a 100 W amplifier sound twice as loud as one with a 40 W amplifier (assuming the same speakers)?

51. Hi-fi amplifiers are often rated with a "headroom factor" in decibels. This indicates how much extra power the amplifier can produce for short periods of time, over and above its nominal rating. What is the maximum output power of a 250 W amplifier with 1.6 dB headroom?

52. If the amplifier of Problem 34 picks up an extraneous signal which is a -10 dBV sine wave at 15 kHz, what is the output?

53. If the amplifier of Problem 39 picks up a high-frequency interference signal at 30 MHz, how much is it attenuated over a normal signal? If this input signal is measured at 2 dBV, what should the output be?

54. If an amplifier has two lag networks, and both are critical at 2 MHz, is the resulting f_2 less than, equal to, or greater than 2 MHz?

55. If an amplifier has two lead networks, and both are critical at 30 Hz, is the resulting f_2 less than, equal to, or greater than 30 Hz?

▷ SPICE Problems

56. Use SPICE to plot the Bode gain response of the circuit in Problem 39.

57. Use SPICE to plot the Bode phase response of the circuit in Problem 34.

58. Use a SPICE program to generate a Bode plot for a lead network comprised of a 1 kΩ resistor and a .1 μF capacitor.

2 Operational Amplifier Internals

Chapter Objectives

After completing this chapter, you should be able to:

❏ Describe the internal layout of a typical op amp.

❏ Describe a simple op amp SPICE model.

❏ Determine fundamental parameters from an op amp data sheet.

❏ Describe an op amp–based comparator, and note where it might be used.

❏ Describe how integrated circuits are constructed.

❏ Define *monolithic planar* construction.

❏ Define *hybrid* construction.

2.1 Introduction

In this chapter we introduce the fundamentals of the operational amplifier, or op amp, as it is commonly known. We shall investigate the construction and usage of a typical general-purpose op amp. Specific as well as generalized internal circuits and associated block diagrams are examined. An initial op amp data sheet interpretation is given as well. Toward the middle of the chapter, the first op amp circuit examples are presented. The chapter finishes with an explanation of semiconductor integration and construction techniques. After finishing this chapter, you should be familiar with the concepts of what an op amp is composed of, how it is manufactured, and an emerging idea of how it might be used in application circuit design.

2.2 What Is an Op Amp?

An operational amplifier is, in essence, a multistage high-gain amplifier treated as a single entity. Normally, op amps have a differential input and a single-ended output. In other words, one input produces an inverted output signal, and the other input produces a noninverted output signal. Often, the op amp is driven from a bipolar power supply (i.e., two supplies, one positive and one negative). Just about any sort of active amplifying device may be used for the individual stages. Op amps can be made entirely from vacuum tubes or discrete bipolar transistors (and of course, they were made that way some years ago). The advances in semiconductor manufacture in the late 1960s and early 1970s eventually made it possible to miniaturize the required components and to place the whole affair on a single silicon chip (hence the term, integrated circuit). Through common use, this is what is generally meant by the term *op amp* today.

As seen in Figure 2.1, a typical op amp has at least five distinct connections; an inverting input (labelled "−"), a noninverting input (labelled "+"), an output, and positive and negative power supply inputs. These power supply connections are sometimes referred to as *supply rails*. Note that a ground connection is not directly given. Rather, a ground connection is implied through the other connections. This triangular symbol and its associated connections are typical, but by no means absolute. There is a wide variety of devices available to the designer which offer such features as differential outputs or unipolar power supply operation. In any case, some form of triangle will be used for the schematic symbol.

It is best to think of op amps as general-purpose building blocks. With them, you can create a wide variety of useful circuits. For general-purpose

Figure 2.1
General op amp symbol

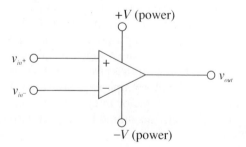

work, designing with op amps is usually much faster and more economical than an all-discrete approach. Likewise, troubleshooting and packaging constraints may be lessened. For more demanding applications, such as those requiring very low-noise, high-output current and/or voltage, or wide bandwidth, manufacturers have created specialized op amps. However, the final word in performance today is still dominated by discrete circuit designs. Also, it is very common to see a mix of discrete and integrated devices in a given circuit. There is certainly no law which states that op amps can be used only with other op amps. Often, a judicious mix of discrete devices and op amps can produce a circuit superior to one made entirely of discrete devices or op amps alone.

Where might you find op amp circuits? In a word, anywhere. They're probably in use in your home stereo or TV (where they help capture incoming signals), in electronic musical instruments (where they can be used to create and modify tones), in a camera (in conjunction with a light metering system), or in medical instruments (where they might be used along with various biosensing devices). The possibilities are almost endless. As you progress through this text, many different real-world circuits will be presented. These are not contrived circuit diagrams; they come right out of real-world tech manuals. For an interesting op amp application, refer to Figure 2.2. This is part of the audio output section of the Commodore Amiga® personal computer. Here, an op amp is used to help properly reconstruct computer-generated sounds, such as artificial speech. You'll see exactly how this is done in Chapter 12.

Block Diagram of an Op Amp

At this point you may be asking yourself, "What's inside the op amp?" The generic op amp consists of three main functional stages. A real op amp may contain more than three distinct stages, but can be reduced to this level for analysis. A generalized discrete representation is given in Figure 2.3. Since the op amp requires a differential input scheme, the first stage is most often a differential amplifier. As seen here, Q_1 and Q_2 comprise a PNP-based differential amplifier. The output of one collector (Q_2 here) is then fed to a high-gain second stage. This stage usually includes a lag network capacitor which plays a major role in setting the op amp's AC characteristics (as examined further in Chapter 5). Q_3 makes up the second stage in the example. It is set in a common emitter configuration for both current and voltage gain. The aforementioned lag capacitor is positioned across Q_3's base-collector junction in order to take advantage of the Miller effect. The third and final section is a class B or class AB follower for the most effective load drive. Q_4 and Q_5 make up our final stage. The twin diodes compensate for the Q_4 and Q_5 V_{BE} drops, and produce a trickle bias current which minimizes distortion. This is a relatively standard class AB stage. Note that the entire circuit is direct-coupled.

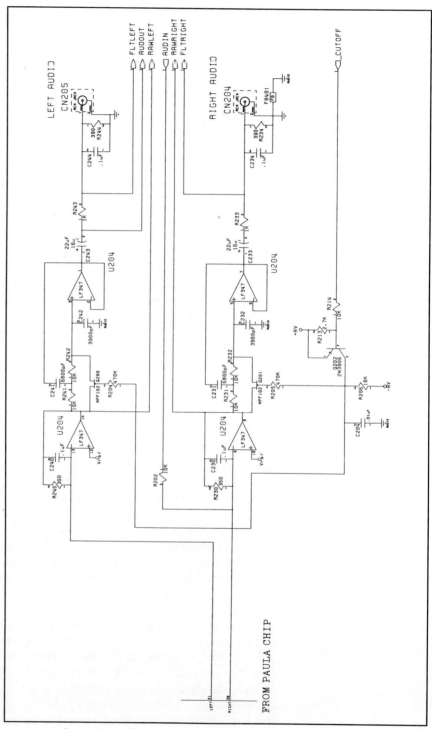

Figure 2.3
General op amp
schematic

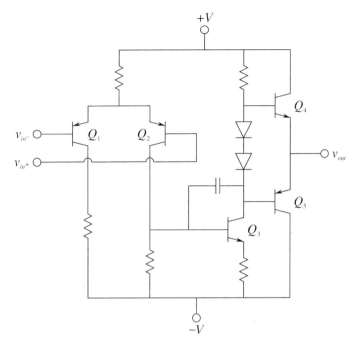

There are no lead networks, and thus the op amp can amplify down to 0 Hz (DC). There are many possible changes which may be seen in a real-world circuit, including the use of Darlington pairs or FETs for the differential amplifier, multiple high-gain stages, and output current limiting for the class B section.

The discrete circuit of Figure 2.3 uses only five transistors and two diodes, although an integrated version may use two to three dozen active devices. Because of the excellent device-matching abilities of single-chip integration, certain techniques are used in favor of standard discrete designs. Internal integrated circuit (IC) current sources are normally made through the use of current mirrors. Current mirror configurations are also employed to create active loads, in order to achieve maximum circuit gain. A typical integrated op amp will contain very few resistors, and usually only one or two lag network capacitors. Due to size limitations and other factors, inductors are virtually never seen in these circuits. A simplified equivalent circuit of the LF351 op amp is shown in Figure 2.4. Note that this device uses JFETs for the diff amp with an active load. The diff amp tail current source and the class AB trickle bias source are shown as simple current sources. In reality, they are a bit more complex, utilizing current mirror arrangements.

One of the most popular op amps over the years has been the 741. The specifications of this device seem rather lackluster by today's standards, but

Figure 2.4
LF351 simplified schematic

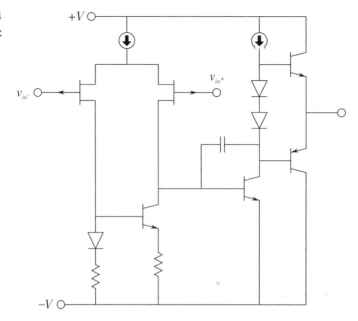

it was one of the first easy-to-use devices produced. As a result, it has found
its way into a large number of designs. Indeed, it is still a wise choice for less
demanding applications, or where parts costs are a major consideration.

A complete schematic of the μA741 is shown in Figure 2.5. Several different
manufacturers make the 741. This version is manufactured by Signetics, and

Figure 2.5
Schematic of the μA741

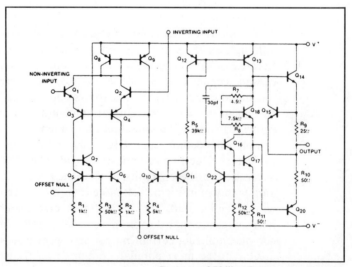

Courtesy of Philips Components-Signetics

may be somewhat different from a 741 made by another company.[1] The circuit contains 20 active devices and about one dozen resistors.

At first glance, this circuit may look hopelessly confusing. A closer look reveals many familiar circuit blocks. First off, you will notice that a number of devices show a shorting connection between their base and collector terminals, such as Q_8, Q_{11}, and Q_{12}. In essence, these are diodes. (They are drawn this way since that is how they are manufactured. It is actually easier to make diodes in this fashion.) For the most part, these diodes are part of current-mirror biasing networks. The bias setup is found in the very center of the schematic, and revolves around Q_9 through Q_{12}. The setup current is found by subtracting two diode drops (Q_{11}, Q_{12}) from the total power supply potential ($V_+ - V_-$), and dividing the result by R_5. For a standard ± 15 V power supply, this works out to

$$I_{bias} = \frac{V_+ - V_- - V_{BE-Q_{11}} - V_{BE-Q_{12}}}{R_5}$$

$$I_{bias} = \frac{30 \text{ V} - 1.4 \text{ V}}{39 \text{ k}\Omega}$$

$$I_{bias} = .733 \text{ mA}$$

This current is reflected into Q_{13}. A close look at Q_{10}, Q_{11} reveals that this portion is not a simple current mirror. By including R_4, the voltage drop across the base-emitter of Q_{10} is decreased, thus producing a current less than .733 mA. This configuration is known as a *Widlar current source*. The derivation of the exact current equation is rather involved, and beyond the scope of this chapter.[2] This current is reflected into Q_8 via Q_9, and establishes the tail current for the differential amplifier. The diff amp stage uses a total of four amplifying transistors in a common-collector/common-base configuration (Q_1 through Q_4). In essence, Q_1 and Q_2 are configured as emitter followers, thus producing high input impedance and reasonable current gain. Q_3 and Q_4 are configured as common base amplifiers, and, as such, produce a large voltage gain. The gain is maximized by the active load comprised of Q_5 through Q_7. The output signal at the collector of Q_4 passes on to a dual-transistor high-gain stage (Q_{16} and Q_{17}). Q_{16} is configured as an emitter follower and buffers Q_{17}, which is set as a common emitter voltage amplifier. Resistor R_{11} serves to stabilize both the bias and gain of this stage (i.e., it is an emitter degeneration or swamping resistor). Q_{17} is directly coupled to the class AB output stage (Q_{14}

---

[1] While the exact internal circuitry may be altered, the various manufacturers' versions will have the same pinouts, and very similar performance specifications.

[2] A complete derivation of the Widlar current source can be found in *Principles of Electronic Circuits*, by S. G. Burns and P. R. Bond, 1987, West Publishing Company.

and Q_{20}). Note the use of a V_{BE} multiplier to bias the output transistors. The V_{BE} multiplier is formed from Q_{18} and resistors R_7 and R_8. Note that this section receives its bias current from Q_{13}, which is part of the central current mirror complex.

Some transistors in this circuit are used solely for protection from overloads. A good example of this is Q_{15}. As the output current increases, the voltage across R_9 will increase proportionally. Note that this resistor is in parallel with the base-emitter junction of Q_{15}. If this potential gets high enough, Q_{15} will turn on, shunting base drive current around the output device (Q_{14}). In this manner, current gain is reduced, and the maximum output current is limited to a safe value. This limiting value may be found via Ohms Law:

$$ I_{limit} = \frac{V_{BE}}{R_9} $$

$$ I_{limit} = \frac{.7 \text{ V}}{25 \text{ }\Omega} $$

$$ I_{limit} = 28 \text{ mA} $$

In a similar fashion, Q_{20} is protected by R_{10}, R_{11}, and Q_{22}. If the output tries to sink too large a current, Q_{22} will turn on, shunting current away from the base of Q_{16}. While individual op amp schematics will vary widely, they generally hold to the basic four-part theme presented here:

1. A central current source/current mirror section to establish proper bias.

2. A differential amplifier input stage with active load.

3. A high-voltage-gain intermediate stage.

4. A class B or AB follower output section.

Fortunately for the designer or repair technician, intimate knowledge of a particular op amp's internal structure is usually not required for successful application of the device. In fact, a few simple models can be used for the majority of cases. One very useful model is given in Figure 2.6. Here the entire multistage op amp is modeled with a simple resistive input network, and a voltage source output. This output source is a dependent source. Specifically, it is a voltage-controlled voltage source. The value of this source is

$$ e_{out} = A_v(v_{in+} - v_{in-}) $$

The input network is specified as a resistance from each input to ground, as well as an input-to-input isolation resistance. For typical op amps these values are normally hundreds of kilo-ohms or more at low frequencies. Due

Figure 2.6
Simplified model

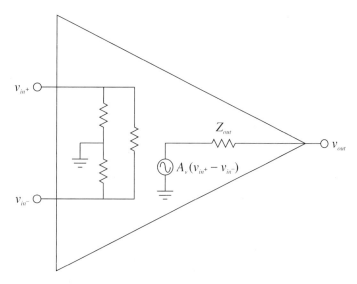

to the differential input stage, the difference between the two inputs is multiplied by the system gain. This signal is presented to the output terminal through the final stage's output impedance. The output impedance will most likely be less than 100 Ω. System voltage gains in excess of 80 dB (10,000) are the norm.

A Simple Op Amp SPICE Model

It is possible to create a great variety of SPICE models for any given op amp. Generally speaking, the more accurate the model is, the more likely it is to be complex. Due to the nature of SPICE, a more complex model requires a greater amount of time for a simulation to be completed. There is always a trade-off between model complexity and computation time. We can create a very simple SPICE model based on the previous section. This model is shown in Figure 2.7 (page 70). It consists of just five nodes. The input section is modeled as a single resistor, R_{in}, between nodes 1 and 2. These two nodes are the noninverting and inverting inputs of the op amp, respectively. The second half of the model consists of a voltage-controlled voltage source and an output resistor. The value of this dependent source is a function of the differential input voltage and the voltage gain. With a minimum of components, the simulation time for this model is very low. In order to use this model, you need only set three parameters: the input resistance, the output resistance, and the voltage gain. An example is shown in Figure 2.8.

This model must be used with great care since it is so simplistic. It is useful as a learning tool for investigating the general operation of an op amp, but

Figure 2.7
Simple SPICE model

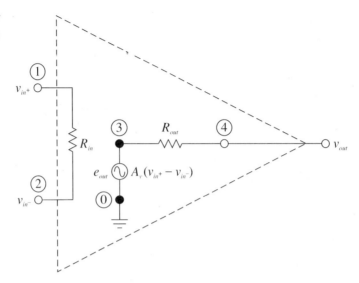

should never be considered as part of a true-to-life simulation. This model makes no attempt to consider the many limitations of the op amp. Since this model in no way imposes output signal swing limits, the effects of saturation will go unnoticed. Similarly, no attempt has been made at modeling the frequency response of the op amp. This is of great concern and we will spend considerable time on this subject in later chapters. Many other effects are also ignored. With so many limitations, you might wonder just where such a model may be used. This model is useful for noncritical simulations given low frequency inputs. You must also recognize the onset of saturation (clipping) yourself. Its primary advantage is that the circuit model is small, and thus computationally fast. Because of this, it is very efficient for students who are new to both op amps and SPICE. Perhaps of equal importance, this model points out the fact that your simulation results can only be as good as the models you use.

Figure 2.8
SPICE file for Figure 2.7

```
VERY SIMPLE OP AMP MODEL
*
* GOOD FOR LOW FREQUENCY USE ONLY
*
*
RIN   1 2    1M OHM
ROUT  3 4    75 OHM
EOUT  3 0 1 2    50K
* NOTE THAT NODES 1 AND 2 ARE THE
* CONTROLLING NODES FOR THE DEPENDENT
* SOURCE
```

Many people fall into the trap that "since the simulation came from a computer, it must be correct." Nothing could be further from the truth. Always remember the old axiom: GIGO (Garbage In = Garbage Out). It can be very instructive to simulate a circuit using differing levels of accuracy and complexity, and then to note how closely the results match the same circuit built in the laboratory.

An Op Amp Data Sheet and Interpretation

Different manufacturers often use special codes and naming conventions to delineate their products from those of other manufacturers, as well as to provide quality level and manufacturing information. A manufacturer's code is usually a letter prefix, while a quality or construction code is a suffix. Common prefix codes include μA (Fairchild), TL (Texas Instruments), LM, LH, and LF (National Semiconductor, with M indicating monolithic construction, H indicating hybrid construction, and F indicating an FET device), NE and SE (Signetics), MC (Motorola), AD (Analog Devices), and CS (Crystal).

Many manufacturers each make a host of standard parts such as the 741. For example, National Semiconductor makes the LM741, while Fairchild made the μA741. These parts are generally considered to be interchangeable, although they may vary in some ways.

Some manufacturers will use the prefix code of the original developer of a part, and reserve their prefix for their own designs. As an example, Signetics produces their version of the 741, which they called μA741 since this op amp was first developed by Fairchild. (Signetics is then referred to as a *second source* for the μA741.)

Suffix codes vary widely between manufacturers. Typical designations for consumer-grade parts are C and CN. The suffix N often means *Not Graded*. Interestingly, the lack of a final suffix often indicates a very high quality part, usually with an extended temperature range. Suffix codes are also used to indicate package styles. This practice is particularly popular as a way to code voltage regulators and other high-power linear ICs, and less common among codes for op amps.

Finally, some manufacturers will use a parallel numbering system for high-grade parts. For example, the commercial-grade device may have a "300 series" part number, with industrial grade given a "200 series" designation, and a military-grade part given a "100 series" number. One example is the LM318 commercial-grade op amp versus its high-grade counterpart, the LM118. Generally, military-specified parts will have a very wide temperature range, with industrial and commercial grades offering progressively narrower ranges.

The data sheet for the LF351 op amp is shown in Figure 2.9 (page 72). Let's look at some of the basic parameters and descriptions. The values are typical of a modern op amp. A complete investigation of all parameters will be given in Chapter 5, once we've gotten a little more familiar with the device.

Figure 2.9   Data sheet for the LF351

Absolute Maximum Ratings

If Military/Aerospace specified devices are required, contact the National Semiconductor Sales Office/Distributors for availability and specifications.

Supply Voltage	±18V
Power Dissipation (Notes 1 and 6)	670 mW
Operating Temperature Range	0°C to +70°C
$T_{j(MAX)}$	115°C
Differential Input Voltage	±30V
Input Voltage Range (Note 2)	±15V
Output Short Circuit Duration	Continuous
Storage Temperature Range	−65°C to +150°C
Lead Temp. (Soldering, 10 sec.)	
Metal Can	300°C
DIP	260°C

	H Package	N Package
θ_{jA}	225°C/W (Still Air)	120°C/W
	160°C/W	
	(400 LF/min Air Flow)	
θ_{jC}	25°C/W	

Soldering Information
Dual-In-Line Package
Soldering (10 sec.)   260°C
Small Outline Package
Vapor Phase (60 sec.)   215°C
Infrared (15 sec.)   220°C
See AN-450 "Surface Mounting Methods and Their Effect on Product Reliability" for other methods of soldering surface mount devices.

ESD rating to be determined.

DC Electrical Characteristics (Note 3)

Symbol	Parameter	Conditions	LF351 Min	LF351 Typ	LF351 Max	Units
V_{OS}	Input Offset Voltage	$R_S = 10\ k\Omega$, $T_A = 25°C$		5	10	mV
		Over Temperature			13	mV
$\Delta V_{OS}/\Delta T$	Average TC of Input Offset Voltage	$R_S = 10\ k\Omega$		10		$\mu V/°C$
I_{OS}	Input Offset Current	$T_j = 25°C$, (Notes 3, 4)		25	100	pA
		$T_j \leq 70°C$			4	nA
I_B	Input Bias Current	$T_j = 25°C$, (Notes 3, 4)		50	200	pA
		$T_j \leq \pm70°C$			8	nA
R_{IN}	Input Resistance	$T_j = 25°C$		10^{12}		Ω
A_{VOL}	Large Signal Voltage Gain	$V_S = \pm15V$, $T_A = 25°C$ $V_O = \pm10V$, $R_L = 2\ k\Omega$	25	100		V/mV
		Over Temperature	15			V/mV
V_O	Output Voltage Swing	$V_S = \pm15V$, $R_L = 10\ k\Omega$	±12	±13.5		V
V_{CM}	Input Common-Mode Voltage Range	$V_S = \pm15V$		+15		V
			±11			V
				−12		V
CMRR	Common-Mode Rejection Ratio	$R_S \leq 10\ k\Omega$	70	100		dB
PSRR	Supply Voltage Rejection Ratio	(Note 5)	70	100		dB
I_S	Supply Current			1.8	3.4	mA

AC Electrical Characteristics (Note 3)

Symbol	Parameter	Conditions	LF351 Min	LF351 Typ	LF351 Max	Units
SR	Slew Rate	$V_S = \pm15V$, $T_A = 25°C$		13		$V/\mu s$
GBW	Gain Bandwidth Product	$V_S = \pm15V$, $T_A = 25°C$		4		MHz
e_n	Equivalent Input Noise Voltage	$T_A = 25°C$, $R_S = 100\Omega$, $f = 1000$ Hz		25		$nV/\sqrt{Hz}$
i_n	Equivalent Input Noise Current	$T_j = 25°C$, $f = 1000$ Hz		0.01		$pA/\sqrt{Hz}$

Note 1: For operating at elevated temperature, the device must be derated based on the thermal resistance, θ_{JA}.

Note 2: Unless otherwise specified the absolute maximum negative input voltage is equal to the negative power supply voltage.

Note 3: These specifications apply for $V_S = \pm15V$ and $0°C \leq T_A \leq +70°C$. V_{OS}, I_B and I_{OS} are measured at $V_{CM} = 0$.

Note 4: The input bias currents are junction leakage currents which approximately double for every 10°C increase in the junction temperature, T_j. Due to the limited production test time, the input bias currents measured are correlated to junction temperature. In normal operation the junction temperature rises above the ambient temperature as a result of internal power dissipation, P_D. $T_j = T_A + \theta_{jA} P_D$ where θ_{jA} is the thermal resistance from junction to ambient. Use of a heat sink is recommended if input bias current is to be kept to a minimum.

Note 5: Supply voltage rejection ratio is measured for both supply magnitudes increasing or decreasing simultaneously in accordance with common practice. From ±15V to ±5V.

Note 6: Max. Power Dissipation is defined by the package characteristics. Operating the part near the Max. Power Dissipation may cause the part to operate outside guaranteed limits.

At the very top of the data sheet is a listing of the absolute maximum ratings. The op amp should never operate at values greater than those presented, as doing so may permanently damage it. Like most general-purpose op amps, the LF351 is powered by a bipolar power supply. The supply rails should never exceed ± 18 V DC. Normally, op amps will be used with ± 15 V supplies. Maximum power dissipation is given as 670 mW. Obviously then, this is a small signal device. In keeping with this, the operating temperature range and maximum junction temperatures are relatively low. We also see that the device can withstand differential input signals of up to 30 V, and single-ended inputs of up to 15 V, without damage. This is also dependent on the negative power-supply rail (see data sheet Note 2). On the output, the LF351 is capable of withstanding a shorted load condition continuously. This makes the op amp a bit more "bullet-proof." The remainder of this section details soldering conditions. Excessive heat during soldering may damage the device.

The second section of the data sheet lists the DC characteristics of the op amp. This table is broken down into five major sections:

1. The parameter symbol

2. The parameter name

3. The conditions under which the parameter is measured

4. The parameter values, either typical or min/max

5. The parameter units

We shall examine a few of these parameters right now. The fourth parameter given is I_B, the input bias current. I_B is the current drawn by the bases (or gates) of the input differential amplifier stage. Since the LF351 utilizes a JFET diff amp, we expect this value to be rather small. For an operating temperature of 25°C, a typical LF351 will draw 50 pA, and a worst-case LF351 no more than 200 pA. If we extend the temperature range out a bit, I_B can extend out to 8 nA. This is a sizable jump, but even 8 nA is a very small value for general-purpose work. Since larger bias currents are normally seen as undesirable, the maximum I_B is the worst-case scenario, hence a minimum I_B is not reported. Also, we see a very high value for input resistance (typically around 10^{12} ohms). Op amps utilizing bipolar input devices will show much higher values for I_B, and much lower values for R_{in}.

Next in line comes A_{vol}. This is the DC voltage gain. Note the test conditions. The power supply is set to ± 15 V, the load is 2 kΩ, and V_{out} is 10 V peak. Normally, we desire as much gain as possible, so the worst-case scenario is the minimum A_{vol}. For 25°C operation this is specified as 25 V/mV, or 25,000. The average device will produce a gain of 100,000. As is typical, once the temperature range is expanded, performance degrades. Over the operating temperature range, the minimum gain may drop to 15,000.

Since the op amp uses a class AB follower for its output stage, we should expect the output compliance to be very close to the power supply rails. The output voltage swing is specified for ±15 V supplies with a 10 kΩ load. The typical device can swing out to ±13.5 V, with a worst-case swing of ±12 V. A reduction in power supply value will naturally cause the maximum output swing to drop. A sizable reduction in the load resistance will also cause a drop in V_{out}, as we shall see a bit later. These maximum output values are caused by the internal stages reaching their saturation limits. When this happens, the op amp is said to be clipping or in saturation. As a general rule of thumb, saturation may be approximated as 1.5 V less than the magnitude of the power supplies.

The last item in the list is the standby current draw, I_S. Note how small this is: only 1.8 mA, 3.4 mA worst case. This is the current the op amp draws from the supply under no signal conditions. When producing output signals, the current draw will rise.

The final section of the data sheet lists certain AC characteristics of the op amp which will be of great concern to us in later sections. Many device parameters change a great deal with frequency, temperature, supply voltage, or other factors. Because of this, data sheets also include a large number of graphs which further detail the op amp's performance. Finally, application hints and typical circuits may round out the basic data sheet.

2.3 A Simple Op Amp Comparator

Now that you have a feel for what an op amp is and what some typical parameters are, let's take a look at an application. The one thing that jumps to most people's attention is the very high gain of the average op amp. The typical LF351 showed A_{vol} at approximately 100,000. With gains this high, it is obvious that even very small input signals can force the output into saturation (clipping). Take a look at Figure 2.10. Here an op amp is being supplied by ±15 V, and is driving a 10 k load. As seen in our model of Figure 2.6, v_{out} should equal the differential input voltage times the op amp's gain, A_{vol}:

$$v_{out} = A_{vol}(v_{in+} - v_{in-})$$

$$v_{out} = 100,000 \times (.1 \text{ V} - 0 \text{ V})$$

$$v_{out} = 10,000 \text{ V}$$

The op amp cannot produce 10,000 V. The data sheet lists a maximum output swing of only ±13.5 V when using ±15 V supplies. The output will be truncated at 13.5 V. If the input signal is reduced to only 1 mV, the output

Figure 2.10
Comparator (single input)

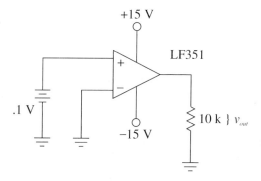

Figure 2.10
Comparator (single input)

will still be clipped at 13.5 V. This holds true even if we apply a signal to the inverting input, as in Figure 2.11.

$$v_{out} = A_{vol}(v_{in+} - v_{in-})$$

$$v_{out} = 100{,}000 \times (.5 \text{ V} - .3 \text{ V})$$

$$v_{out} = 20{,}000 \text{ V}$$

$$v_{out} = 13.5 \text{ V} \qquad \text{(due to clipping)}$$

For any reasonable set of inputs, as long as the noninverting signal is greater than the inverting signal, the output will be positive saturation. If you trade the input signals so that inverting signal is the larger, the converse will be true. As long as the inverting signal is greater than the noninverting signal, the output will be negative saturation. If the inverting and noninverting signals are identical, v_{out} should be 0 V. In the real world, this will not happen. Due to minute discrepancies and offsets in the diff amp stage, either positive or negative saturation will result. You have no quick way of knowing in which

Figure 2.11
Comparator (dual input)

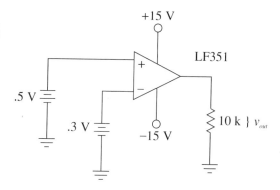

direction it will go. It is for this reason that it is impractical to amplify a very small signal, say around 10 μV. You might then wonder, "What is the use of this amplifier if it always clips? How can I get it to amplify a simple signal?" Well, for normal amplification uses, we will have to add on some extra components, and through the use of negative feedback (next chapter), we will create some very well-controlled, useful amplifiers. This is not to say that our barren op amp circuit is useless. Quite to the contrary, we have just created a comparator.

A comparator has two output states: high and low. In other words, it is a digital, logical output. Our comparator has a high-state potential of 13.5 V, and a low-state potential of -13.5 V. The input signals, in contrast, are continuously variable analog potentials. A comparator, then, is an interface between analog and digital circuitry. One input will be considered the reference, while the other input will be considered the sensing line. Note that the differential input signal is the difference between the sensing input and the reference input. When the polarity of the differential input signal changes, the logical output of the comparator changes state.

Example 2.1   Figure 2.12 shows a light leakage detector which might be used in a photographer's dark room. This circuit utilizes a cadmium sulfide (CdS) cell which is used as a light-sensitive resistor. The inverting input of the op amp is being used as the reference input, with a 1 V DC level. The noninverting input is being used as the sensing input. Under normal (no light) conditions, the CdS cell acts as a very high resistance, perhaps 1 M. Under these conditions, a voltage divider is set up with the 10 k resistor, producing about .15 V at the noninverting input. Remember, no loading of the divider occurs since the LF351

Figure 2.12
Light alarm

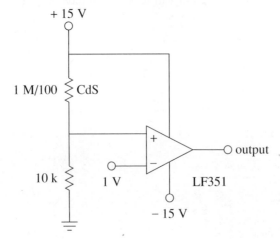

utilizes a JFET input. Since the noninverting input is less than the inverting input, the comparator's output is negative saturation, or approximately − 13.5 V. If the ambient light level rises, the resistance of the CdS cell drops, thus raising the signal applied to the noninverting input. Eventually, if the light level is high enough, the noninverting input signal will exceed the 1-volt reference, and the comparator's output will move to positive saturation (about + 13.5 V). This signal could then be used to trigger some form of audible alarm. A real-world circuit would need the flexibility of an adjustable reference in place of the fixed 1-volt reference. By swapping the CdS cell and the 10 k resistor, and adjusting the reference, an inverse circuit (i.e., an alarm which senses darkness) may be produced.

Circuits of this type can be used to sense a variety of over-level/under-level conditions, including temperature and pressure. All that is needed is an appropriate sensing device. Comparators can also be used with AC input signals.

Example 2.2   Sometimes, it is necessary to square up an AC signal for further processing. That is, we must turn it into an equivalent pulse waveform. One example of this might be a frequency counter. A frequency counter works by tallying the number of high-to-low or low-to-high transitions in the input signal over a specific length of time. For accurate counts, good edge transitions are required. Since a simple sine wave changes relatively slowly compared to a square wave of equal frequency, some inaccuracy may creep into the readings. We can turn the input into a pulse-type output by running it through the comparator of Figure 2.13. Note that the reference signal is adjustable from − 15 to + 15 V. Normally, the reference is set for 0 volts. Whenever the input is greater than the reference, the output will be positive saturation. When the input is less than the reference, the output will be negative saturation. By making the

Figure 2.13
"Square-up" circuit

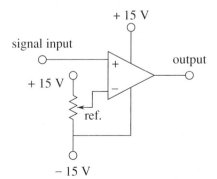

Figure 2.14
Output of "square-up"
circuit

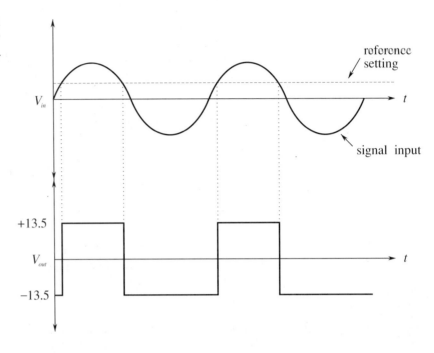

reference adjustable, we have control over the output duty cycle, and can also compensate for DC offsets on the input signal. A typical input/output signal set is given in Figure 2.14.

There are a few limitations with our simple op amp comparator. For very fast signal changes, a typical op amp will not be able to accurately track its output. Also, the output signal range is rather wide and is bipolar. It is not at all compatible with normal TTL logic circuits. Extra limiting circuitry is required for proper interfacing. To help reduce these problems, a number of circuits have been specially optimized for comparator purposes. We shall take a closer look at a few of them in Chapter 7.

2.4 Op Amp Manufacture

Op amps are generally manufactured in one of two ways: the device is either a *hybrid*, or is *monolithic*. In either case, the circuit can contain hundreds of components. The resulting op amp will be packaged in a variety of styles,

Plate 1

Entry to the clean room. Since the scale of integrated circuits is so small, even the smallest particles of dust and debris must be avoided. Here, technicians don special suits to avoid contaminating the work environment.

Plate 2

A technician checks diffusion tubes in the clean room.

Plate 3

Wafer probe station. A technician verifies a silicon wafer with the help of a microscope.

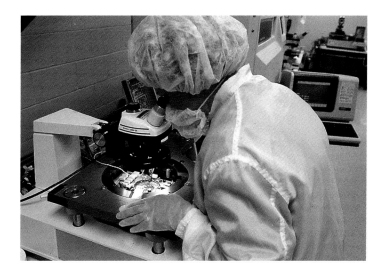

Plate 4

CAD layout equipment. Computer workstations are used to lay out the various integrated circuit components.

Plate 5

Integrated circuit layout, as produced by a CAD system.

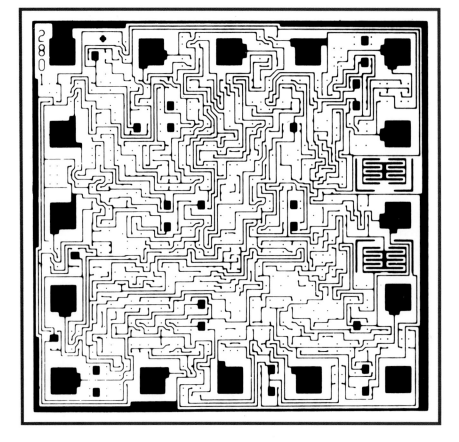

Plate 6

The scanning electron microscope (SEM) is used for close inspection of the integrated circuit. The SEM can resolve even the finest details.

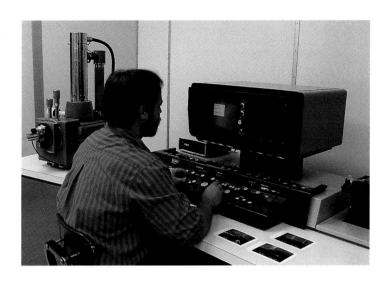

Plate 7

Wafer inspection system in the clean room.

Plate 8

Metal deposition. Different integrated circuit fabrication techniques require various forms of metal deposition (for example, for interconnections).

Plate 9

Ion implanter in the
clean room.

Plate 10

A 4-inch wafer. This
completed wafer will be
broken into separate
chips. Each chip can
then be mounted into
any of a variety of
different packages.
Testing is performed
at each stage of
production.

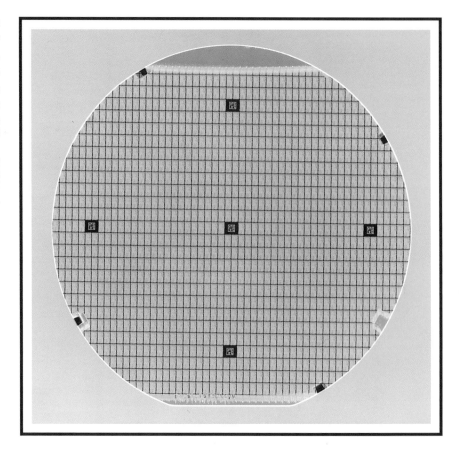

Photos courtesy of Cherry Semiconductor Corp.
2000 South County Trail
East Greenwich RI 02818

—— Figure 2.15   Package styles

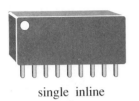

single inline   dual inline   can   flat pack

including plastic or ceramic dual inline packages (DIPs) and single inline packages (SIPs), multi-lead cans, flat packages, and surface-mount forms. Some examples are shown in Figure 2.15. In each type, the circuitry is completely encased and not accessible to the designer or technician. If one of the components should fail, the entire op amp is replaced. The design and layout of the integrated circuit itself is normally carried out with the use of special CAD workstations. These allow the designers to simulate portions of the circuit, and to create the outlines and interconnections for the various components to be formed.

Monolithic Construction

The term *monolithic* is from the Greek, meaning literally "single stone." In this process, all circuit elements are created and interconnected using a single slab of silicon (or other suitable material). Normally, several op amps are made from a single silicon wafer. Each wafer may be a few inches in diameter, with each op amp circuit chip comprising perhaps a square 1 millimeter by 1 millimeter in area. A single transistor can easily be smaller than 15 micrometers by 20 micrometers. Since the scale of construction is so small, special *clean rooms* are required in order to remove tiny airborne particles of dust and grit, which could interfere with the production of these super-small components. Workers in clean rooms are required to wear special suits as well.

Figure 2.16 (page 80) outlines the major steps in the chip manufacturing process. This process starts with the preparation of a p-type silicon wafer. This is referred to as the *substrate*. After it has been cleaned and polished, an n-type *epitaxial* region is *diffused* into the p-type base. "Epitaxial" is from the Greek, roughly meaning "to arrange upon." It is within this thin epitaxial region that the circuit elements will be formed, with the remainder of the substrate lending mechanical support to the structure. The term *diffusion* refers to the manner in which the semiconductor material becomes doped. In essence, the base material is surrounded by a high concentration of dopant, usually

Figure 2.16   Chip manufacturing process

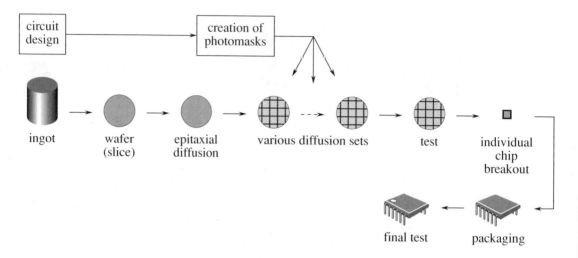

gaseous, as heat is applied. The low-concentration wafer material will be in-filtrated by the high-concentration doping material. Diffusion is a relatively accurate and inexpensive means of controlling the semiconductor's properties.

Once the n-type region is produced, the wafer will undergo an oxidizing process which will leave the top surface covered with silicon dioxide. This layer prevents impurities from entering the n-type region. At this point, a series of steps will be used to create *wells* or alternating deposits of p and n material. These deposits will form the various active and passive components. Normally, this is done through a photolithographic process. This involves the use of light-sensitive materials and *masks*. Conceptually, the process is not very much different from the way in which printed circuit boards are often made. In essence, specific areas of the silicon dioxide layer will be stripped away, thus exposing the epitaxial region, and allowing diffusion of other acceptor/donor impurities to take place. Since the silicon dioxide serves as an effective barrier to diffusion, only areas cleared of silicon dioxide will be affected by the diffusion process. In this manner, specific areas can be singled out and selectively doped to create specific components. This is detailed below and in Figure 2.17.

In order to selectively remove the silicon dioxide, the top surface is coated with a light-sensitive material called *photoresist*. Above this is placed a mask. This mask is much like a black and white negative; some areas are clear, and some areas are opaque. The resulting sandwich is then exposed to ultraviolet light. The clear areas of the mask will allow the light to pass through to cause a chemical change in the photoresist. A solution is then used to wash away the unexposed photoresist. At this point, a second solution is used to wash away the silicon dioxide. This solution will not affect the exposed photoresist,

Figure 2.17   Diffusion process (one run)

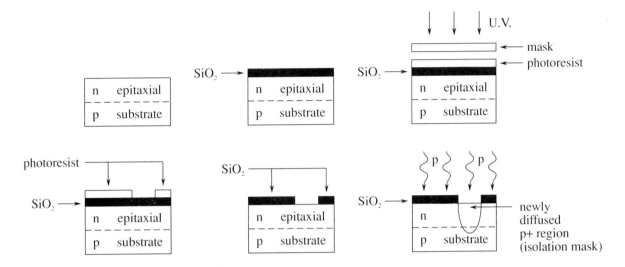

and thus, the silicon dioxide beneath it is unaffected. After this protective layer of photoresist is removed, all that remains on the top surface of the wafer is alternating patches of silicon dioxide. The wafer can now be led through another diffusion process.

The process of oxidizing, masking, and diffusing will be repeated several times. The initial run will be produced with an isolation mask. This is used to separate the various components. Normally, a base mask will be used next, followed by the emitter mask. The final masks will be used for contacts and interconnections. In this way, n-type material can be placed next to, or completely within, p-type material. The adjoining areas are, of course, PN junctions. Since all circuit elements are laid out lengthwise on a thin strip, this method of manufacturing is referred to as a *planar* process.

Once the final mask is completed, the wafer will be inspected. The individual chips will then be broken out of the wafer and mounted in the desired package. Leads will be connected to the chip with fine *angel hair* wire, and then the package will be sealed. Part numbers and date codes will also be imprinted. It is now ready for final testing and inspection.

Several key points in this process are illustrated in the series of photos found in Plates 1–10 in the color insert in this chapter. Included are various steps in the process of designing and manufacturing an integrated circuit, as well as the final chip.

Virtually all general-purpose op amps today use a planar monolithic process. Some of the advantages of monolithic construction are its relative simplicity and low per-part cost.

Hybrid Construction

Hybrids are usually used where a complete monolithic solution is impractical. This is usually the case for special-purpose devices, such as those requiring very high output current, very wide bandwidth, or which are very complex or sensitive. Hybrids, as the name suggests, are a collection of interconnected smaller circuit elements. A typical hybrid may contain two or three smaller monolithic chips and assorted miniaturized passive and/or power components. Passive components may be further integrated by using either a thin- or thick-film chip process. (A discussion of thin- and thick-film chip techniques is beyond the scope of this text.) Due to the complexity of a hybrid chip, it is normally more expensive than its monolithic cousins. Although the IC itself may be more expensive, the complete application may very well wind up being less costly to produce, since the cost of other components are effectively absorbed within the hybrid IC. One place where hybrids are often used is in consumer stereo music systems. A hybrid power amplifier IC offers the convenience of a single IC solution with the capabilities of a discrete transistor approach. As an op amp user, it makes little difference whether the device is hybrid or monolithic when it comes to circuit analysis or design.

Summary

Op amps are presently in wide use in just about every aspect of linear electronics. An op amp is a multistage amplifier treated as a single entity. The first stage usually utilizes a differential amplifier which can be made with either bipolar or FET devices. The following stage(s) create a large voltage gain. The final stage is a class B voltage follower. The resulting op amp typically has a high input impedance, a low output impedance, and voltage gains in excess of 10,000. The op amp operates from a bipolar power supply, usually around ± 15 V. Externally, it has connections for the inverting and noninverting inputs, the single-ended output, and the power supplies. The op amp may be packaged in a variety of forms, including DIPs, SIPs, cans, flat packs and surface mount.

The general-purpose op amp is manufactured using a monolithic structure and a photolithographic process. Several chips are created from a single master wafer. Creation is a multistep process involving the selective doping of specific areas on the chip through diffusion. The monolithic technique is relatively inexpensive and accurate. Since integration allows for very tight part matching and consistency, certain circuit design techniques are favored, including the use of current mirrors and active loads.

Finally, with very little supporting circuitry, simple op amps can make effective comparator circuits. A comparator is, in essence, a bridge between the analog and digital worlds.

— Self Test Questions

1. What is an op amp?
2. Give several examples of where op amps might be used.
3. What is the typical stage layout of an op amp?
4. What comprises the first stage of a typical op amp?
5. What comprises the final stage of a typical op amp?
6. How does integrated circuit design differ from discrete design?
7. What is a comparator, and how might it be used?
8. What is meant by monolithic planar construction?
9. What is a mask, and how is it used in the construction of IC op amps?
10. What is the process of diffusion, and how does it relate to the construction of IC op amps?
11. What are the advantages of monolithic IC construction?
12. How does hybrid construction differ from monolithic construction?

— Problem Set

1. For the circuit in Figure 2.18, find v_{out} for the following inputs:
   a. 0 V   b. −1 V   c. +2 V

——— Figure 2.18

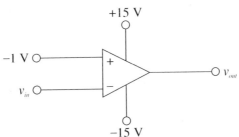

2. For the circuit in Figure 2.19, find v_{out} for the following inputs:
   a. 0 V   b. −4 V   c. +5 V   d. −.5 V

──────────── **Figure 2.19**

+15 V

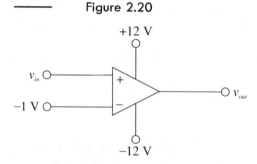

3. For the circuit in Figure 2.20, find v_{out} for the following inputs:
   a. 0 V   b. -2 V   c. $+1$ V   d. $-.5$ V

4. Sketch v_{out} if $v_{in}(t) = 1 \sin 60t$ in Figure 2.18.

5. Sketch v_{out} if $v_{in}(t) = 2 \sin 20t$ in Figure 2.19.

6. Sketch v_{out} if $v_{in}(t) = 3 \sin 10t$ in Figure 2.20.

──────────── **Figure 2.20**

+12 V

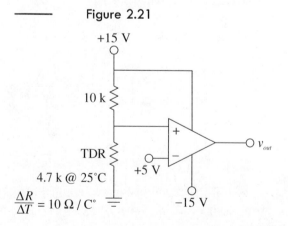

7. A temperature dependent resistor (TDR) is used in the comparator of Figure 2.21. At what temperature will the comparator change state?

──────────── **Figure 2.21**

+15 V

10 k

TDR

4.7 k @ 25°C

$\dfrac{\Delta R}{\Delta T} = 10 \ \Omega/C°$

+5 V

V_{out}

−15 V

8. What is the value of the light dependent resistor (LDR) at the comparator trip point in Figure 2.22?

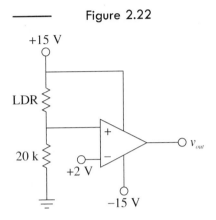

Figure 2.22

9. What reference voltage is required for a 50°C trip point in Figure 2.21?

10. A thermal fuse is a device found in common items such as coffee
makers. Normally, its resistance is very low (ideally 0 ohms). With the
application of excessive heat, the fuse opens, presenting a very high
resistance (ideally infinite). Explain the operation of the circuit in Figure
2.23. Is the output voltage normally high or low?

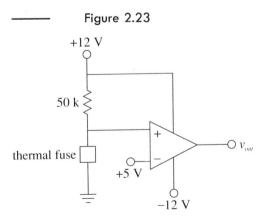

Figure 2.23

11. A strain gauge is a device which can be used to measure the amount of
bend or deflection in a part which it is attached to. It can be connected
so that its resistance rises as the bend increases. What resistance is required to trip the comparator of Figure 2.24?

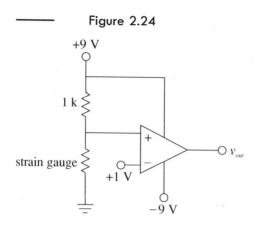

Figure 2.24

12. Determine the current flowing out of the collector of Q_{13} in Figure 2.5.

SPICE Problems

13. Alter the simple SPICE model presented in Figure 2.8 to include saturation effects. (*Hint*: consider a device which limits voltage.)

14. Describe the circuit of Figure 2.3 as a SPICE input file.

3

Negative Feedback

After completing this chapter, you should be able to:

❏ Give examples of how negative feedback is used in everyday life.

❏ Discuss the four basic feedback connections, detailing their similarities and differences.

❏ Detail which circuit parameters negative feedback will alter, and how.

❏ Discuss which circuit parameters are not altered by negative feedback.

❏ Define the terms *sacrifice factor*, *gain margin*, and *phase margin*, and relate them to a Bode plot.

❏ Discuss in general the limits of negative feedback in practical amplifiers.

3.1 Introduction

As we saw in the last chapter, op amps are very useful devices. However, in many applications, the device's gain is simply too large, and its bandwidth too narrow, for effective use. In this chapter we shall explore the concept of negative feedback. This concept is realized by *feeding* a portion of the output signal *back* to the input of the system. The proper use of negative feedback will allow us to exercise fine control over the performance of electronic circuits. As a matter of fact, negative feedback is so useful to us that we will seldom use op amps without it. Negative feedback is not tied solely to op amps though, as almost any electronic circuit may benefit from its application. As with most things, there are disadvantages as well. A successful design will minimize the disadvantages and capitalize on the positive aspects. We shall begin with the basic concepts of what negative feedback is and does, and then fine-tune our viewing by examining its four specific variants. We will look at specific examples of how negative feedback is applied to op amps, and conclude with a discussion of its practical limits.

3.2 What Negative Feedback Is and Why We Use It

People use negative feedback every day of their lives. In fact, we probably couldn't get along without it. Simply put, negative feedback is a very rudimentary part of intelligence. In essence, negative feedback lets something correct for mistakes. It tends to stabilize operations and reduce change. Negative feedback relies on a loop concept. In human terms, it is akin to knowing what you are doing, and being able to correct for mistakes as they happen. You are constantly evaluating and correcting your actions in order to achieve a desired goal. This may be stated as letting the input know what the output is doing. A good example of this is your ability to maintain a constant speed while driving along the highway. You have a desired result, or set point, in mind, say 60 mph. As you drive, you constantly monitor the speedometer. If you glance down and see that you're zipping along at 70 mph, you think "Ooops, I'm going a bit too fast" and lift your foot slightly off of the gas pedal. On the other hand, if you're only going 40 mph, you will depress the gas pedal further. The faster and more accurate your updates are, the better you will be at maintaining an exact speed.

In contrast to negative feedback is positive feedback, which reinforces change. If you were to correct your speed by saying "Hmmm, I'm going 70 mph, I'd better step on the gas," you'd be using positive feedback. Other examples of positive feedback include the "acoustic squeal" often heard over public address systems, and thermal runaway effects seen in discrete devices. When positive feedback is applied to normal amplifiers, they *oscillate*. That is, they produce their own signals without any input applied.

3.3 Basic Concepts

Seeing the usefulness of negative feedback, it would be nice if we could apply the concept to our electronic circuits. The basic idea is quite simple, really. What we will do is sample a piece of the output signal, and then add it to the input signal out of phase. By adding it out of phase, the circuit will see the difference between the input and the output. If the output signal is too large, the difference will be negative. Conversely, the difference signal will be positive if the output is too small. This signal is then multiplied by the circuit gain, and cancels the output error. Thus, the circuit will be presented with the undesired errors in such a way that it will force the output to compensate (move in the opposite direction). This process is done continuously: the only time lags involved are the propagation delays of the circuits used. Since the sampled output signal is effectively subtracted from the input signal, negative feedback is sometimes referred to as degenerative or destructive feedback. This subtraction can be achieved in a variety of ways. A differential amplifier is tailor-made for this task since it has one inverting input and one noninverting input. (Note: If the error is presented in phase, the circuit magnifies the errors and positive feedback results.)

To see an example of how this works, refer to Figure 3.1. The triangle represents an amplifying circuit. It has a gain of A. The output signal is also presented to the input of the feedback network represented by the box. This network scales the output signal by a factor, β. The feedback network ranges from very simple to complex. It may contain several resistors, capacitors, diodes, and what not, or it may be as simple as a single piece of wire. In any case, this scaled output signal is referred to as the *feedback signal* and is effectively subtracted from the input signal. This combination, called the *error signal*, is then fed to the amplifier, where it is boosted and appears at the output. The process repeats like this forever (or at least until the power is switched off). Let's assume that for some reason (perhaps a temperature change) the amplifier's gain were to rise. This should make the output signal increase by a similar percentage, but it doesn't. Here's why: As the output signal tries to rise, the feedback signal tracks with it. Now that there is a larger feedback signal, the error signal will become smaller (remember, error = input − feedback). This smaller signal is multiplied by the gain of the amplifier, thus producing a smaller output signal which almost completely offsets the original

Figure 3.1
Negative feedback

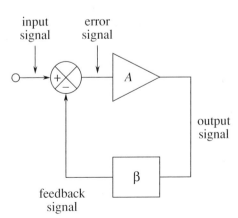

positive change. Note that if the output signal were too small, the error signal would increase, thus bringing the output back up to a normal level. When everything is working right, the feedback and input signals are almost the same size. (Actually, the feedback signal is somewhat smaller in magnitude.)

The Effects of Negative Feedback

Besides smoothing out gain anomalies, negative feedback can reduce the effect of device nonlinearities, thus producing a reduction in static forms of distortion such as THD (Total Harmonic Distortion). Basically, these nonlinearities can be viewed as a string of small gain errors. As such, they produce appropriate error signals and are compensated for in the above manner. Negative feedback can also increase the bandwidth of the system. It can increase the upper cutoff frequency f_2 and decrease the lower cutoff frequency f_1 (assuming the system has one). Also, we can exercise control over the input and output impedances of the circuit. It is possible to increase or decrease the impedances. As you might have guessed, we don't receive these benefits for nothing. The downside to negative feedback is that you lose gain. Effectively, you get to trade off gain for an increase in bandwidth, a decrease in distortion, and control over imped- ances. The more gain you trade off, the greater your rewards in the other three areas. In the case of our op amp, this is a wise trade-off since we already have more gain than we need for typical applications. This give-and-take is a very important idea, so remember "BIG D." That stands for **B**andwidth, **I**mpedance, **G**ain, and **D**istortion.

At this point, we need to define a few terms. *Closed loop* refers to the characteristics of the system when feedback exists. For example, *closed-loop gain* is the gain of the system with feedback, while *closed-loop frequency response* refers to the new system break points. Generalized closed-loop quantities will be shown with the subscript *cl*. Similarly, we will denote impedances, gain, and the like for specific feedback variants with a two-letter subscript abbre- viating the exact feedback configuration. One possibility for closed-loop gain would be A_{sp}. *Open loop* refers to the characteristics of the amplifier itself. To remember this, think of disconnecting or opening the path through the feedback network. Once the path is broken, the amplifier is on its own. *Open-loop gain* then, refers to the gain of the amplifier by itself, with no feedback. All open- loop quantities will be shown with the subscript *ol*. The symbol for open-loop gain will be A_{ol}. The term *loop gain* refers to the ratio between the open- and closed-loop gains. It may also be computed from their difference in decibels on a Bode plot. Loop gain indicates how much gain we have given up or sacrificed in order to enhance the operation of the system. Consequently, loop gain is often called *sacrifice factor*, and given the symbol S. Generally, trade- offs are proportional to the sacrifice factor. For example, if we cut the gain in half, we will generally double the bandwidth, and halve the distortion. An example which illustrates this can be seen in Figure 3.2. Note how the loop

Figure 3.2

Response with and without feedback

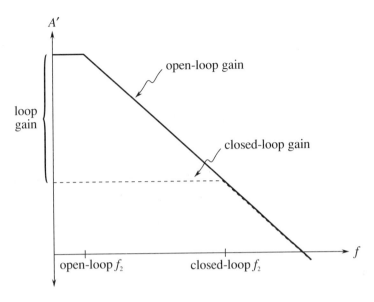

gain decreases with increasing frequency. This means that the effects of feedback at higher frequencies are not as great.

Up to this point we have made one simple assumption about our system, and that is that it exhibits no "extra" phase change beyond the desired inversion. As we saw in the first chapter, though, all circuits do produce phase changes as the input frequency is increased. If this extra phase change reaches $-180°$ while the gain is greater than unity (0 dB), our negative feedback will turn into positive feedback. (The inversion $= -180°$, plus this extra $-180°$, places us at $-360°$. The net result is an in-phase signal.) If this were to happen, our amplifier would no longer be stable. In fact, it may very well turn into a high-frequency oscillator. (You will see how to do this on purpose when you cover Chapter 9.) As the input frequency is raised, the phase will eventually exceed $-180°$ and the gain will drop to a fraction (<0 dB). The real key here is making sure that the phase never reaches $-180°$ when the gain falls to 1. Stated another way, when the phase hits $-180°$, the gain should be a fraction. Still another way of looking at this is to make sure that the curve never encloses the -1 point on a Nyquist plot. Generally, the farther you are from this "danger zone," the better. In other words, you're better off if the extra phase at the unity gain point is $-90°$ rather than $-170°$. Both are stable, but the first one gives you some breathing room.

Two measures of just how safe the circuit is are given by the *gain* and *phase margins*. Phase margin indicates the difference between the actual phase at unity gain and $-180°$. In the example above, the first circuit would have a phase margin of $180°$ while the second would have a margin of $10°$. Gain margin is the difference between the actual gain in dB at the $-180°$ phase point, and 0 dB. If our gain were -9 dB at $-180°$, the gain margin would be 9 dB (i.e., we have 9 dB "to spare"). Reasonable values for gain and phase margin are >6 dB and $>45°$. Gain and phase margin are depicted in Figure 3.3. It is possible to guarantee safe margins if the amplifier's open-loop response maintains a 20 dB-per-decade rolloff up to the unity gain frequency, f_{unity}. This means that there is only one dominating lag network, which will add a maximum phase shift of $-90°$. Even if the second network coincided with f_{unity}, it would add $-45°$ at most. This would still leave us with a 45° phase margin. (Note that if we had several secondary networks critical at f_{unity}, the phase could exceed $-180°$, however the slope would no longer be 20 dB per decade in reality.) It is for this reason that the general-purpose op amps examined in Chapter 2 included a compensating capacitor. No matter how much feedback we wish to use, our circuits will always end up being stable. For the best circuit performance, it is possible to use amplifiers that do not have the "constant rolloff" characteristic. The possibility exists that they may go into oscillation or become unstable if you are not careful and ignore the margins.

Figure 3.3
Gain and phase margin
graphically determined
from Bode plot

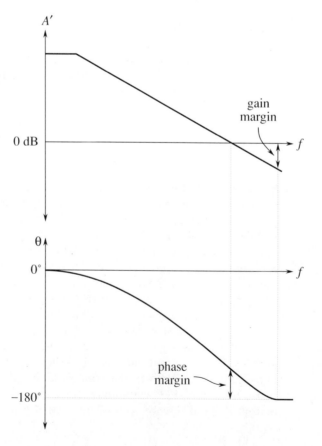

3.4 The Four Variants of Negative Feedback

Negative feedback can be achieved via four different forms of connections. They differ in how the input and output impedances are changed. We have basically two choices when it comes to connecting the input and output of the amplifier to the output and input of the feedback network. We can produce either a series connection or a parallel connection. This yields four possibilities total. Each connection will produce a specific effect on the input or output impedance of the system. As you might guess, parallel connections decrease the impedance and series connections increase it. A high input impedance is desirable for maximum voltage transfer, while a low impedance is required for maximum current transfer. As a memory aid, think of volt meters and ammeters. For the smallest loading effect, volt meters should exhibit a high impedance and ammeters a very low one. In the case of the output connection, a low source impedance is required for the best voltage transfer, and a high source impedance for current transfer. You should think of the ideal voltage source (zero Z_{out}) and the ideal current source (infinite Z_{out}) here. Consequently, if we were to connect our feedback network in series with the amplifier's input, and in parallel with its output, we would have an increase in Z_{in} and a decrease in Z_{out}. This means that our system would be very good at sensing an input voltage, and ideal for producing a voltage. We would have created a voltage-controlled voltage source (VCVS), the ideal voltage amplifier. So that you can get a good idea of the possibilities, all four types are summarized below.

Type (in-out)	Z_{in}	Z_{out}	Model	Idealization	Transfer Ratio
Series-parallel	high	low	VCVS	voltage amplifier	$\dfrac{v_{out}}{v_{in}}$ (voltage gain)
Series-series	high	high	VCCS	voltage-to-current transducer	$\dfrac{i_{out}}{v_{in}}$ (transconductance)
Parallel-parallel	low	low	CCVS	current-to-voltage transducer	$\dfrac{v_{out}}{i_{in}}$ (transresistance)
Parallel-series	low	high	CCCS	current amplifier	$\dfrac{i_{out}}{i_{in}}$ (current gain)

Generally speaking, the input and output impedances will be raised or lowered from the nonfeedback value by the sacrifice factor. Note that by using the proper form of feedback, we can achieve any of the possible models. This greatly enhances our ability to deal with specific applications. Much work in our field relies on optimal voltage transfer, therefore series-parallel is often used. A variation on parallel-parallel is also used frequently, as we shall see.

Series-Parallel Connections

The series-parallel (SP) connection makes for the ideal voltage amplifier. A generalized block diagram is shown in Figure 3.4. You can tell that it has a series input because there is no input-current node. Contrast this with the output: note that the op amp's output current splits into two paths, one through the load, the other into the feedback network. This output node clearly denotes the feedback's parallel output connection.

Let's take a look at exactly how SP negative feedback alters the system gain, impedances, and frequency response. First, let's examine the closed-loop gain (A_{sp}). Our amplifier block produces a gain A, and could be a diff amp, op amp, or other multistage possibility. The feedback network is typically a voltage divider and produces a loss β. The signal presented to the inverting input of the amp is the feedback signal, and is equal to $v_{out}\beta$. Note that the source's signal, v_{in}, is applied to the noninverting input. Therefore, the differential input voltage (usually referred to as v_{error}) equals $v_{in} - v_{feedback}$. We also know from previous work with differential amplifiers that $v_{out} = v_{error}A_{ol}$ (where A_{ol} is the amplifier's open-loop gain). In other words, we know:

$$v_{in} = v_{error} + v_{feedback} \tag{3.1}$$

$$v_{feedback} = v_{out}\beta \tag{3.2}$$

$$v_{out} = v_{error}A_{ol} \tag{3.3}$$

Figure 3.4   Series-parallel connection

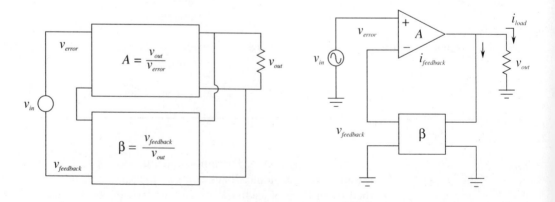

and by definition,

$$A_{sp} = \frac{v_{out}}{v_{in}} \tag{3.4}$$

Substituting Equation 3.3 into Equation 3.2,

$$v_{feedback} = v_{error}\beta A_{ol} \tag{3.5}$$

Substituting Equation 3.5 into Equation 3.1, and simplifying,

$$v_{in} = v_{error}(1 + \beta A_{ol}) \tag{3.6}$$

Finally, substituting Equations 3.6 and 3.3 into Equation 3.4 and simplifying, yields

$$A_{sp} = \frac{A_{ol}}{1 + \beta A_{ol}} \tag{3.7}$$

Since the fundamental definition of sacrifice factor, S, is A_{ol}/A_{cl}, we may also say $A_{sp} = A_{ol}/S$ and therefore, for SP,

$$S = 1 + \beta A_{ol} \tag{3.8}$$

Equation 3.7 is our general gain equation. However, if we can make $\beta A_{ol} \gg 1$, we can ignore the "$+1$" in the denominator and further simplify this as

$$A_{sp} = \frac{1}{\beta} \tag{3.9}$$

This seemingly innocent equation packs a rather hefty punch. It is telling us that the open-loop gain of the amplifier does **not** play a role in setting the system gain as long as the open-loop gain is very large. The system gain is controlled solely by the feedback network. Consequently, our amplifier can exhibit large gain changes in its open-loop response, but the closed-loop response will remain essentially constant. For this reason, we will achieve identical closed-loop gains for op amps which exhibit sizable differences in their open-loop gains. Since signal distortion is produced by nonlinearities which can be viewed as dynamic gain changes, our closed-loop distortion drops as well. Also, it is this very effect that extends our closed-loop frequency response. Imagine that our amplifier exhibits a gain of 10,000 at its upper-break frequency of 100 Hz. If the feedback factor equals .1, our exact gain is

$$A_{sp} = \frac{10,000}{1 + .1 \times 10,000} = 9.99$$

If we were to measure the amplifier's open-loop gain one decade up, at 1 kHz, it should be around 1000 (assuming 20 dB/decade loss). The closed-loop gain now equals

$$A_{sp} = \frac{1000}{1 + .1 \times 1000} = 9.9$$

As you can see, the closed-loop gain changed only about 1% despite the fact that the open-loop gain dropped by a factor of 10. If we continue to raise the frequency, A_{sp} will equal 9.09 at 10 kHz. Finally, at 100 kHz, a sizable drop is seen since the gain falls to 5. At this point, our assumption of $\beta A_{ol} \gg 1$ falls apart. Note, however, that our loss relative to the midband gain is only 6 dB. We have effectively stretched out the bandwidth of the system. Actually, this calculation is somewhat oversimplified, since we have ignored the extra phase lag produced by the amplifier above the open-loop break frequency. If we assume that the open-loop response is dominated by a single lag network (and it should be, in order to guarantee stability, remember?), a phase-sensitive version of Equation 3.7 would be

$$A_{sp} = \frac{-jA_{ol}}{1 - jA_{ol}\beta}$$

This extra phase will reach its maximum of $-90°$ approximately one decade above the open-loop break frequency. Consequently, when we find the magnitude of gain at 100 kHz, it's not

$$A_{sp} = \frac{10}{1 + 1}$$

but rather

$$A_{sp} = \frac{10}{\sqrt{1^2 + 1^2}}$$

which equals 7.07, for a -3 dB relative loss.

A simpler way of stating all of this is: The new upper-break frequency is equal to the open-loop upper break times the sacrifice factor, S. Since S is the loop gain, it is equal to A_{ol}/A_{sp}. Note that our low frequency $S = 10,000/10$, or 1000. Therefore, our closed-loop break equals 1000×100 Hz, or 100 kHz.

A very important item to notice here is that there is an inverse relation between closed-loop gain and frequency response. Systems with low gains will have high upper breaks, while high-gain systems will suffer from low upper breaks. This sort of trade-off is very common. While most diff amps and op amps do not have lower break frequencies, circuits that do will see an extension of their lower response in a similar manner (i.e., the lower break will be reduced by S). In order to achieve both high gain and wide bandwidth, it may be necessary to cascade multiple low-gain stages.

Example 3.1 Assume that you have an amplifier connected as in Figure 3.4. The open-loop gain (A_{ol}) of the amp is 200 and its open-loop upper break frequency (f_{2-ol}) is 10 kHz. If the feedback factor (β) is .04, what are the closed-loop gain (A_{sp}) and break frequency (f_{2-sp})?

To find A_{sp},

$$A_{sp} = \frac{A_{ol}}{1 + \beta A_{ol}}$$

$$A_{sp} = \frac{200}{1 + .04 \times 200}$$

$$A_{sp} = 22.22$$

Or, using the approximation,

$$A_{sp} = \frac{1}{\beta}$$

$$A_{sp} = \frac{1}{.04}$$

$$A_{sp} = 25$$

which is reasonably close to the general equations answer. (Note that there is no need to include phase effects since we are looking for the midband gain.) The approximation is more accurate when A_{ol} is larger.

To find f_{2-sp}, first find the sacrifice factor, S:

$$S = \frac{A_{ol}}{A_{sp}}$$

$$S = \frac{200}{22.22}$$

$$S = 9$$

$$f_{2-sp} = f_{2-ol}S$$

$$f_{2-sp} = 10\,\text{kHz} \times 9$$

$$f_{2-sp} = 90\,\text{kHz}$$

One interesting result to note is that the product of the gain and upper break frequency will always equal a constant value, assuming a 20 dB per decade rolloff. Our open-loop product is 200×10 kHz, or 2 MHz. Our closed-loop product is 22.22×90 kHz, which is 2 MHz. If we choose any other feedback factor, the resulting A_{sp} and $f_{2\text{-}sp}$ will also produce a product of 2 MHz. (Try it and see.) The reason for this is simple: A 20 dB per decade rolloff means that the gain drops by a factor of 10 when the frequency is increased by a factor of 10. There is a perfect 1:1 inverse relationship between the two parameters. No matter how much you increase one parameter, the other one will decrease by the exact same amount. Thus, the product is a constant.

At this point you may be asking yourself, "What exactly is in that feedback network and how do I figure out β?" Usually, the feedback network just needs to produce a loss—it has to scale v_{out} down to $v_{feedback}$. The simplest item for the job would be a resistive voltage divider. (It is possible to have complex frequency dependent or nonlinear elements in the network, as we shall see in the future.) An example is presented in Figure 3.5. If you study this diagram for a moment, you will notice that the feedback factor β is really nothing more than the voltage divider loss. v_{out} is the input to the feedback network and appears across $R_f + R_i \parallel Z_{in}$. The output of the network is $v_{feedback}$, which appears across R_i. Simply put, the ratio is

$$\beta = \frac{R_i \parallel Z_{in}}{R_f + R_i \parallel Z_{in}}$$

If Z_{in} is large enough to ignore, as in most op amps, this simplifies to

$$\beta = \frac{R_i}{R_f + R_i}$$

Figure 3.5
Simple voltage divider
for β

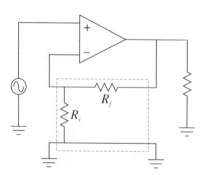

By substituting this equation into our approximate gain Equation 3.9, we find

$$A_{sp} = \frac{R_f + R_i}{R_i} = \frac{R_f}{R_i} + 1 \qquad (3.10)$$

Note that the values of R_f and R_i are not really important; rather, their ratio is. We would arrive at the same gain if $R_f = 10$ k and $R_i = 1$ k, or $R_f = 20$ k and $R_i = 2$ k. We obviously have quite a bit of latitude when designing circuits for a specific gain, but we do face a few practical limits. If the resistors are too small we will run into problems with op amp output current. On the other hand, if the resistors are too large, excessive noise, offset, drift, and loading effects will result. As a guideline for general-purpose circuits, $R_f + R_i$ is usually in the range of 10 k to 100 k.

Example 3.2   Let's say that the microphone that you use for acoustic instruments produces a signal that is just too weak for you to record without excessive tape hiss. After a little experimentation in lab, you discover that you need about 20 dB of voltage gain before you can successfully capture a softly picked guitar. Using Figure 3.6 as a guide, design this amplifier with the following device: $A_{ol} = 50,000$, $Z_{in\text{-}ol} = 600$ kΩ.

First, note that our A_{ol} and $Z_{in\text{-}ol}$ values are more than sufficient for us to use the approximation formulas. Since our formulas all deal with ordinary gain, we must convert 20 dB:

$$A_{sp} = \log^{-1} \frac{A'_{sp}}{20}$$

$$A_{sp} = \log^{-1} \frac{20 \text{ dB}}{20}$$

$$A_{sp} = 10$$

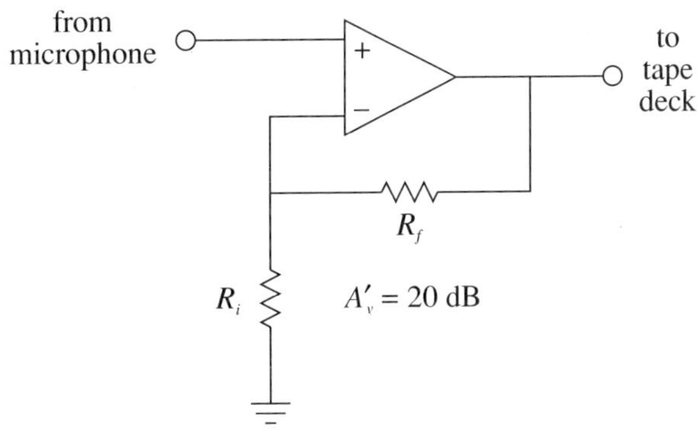

Figure 3.6
Microphone amplifier

By rearranging Equation 3.10, we find that

$$\frac{R_f}{R_i} = A_{sp} - 1$$

$$\frac{R_f}{R_i} = 9$$

and see that R_f must be 9 times larger than R_i. There is no single right answer here; there are many possibilities. One viable solution is simulated using SPICE in Figure 3.7. This circuit uses $R_f = 9\,k$ and $R_i = 1\,k$. The op amp model is the simple dependent source version examined in Chapter 2. The input signal is set to .1 V DC for simplicity. Both the output and feedback potentials are presented. The results of this simulation verify our hand calculations. In order to note the sensitivity of the design, you can alter certain parameters of the input file, and rerun the simulation. Two of the more interesting quantities are the absolute values of the feedback resistors and the open-loop gain of the op amp. You will note that, as these quantities are lowered, our approximation formulas become less accurate. With the given values, the approximations deviate from the SPICE results by less than 1%.

Figure 3.7
SPICE analysis of the simple op amp model of Example 3.2

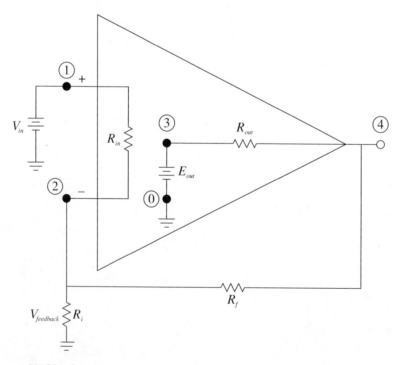

a. SPICE schematic

——————   Figure 3.7   (Continued)

```
*************************************** **** AMIGASPICE 5.0 ****************************************

****      INPUT LISTING                   TEMPERATURE=27.000 DEG C

***************************************************************************************************

*
*The op amp's ROUT is set to 75 ohms, and a .1 V DC input is used.
*The feedback resistors are 1K for Ri and 9K for Rf.
*Note that you can change RIN, ROUT, and the gain of EOUT with little
*change in the final output values. Rf and Ri may also be changed, as
*long as their ratio remains at 9:1
*
RIN    1   2   600K
ROUT   3   4   75
EOUT   3   0   1  2  50K
Ri     0   2   1k
Rf     4   2   9K
VIN    1   0   DC
*
*Set the input to .1 V DC with no sweeps, then print the desired
*potentials. V(4) is the output, V(2) is the feedback voltage.
*
.DC VIN 0.1 0.1 0.1
.PRINT DC V(4) V(2)
.END

*************************************** **** AMIGASPICE 5.0 ****************************************

****      DC TRANSFER CURVES               TEMPERATURE=27.000 DEG C

***************************************************************************************************

VIN         V(4)       V(2)
1.000E-01   9.998E-01  9.998E-02

JOB CONCLUDED
```

b. SPICE input and output listings

——————|

Series-Parallel Impedance Effects

As noted earlier, negative feedback affects the closed-loop input and output impedances of our system. Series connections increase the impedance and parallel connections decrease the impedance.

Let's see exactly how this works in the SP case. First, we must distinguish between the Z_{in} of the amplifier itself and the Z_{in} of the system with feedback.

Figure 3.8
Series-parallel input
impedance

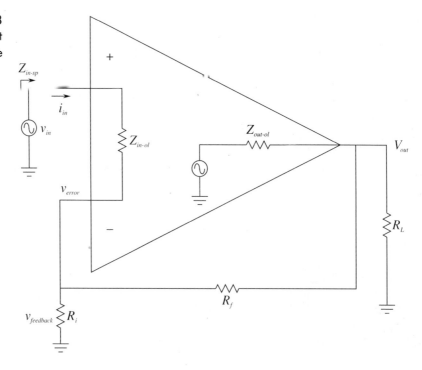

We shall call them $Z_{in\text{-}ol}$ and $Z_{in\text{-}sp}$, respectively. Figure 3.8 shows this difference by using a simple model of the amplifier. By definition,

$$Z_{in\text{-}sp} = \frac{v_{in}}{i_{in}}$$

The idea here is to notice that the source needs to supply only enough signal current to develop the v_{error} drop across the amplifier's $Z_{in\text{-}ol}$. As far as the v_{in} signal source is concerned, $v_{feedback}$ is a voltage source, not a voltage drop. Therefore, $i_{in} = v_{error}/Z_{in\text{-}ol}$. We may now say

$$Z_{in\text{-}sp} = \frac{v_{in}}{\dfrac{v_{error}}{Z_{in\text{-}ol}}} = \frac{Z_{in\text{-}ol}v_{in}}{v_{error}}$$

Since v_{error} ideally equals v_{out}/A_{ol},

$$Z_{in\text{-}sp} = \frac{Z_{in\text{-}ol}A_{ol}v_{in}}{v_{out}} = \frac{Z_{in\text{-}ol}A_{ol}}{A_{sp}}$$

Figure 3.9
Common-mode input
impedance

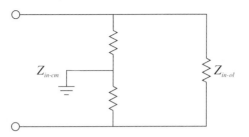

Sacrifice factor S is defined as A_{ol}/A_{sp}, so

$$Z_{in\text{-}sp} = Z_{in\text{-}ol}S$$

This is our ideal SP input impedance. Obviously, even moderate open-loop Z_{in}'s with moderate sacrifice factors can yield high closed-loop Z_{in}'s. The upper limit to this will be the impedance seen from each input to ground. In the case of a typical op amp, this is sometimes referred to as the common-mode input impedance, $Z_{in\text{-}cm}$, and can be very high (perhaps hundreds of megohms). This is the impedance presented to common-mode signals. This value effectively appears in parallel with our calculated $Z_{in\text{-}sp}$, above. An example is shown in Figure 3.9. Note that since $Z_{in\text{-}cm}$ is measured with the inputs of the op amp in parallel, each input has approximately twice the value to ground. In the case of a discrete amplifier, you would be most concerned with the noninverting input's Z_{in}. In any case, since S drops as the frequency increases, $Z_{in\text{-}sp}$ decreases as well. At very high frequencies, input and stray capacitances dominate, and the real system input impedance may be a small fraction of the low frequency value. Negative feedback cannot reduce effects which live outside of the loop.

Now for our $Z_{out\text{-}sp}$. Refer to Figure 3.10 on page 104. $Z_{out\text{-}sp}$ is the output impedance. In order to find it, we will drive the amplifier's output with a voltage source and reduce all other independent voltage sources to zero. (We have no current sources. Remember, this is a "paper" analysis technique and may not work in lab due to other factors.) By figuring out the resulting output current, we can find $Z_{out\text{-}sp}$ (by definition, $Z_{out\text{-}sp} = v_{out}/i_{out}$).

First, notice that i_{out} is made of two pieces, $i_{feedback}$ and i_{amp}. If we can find the two impedances associated with these parts, we can simply perform a parallel equivalent in order to determine $Z_{out\text{-}sp}$. The $i_{feedback}$ portion is very easy to determine. Ignoring the inverting input's loading effects on R_i, this impedance is just $R_f + R_i$. Finding the output impedance of the amplifier itself is a little more involved. i_{amp} is found by taking the drop across $Z_{out\text{-}ol}$ and using Ohm's Law. The voltage across $Z_{out\text{-}ol}$ is the difference between v_{out} and the signal created by the feedback path to the inverting input. This signal is

──────　**Figure 3.10**
Series-parallel output
impedance

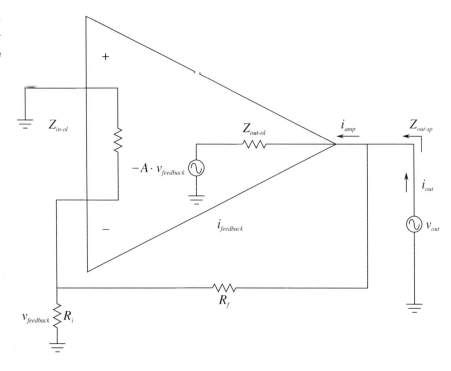

$-A_{ol}v_{feedback}$, so

$$i_{amp} = \frac{v_{out} - (-A_{ol}v_{feedback})}{Z_{out\text{-}ol}}$$

Since $v_{feedback} = v_{out}\beta$,

$$i_{amp} = \frac{v_{out} + A_{ol}\beta v_{out}}{Z_{out\text{-}ol}}$$

$$i_{amp} = \frac{v_{out}(1 + A_{ol}\beta)}{Z_{out\text{-}ol}}$$

By using Equation 3.8, this may be simplified to

$$i_{amp} = \frac{S v_{out}}{Z_{out\text{-}ol}}$$

Since

$$Z_{out\text{-}amp} = \frac{v_{out}}{i_{amp}}$$

we may say

$$Z_{out\text{-}amp} = \frac{Z_{out\text{-}ol}}{S}$$

This is the part of $Z_{out\text{-}sp}$ contributed by i_{amp}. To find $Z_{out\text{-}sp}$, just combine the two pieces in parallel:

$$Z_{out\text{-}sp} = \frac{Z_{out\text{-}ol}}{S} \parallel (R_f + R_i)$$

With op amps, $R_f + R_i$ is much larger than the part contributed by i_{amp}, and can be ignored. For example, a typical device may have $Z_{out\text{-}ol} = 75\ \Omega$. Even a very modest sacrifice factor will yield a value many times smaller than a typical $R_f + R_i$ combo (generally over 1 kΩ). Discrete circuits using common emitter or common base connections will have larger $Z_{out\text{-}ol}$ values, and therefore the feedback path may produce a sizable effect. As in the case of input impedance, $Z_{out\text{-}sp}$ is a function of frequency. Since S decreases as the frequency rises, $Z_{out\text{-}sp}$ will increase.

Example 3.3
An op amp has the following open-loop specs: $Z_{in} = 300$ kΩ, $Z_{out} = 100\ \Omega$, and $A = 50{,}000$. What are the low frequency system input and output impedances if the closed-loop gain is set to 100?
First we must find S:

$$S = \frac{A_{ol}}{A_{sp}}$$

$$S = \frac{50{,}000}{100} = 500$$

We may now find the approximate solutions:

$$Z_{in\text{-}sp} = SZ_{in\text{-}ol}$$

$$Z_{in\text{-}sp} = 500 \times 300\ \text{k}\Omega = 150\ \text{M}\Omega$$

$$Z_{out\text{-}sp} = \frac{Z_{out\text{-}ol}}{S}$$

$$Z_{out\text{-}sp} = \frac{100}{500} = .2\ \Omega$$

The effects are quite dramatic. Note that with such high Z_{in} values, op amp circuits may be used in place of FETs in some applications. One example is the front end amplifier/buffer in an electrometer (electrostatic voltmeter).

Distortion Effects

As noted earlier, negative feedback lowers static forms of distortion, such as THD. The question, as always, is by how much. Since something like harmonic distortion is internally generated, we can model it as a voltage source in series with the output source.

An example of this model can be seen in Figure 3.11. v_{dist} is the distortion generator. In the case of a simple input sine wave, v_{dist} will contain harmonics at various amplitudes (sine waves at integer multiples of the input frequency). These amplitudes are directly related to the input signal's amplitude. If we assume that Z_{out} is small enough to ignore, v_{dist} will appear at the output of the open-loop circuit. Thus, the total output voltage is the desired $Av_{in} + v_{dist}$. When we add feedback, as in Figure 3.12 (page 107), this distortion signal is fed back to the inverting input, and since it is now out of phase, it partially cancels the internally generated distortion. Thus the SP distortion signal ($v_{dist-sp}$) is much smaller. The SP output signal is

$$v_{out} = A_{ol}v_{error} + v_{dist}$$

$$v_{out} = A_{ol}(v_{in} - v_{feedback}) + v_{dist}$$

$$v_{out} = A_{ol}(v_{in} - \beta v_{out}) + v_{dist}$$

Figure 3.11
Distortion model
(open-loop)

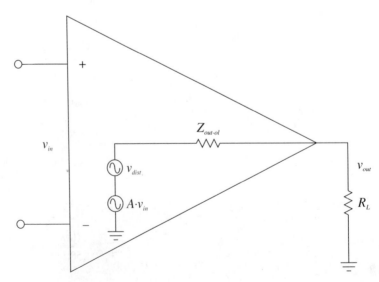

Figure 3.12
Distortion model
(closed-loop)

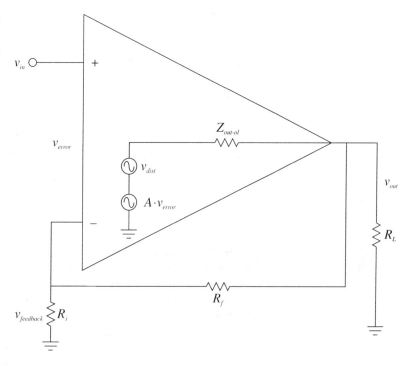

We now perform some algebra in order to get this into a nicer form and solve for v_{out}:

$$v_{out} = A_{ol}v_{in} - A_{ol}\beta v_{out} + v_{dist}$$

$$v_{out} + A_{ol}\beta v_{out} = A_{ol}v_{in} + v_{dist}$$

$$v_{out}(1 + A_{ol}\beta) = A_{ol}v_{in} + v_{dist}$$

Remember that $1 + A_{ol}\beta = S$, so

$$v_{out}S = A_{ol}v_{in} + v_{dist}$$

$$v_{out} = \frac{A_{ol}v_{in}}{S} + \frac{v_{dist}}{S}$$

Since A_{ol}/S is just A_{sp}, this reduces to

$$v_{out} = A_{sp}v_{in} + \frac{v_{dist}}{S}$$

The internally generated distortion is reduced by the sacrifice factor. As you can see, large sacrifice factors can drastically reduce distortion. An amplifier with 10% THD and a sacrifice factor of 100 produces an effective distortion of only .1%. This analysis does assume that the open-loop distortion is not overly grotesque. If the distortion is large, we cannot use this superposition approach. (Remember, superposition assumes that the circuit is essentially linear.) Also, we are ignoring any additional distortion created by feeding this distortion back into the amplifier. For any reasonably linear amplifier, this extra distortion is only a small part of the total.

As we have seen, the sacrifice factor is a very useful item. Our gain, distortion, and Z_{out} are all reduced by S, while f_2 and Z_{in} are increased by S.

Noise

It is possible to model noise effects in much the same way as we just modeled distortion effects. By doing so, you will discover that noise can also be decreased by a large amount. Unfortunately, there is one major flaw: Unlike our distortion generator, a noise generator will produce a signal that is not dependent on the input signal. The net result is that, while the noise level does drop by the sacrifice factor, so does the desired output signal. The signal-to-noise ratio at the output is unchanged. In contrast, the distortion signal is proportional to the input signal, so that when the desired signal is cut by S, the distortion signal sees a further cut by S (i.e., the distortion drops by S relative to the desired signal). As a matter of fact, it is quite possible that the noise produced in following stages may add up to more noise than the circuit has without feedback. Sad but true, negative feedback doesn't help us much when it comes to signal-to-noise ratio.

Parallel-Series Connections

The parallel-series (PS) connection is the opposite of the series-parallel form. PS negative feedback is used to make the ideal current amplifier. Its gain is dimensionless, but for convenience, it is normally given the units A/A (amps per amp). It produces a low Z_{in} (perfect for sensing i_{in}) and a high Z_{out} (making for an ideal current source).

An example of PS is shown in Figure 3.13 on page 109. The signal source's current splits in two, with part going into the amplifier, and part going through the feedback network. This is how you know that you have a parallel input connection. The output current, on the other hand, passes through the load and then enters the feedback network, indicating a series output connection. Note that this general model is an inverting type, and that the load is floating (i.e., not ground-referenced). It is possible, though, to use ground-referenced loads with some additional circuitry.

Figure 3.13 Parallel-series connection

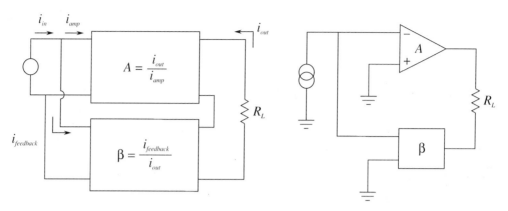

Since PS is used for current amplification, let's see how we can find the current gain. Figure 3.14 will help us along. PS current gain is defined as

$$A_{ps} = \frac{i_{out}}{i_{in}} \tag{3.11}$$

The signal source's current splits into two paths, so

$$i_{in} = i_{amp} + i_{feedback} \tag{3.12}$$

Since i_{amp} multiplied by the amplifier's open-loop current gain is i_{out}, we can also say

$$i_{amp} = \frac{i_{out}}{A_{ol}} \tag{3.13}$$

Figure 3.14
Parallel-series analysis

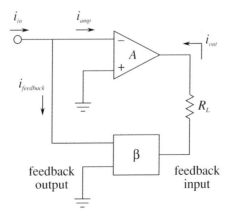

The feedback network is nothing more than a current divider, where the feedback network's output ($i_{feedback}$) is β times smaller than its input (i_{out}). Don't let the arbitrary current direction fool you—the feedback "flow" is still from right to left as always. In this case, the amplifier is sinking current instead of sourcing it. Since β is just a fraction, we may say

$$\beta = \frac{i_{feedback}}{i_{out}}$$

or

$$i_{feedback} = \beta i_{out}$$

$$i_{feedback} = A_{ol}\beta i_{amp} \qquad (3.14)$$

By substituting Equation 3.14 into Equation 3.12 we see that

$$i_{in} = i_{amp} + A_{ol}\beta i_{amp}$$

or

$$i_{in} = i_{amp}(1 + A_{ol}\beta) \qquad (3.15)$$

After substituting this result into Equation 3.11 we find that

$$A_{ps} = \frac{i_{out}}{i_{amp}(1 + A_{ol}\beta)}$$

or, with the help of Equation 3.13,

$$A_{ps} = \frac{A_{ol}}{1 + A_{ol}\beta} \qquad (3.16)$$

To make a long story short, the open-loop gain is reduced by the sacrifice factor. (Where have we seen this before?) The one item that you should note is that we have used only current gains in our derivation (compared to voltage gains in the SP case). It is possible to perform a derivation using the open-loop voltage gain; however, the results are basically the same, as you might have guessed. Once again, our approximation for gain can be expressed as $1/\beta$.

Example 3.4   One day while you are working in the lab, a co-worker walks in and asks you to measure the output current of a new circuit that he's just built. If all is working correctly, this circuit should produce 100 μA. Unfortunately, your

Figure 3.15
Current amplifier for
Example 3.4

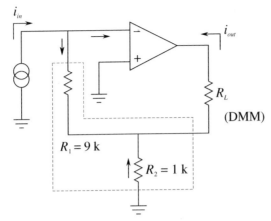

hand-held DMM will accurately measure down to only 1 mA. It is 10 times less sensitive than it needs to be. Being the resourceful person that you are, you refer to Figure 3.15. Does this amplifier have the extra current gain you need?

If you are using an op amp, you can assume that $A_{ps} = 1/\beta$. The question then becomes, "What is β?" β is the current divider ratio. The resistors R_1 and R_2 make up the current divider. The output current splits between R_1 and R_2, where the R_1 path is $i_{feedback}$. According to the current divider rule,

$$\beta = \frac{R_2}{R_1 + R_2}$$

Consequently,

$$A_{ps} = \frac{R_1 + R_2}{R_2}$$

or

$$A_{ps} = \frac{R_1}{R_2} + 1$$

This equation looks a lot like the one in our SP example, doesn't it? Solving for A_{ps} yields

$$A_{ps} = \frac{9\,k}{1\,k} + 1$$

$$A_{ps} = 10$$

Yes, this circuit has just the gain you need. All you have to do is connect your co-worker's circuit to the input, and replace the load resistor with your hand-held DMM. Note that most DMMs are floating instruments and are not tied to ground (unlike an oscilloscope), so using one as a floating load presents no problems. The accuracy of the gain (and thus your measurement) depends on the accuracy of the resistors and the relative size of the op amp's input bias current. Therefore it would be advisable to use a bi-FET type device (i.e., an op amp with an FET diff amp front end) and precision resistors.

Parallel-Series Impedance Effects

As you have seen, the gain derivation for PS is similar to that for SP. The same is true for the impedance equations. First, we'll take a look at input impedance with Figure 3.16. As always, we start with our base definition:

$$Z_{in\text{-}ps} = \frac{v_{in}}{i_{in}}$$

Recalling Equation 3.15, this can be rewritten as

$$Z_{in\text{-}ps} = \frac{v_{in}}{i_{amp}(1 + A_{ol}\beta)} \tag{3.17}$$

v_{in} is merely the voltage that appears from the inverting input-to-ground. By using Ohm's Law, we can say

$$v_{in} = i_{amp}Z_{in\text{-}ol}$$

Figure 3.16
Parallel-series input impedance

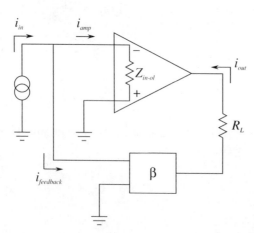

where $Z_{in\text{-}ol}$ is the open-loop input impedance. Finally, substituting this into Equation 3.17 gives

$$Z_{in\text{-}sp} = \frac{i_{amp}Z_{in\text{-}ol}}{i_{amp}(1 + A_{ol}\beta)}$$

$$Z_{in\text{-}sp} = \frac{Z_{in\text{-}ol}}{1 + A_{ol}\beta}$$

The input impedance is lowered by the sacrifice factor. (If you're starting to wonder whether everything is altered by the sacrifice factor, the answer is, *yes*. This is, of course, an approximation.) At this point, what happens to the output impedance shouldn't be too surprising. For this proof, refer to Figure 3.17. We shall use the same general technique to find Z_{out} that we did with the SP configuration. In this case, we replace the load with a current source, and then determine the resulting output voltage. Note that the input signal current source is opened. The output current drives the feedback network and produces the feedback current. The feedback current is then multiplied by the amplifier's open-loop current gain. Because the feedback current enters the inverting input, the internal source is sinking current. Since we are driving the

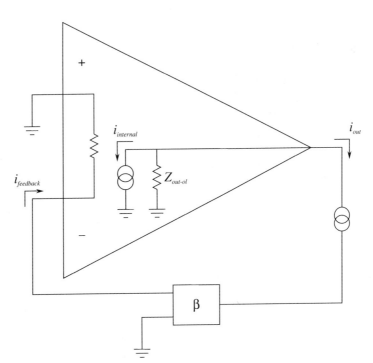

Figure 3.17
Parallel-series output
impedance

circuit from the output,

$$Z_{out-ns} = \frac{v_{out}}{i_{out}}$$

v_{out} is found through Ohm's Law:

$$v_{out} = Z_{out-ol}(i_{out} + i_{internal})$$

We now expand on our currents.

$$i_{internal} = i_{feedback}A_{ol}$$

$$i_{internal} = i_{out}\beta A_{ol}$$

$$v_{out} = Z_{out-ol}(i_{out} + i_{out}\beta A_{ol})$$

$$v_{out} = Z_{out-ol}i_{out}(1 + A_{ol}\beta)$$

$$v_{out} = Z_{out-ol}i_{out}S$$

Finally, we see that

$$Z_{out-ps} = \frac{Z_{out-ol}i_{out}S}{i_{out}}$$

$$Z_{out-ps} = Z_{out-ol}S$$

As expected, the series output connection increased Z_{out} by the sacrifice factor. The remainder of the PS equations are essentially those used in our earlier SP work. Once again, the bandwidth will be increased by S, and the distortion will be reduced by S. This is also true for the parallel-parallel and series-series connections. As a matter of fact, the Z_{in} and Z_{out} relations are just what you might expect. The proofs are basically the same as those already presented, so we won't go into them. Suffice to say that parallel connections reduce the impedance by S, and series connections increase it by S.

Parallel-Parallel Connections

Unlike our two earlier examples, the concept and modeling of gain is not quite as straightforward in the parallel-parallel (PP) case. Parallel-parallel does not produce gain, so to speak. It is neither a voltage amplifier, nor a current amplifier. Instead, this connection is used as a *transducer*. It turns an input current into an output voltage. Consequently, our normal gain and feedback factors have units associated with them. Since our output quantity is measured

Figure 3.18   Parallel-parallel connection

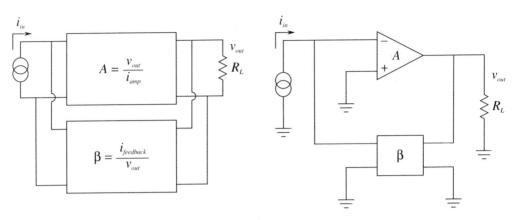

in volts and our input quantity is in amps, the appropriate units for our factors are ohms and siemens. To be specific, we refer to our gain as a transresistance value. Our general model is shown in Figure 3.18. Note that the A and β factors are measured in volts per amp and amps per volt, respectively. Our only new item of interest here is to find the gain with feedback. By definition, our gain in Figure 3.19 is

$$A_{pp} = \frac{v_{out}}{i_{in}}$$

As in the PS case, i_{in} splits into two paths:

$$i_{in} = i_{amp} + i_{feedback}$$

Figure 3.19
Parallel-parallel analysis

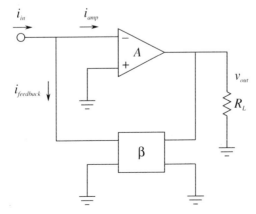

Let's write $i_{feedback}$ in terms of i_{amp}:

$$i_{feedback} = \beta v_{out}$$

$$i_{feedback} = A_{ol}\beta i_{amp}$$

Also, we can rewrite v_{out} as

$$v_{out} = i_{amp}A_{ol}$$

(remember, A_{ol} is measured in volts per amp), so

$$A_{pp} = \frac{i_{amp}A_{ol}}{i_{amp} + A_{ol}\beta i_{amp}}$$

or

$$A_{pp} = \frac{A_{ol}}{1 + A_{ol}\beta}$$

It is imperative to remember the units here. Since β is measured in amps per volt, its units cancel those of A_{ol} in the denominator. As a result, the numerator is divided by a pure (unitless) number. This means that the final units for A_{pp} are the same as those of A_{ol}. This all works well with your own discrete designs, but with op amps, the transresistance is never specified. Fortunately, we can make a few approximations and create some very useful circuits utilizing PP feedback and op amps. These design and analysis shortcuts are presented in the next chapter, along with practical applications.

Series-Series Connections

As with the PP connection, the series-series (SS) connection is used to create a transducer. Series-series creates the ideal voltage-to-current converter. This idea should not be new to you. After all, the basic model of an FET is that of a voltage-controlled current source. As in the FET case, we are concerned with the system's transconductance. The basic model for SS feedback is shown in Figure 3.20 on page 117. Again, note the units of the gain and feedback factors. In order to find the gain of Figure 3.21 (page 117), we start with the base definition

$$A_{ss} = \frac{i_{out}}{v_{in}}$$

We may also say $i_{out} = v_{error}A_{ol}$. (Remember, A_{ol}'s units are amps per volt.) As

Figure 3.20   Series-series connection

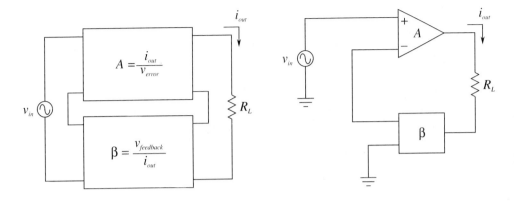

Figure 3.21
Series-series analysis

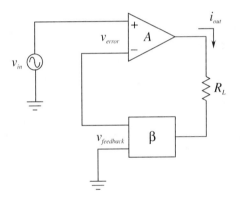

with the SP case, v_{in} is made up of two pieces:

$$v_{in} = v_{error} + v_{feedback}$$

$v_{feedback}$ is derived from i_{out}. Expanding on this, we find

$$v_{feedback} = \beta i_{out}$$

$$v_{feedback} = A_{ol}\beta v_{error}$$

$$v_{in} = v_{error} + A_{ol}\beta v_{error}$$

$$v_{in} = v_{error}(1 + A_{ol}\beta)$$

Finally,

$$A_{ss} = \frac{v_{error} A_{ol}}{v_{error}(1 + A_{ol}\beta)}$$

$$A_{ss} = \frac{A_{ol}}{1 + A_{ol}\beta}$$

This is as expected. Once again, the gain is reduced by the sacrifice factor. Like its PP counterpart, the denominator units cancel, thus leaving the final units the same as A_{ol}'s. The units for transconductance are, of course, siemens. This makes for a reasonably efficient application in discrete circuits, although, once again, op amps are not generally rated with a transconductance. The high gain of op amps will allow us to make certain idealizations and approximations in the next chapter.

3.5 Limitations On The Use of Negative Feedback

From the foregoing discussion, you may well think that negative feedback can do just about anything, short of curing a rainy day. Such is not the case. Negative feedback *can* drastically lower distortion and increase bandwidth. It *can* have a very profound effect on input and output impedance. And yes, it certainly does stabilize our gains.

However, there are limitations to the usefulness of negative feedback. The first fact that you should take note of is that S is a function of frequency, as was graphically depicted in Figure 3.2. Since the amount of change seen in impedances and distortion is a function of S, it follows that these changes must be a function of frequency. Since S drops as the frequency increases, the effects of negative feedback diminish as well. For example, if an SP amplifier has an open-loop Z_{in} of 200 kΩ and the low frequency S is 500, the resulting Z_{in} with feedback is 100 MΩ. If we increase the input frequency past the open-loop f_2, the open-loop gain drops, and thus S drops. One decade up, S will only be 50, so the Z_{in} with feedback will only be 10 MΩ. If this amplifier has a lower break frequency (f_1), S will also drop as the frequency is reduced (below f_1). The same sort of effect occurs with distortion; however, the harmonics each see a different S, so the calculation is a bit more involved. Along with the reduction in gain, there is also a change in phase. If the phase around the feedback loop varies from $-180°$, incomplete cancellation takes place, and thus, the effects of feedback are lessened. The bottom line is that the effects of negative feedback weaken as we approach the frequency extremes.

The other fact that must be kept in mind is that negative feedback does not change specific fundamental characteristics of the amplifier. Negative feedback cannot get a circuit to do something beyond its operational parameters. For example, feedback has no effect on *slew rate* (the maximum rate of output signal change, and an item which we will examine in a later chapter). Actually, when an amplifier slews, feedback is effectively blocked. An accurate output signal is no longer sent back to the input, but the amplifier can't correct for the errors any faster than it already is. In a similar manner, even though feedback can be used to lower the output impedance of a system, it does not enable the system to produce more output current.

One possible problem with negative feedback is really the fault of the designer. It can be very tempting to sloppily design an amplifier with poor characteristics and then correct for them with large amounts of feedback. No matter how much feedback is used, the result will never be as good as a system that was designed carefully from the start. An example of this effect can be seen with TIMD (Transient Inter-Modulation Distortion). TIMD is a function of nonlinearities in the first stages of an amplifier, and the excessive application of negative feedback will not remove it. On the other hand, if the initial stages of the system are properly designed, TIMD is not likely to be a problem.[1]

Summary

We have seen that negative feedback can enhance the performance of amplifier circuits. This is done by sampling a portion of the output signal and summing it out of phase with the input signal. In order to maintain stability, the gain of the amplifier must be less than unity by the time its phase reaches $-180°$. There are four basic variants of negative feedback: series-parallel, parallel-series, parallel-parallel, and series-series. In all cases, gain and distortion are lowered by the sacrifice factor S, while the bandwidth is increased by S. Parallel connections reduce impedance by S while series connections increase the impedance by S. At the input, parallel connections are current-sensing, while series connections are voltage-sensing. At the output, parallel connections produce a voltage-source model, while series connections produce a current-source model. The sacrifice factor is the ratio between the open- and closed-loop gains. It is a function of frequency, and, therefore, the effects of negative feedback lessen at the frequency extremes.

---

[1] See E. M. Cherry, K. P. Dabke, "Transient Intermodulation Distortion- Part 2: Soft Nonlinearity," *Journal of the Audio Engineering Society*, Vol. 34 No. 1/2, pp. 19–35.

---   **Self Test Questions**

1. Give two examples of how negative feedback is used in everyday life.
2. What circuit parameters will negative feedback alter, and to what extent?
3. What is meant by the term *sacrifice factor*?
4. What is the usage of the *gain and phase margins*?
5. Name the different negative feedback connections (i.e., variants or forms).
6. How might negative feedback accidentally turn into positive feedback?
7. What circuit parameters won't negative feedback affect?
8. In practical amplifiers, when does negative feedback "stop working," and why?

---   **Problem Set**

▷ **Analysis Problems**

1. An amplifier's open-loop gain plot is given in Figure 3.22. If the amplifier is set up for a closed-loop gain of 100, what is the sacrifice factor (S) at low frequencies? What is S at 1 kHz?

——— **Figure 3.22**

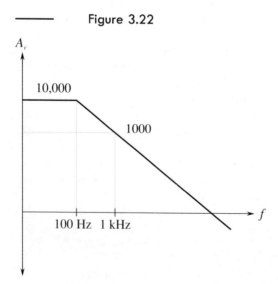

2. If the amplifier in Problem 1 has an open-loop THD of 5%, what is the closed-loop THD at low frequencies? Assuming that the open-loop THD doesn't change with frequency, what is the closed-loop THD at 1 kHz?

3. If an amplifier has an open-loop response as given in Figure 3.22, and a feedback factor (β) of .05 is used, what is the exact low-frequency closed-loop gain? What is the approximate low-frequency gain? What is the approximate gain at 1 kHz?

4. Using the open-loop response curve in Figure 3.23, determine exact and approximate values of β for a closed-loop gain of 26 dB.

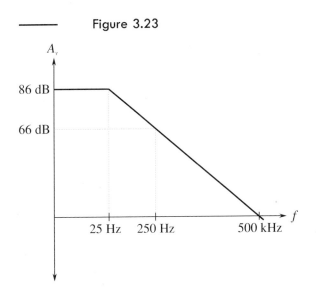

Figure 3.23

5. Determine the closed-loop f_2 for the circuit of Problem 4.

6. What is the maximum allowable phase shift at 500 kHz for stable usage of negative feedback in Figure 3.23?

7. Determine the gain and phase margins for the amplifier response given in Figure 3.24. Is this amplifier a good candidate for negative feedback?

8. Given the Nyquist plot in Figure 3.25, would this amplifier be stable if negative feedback were used?

9. Determine the closed-loop (midband) gain in Figure 3.26.

10. What is the closed-loop Z_{in} in Figure 3.26? What is Z_{out}?

11. What is the low-frequency sacrifice factor in Figure 3.26?

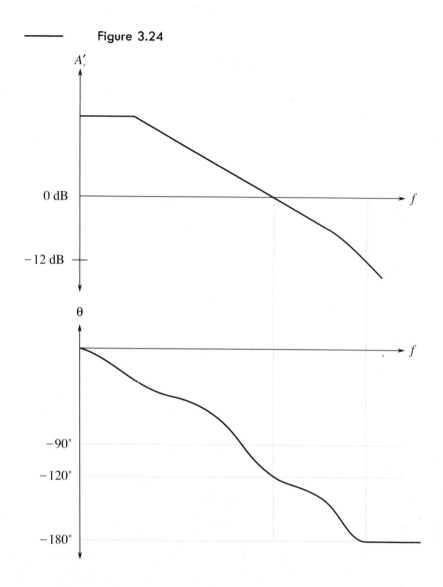

Figure 3.24

12. How much distortion reduction can we hope for in Figure 3.26?

13. How much of a signal-to-noise improvement can we expect in Figure 3.26?

14. Assuming $v_{in}(t) = .1 \sin 2\pi 500t$, what is $v_{out}(t)$ in Figure 3.26?

15. If the circuit in Figure 3.26 had an open-loop f_1 of 10 Hz, what would the closed-loop f_1 be?

16. Determine an appropriate pair of resistors to set β to .1 in Figure 3.26.

Figure 3.25

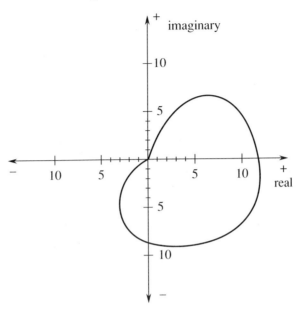

Figure 3.26

$f_2 = 200$ Hz
$Z_{in} = 10$ kΩ
$Z_{out} = 500$ Ω

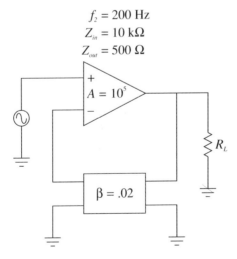

Figure 3.27

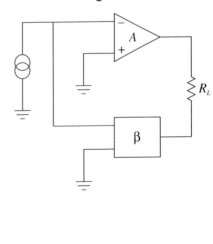

17. In Figure 3.27, determine the appropriate value of closed-loop transconductance such that 1 V in will produce .5 mA through the load.

18. If the circuit in Figure 3.27 has open-loop Z_{in} and Z_{out} values of 20 kΩ and 600 Ω, respectively, what are the closed-loop impedance values if $S = 300$?

19. If the open-loop distortion in Problem 18 is 7%, what will the distortion be after feedback is applied?

20. The circuit in Figure 3.28 has an open-loop current gain of 100,000. We desire a 3 mA load current for an input current of .1 mA. What must β be? What is S?

Figure 3.28

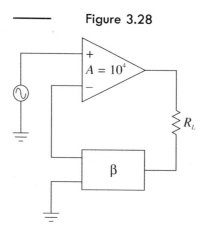

21. If the $Z_{in\text{-}ol}$ of the amplifier in Problem 20 is 50 kΩ, what is the value after feedback is applied?

22. Assuming that the open-loop f_2 is 20 Hz in Problem 20, what is the closed-loop f_2?

23. Determine the input and output impedances in Figure 3.29 if $\beta = .04$.

Figure 3.29

$$f_2 = 1 \text{ kHz} \qquad A_{ol} = 5000$$
$$Z_{in} = 40 \text{ k}\Omega \qquad Z_{out} = 1 \text{ k}\Omega$$

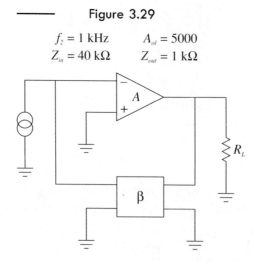

24. Assuming that the open-loop distortion is 11% in Problem 23, what is the closed-loop distortion?

25. If we desire a .1 V output for an input current of 1 mA in the circuit of Figure 3.29, what must β be (approximately)?

▷ Challenge Problems

26. If the feedback network of Figure 3.29 produces a phase shift of $-200°$ at 4 kHz, what effect will this have on circuit operation in Problem 23?

27. Consider the circuit of Figure 3.6. In general, what effect would the following alterations to the feedback network have on the closed-loop system response?
   a. Placing a capacitor across R_f
   b. Placing a capacitor across R_i
   c. Placing a rectifying diode across R_f (both polarities)

▷ SPICE Problems

28. Rerun the SPICE simulation of Figure 3.7 using the following open-loop gains: 1 k, 10 k, 100 k. What can you conclude from the results?

29. Verify the results of Problem 14 using SPICE. It is possible to extend the basic op amp model with a lag network in order to mimic $f_{2\text{-}ol}$.

30. Verify the stability of the circuit used in Problem 14 with regard to the open-loop gain. Run simulations with the given A_{ol}, and with values one decade above and below. Compare the resulting simulations and determine the maximum deviation of A_{sp}.

31. Create a simple SPICE model suitable for a circuit such as the one in Figure 3.27, using a dependent current source.

4 Basic Op Amp Circuits

After completing this chapter, you should be able to:

❑ Relate each op amp circuit to its general feedback form.

❑ Detail the general op amp circuit analysis idealizations.

❑ Solve inverting and noninverting voltage amplifier circuits for a variety of parameters, including gain and input impedance.

❑ Solve voltage/current transducer circuits for a variety of parameters.

❑ Solve current-amplifier circuits for a variety of parameters.

❑ Define the term *virtual ground*.

❑ Analyze differential amplifiers.

❑ Analyze inverting and noninverting summing amplifiers.

❑ Discuss how output current capability can be increased.

❑ Outline the circuit modifications required for operation from a single-polarity power supply.

4.1 Introduction

In this chapter we will be examining some common uses of op amps. While Chapter 3 focused on the theory of negative feedback in a more abstract way, this chapter zeroes in on the practical results of using negative feedback with op amps. The main theme is the design and analysis of simple small-signal linear amplifiers. We will also be looking at some convenient approximation methods of analysis which can prove quite efficient. By the end of the chapter, you will be able to design single and multistage voltage amplifiers, voltage followers, and even amplifiers which sense current and/or produce constant current output. The voltage amplifiers can be inverting or noninverting. Simple differential amplifiers are examined, too. The chapter wraps up with sections on using op amps with single-polarity power supplies, and on how to increase the available output current.

Since this is an introductory design section, the details of high-frequency response, noise, offsets, and other important criteria are ignored. These items await a detailed analysis in the following chapter. For the most part, all calculations and circuit operations in this chapter are assumed to be in the midband region.

4.2 Inverting and Noninverting Amplifiers

As noted in our earlier work, negative feedback can be applied in one of four ways. The parallel input form inverts the input signal, and the series input form doesn't. Since these forms were presented as current-sensing and voltage-sensing, respectively, you might get the initial impression that all voltage amplifiers must be noninverting: This is not the case. With the simple inclusion of one or two resistors, for example, we can make inverting voltage amplifiers or noninverting current amplifiers. Virtually all topologies are realizable. First we will look at the controlled-voltage source forms (those using SP and PP negative feedback).

For analysis, you can use the classic treatment given in Chapter 3; however, due to some rather nice characteristics of the typical op amp, approximations will be shown. These approximations are only valid in the midband, and say nothing of the high-frequency performance of the circuit. Therefore, they are not suitable for general-purpose discrete work. The idealizations for the approximations are:

1. The input current is virtually zero (Z_{in} is infinite).

2. The potential difference between the inverting and noninverting inputs is virtually zero (loop gain is infinite). This signal is also called the error signal.

Also note that, for simplicity, the power supply connections are not shown in most of the diagrams.

The Noninverting Voltage Amplifier

The noninverting voltage amplifier is based upon SP negative feedback. An example is given in Figure 4.1. Note the similarity to the generic SP circuits of Chapter 3. Recalling the basic action of SP negative feedback, we expect a very high Z_{in}, very low Z_{out}, and a reduction in voltage gain. Idealization 1 states that Z_{in} must be infinite. Since we already know that op amps have low Z_{out}'s, the second idealization is also taken care of. Now, let's take a took at voltage gain:

$$A_v = \frac{v_{out}}{v_{in}}$$

Since ideally $v_{error} = 0$,

$$v_{in} = v_{R_i}$$

Also,

$$v_{out} = v_{R_i} + v_{R_f}$$

$$A_v = \frac{v_{R_i} + v_{R_f}}{v_{R_i}}$$

Expansion gives

$$A_v = \frac{R_i i_{R_i} + R_f i_{R_f}}{R_i i_{R_i}}$$

——————

Figure 4.1
Noninverting voltage
amplifier

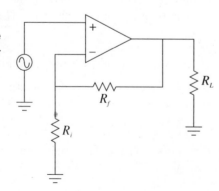

Since $i_{in} = 0$,

$$i_{R_f} = i_{R_i}$$

So, finally we arrive at

$$A_v = \frac{R_i + R_f}{R_i}$$

or
$$A_v = 1 + \frac{R_f}{R_i} \tag{4.1}$$

Now that's convenient. The gain of this amplifier is set by the ratio of two resistors. The larger R_f is relative to R_i, the larger the resulting gain. (Remember, this is an approximation.) The closed-loop gain can never exceed the open-loop gain, and eventually A_v will fall off as frequency increases. Note that the calculation ignores the effect of the load impedance. Obviously, if R_L is too small, the excessive current draw will cause the op amp to clip.

Example 4.1   What are the input impedance and gain of the circuit in Figure 4.2?

Figure 4.2
Noninverting circuit for
Example 4.1

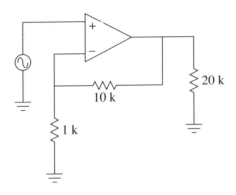

First, Z_{in} is ideally infinite. Now to find the gain:

$$A_v = 1 + \frac{R_f}{R_i}$$

$$A_v = 1 + \frac{10 \text{ k}}{1 \text{ k}}$$

$$A_v = 11$$

The opposite process of amplifier design is just as straightforward.

Example 4.2   Design an amplifier with a gain of 26 dB and an input impedance of 47 k.
To find the gain, first turn 26 dB into ordinary form. This is a voltage gain of about 20:

$$A_v = 1 + \frac{R_f}{R_i}$$

$$\frac{R_f}{R_i} = A_v - 1$$

$$\frac{R_f}{R_i} = 19$$

At this point, choose a value for one of the resistors, and solve for the other one. For example, all of the following are valid:

$$R_i = 1\text{ k}, \qquad R_f = 19\text{ k}$$
$$R_i = 2\text{ k}, \qquad R_f = 38\text{ k}$$
$$R_i = 500, \qquad R_f = 9.5\text{ k}$$

A reasonable range is $100\text{ k} > R_i + R_f > 10\text{ k}$. The accuracy of this gain will depend on the accuracy of the resistors.

Finding Z_{in} is deceptively simple. Since Z_{in} is assumed to be infinite, all you need to do is place a 47 k resistor in parallel with the input. The resulting circuit is shown in Figure 4.3. If a specific Z_{in} is not required, a resistor in this position is not required. There is one exception to this rule: If the driving source is not directly coupled to the op amp input (e.g., it is capacitively coupled), a resistor will be required to establish a DC return path to ground.

Figure 4.3
Noninverting design for
Example 4.2

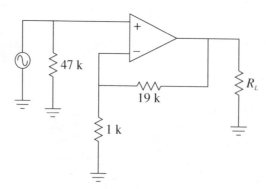

Without a DC return path, the input section's diff amp stage will not be properly biased. This point is worth remembering as it can save you a great deal in future headaches. For example, in lab a circuit like the one in Figure 4.2 may work fine with one function generator, but not with another. This would be the case if the second generator used an output coupling capacitor and the first one didn't.

Example 4.3 Design a voltage follower (i.e., ideally infinite Z_{in} and a voltage gain of 1). The Z_{in} requirement is straightforward enough. As for the second requirement, what ratio of R_f to R_i will yield a gain of 1?

$$A_v = 1 + \frac{R_f}{R_i}$$

$$\frac{R_f}{R_i} = A_v - 1$$

$$\frac{R_f}{R_i} = 0$$

This says that R_f must be 0. Practically speaking, this means that R_f is replaced with a shorting wire. What about R_i? Theoretically, almost any value will do. As long as there's a choice, consider infinity. Zero divided by infinity is certainly zero. The practical benefit of choosing $R_i = \infty$ is that you can delete R_i. The resulting circuit is shown in Figure 4.4. Remember, if the source is not directly coupled, a DC return resistor will be needed. The value of this resistor has to be large enough to avoid loading the source.

Figure 4.4
Voltage follower for
Example 4.3

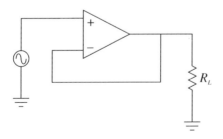

As you can see, designing with op amps can be much faster than its discrete counterpart. As a result, your efficiency as a designer or repair technician can improve greatly. You can be free to concentrate on the system, rather than on the specifics of an individual biasing resistor. In order to make multistage amplifiers, just link individual stages together.

Example 4.4 What is the input impedance of the circuit in Figure 4.5? What is v'_{out}?

As in any multistage amplifier, the input impedance to the first stage is the system Z_{in}. The DC return resistor sets this at 100 k.

To find v'_{out}, we need to find the gain (in decibels):

$$A_{v1} = 1 + \frac{R_f}{R_i}$$

$$A_{v1} = 1 + \frac{14 \text{ k}}{2 \text{ k}}$$

$$A_{v1} = 8$$

$$A'_{v1} = 18 \text{ dB}$$

$$A_{v2} = 1 + \frac{R_f}{R_i}$$

$$A_{v2} = 1 + \frac{18 \text{ k}}{2 \text{ k}}$$

$$A_{v2} = 10$$

$$A'_{v2} = 20 \text{ dB}$$

$$A'_{vt} = 18 \text{ dB} + 20 \text{ dB} = 38 \text{ dB}$$

Figure 4.5 Multistage circuit for Example 4.4

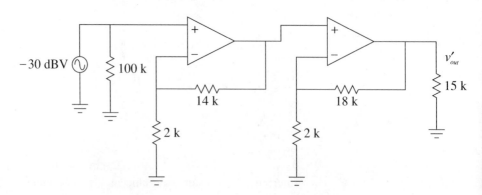

$$v'_{out} = A'_{vt} + v'_{in}$$

$$v'_{out} = 38 \text{ dB} + -30 \text{ dBV}$$

$$v'_{out} = 8 \text{ dBV}$$

Since 8 dBV translates to about 2.5 V, there is no danger of clipping.

The Inverting Voltage Amplifier

The inverting amplifier is based on the PP negative feedback model. The base form is shown in Figure 4.6. By itself, this form is current-sensing, not voltage-sensing. In order to achieve voltage sensing, an input resistor, R_i, is added. See Figure 4.7. Here's how the circuit works: Since V_{error} is virtually zero, the inverting input potential must equal the noninverting input potential. This means that the inverting input is at a *virtual ground*. The signal here is so small that it is negligible. Because of this we may also say that the impedance seen looking into this point is zero. This last point may cause a bit of confusion. You may ask "How can the impedance be zero if the current into the op amp

Figure 4.6
A basic parallel-parallel amplifier

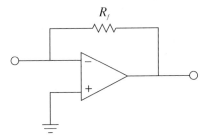

Figure 4.7
Inverting voltage amplifier

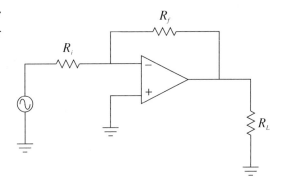

Figure 4.8
Analysis of the inverting
amplifier from Figure 4.7

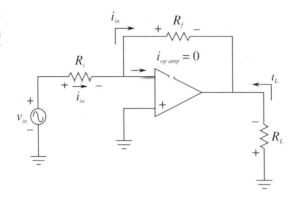

is zero?" The answer lies in the fact that all of the entering current will be drawn through R_f, thus bypassing the inverting input.

Refer to Figure 4.8 for the detailed explanation. Since the right end of R_i is at virtual ground, all of the input voltage drops across it, creating i_{in}, the input current. This current cannot enter the op amp, and instead will pass through R_f. Since a positive signal is presented to the inverting input, the op amp will sink output current, thus drawing i_{in} through R_f. The resulting voltage drop across R_f is the same magnitude as the load voltage. This is true because R_f is effectively in parallel with the load. Note that both elements are tied to the op amp's output and to (virtual) ground. There is a change in polarity since we reference the output signal to ground. In short, v_{out} is the voltage across R_f, inverted:

$$A_v = \frac{v_{out}}{v_{in}}$$

$$v_{in} = i_{in}R_i$$

$$v_{out} = -v_{R_f}$$

$$v_{R_f} = i_{in}R_f$$

Substitution yields

$$A_v = \frac{-i_{in}R_f}{i_{in}R_i}$$

$$A_v = -\frac{R_f}{R_i} \qquad\qquad (4.2)$$

Again, we see that the voltage gain is set by resistor ratio. Again, there is an allowable range of values.

The foregoing discussion leads up to the derivation of input impedance. Since all of the input signal drops across R_i, it follows that all the driving source "sees" is R_i. Quite simply, R_i sets the input impedance. Unlike the noninverting voltage amp, there is a definite interrelation between Z_{in} (R_i) and A_v ($-R_f/R_i$). This indicates that it is very hard to achieve both high gain and high Z_{in} with this circuit.

Example 4.5 Determine the input impedance and output voltage for the circuit shown in Figure 4.9.

Figure 4.9
Inverting amplifier for Example 4.5

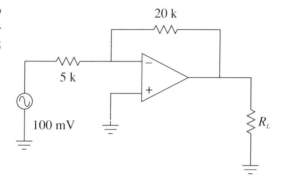

The input impedance is set by R_i. $R_i = 5$ k, therefore $Z_{in} = 5$ k. The output voltage is found as follows.

$$v_{out} = v_{in} A_v$$

$$A_v = -\frac{R_f}{R_i}$$

$$A_v = -\frac{20 \text{ k}}{5 \text{ k}}$$

$$A_v = -4$$

$$v_{out} = 100 \text{ mV} \times (-4)$$

$$v_{out} = -400 \text{ mV} \qquad \text{(i.e., inverted)}$$

———— **Example 4.6** Design an inverting amplifier with a gain of 10 and an input impedance of 15 k.

The input impedance tells us what R_i must be:

$$Z_{in} = R_i$$

$$R_i = 15 \text{ k}$$

Knowing R_i, solve for R_f:

$$A_v = -\frac{R_f}{R_i}$$

$$R_f = R_i(-A_v)$$

$$R_f = 15 \text{ k} \times (-(-10))$$

$$R_f = 150 \text{ k}$$

A SPICE simulation of the result is shown in Figure 4.10, along with its schematic. This simulation uses the simple dependent source model presented in Chapter 2. The input is set at .1 V DC for simplicity. Note that the output potential is negative, indicating the inverting action of the amplifier. Also, note that the virtual ground approximation is borne out quite well, with the inverting input potential measuring in the microvolt region.

———— **Figure 4.10**
SPICE simulation of the
simple op amp model for
Example 4.6

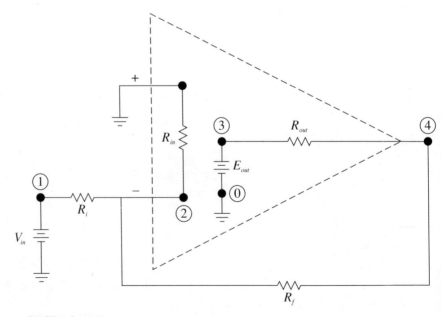

a. SPICE schematic

——— Figure 4.10 (Continued)

```
******************************** AMIGASPICE 5.0 *******************************

****    INPUT LISTING                TEMPERATURE=27.000 DEG C

*******************************************************************************
*
*The op amp's ROUT is set to 75 ohms, and a .1 V DC input is used.
*The feedback resistors are 15 K for Ri and 150 K for Rf.
*
*Note that you can change RIN, ROUT, and the gain of EOUT with little
*change in the final output values. Rf and Ri may also be changed, so long
*as their ratio remains at 10:1.
*
RIN    0   2   600K
ROUT   3   4   75
EOUT   3   0   0   2   50K
Ri     1   2   15K
Rf     4   2   150K
VIN    1   0   DC
*
*Set the input to .1 V DC with no sweeps, then print the desired
*potentials. V(4) is the output, V(2) is a virtual ground.
*Note how small V(2) is, thereby pointing out the virtual ground
*approximation. Also, note the inversion at the output, V(4).
*
.DC VIN 0.1 0.1 0.1
.PRINT DC V(4) V(2)
.END

******************************** AMIGASPICE 5.0 *******************************

****    DC TRANSFER CURVES           TEMPERATURE=27.000 DEG C

*******************************************************************************

VIN         V(4)         V(2)
1.000E-01   -9.998E-01   2.001E-05
JOB CONCLUDED
```

b. SPICE output listing

——— Example 4.7   The circuit of Figure 4.11 (page 138) is a preamplifier stage for an electronic music keyboard. Like most musicians' preamplifiers, this one offers adjustable gain. This is done by following the amplifier with a pot. What are the maximum and minimum gain values?

Note that the gain for the pre-amp is the product of the op amp gain and the voltage divider ratio produced by the pot. For maximum gain, use the pot in its uppermost position. Since the pot acts as a voltage divider, the uppermost

Figure 4.11
Musical instrument
preamplifier for
Example 4.7

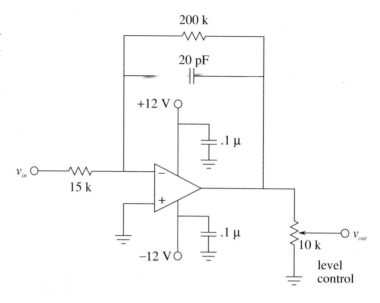

position provides no divider action (its gain is unity). For midband frequencies, the 20 pF may be ignored.

$$A_{v(max)} = -\frac{R_f}{R_i}$$

$$A_{v(max)} = -\frac{200\ k}{15\ k}$$

$$A_{v(max)} = -13.33$$

$$A'_{v(max)} = 22.5\ dB$$

For minimum gain, the pot is dialed to ground. At this point, the divider action is infinite, and thus the minimum gain is zero (resulting in silence).

Z_{in} for the system is about 15 K. As far as the extra components are concerned, the 20 pF capacitor is used to decrease high-frequency gain. The two .1 μF bypass capacitors across the power supply lines are very important. Virtually all op amp circuits use bypass capacitors. Due to the high gain nature of op amps, it is essential to have good AC grounds at the power supply pins. At higher frequencies the inductance of power supply wiring may produce a sizable impedance. This impedance may create a positive feedback loop which wouldn't exist otherwise. Without the bypass capacitors, the circuit may oscillate, or produce spurious output signals. The precise values for the capacitors are usually not critical, with .1 to 1 μF being typical.

The Inverting Current-to-Voltage Transducer

As previously mentioned, the inverting voltage amplifier is based on PP negative feedback, with an extra input resistor used to turn the input voltage into a current. What happens if that extra resistor is left out, and a circuit like Figure 4.6 is used? Without the extra resistor, the input is at virtual ground, thus setting Z_{in} to zero. This is ideal for sensing current. This input current will pass through R_f and produce an output voltage as outlined above. The characteristic of transforming a current to a voltage is measured by the parameter transresistance. By definition, the transresistance of this circuit is the value of R_f. To find v_{out}, multiply the input current by the transresistance:

$$v_{out} = i_{in}R_f \tag{4.3}$$

This circuit will invert polarity, as well.

Example 4.8   Design a circuit based on Figure 4.6 if an input current of 50 μA is to produce an output of 4 V.

The transresistance of the circuit, R_f, is found as follows.

$$R_f = \frac{v_{out}}{i_{in}}$$

$$R_f = \frac{4 \text{ V}}{50 \text{ } \mu\text{A}}$$

$$R_f = 80 \text{ k}$$

The input impedance is assumed to be zero.

At first glance, the circuit applications of this topology seem very limited. In reality, there are a large number of linear integrated circuits which produce their output in current form.[1] In many cases, this signal must be turned into a voltage in order to properly interface with other circuit elements. The current-to-voltage transducer is widely used for this purpose.

---

[1] Most notably, operational transconductance amplifiers and digital-to-analog convertors, which we shall examine in Chapters 6 and 12, respectively.

The Noninverting Voltage-to-Current Transducer

This circuit topology utilizes SS negative feedback. It senses an input voltage and produces a current. A conceptual comparison can be made to the FET (a voltage controlled current source). Instead of circuit gain, we are interested in transconductance. In other words, how much input voltage is required to produce a given output current? The op amp circuit presented in this discussion drives a floating load. That is, the load is not referenced to ground. This can be convenient in some cases, and a real pain in others. With some added circuitry, it is possible to produce a grounded-load version, although space precludes us from examining it here.

A typical voltage-to-current circuit is shown in Figure 4.12. Since this uses series-input feedback, we may immediately assume that Z_{in} is infinite. The voltage-to-current ratio is set by feedback resistor R_i. Since v_{error} is assumed to be zero, all of v_{in} drops across R_i, creating current i_{R_i}. The op amp is assumed to have zero input current, so all of i_{R_i} flows through the load resistor, R_L. By adjusting R_i, the load current may be varied:

$$i_{load} = i_{R_i}$$

$$i_{R_i} = \frac{v_{in}}{R_i}$$

$$i_{load} = \frac{v_{in}}{R_i}$$

By definition, transconductance (denoted gm) is calculated as

$$gm = \frac{i_{load}}{v_{in}}$$

$$gm = \frac{1}{R_i} \tag{4.4}$$

Figure 4.12
Voltage-to-current
transducer

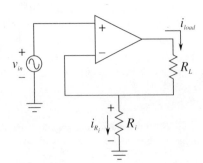

So, the transconductance of the circuit is set by the feedback resistor. As usual, there are practical limits to the size of R_i. If R_i and R_L are too small, the possibility exists that the op amp will "run out" of output current and go into saturation. At the other extreme, the product of the two resistors and the i_{load} cannot exceed the power supply rails. As an example, if R_i plus R_L is 10 k, i_{load} cannot exceed about 1.5 mA if standard ± 15 V supplies are used.

Example 4.9   Given an input voltage of .4 V in the circuit of Figure 4.13, what is the load current?

Figure 4.13
Voltage-to-current
transducer for
Example 4.9

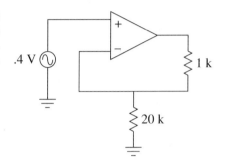

First, we find the transconductance.

$$gm = \frac{1}{R_i}$$

$$gm = \frac{1}{20\ \text{k}}$$

$$gm = 50\ \mu\text{S}$$

We now can solve for the load current:

$$i_{load} = gm \times v_{in}$$

$$i_{load} = 50\ \mu\text{S} \times .4\ \text{V}$$

$$i_{load} = 20\ \mu\text{A}$$

There is no danger of current overload in the example above as the average op amp can produce about 20 mA, maximum. The output current will be 20 μA regardless of the value of R_L, up to clipping. There is no danger of clipping in this situation either. The voltage seen at the output of the op amp to ground is

$$v_{max} = (R_i + R_t)i_{load}$$

$$v_{max} = (20\ \text{k} + 1\ \text{k}) \times 20\ \mu\text{A}$$

$$v_{max} = 420\ \text{mV}$$

That's well below clipping level.

Example 4.10   The circuit of Figure 4.14 can be used to make a high-input-impedance DC voltmeter. The load in this case is a simple meter movement. This particular meter requires 100 μA for full scale deflection. If we want to measure voltages up to 10 V, what size must R_i be?

First, we must find the transconductance.

$$gm = \frac{i_{load}}{v_{in}}$$

$$gm = \frac{100\ \mu\text{A}}{10\ \text{V}}$$

$$gm = 10\ \mu\text{S}$$

By rewriting Equation 4.4 in a different form, we can find the size of R_i:

$$R_i = \frac{1}{gm}$$

$$R_i = \frac{1}{10\ \mu\text{S}}$$

$$R_i = 100\ \text{k}$$

The meter deflection is assumed to be linear. For example, if the input signal is only 5 V, the current produced is halved to 50 μA, which should produce

Figure 4.14
DC voltmeter for
Example 4.10

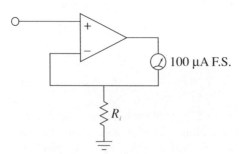

100 μA F.S.

half-scale deflection. The accuracy of this electronic voltmeter depends on the accuracy of R_i, and the linearity of the meter movement. Note that this little circuit can be quite convenient in lab if it is powered by batteries. In order to change scales, new values of R_i can be swapped in with a rotary switch. For a 1-volt scale, R_i equals 10 k. Note that for higher input ranges some form of input attenuator is needed. This is due to the fact that most op amps may be damaged if input signals larger than the supply rails are used.

The Inverting Current Amplifier

The inverting current amplifier uses PS negative feedback. As in the voltage-to-current transducer, the load is floating. The basic circuit is shown in Figure 4.15. Due to the parallel negative-feedback connection at the input, the circuit input impedance is assumed to be zero. This means that the input point is at virtual ground. Since the current into the op amp is negligible, all input current flows through R_i to node A. Effectively, R_i and R_f are in parallel (they both share node A and ground—actually virtual ground for R_i). Therefore, v_{R_i} and v_{R_f} are the same value. This means that a current is flowing through R_f, from ground to node A. These two currents join to form the load current. In this manner, current gain is achieved. The larger i_{R_f} is relative to i_{in}, the more current gain there is.

$$A_i = \frac{i_{out}}{i_{in}}$$

$$i_{out} = i_{R_f} + i_{R_i} \tag{4.5}$$

$$i_{R_i} = i_{in}$$

$$i_{R_f} = \frac{v_{R_f}}{R_f}$$

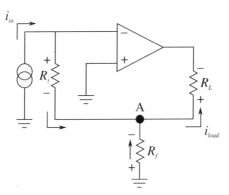

Figure 4.15
Inverting current amplifier

Since v_{R_f} is the same value as v_{R_i},

$$i_{R_f} = \frac{v_{R_i}}{R_f} \qquad (4.6)$$

$$v_{R_i} = i_{in}R_i \qquad (4.7)$$

Substitution of Equation 4.7 into Equation 4.6 yields

$$i_{R_f} = \frac{i_{in}R_i}{R_f}$$

Substituting into Equation 4.5 produces

$$i_{out} = i_{in} + \frac{i_{in}R_i}{R_f}$$

$$i_{out} = i_{in}\left(1 + \frac{R_i}{R_f}\right)$$

$$A_i = 1 + \frac{R_i}{R_f} \qquad (4.8)$$

As you might expect, the gain is a function of the two feedback resistors. Note the similarity of this result to our previous result for the noninverting voltage amplifier.

───────── **Example 4.11**   What is the load current in Figure 4.16?

$$i_{out} = A_i i_{in}$$

$$A_i = 1 + \frac{R_i}{R_f}$$

$$A_i = 1 + \frac{33\ k}{1\ k}$$

$$A_i = 34$$

$$i_{out} = 34 \times 5\ \mu A$$

$$i_{out} = 170\ \mu A$$

Figure 4.16
Current amplifier for
Example 4.11

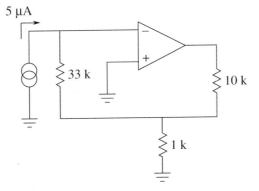

We need to check to make sure that this current doesn't cause output clipping. A simple Ohm's Law check is all that's needed:

$$v_{max} = i_{out}R_L + i_{in}R_i$$

$$v_{max} = 170 \ \mu A \times 10 \ k + 5 \ \mu A \times 33 \ k$$

$$v_{max} = 1.7 \ V + .165 \ V$$

$$v_{max} = 1.865 \ V$$

so output clipping will not be a problem.

Example 4.12 Design an amplifier with a current gain of 50. The load is approximately 200 k. Assuming a typical op amp ($i_{out(max)} = 20$ mA with ± 15 V supplies), what is the maximum load current obtainable?

$$A_i = 1 + \frac{R_i}{R_f}$$

$$\frac{R_i}{R_f} = A_i - 1$$

$$\frac{R_i}{R_f} = 50 - 1 = 49$$

Therefore, R_i must be 49 times larger than R_f. Possible solutions include:

$$R_i = 49 \ k, \qquad R_f = 1 \ k$$

$$R_i = 98 \ k, \qquad R_f = 2 \ k$$

$$R_i = 24.5 \ k, \qquad R_f = 500$$

Maximum load current can be no greater than the op amp's maximum output of 20 mA, but it may be less. We need to determine the current at

Figure 4.17
Current amplifier design
for Example 4.12

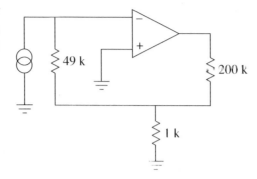

clipping. Due to the large size of the load resistance, virtually all of the output potential will drop across it. Ignoring the extra drop across the feedback resistors will introduce a maximum of 1% error (that's the worst case, assuming resistor set number 2).

With 15 V rails, a typical op amp will clip at 13.5 V. The resulting current is found through Ohm's Law:

$$i_{max} = \frac{13.5 \text{ V}}{200 \text{ k}}$$

$$i_{max} = 67.5 \ \mu\text{A}$$

Another way of looking at this is to say that the maximum allowable input current is $(67.5 \ \mu\text{A})/50$, or $1.35 \ \mu\text{A}$. One possible solution is shown in Figure 4.17.

Summing Amplifiers

It is very common in circuit design to combine several signals into a single common signal. One good example of this is in the broadcast and recording industries. The typical modern music recording will require the use of perhaps dozens of microphones, yet the final product consists of only two stereo output signals. If signals are joined haphazardly, excessive interference, noise, and distortion may result. The ideal summing amplifier would present each input with an isolated load not affected by other channels.

The most common form of summing amplifier is really nothing more than an extension of the inverting voltage amplifier. Since the input to the op amp is at virtual ground, it makes an ideal current-summing node. Instead of placing a single input resistor at this point, several input resistors may be used. Each input source drives its own resistor and there is very little effect from neighboring inputs. The virtual ground is the key. A general summing amplifier is shown in Figure 4.18. The input impedance for the first channel is R_{i1}, and

Figure 4.18
Summing amplifier

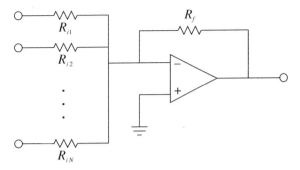

it's voltage gain is $-R_f/R_{i1}$. For channel 2, the input impedance is R_{i2}, with a gain of $-R_f/R_{i2}$. In general, for channel N, we have:

$$Z_{inN} = R_{iN}$$

$$A_{vN} = -\frac{R_f}{R_{iN}}$$

The output signal is the sum of all inputs multiplied by their associated gains:

$$v_{out} = v_{in1}A_{v1} + v_{in2}A_{v2} + \cdots + v_{inN}A_{vN}$$

which is written more conveniently as

$$v_{out} = \sum_{i=1}^{n} v_{in_i}A_{v_i}$$

A summing amplifier may have equal gain for each input channel. This is referred to as an equal weighted configuration.

Example 4.13   What is the output of the summing amplifier in Figure 4.19, with the given DC input voltage?

Figure 4.19
Summing amplifier for
Example 4.13

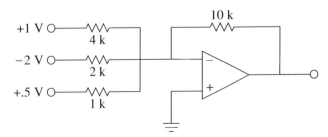

The easy way to approach this is simply to treat the circuit as three inverting voltage amplifiers, and then add the results to get the final output.

channel 1:

$$A_v = -\frac{R_f}{R_i}$$

$$A_v = -\frac{10\text{ k}}{4\text{ k}}$$

$$A_v = -2.5$$

$$v_{out} = -2.5 \times 1\text{ V}$$

$$v_{out} = -2.5\text{ V}$$

channel 2:

$$A_v = -\frac{R_f}{R_i}$$

$$A_v = -\frac{10\text{ k}}{2\text{ k}}$$

$$A_v = -5$$

$$v_{out} = -5 \times -2\text{ V}$$

$$v_{out} = 10\text{ V}$$

channel 3:

$$A_v = -\frac{R_f}{R_i}$$

$$A_v = -\frac{10\text{ k}}{1\text{ k}}$$

$$A_v = -10$$

$$v_{out} = -10 \times .5\text{ V}$$

$$v_{out} = -5\text{ V}$$

The final output is found via summation:

$$v_{out} = -2.5\text{ V} + 10\text{ V} + (-5\text{ V})$$

$$v_{out} = 2.5\text{ V}$$

If the inputs were AC signals, summation would not be quite so straight-forward. Remember, AC signals of differing frequency and phase do not add coherently. You can perform a calculation similar to the preceding to find the peak value; however, an RMS calculation is needed for the effective value (i.e., square root of the sum of the squares).

For use in the broadcast and recording industries, summing amplifiers also require some form of volume control for each input channel and a master volume control as well. This allows the levels of various microphones or instruments to be properly balanced. Theoretically, individual-channel gain control may be produced by replacing each input resistor with a potentiometer. By adjusting R_i, the gain may be directly varied. In practice, there are a few problems with this arrangement. First of all, it is impossible to turn a channel completely off. The required value for R_i would be infinite. Second, since R_i sets the input impedance, a variation in gain will produce a Z_{in} change. This change may overload or alter the characteristics of the driving source. One possible solution is to keep R_i at a fixed value and place a potentiometer before it, as in Figure 4.20. The pot produces a gain from 1 through 0. The R_f/R_i combination is then set for maximum gain. As long as R_i is several times larger than the pot's value, the channel's input impedance will stay relatively constant. The effective Z_{in} for the channel is R_{pot} in parallel with R_i, at a minimum, up to R_{pot}.

As far as a master volume control is concerned, it is possible to use a pot for R_f. Without a limiting resistor, though, a very low master gain runs the risk of overdriving the op amp due to the small effective R_f value. This technique also causes variations in offset potentials and circuit bandwidth. A technique which achieves higher performance uses a stage with a fixed R_f value, followed with a pot, as in Figure 4.20.

Still another application of the summing amplifier is the *level shifter*. A level shifter is a two-input summing amplifier. One input is the desired AC signal, and the second input is a DC value. The proper selection of DC value lets you place the AC signal at a desired DC offset. There are many uses for such a circuit. One possible application is the DC offset control available on many signal generators.

Figure 4.20
Audio mixer

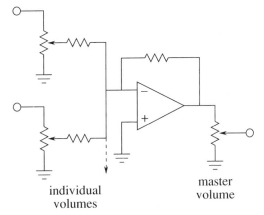

individual
volumes

master
volume

The Noninverting Summing Amplifier

Besides the inverting form, summing amplifiers may also be produced in a noninverting form. Noninverting summers generally exhibit superior high-frequency performance when compared to the inverting type. One possible circuit is shown in Figure 4.21. In this example three inputs are shown, although more could be added. Each input has an associated input resistor. Note that it is not possible to simply wire several sources together in hopes of summing their respective signals. This is because each source will try to bring its output to a desired value which will be different from the values created by the other sources. The resulting imbalance may cause excessive (and possibly damaging) source currents. Consequently, each source must be isolated from the others through a resistor.

In order to understand the operation of this circuit it is best to break it into two parts: the input source/resistor section, and the noninverting amplifier section. The input signals will combine to create a total input voltage, v_t. By inspection, you should see that the output voltage of the circuit will equal v_t times the noninverting gain, or

$$v_{out} = v_t \left(1 + \frac{R_f}{R_i} \right)$$

All that remains is to determine v_t. Since each of the input channels contributes to v_t in a similar manner, the derivation of the contribution from a single channel will be sufficient.

Unlike the inverting summer, the noninverting summer does not take advantage of the virtual-ground summing node. The result is that individual channels will affect each other. The equivalent circuit for channel 1 is redrawn in Figure 4.22. Using superposition, we would first replace the input generators

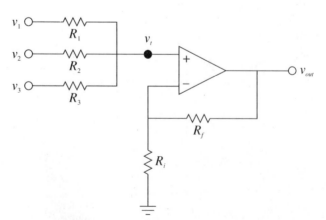

Figure 4.21
Noninverting summing
amplifier

Figure 4.22
Channel 1 input equivalent
circuit

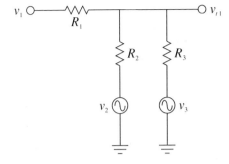

of channels 2 and 3 with short circuits. The result is a simple voltage divider between v_1 and v_{t1}.

$$v_{t1} = v_1 \frac{R_2 \parallel R_3}{R_1 + R_2 \parallel R_3}$$

In a similar manner, we can derive the portions of v_t due to channel 2,

$$v_{t2} = v_2 \frac{R_1 \parallel R_3}{R_2 + R_1 \parallel R_3}$$

and due to channel 3:

$$v_{t3} = v_3 \frac{R_1 \parallel R_2}{R_3 + R_1 \parallel R_2}$$

v_t is the summation of these three portions.

$$v_t = v_{t1} + v_{t2} + v_{t3}$$

Thus, by combining these elements, we find that the output voltage is

$$v_{out} = \left(1 + \frac{R_f}{R_i}\right)\left(v_1 \frac{R_2 \parallel R_3}{R_1 + R_2 \parallel R_3} + v_2 \frac{R_1 \parallel R_3}{R_2 + R_1 \parallel R_3} + v_3 \frac{R_1 \parallel R_2}{R_3 + R_1 \parallel R_2}\right)$$

For convenience and equal weighting, the input resistors are often all set to the same value. This results in a circuit which averages all of the inputs together. Doing so simplifies the equation to

$$v_{out} = \left(1 + \frac{R_f}{R_i}\right)\frac{v_1 + v_2 + v_3}{3}$$

Figure 4.23
Buffered and isolated
noninverting summing
amplifier

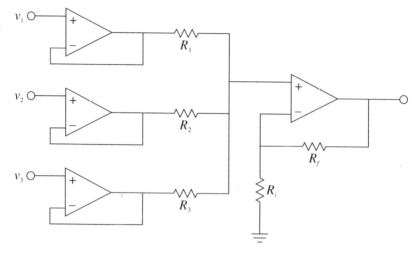

or in a more general sense,

$$v_{out} = \left(1 + \frac{R_f}{R_i}\right) \frac{\sum_{i=1}^{n} v_i}{n}$$

where n is the number of channels.

One problem still remains with this circuit, and that is interchannel isolation or *crosstalk*. This can be eliminated by individually buffering each input, as shown in Figure 4.23.

Example 4.14 A noninverting summer such as the one shown in Figure 4.21 is used to combine three signals: $v_1 = 1$ V DC, $v_2 = -.2$ V DC, and v_3 is a 2 V peak 100 Hz sine wave. Determine the output voltage if $R_1 = R_2 = R_3 = R_f = 20$ k and $R_i = 5$ k.

Since all of the input resistors are equal, we can use the general form of the summing equation.

$$v_{out} = \left(1 + \frac{R_f}{R_i}\right) \frac{v_1 + v_2 + \cdots + v_n}{n}$$

$$v_{out} = \left(1 + \frac{20\text{ k}}{5\text{ k}}\right) \frac{1\text{ V DC} + -.2\text{ V DC} + 2\sin 2\pi 100 t}{3}$$

$$v_{out} = 5 \frac{.8\text{ V DC} + 2\sin 2\pi 100 t}{3}$$

$$v_{out} = 1.33\text{ V DC} + 3.33\sin 2\pi 100 t$$

So we see that the output is a 3.33 V peak sine wave riding on a 1.33 V DC offset.

The Differential Amplifier: A Combination of Inverting and Noninverting Voltage Amplifiers

As long as the op amp is based on a differential input stage, there is nothing preventing you from making a diff amp with it. The applications of a unit based on op amps are the same as the discrete version examined in Chapter 1. In essence, the differential amplifier configuration is a combination of the inverting and noninverting voltage amplifiers. A candidate is seen in Figure 4.24. The analysis is identical to that of the two base types, and superposition is used to combine the results. The obvious problem for this circuit is that there is a large mismatch between the gains if lower values are used. Remember, for the inverting input the gain magnitude is R_f/R_i, while the noninverting input sees $(R_f/R_i) + 1$. For proper operation, the gains of the two halves should be identical. Since the noninverting input has a slightly higher gain, a simple voltage divider can be used to compensate. This is shown in Figure 4.25. The

Figure 4.24
Differential amplifier candidate

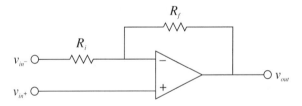

Figure 4.25
Differential amplifier with compensation for mismatched gains

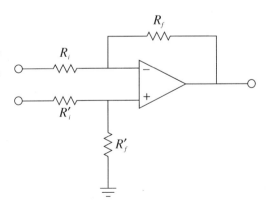

ratio should be the same as the R_f/R_i ratio. The target gain is R_f/R_i, the present gain is $1 + (R_f/R_i)$, which may be written as $(R_f + R_i)/R_i$. To compensate, a gain of $R_f/(R_f + R_i)$ is used.

$$A_{v+} = \frac{R_f + R_i}{R_i} \quad \frac{R_f}{R_f + R_i}$$

$$A_{v+} = \frac{R_f(R_f + R_i)}{R_i(R_f + R_i)}$$

$$A_{v+} = \frac{R_f}{R_i}$$

For a true differential amplifier, R_i' is set to R_i, and R_f' is set to R_f. A small potentiometer is typically placed in series with R_f' in order to compensate for slight gain imbalances due to component tolerances. This makes it possible for the circuit's common-mode rejection ratio to reach its maximum value. Another option for a simple difference amplifier is to set R_i' plus R_f' equal to R_i. Doing so will maintain roughly equal input impedance between the two halves if two different input sources are used.

Once the divider is added, the output voltage is found by multiplying the differential input signal by R_f/R_i.

──────── **Example 4.15**   Design a simple difference amplifier with an input impedance of 10 k per leg, and a voltage gain of 26 dB.

First of all, converting 26 dB into ordinary form yields 20. Since R_i sets Z_{in}, set $R_i = 10$ k, from the specifications.

$$A_v = \frac{R_f}{R_i}$$

$$R_f = A_v R_i$$

$$R_f = 20 \times 10 \text{ k}$$

$$R_f = 200 \text{ k}$$

For equivalent inputs,

$$R_i' + R_f' = R_i$$

$$R_i' + R_f' = 10 \text{ k}$$

Given that $A_v = 20$,

$$R_f' = 20 \times R_i'$$

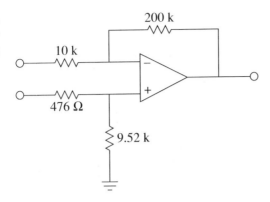

Figure 4.26
Difference amplifier for
Example 4.15

Therefore,

$$21 \times R'_i = 10 \text{ k}$$
$$R'_i = 476 \ \Omega$$
$$R'_f = 20 \times R'_i$$
$$R'_f = 9.52 \text{ k}$$

The final result is shown in Figure 4.26. As you will see later in Chapter 6, the differential amplifier figures prominently in another useful circuit, the *instrumentation amplifier*.

4.3 Single-Supply Biasing

Up to this point, all of the example circuits have used a bipolar power supply, usually ± 15 V. Sometimes this is not practical. For example, a small amount of analog circuitry may be used along with a predominantly digital circuit that runs off a unipolar supply. It may not be economical to create an entire negative supply just to run one or two op amps. While it is possible to buy op amps which have been specially designed to work with unipolar supplies,[2] the addition of simple bias circuitry will allow almost any op amp to run from a unipolar supply. This supply can be up to twice as large as the bipolar counterpart. In other words, a $+30$ V unipolar supply produces results similar to

---

[2] Examples include the LM324 and TLC270 series.

a ± 15 V bipolar unit. We will look at examples using both the noninverting and inverting voltage amplifiers.

The idea is to bias the input at one-half of the total supply potential. This can be done with a simple voltage divider. A coupling capacitor can be used to isolate this DC potential from the driving stage. For proper operation, the op amp's output should also be sitting at one-half of the supply, implying that the circuit gain must be unity. This may appear to be a very limiting factor, but in reality, it isn't. Remember that the gain need only be unity for DC. The AC gain can be just about any gain you'd like.

An example using the noninverting voltage amplifier is shown in Figure 4.27. In order to set DC gain to unity without affecting the AC gain, capacitor C_3 is placed in series with R_i. R_1 and R_2 establish the 50% bias point. Their parallel combination sets the input impedance, too. Resistors R_3 and R_4 are used to prevent destructive discharge of the coupling capacitors C_1 and C_2 into the op amp. They may not be required, but if present, typically run around 1 k and 100 Ω respectively. The inclusion of the capacitors produces three lead networks. A standard frequency analysis and circuit simplification shows that the approximate critical frequencies are:

$$f_{in} = \frac{1}{2\pi C_1 R_1 \parallel R_2}$$

$$f_{out} = \frac{1}{2\pi C_2 R_L}$$

$$f_{fdbk} = \frac{1}{2\pi C_3 R_i}$$

Figure 4.27
Single-supply bias in a noninverting amplifier

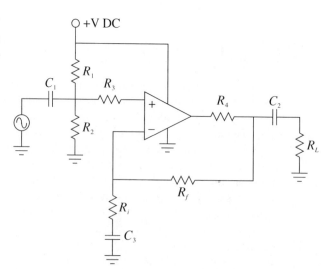

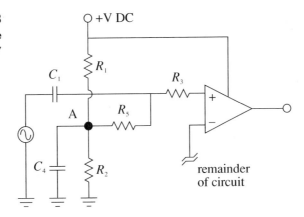

Figure 4.28
Improved bias for the
circuit in Figure 4.27

The input bias network can be improved by using the circuit of Figure 4.28, which reduces the hum and noise transmitted from the power supply into the op amp's input. It does so by creating a low impedance at node A. This, of course, does not affect the DC potential. R_5 now sets the input impedance of the circuit.

The important points to remember here are that voltage gain is still $1 + (R_f/R_i)$ in the midband, Z_{in} is now set by the biasing resistors R_1 and R_2, or R_5 (if used), and that frequency response is no longer flat down to 0 Hz.

A single-supply version of the inverting voltage amplifier is shown in Figure 4.29. It uses the same basic techniques as the noninverting form. The bias setup

Figure 4.29
Single-supply inverting
amplifier

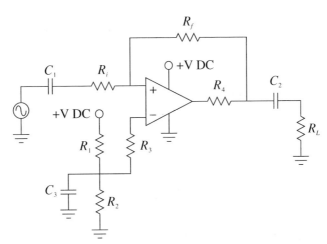

uses the optimized low-noise form. Note that there is no change in input impedance; it is still set by R_i. The approximate lead network critical frequencies are found as follows.

$$f'_{in} = \frac{1}{2\pi C_1 R_i}$$

$$f_{out} = \frac{1}{2\pi C_2 R_L}$$

$$f_{bias} = \frac{1}{2\pi C_3 R_1 \parallel R_2}$$

Note the general similarity between the circuits of Figures 4.28 and 4.29. A simple redirection of the input signal creates one form from the other.

4.4 Current Boosting

As previously noted, general-purpose op amps produce a maximum output current of around 20 mA. This is sufficient for a wide variety of uses. If the load is less than about 1 kΩ, the op amp will start to clip on the higher output signals. The average op amp cannot drive low impedance loads. Two examples of applications whose loads are inappropriate include distribution amplifiers and small audio power amplifiers, such as headphone amplifiers. This is most unfortunate as we have already seen how useful these devices can be. There is a way out though. It is possible to include a current gain stage right after the op amp. All that is needed is a simple class B or class AB push-pull follower. This follower will be able to produce the higher current required by low impedance loads. The op amp only needs to drive the follower stage. In order to increase system linearity and lower distortion, the follower can be placed inside the op amp's feedback loop. Since the follower is noninverting, there is no problem with maintaining correct feedback (assuming that the power devices used have a wider bandwidth than the op amp). An example of this is shown in Figure 4.30. This particular circuit is from the JBL model 5234 electronic crossover and distribution amplifier. This output circuit needs to drive relatively low impedances through long cable runs (perhaps several hundred feet). The excessive capacitance resulting from long cable runs increases the current demand above that of a purely resistive load.

Circuits like the one in Figure 4.30 can produce currents of several hundred milliamps or more. Many times, small resistors are placed in the emitter or

Figure 4.30
Current boosting

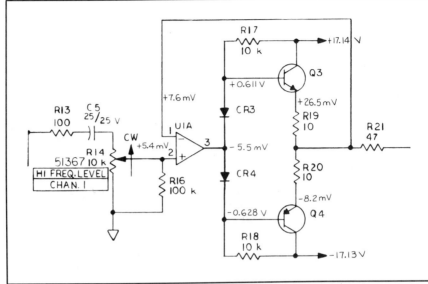

Courtesy of JBL Professional

collector as a means of limiting maximum current or reducing distortion. The maximum output current limitation is a function of the class B devices. For very high current demands, Darlingtons or multistage designs may be required. It is even possible to provide voltage gain stages. Indeed, several consumer audio power amplifiers have been designed in exactly this way. In essence, the designers produce a discrete power amplifier, and then "wrap it" within an op amp feedback loop.

Summary

In this chapter we have explored a variety of basic op amp circuits and learned a few analysis shortcuts. The basic assumptions are that the error voltage (differential input voltage) is zero, and that the op amp's input current is zero. In all circuits, the gain or transfer parameter is a function of just one or two resistors. Circuits can be made which produce voltage gain, current gain, voltage-to-current conversion, or current-to-voltage conversion. The most popular op amp circuits are the noninverting voltage amplifier and the inverting voltage amplifier. These are based on SP and PP negative feedback, respectively. The noninverting type shows an ideally infinite input impedance, while the inverting type has its input impedance set by one of the feedback resistors. A variation of the inverting voltage amplifier is the summing amplifier. This adds its several input channels together in order to arrive at its single output signal. The input node is at virtual ground. The differential amplifier is basically the simultaneous use of both the inverting and noninverting voltage amplifier forms.

The voltage-to-current transducer is based on SS feedback. Its transconductance is set by a single feedback resistor. In a similar manner, the current-to-voltage transducer is based on PP feedback and has a single feedback resistor to set its transresistance. The current amplifier is based on PS feedback.

Although op amps are designed to run off bipolar power supplies, they can by used with unipolar supplies. Extra circuitry is needed for the proper bias. There is no restriction on AC gain; however, DC gain must be set to unity. Since lead networks are introduced, the system gain cannot be flat down to 0 Hz.

Finally, if higher output current requirements need to be met, it is possible to boost the op amp's capabilities with a discrete output stage. This stage is typically a class B or class AB push-pull follower. In order to lower system distortion, the follower is kept within the op amp's feedback loop.

Self Test Questions

1. What forms of feedback are used for the inverting and noninverting voltage amplifiers?

2. What forms of feedback are used for the current-to-voltage and voltage-to-current transducers?

3. What form of feedback is used for the inverting current amplifier?

4. What are the op amp analysis idealizations?

5. What is virtual ground?

6. What is a summing amplifier?

7. How can output current be increased?

8. What circuit changes are needed in order to bias an op amp with a unipolar supply?

9. What operational parameters change when a circuit is set up for single-supply biasing?

10. How might a circuit's gain be controlled externally?

11. What is meant by the term "floating load?"

Problem Set

▷ **Analysis Problems**

1. What is the voltage gain in Figure 4.31? What is the input impedance?

——— **Figure 4.31**

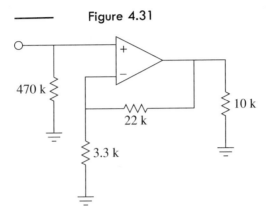

2. What is the voltage gain for the first stage of Figure 4.32? What is the input impedance?

——— **Figure 4.32**

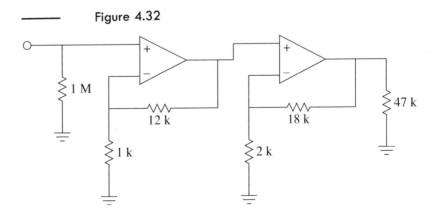

3. What is the voltage gain for the second stage of Figure 4.32? What is the input impedance?

4. What is the system voltage gain in Figure 4.32? What is the input impedance?

5. If the input to Figure 4.32 is -52 dBV, what is v'_{out}?

6. What is the voltage gain in Figure 4.33? What is the input impedance?

—————— **Figure 4.33**

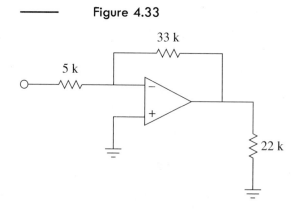

7. If the input voltage to the circuit of Figure 4.33 is 100 mV, what is v_{out}?

8. What is the system input impedance in Figure 4.34? What is the system gain?

—————— **Figure 4.34**

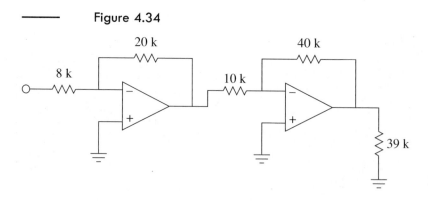

9. Redesign Figure 4.34 for an input impedance of 20 k.

10. Given an input current of 2 μA, what is the output voltage in Figure 4.35?

Figure 4.35

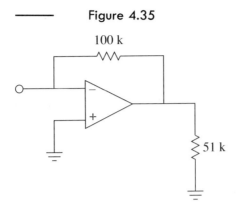

11. What is the meter deflection in Figure 4.36 if the input voltage is 1 V?

12. What input voltage will produce full-scale deflection in Figure 4.36?

13. Determine a new value for the 10 k resistor in Figure 4.36 such that a .1 V input will produce full-scale deflection.

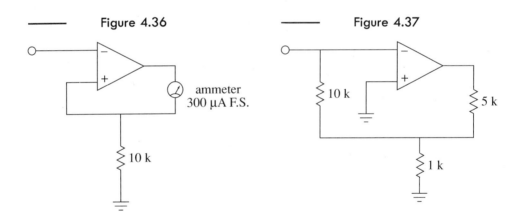

Figure 4.36   Figure 4.37

14. What is the current gain in Figure 4.37?

15. What is the maximum input current in Figure 4.37, assuming the circuit is running off ±15 V supplies, and the op amp has a maximum output current of 25 mA?

16. If the differential input signal is 300 mV in Figure 4.38 (page 164), what is v_{out}?

Figure 4.38

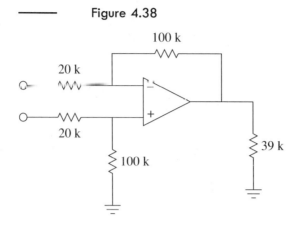

17. Determine new values for the voltage divider resistors in Figure 4.38, such that the resulting input impedance is balanced.

⊳ Design Problems

18. Design a noninverting amplifier with a voltage gain of 32 dB and an input impedance of 200 k.

19. Design a voltage follower with a gain of 0 dB.

20. Design an inverting amplifier with a voltage gain of 14 dB and an input impedance of 15 k.

21. Design a current-to-voltage transducer such that a 20 μA input current will produce an output of 1 V.

22. Design a voltage-to-current transducer such that a 100 mV input will produce an output of 1 mA.

23. Design a current amplifier with a gain of 20.

24. Design a differential amplifier with a gain of 18 dB and a balanced input impedance of 25 k per input.

25. Design a voltage-to-current transducer with a transconductance of 1 mS. If v_{in} is 200 mV, what is i_{out}?

26. Design a current-to-voltage transducer with a transresistance of 10 k. If the input current is .5 mA, what is v_{out}?

27. Redesign the circuit of Figure 4.31 for single-supply operation. (Don't bother calculating capacitor values.)

28. Redesign the circuit of Figure 4.33 for single-supply operation. (Don't bother calculating capacitor values.)

29. Design a summing amplifier such that channel 1 has a gain of 10, channel 2 has a gain of 15, and channel 3 has a gain of 5. The minimum channel input impedance should be 1 k.

30. Determine capacitor values for Problem 27 if the lower break frequency f_1 is set to 20 Hz.

31. Determine capacitor values for Problem 28 if the lower break frequency f_1 is set to 10 Hz.

▷ **Challenge Problems**

32. Design a three-channel summing amplifier with the following characteristics:

$$\text{channel 1:}\quad Z_{in} \geq 10 \text{ k}, A_v = 6 \text{ dB}$$
$$\text{channel 2:}\quad Z_{in} \geq 22 \text{ k}, A_v = 10 \text{ dB}$$
$$\text{channel 3:}\quad Z_{in} \geq 5 \text{ k}, A_v = 16 \text{ dB}$$

33. Assuming 10% resistor values, determine the production gain range for Figure 4.31.

34. Assuming 5% resistor values, determine the highest gain produced in Figure 4.32.

35. Design an inverting amplification circuit with a gain of at least 40 dB and an input impedance of at least 100 k. No resistor used may be greater than 500 k. Multiple stages are allowed.

36. Redesign the circuit of Figure 4.36 as a voltmeter with 500 mV, 2 V, 5 V, 20 V, and 50 V ranges.

37. Assuming 1% precision resistors and a meter accuracy of 5%, what range of input values can produce a full-scale reading of 2 V for the circuit of Problem 36?

38. Design an amplifier with a gain range from -10 dB to $+20$ dB and an input impedance of at least 10 k.

39. Design an amplifier with a gain range from 0 to 20. The input impedance should be at least 5 k.

40. What is the input impedance in Figure 4.39? What is A_v?

————— Figure 4.39

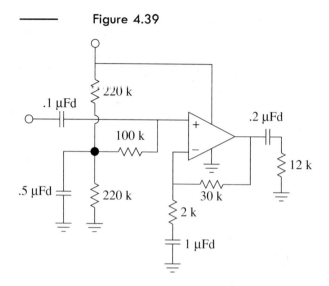

41. How much power supply ripple attenuation does the input biasing net-work of Figure 4.39 produce? (Assume $f_{ripple} = 120$ Hz.)

42. Assume that the circuit of Figure 4.40 utilizes a standard 20 mA output op amp. If the output devices are rated for a maximum collector current of 5 A and a β of 50, what is the maximum load current obtainable?

————— Figure 4.40

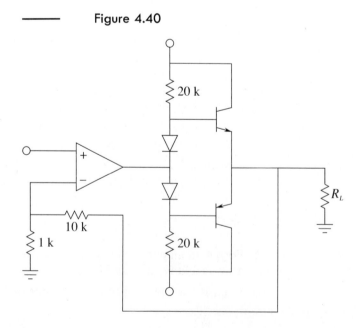

43. What are the voltage gain and input impedance in Figure 4.40?

44. Given a summer based on Figure 4.21, sketch the output waveform if $R_1 = R_2 = 10$ k, $R_3 = R_f = 30$ k, $R_i = 15$ k, $v_1 = .3$ V DC, $v_2 = .1 \sin 2\pi 50t$, and $v_3 = -.2 \sin 2\pi 200t$.

▷ SPICE Problems

45. Simulate the operation of the circuit in Figure 4.9. Verify the output voltage and the virtual ground at the inverting input.

46. Use SPICE to verify the maximum and minimum gains of the circuit in Figure 4.11.

47. Use SPICE to verify the load current and the output voltage of the circuit in Figure 4.16.

48. Verify the output potential of the circuit in Figure 4.19.

49. Simulate the output voltage of the circuit of Figure 4.34 for the following inputs. Also, note the potential at the output of the first stage. How might your op amp model affect the results?

    a. $v_{in} = .1$ V DC   b. $v_{in}(t) = 1 \sin 2\pi 10t$ c. $v_{in} = 5$ V DC

50. Use SPICE to simulate the circuit in Figure 4.26. Determine the output potential for the following inputs.

    a. $v_{in+}(t) = .1 \sin 2\pi 10t$, $v_{in-}(t) = .1 \sin 2\pi 10t$

    b. $v_{in+}(t) = .1 \sin 2\pi 10t$, $v_{in-}(t) = -.1 \sin 2\pi 10t$

51. Simulate the circuit of Figure 4.40, and determine the output of the circuit and op amp for inputs of .1 V DC and 1 V DC.

5

Practical Limitations of Op Amp Circuits

Chapter Objectives

After completing this chapter, you should be able to:

❑ Define *gain-bandwidth product* and describe its use in circuit design and analysis.

❑ Determine upper and lower break frequencies in a multistage circuit.

❑ Define *slew rate* and *power bandwidth,* and calculate their effect on circuit performance.

❑ Understand the difference between *power bandwidth* and *small signal bandwidth.*

❑ Detail the differences between compensated, noncompensated, and decompensated op amps.

❑ Calculate the DC offset of an op amp circuit, and understand how to minimize it.

❑ Calculate the DC drift of an op amp circuit, and understand how to minimize it.

❑ Discuss which factors affect the noise performance of an op amp circuit.

❑ Calculate the noise voltage of an op amp circuit.

❑ Analyze the CMRR, PSRR, and S/N performance of an op amp circuit.

5.1 Introduction

Up to now, we have treated the op amp as an ideal device. While these idealizations are very useful in their place, it is imperative that closer examination follow. Without further knowledge, it is impossible to accurately predict a circuit's performance for very high or low frequencies, to judge its noise characteristics, or to determine its stability with temperature or power supply variations. Armed with this information, you can optimize circuit performance for given applications. A major part of such optimization is being able to determine the most desirable op amp for the job. The function of this chapter is to delve deeper into the specifics of individual op amps, and to present methods for determining system parameters such as frequency response, noise level, offsets, and drift. We will be primarily interested in investigating the popular inverting and noninverting voltage amplifier topologies.

5.2 Frequency Response

In Chapter 4, a number of equations were presented for the various amplifier topologies. These enable you to find the circuit gain, among other things. These equations are, of course, valid only in the midband region of the amplifier. They say nothing of the amplifier response at the frequency extremes. In Chapter 1, you found that all amplifiers eventually roll off their gain as the input frequency increases. Some amplifiers also exhibit a rolloff as the input frequency is decreased. Op amp circuits are no exception. There are two things we can say about the average op amp circuit's frequency response: 1) If there are no coupling or other lead-network capacitors, the circuit gain will be flat from midband down to DC; and 2) There will eventually be a well-controlled high-frequency rolloff that is usually very easy to find. The first statement should come as no great surprise, but you may well wonder about the second. For general-purpose op amps, the high-frequency response may be determined with a parameter called the *gain-bandwidth product* (GBW).

5.3 The Gain-Bandwidth Product

The open-loop frequency response of a general-purpose op amp is shown in Figure 5.1a. While the exact frequency and gain values will differ from model to model, all devices will exhibit this same general shape and 20 dB per decade rolloff slope. This is because the lag break frequency (referred to as f_c) is determined by a single capacitor called the *compensation capacitor*. This capacitor is usually in the Miller position (straddling input and output) of an intermediate stage, such as C in Figure 5.1b. Although this capacitor is rather small, the Miller effect drastically increases its apparent value. The resulting critical frequency is very low, often in the range of 10 to 100 Hz. The other circuit lag networks caused by stray or load capacitances are much higher, usually over 1 MHz. As a result, a constant 20 dB per decade rolloff is maintained from the lag break frequency up to very high frequencies. The remaining lag networks will not affect the open-loop response until the gain has already dropped below 0 dB.

This type of frequency response curve has two benefits: 1) The most important benefit is that it allows you to set almost any gain you desire with stability. Since only a single network is active, satisfactory gain and phase margins will be maintained. Therefore, your negative feedback never turns into positive feedback (as noted in Chapter 3). 2) The product of any break frequency and its corresponding gain is a constant. In other words, the gain decreases at the same rate at which the frequency increases. In Figure 5.1a, the product is 1 MHz. As you might have guessed, this parameter is the gain-bandwidth product of the op amp. The GBW is also referred to as f_{unity} (the frequency at which the open-loop gain equals one). You will find both terms used on manufacturers' spec sheets.

As you already know, operating an op amp with negative feedback lowers the midband gain. To a first approximation, this gain will continue until it reaches the open-loop response. At this point, the closed-loop response will follow the open-loop rolloff. (Remember, this is because of the reduction in loop gain, as seen in Chapter 3.) This effect is shown in Figure 5.2 on page 172. Knowing the GBW and the gain, you can quickly determine the associated break frequency. For the inverting and noninverting voltage amplifiers,

$$f_2 = \frac{\text{GBW}}{A_{noise}} \qquad (5.1)$$

The use of *noise gain* versus ordinary voltage gain simplifies things and actually makes the results a bit more accurate. Noise gain is the same for both inverting and noninverting voltage amplifiers. The use of noise gain helps us to take into account the true (non-ideal) feedback effects and circuit imperfections. An example of these limitations is that the open-loop gain of an op amp is never infinite. To find the noise gain for any circuit, short all voltage sources and open all current sources. The only item remaining for each source will be its internal resistance. At this point, simplify the circuit as required,

Figure 5.1 Gain curve

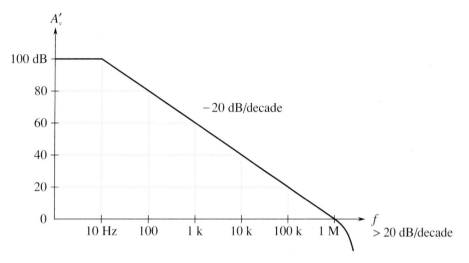

a. Open-loop frequency response

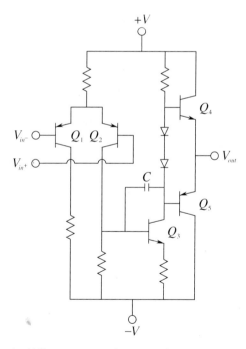

b. Miller compensation capacitor

Figure 5.2
Comparison of open-loop
and closed-loop responses

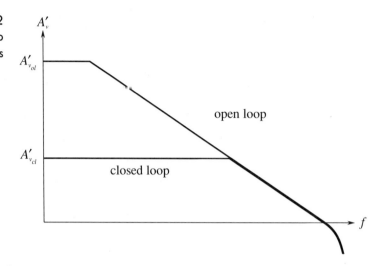

and find the gain from the noninverting input to the output of the op amp. This gain is the noise gain. For the standard inverting and noninverting voltage amplifiers, we find

$$A_{noise} = 1 + \frac{R_f}{R_i}$$

Noise gain is the same as ordinary voltage gain for the noninverting voltage amplifier, but is one unit larger than the inverting amplifier's ordinary gain (R_f/R_i). The deviation is noticeable only at lower gains. This does imply though that, for the same gain, noninverting amplifiers will exhibit a higher break frequency than inverting types. Thus, for maximum bandwidth with low gain circuits, the noninverting form is generally preferred. The worst case occurs with an ordinary voltage gain of 1. For the noninverting configuration, the noise gain will also equal 1, and the closed-loop bandwidth will equal f_{unity}. On the other hand, an inverting amplifier with a voltage gain of 1 will produce a noise gain of 2, and will exhibit a small signal bandwidth of $f_{unity}/2$. Never use the gain in decibel form for this calculation!

Example 5.1   Using a 741 op amp, what is the upper break frequency for a noninverting amplifier with a gain of 20 dB?
A 741 data sheet shows a typical GBW of 1 MHz. The noise gain for a noninverting amplifier is the same as its ordinary gain. Converting 20 dB into

ordinary form yields a gain of 10:

$$f_2 = \frac{GBW}{A_{noise}}$$

$$f_2 = \frac{1 \text{ MHz}}{10}$$

$$f_2 = 100 \text{ kHz}$$

So, the gain is constant at 10 up to 100 kHz. Above this frequency, the gain rolls off at 20 dB per decade.

Example 5.2   Sketch the frequency response of the circuit in Figure 5.3. This is an inverting voltage amplifier. The gain is

$$A_v = -\frac{R_f}{R_i}$$

$$A_v = -\frac{10 \text{ k}}{2 \text{ k}}$$

$$A_v = -5$$

$$A'_v = 14 \text{ dB}$$

The noise gain is

$$A_{noise} = 1 + \frac{R_f}{R_i}$$

$$A_{noise} = 1 + \frac{10 \text{ k}}{2 \text{ k}}$$

$$A_{noise} = 6$$

Figure 5.3
Circuit for Example 5.2

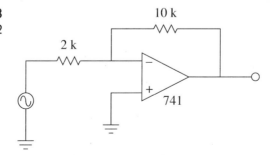

From a data sheet, the GBW for a 741 is found to be 1 MHz.

$$f_2 = \frac{GBW}{A_{noise}}$$

$$f_2 = \frac{1 \text{ MHz}}{6}$$

$$f_2 = 167 \text{ kHz}$$

The resulting gain Bode plot is shown in Figure 5.4a. Note that if a "faster" op amp is used (one with a higher GBW, like the LF351), the response will extend further. As you might guess, faster op amps are more expensive op amps. A SPICE schematic and simulation of the Bode gain response for this circuit are shown in Figures 5.4b and c. A subcircuit is used for the 741 op

Figure 5.4
SPICE analysis for the
inverting 741 op amp of
Figure 5.3

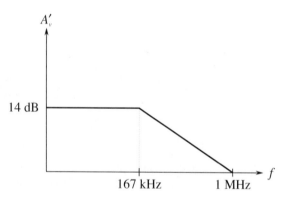

a. Bode plot for the circuit

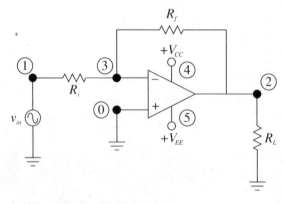

b. Schematic for SPICE simulation

```
*******************************AMIGASPICE 2.3/2G.6 GWS*******************************
***   INPUT LISTING                    TEMPERATURE=27.000 DEG C
**********************************************************************************
*****************************Start of UA741 op amp*****************************
.SUBCKT UA741   1    2    3   4   5    6
*               IN+  IN- GND V+ V- OUT
Q1    11   1   13 UA741QA
Q2    12   2   14 UA741QB
RC1    4  11   5.305165E+03
RC2    4  12   5.305165E+03
C1    11  12   5.459553E-12
RE1   13  10   2.151297E+03
RE2   14  10   2.151297E+03
IEE   10   5   1.666000E-05
CE    10   3   3.000000E-12
RE    10   3   1.200480E+07
GCM    3  21   10 3 5.960753E-09
GA    21   3   12 11 1.884655E-04
R2    21   3   1.000000E+05
C2    21  22   3.000000E-11
GB    22   3   21 3 2.357851E+02
RO2   22   3   4.500000E+01
D1    22  31   UA741DA
D2    31  22   UA741DA
EC    31   3   6 3 1.0
RO1   22   6   3.000000E+01
D3     6  24   UA741DB
VC     4  24   2.803238E+00
D4    25   6   UA741DB
VE    25   5   2.803238E+00
.ENDS
*Diode and transistor models for the 741 op amp.
.MODEL UA741DA D (IS=9.762287E-11)
.MODEL UA741DB D (IS=8.000000E-16)
.MODEL UA741QA NPN (IS=8.000000E-16 BF=9.166667E+01)
.MODEL UA741QB NPN (IS=8.309478E-16 BF=1.178571E+02)
******************************End of UA741 op amp******************************
*
****************************Main circuit description****************************
RL     0  2 100K
RI     1  3 2K
RF     3  2 10K
VCC    4  0 DC 15
VEE    5  0 DC -15
XOA1   0  3 0  4 5 2   UA741
VIN    1  0 AC 1
****************************Analysis directives****************************
*Calculate 10 points per decade from 2 KHz to 2 MHz.
.AC DEC 10 2K 2MEG
*Plot the output amplitude versus frequency (node 2).
.PLOT AC VDB(2)
.END
```

c. SPICE simulation

(Continued)

Figure 5.4 *(Part c, continued)*

```
*****************************AMIGASPICE 2.3/2G.6 GWS*************************************
****                        AC ANALYSIS        TEMPERATURE=27.000 DEG C
***************************************************************************************
FREQ         VDB(2)
                       -1.000D+01        0.000D-01        1.000D+01        2.000D+01
             ----------------------------------------------------------------------
2.000D+03    1.398D+01   .                .                *               .
2.518D+03    1.398D+01   .                .                *               .
3.170D+03    1.398D+01   .                .                *               .
3.991D+03    1.398D+01   .                .                *               .
5.024D+03    1.398D+01   .                .                *               .
6.325D+03    1.397D+01   .                .                *               .
7.962D+03    1.397D+01   .                .                *               .
1.002D+04    1.397D+01   .                .                *               .
1.262D+04    1.396D+01   .                .                *               .
1.589D+04    1.394D+01   .                .                *               .
2.000D+04    1.392D+01   .                .                *               .
2.518D+04    1.389D+01   .                .                *               .
3.170D+04    1.384D+01   .                .                *               .
3.991D+04    1.376D+01   .                .               *                .
5.024D+04    1.364D+01   .                .              *                 .
6.325D+04    1.345D+01   .                .             *                  .
7.962D+04    1.317D+01   .                .            *                   .
1.002D+05    1.276D+01   .                .          *                     .
1.262D+05    1.218D+01   .                .        *                       .
1.589D+05    1.140D+01   .                .      *                         .
2.000D+05    1.038D+01   .                .    *                           .
2.518D+05    9.132D+00   .                .  *                             .
3.170D+05    7.674D+00   .              *                                  .
3.991D+05    6.035D+00   .            *                                    .
5.024D+05    4.249D+00   .          *                                      .
6.325D+05    2.338D+00   .        *                                        .
7.962D+05    3.158D-01   .      *                                          .
1.002D+06   -1.821D+00   .    *                                            .
1.262D+06   -4.090D+00   .  *                                              .
1.589D+06   -6.517D+00   . *                                               .
2.000D+06   -9.135D+00   . *                                               .
             ----------------------------------------------------------------------
```

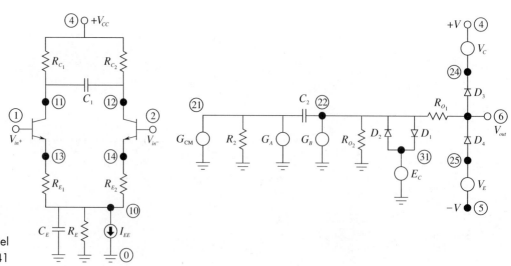

d. SPICE model
   for the 741

amp, and a 100 k resistor is used for the load. The nodes for the simulation are assigned as follows:

0. Ground (as usual)

1. Junction of the input source and R_i (2 k)

2. The output of the op amp

3. The inverting input of the op amp

4. The positive power supply (V_{CC})

5. The negative power supply (V_{EE})

The simple dependent source model of Chapter 2 cannot be used here since it does not have the proper frequency response. The model presented here is fairly complex, and is quite accurate. The op amp model is redrawn in Figure 5.4d for closer inspection. It is comprised of two basic parts, a diff amp input portion, and a dependent source output section. The input portion utilizes a pair of NPN transistors with simple resistors for the loads (R_{C_1} and R_{C_2}). Resistors R_{E_1} and R_{E_2} serve as swamping or emitter-degeneration resistors. The tail-current source is set by the independent source I_{EE}. The non-ideal internal impedance and frequency limitations of this current source are taken into account by R_E and C_E, while C_1 helps to model the high-frequency loading of the diff amp's output. The output portion revolves around a series of voltage-controlled current sources. G_{CM} models common-mode gain, while G_A models the ordinary gain, and R_2 serves as the combined internal impedance of these sources. C_2 is the system compensation capacitor and has a value of 30 pF. R_{O_1} and R_{O_2} serve to model the output impedance of the op amp. Diodes D_1 through D_4 and voltage sources E_C, V_C, and V_E model the limits of the op amp's class AB output stage.

In spite of their accuracy, models such as this are time consuming and tedious to recreate. Fortunately, many manufacturers offer SPICE models for their components in diskette form. To use these models, all you need to do is clip out the desired op amps with your text editor and paste them into your SPICE input file. This is exactly how the input file for Figure 5.4b was created.

The op amp model uses typical, rather than worst-case, values. Concerning the simulation results, note that the low frequency gain agrees with the hand calculation of approximately 14 dB. The 3 dB down point (f_2) also agrees with the calculated break of approximately 167 kHz.

Example 5.3   Determine the minimum acceptable f_{unity} for the circuit of Figure 5.5 (page 178) if response should extend to at least 50 kHz.

Figure 5.5
Circuit for Example 5.3

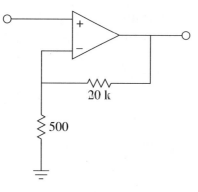

$$A_{noise} = 1 + \frac{R_f}{R_i}$$

$$A_{noise} = 1 + \frac{20\text{ k}}{500}$$

$$A_{noise} = 41$$

$$f_{unity} = A_{noise} f_2$$

$$f_{unity} = 41 \times 50\text{ kHz}$$

$$f_{unity} = 2.05\text{ MHz}$$

For this application, a stock 741 would not be fast enough; however, a 351 would be fine. From the foregoing, it is apparent that there is a direct trade-off between circuit gain and high-frequency performance for a given device. For an application requiring both high gain and wide bandwidth, a multistage approach should be considered.

Multistage Considerations

By combining two or more wide-bandwidth, low-gain stages, a single high-gain, wide-bandwidth system may be produced. While the overall system gain will be simply the combination of the individual stage gains, the upper break frequency calculation can be a little tricky. Chances are, in a multistage op amp design, all stages will not exhibit the same upper break frequency. In this case the system's upper break is approximately equal to the lowest of the stage f_c's. In other words, the system is treated as though it were a discrete stage with multiple lag networks. On the other hand, if two or more stages are dominant, this approximation can lead to a sizable error. This is best illustrated with a quick example: Imagine two stages exhibiting a 100 kHz break. If each stage produces a 3 dB loss at 100 kHz, it is obvious that the cascaded system must be producing a 6 dB loss at 100 kHz. Therefore, the system's critical frequency (3 dB point) must be somewhat lower than 100 kHz (to be exact, it is the frequency at which each stage produces a 1.5 dB loss). Taking this a step further, if we cascade three identical stages, the total loss at 100 kHz will be 9 dB. The system break will be the point at which each of the three stages produce a 1 dB loss. The more identical stages that are added, the lower the effective break becomes. If we make a few assumptions about the exact shape of the rolloff curve, we can reduce this to a simple equation. In Chapter 1, we derived the general equation describing the amplitude response of a lead network (Equation 1.7). In a similar vein, the response for a lag network may be determined to be

$$A_v = \frac{1}{\sqrt{1 + \dfrac{f^2}{f_c^2}}} \tag{5.2}$$

where f is the frequency of interest and f_c is the critical frequency.

It is more convenient to write this equation in terms of a *normalized frequency of interest*. Instead of being expressed in Hertz, the frequency of interest is represented as a factor relative to f_c. If we call this normalized frequency k_N, we may rewrite the amplitude response equation as

$$A_v = \frac{1}{\sqrt{1 + k_N^2}} \tag{5.3}$$

We now solve for k_N:

$$\frac{1}{A_v} = \sqrt{1 + k_N^2}$$

$$k_N^2 + 1 = \frac{1}{A_v^2}$$

$$k_N^2 = \frac{1}{A_v^2} - 1$$

$$k_N = \sqrt{\frac{1}{A_v^2} - 1} \tag{5.4}$$

We will now find the gain contribution of each stage. If all stages are critical at the same frequency, each stage must produce the same gain as the other stages at any other frequency. Since the combined gain of all stages must, by definition, be -3 dB or .707 at the system's break frequency, we can find the gain of each stage at this new frequency:

$$A_v^n = .707 \tag{5.5}$$

where n is the number of stages involved.

We may rewrite this as

$$A_v = .707^{1/n} \tag{5.6}$$

Combining Equation 5.6 with Equation 5.4 yields

$$k_N = \sqrt{\frac{1}{(.707^{1/n})^2} - 1}$$

$$k_N = \sqrt{2^{1/n} - 1} \tag{5.7}$$

Since k_N is nothing more than a factor, this may be rewritten into a final convenient form:

$$f_{2\text{-}system} = f_2 k_N$$

$$f_{2\text{-}system} = f_2 \sqrt{2^{1/n} - 1} \tag{5.8}$$

where n is the number of identical stages.

Example 5.4   Assuming that all stages in Figure 5.6 use 741s, what is the system gain and upper break frequency?

stage 1:
$$A_v = 1 + \frac{R_f}{R_i}$$

$$A_v = 1 + \frac{14\,k}{2\,k}$$

$$A_v = 8$$

$$A_{noise} = 1 + \frac{R_f}{R_i}$$

$$A_{noise} = 8$$

Figure 5.6
Multistage circuit for
Example 5.4

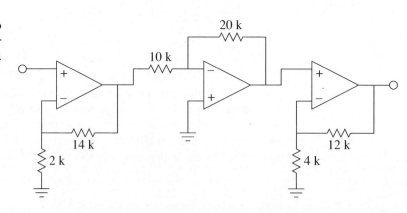

$$f_2 = \frac{GBW}{A_{noise}}$$

$$f_2 = \frac{1 \text{ MHz}}{8}$$

$$f_2 = 125 \text{ kHz}$$

stage 2:
$$A_v = -\frac{R_f}{R_i}$$

$$A_v = -\frac{20 \text{ k}}{10 \text{ k}}$$

$$A_v = -2$$

$$A_{noise} = 1 + \frac{R_f}{R_i}$$

$$A_{noise} = 3$$

$$f_2 = \frac{GBW}{A_{noise}}$$

$$f_2 = \frac{1 \text{ MHz}}{3}$$

$$f_2 = 333 \text{ kHz}$$

stage 3:
$$A_v = 1 + \frac{R_f}{R_i}$$

$$A_v = 1 + \frac{12 \text{ k}}{4 \text{ k}}$$

$$A_v = 4$$

$$A_{noise} = 1 + \frac{R_f}{R_i}$$

$$A_{noise} = 4$$

$$f_2 = \frac{GBW}{A_{noise}}$$

$$f_2 = \frac{1 \text{ MHz}}{4}$$

$$f_2 = 250 \text{ kHz}$$

system:
$$A_v = 8 \times (-2) \times 4$$
$$A_v = -64$$

Compare f_2 values to identify the dominant stage. The dominant break here is 125 kHz (stage 1).

The system has a gain of 64 and an upper break of 125 kHz. If this level of performance is to be achieved with a single op amp, it will need a gain-bandwidth product of 125 kHz × 64, or 8 MHz.

Example 5.5 A three-stage amplifier uses identical noninverting voltage stages with gains of 10 each. If the op amps used have an f_{unity} of 4 MHz, what is the system gain and upper break?

Since these are noninverting amplifiers, the noise gain equals the signal gain. The break frequency for each stage is

$$f_2 = \frac{f_{unity}}{A_{noise}}$$
$$f_2 = \frac{4 \text{ MHz}}{10}$$
$$f_2 = 400 \text{ kHz}$$

Since the three stages are identical, the system will roll off before 400 kHz.

$$f_{2\text{-}system} = f_2 \sqrt{2^{1/n} - 1}$$
$$f_{2\text{-}system} = 400 \text{ kHz} \sqrt{2^{1/3} - 1}$$
$$f_{2\text{-}system} = 203.9 \text{ kHz}$$

Note that the system response in this case is reduced about an octave from the single-stage response.

Example 5.6 Using only LF351 op amps, design a circuit with an upper break of 500 kHz and a gain of 26 dB.

A gain of 26 dB translates to an ordinary gain of 20. Assuming a noninverting voltage stage, a single op amp would require an f_{unity} of:

$$f_{unity} = f_2 A_v \qquad (A_{noise} = A_v \text{ for the noninverting form})$$
$$f_{unity} = 500 \text{ kHz} \times 20$$
$$f_{unity} = 10 \text{ MHz}$$

Figure 5.7
Completed design for
Example 5.6

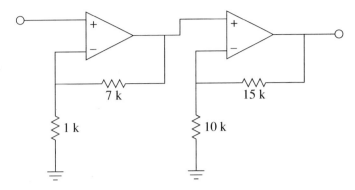

Figure 5.7
Completed design for
Example 5.6

Since the 351 has a typical f_{unity} of 4 MHz, at least two stages are required. There are many possibilities. One option is to set one stage as the dominant stage and set its gain to produce the desired f_2. The second stage will then be used to make up the difference in gain to the desired system gain.

stage 1:
$$f_2 = \frac{f_{unity}}{A_v}$$

$$A_v = \frac{f_{unity}}{f_2}$$

$$A_v = \frac{4\ \text{MHz}}{500\ \text{kHz}}$$

$$A_v = 8$$

stage 2: To achieve a final gain of 20, stage 2 requires a gain of 2.5. Its f_2 is:

$$f_2 = \frac{f_{unity}}{A_v}$$

$$f_2 = \frac{4\ \text{MHz}}{2.5}$$

$$f_2 = 1.6\ \text{MHz}$$

Note that if this frequency had worked out to less than 500 kHz, three or more stages would be required. To set the resistor values, the rules of thumb presented in Chapter 4 may be used. One possible solution is shown above in Figure 5.7.

Low Frequency Limitations

As mentioned earlier, standard op amps are direct-coupled. That is, their gain response extends down to 0 Hz. Consequently, many op amp circuits have no lower frequency limit. They will amplify DC signals just as easily as AC signals. Sometimes it is desirable to introduce a low-frequency rolloff. Two instances are single-supply biasing (see Chapter 4) and interference rejection (the removal of undesired signals, such as low-frequency rumble). In both cases, the circuit designer produces a low-frequency rolloff (lead network) by introducing coupling capacitors. For single-supply circuits, these capacitors are a necessary evil. Without them, stages would quickly overload from the large DC input. Also, signal sources and loads may be very intolerant of the DC bias potential. The result could be gross distortion or component failure. Even if a circuit uses a normal bipolar supply, a lead network may be used to reduce interference signals. For example, a well-chosen coupling capacitor can reduce 60 Hz hum interference while hardly affecting the quality of a voice transmission. Generally, these coupling capacitors can be simplified into the straightforward lead networks discussed in Chapter 1. (Remember, for lead networks, the highest critical frequency is the dominant one.) Also, if multiple networks are dominant, the resulting critical frequency will be higher than the individual break frequency. The relationship is the mirror image of Equation 5.8. The proof for the following equation is very similar to that of Equation 5.8, and is left as an exercise.

$$f_{1\text{-}system} = \frac{f_1}{\sqrt{2^{1/n} - 1}} \tag{5.9}$$

If you decide to add coupling capacitors in order to reduce interference, remember that the op amp will need a DC-return resistor. An example is shown in Figure 5.8. The 100 k resistor is needed to ensure that the inverting input's half of the diff amp stage is properly biased. Note that for a typical op amp,

Figure 5.8
DC-return resistor (100 k)

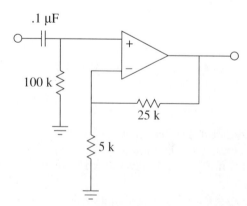

.1 µF

100 k

25 k

5 k

this resistor also ends up setting the input impedance. Assuming a relatively low source impedance, the lead network simplification boils down to the .1 μF capacitor along with the 100 k resistor. The critical frequency is:

$$f_c = \frac{1}{2\pi RC}$$

$$f_c = \frac{1}{2\pi \times 100\ \text{k} \times .1\ \mu\text{F}}$$

$$f_c = 15.9\ \text{Hz}$$

To sum up, when using general-purpose op amps, if no signal-coupling capacitors are being used, the gain response extends back to 0 Hz. If coupling capacitors are used, general lead-network analysis techniques can be used to find the critical frequencies.

5.4 Slew Rate and Power Bandwidth

As noted in the previous section, general-purpose op amps contain a compensation capacitor which is used to control the open-loop frequency response. The signal developed across this capacitor will be amplified in order to create the final output signal. In essence, this capacitor serves as the load for the preceding stage inside the op amp. Like all stages, this one has a finite current-output capability. The compensation capacitor thus can be charged no faster than a rate determined by the standard capacitor charge equation:

$$i = C\frac{dv}{dt}$$

$$\frac{dv}{dt} = \frac{i}{C}$$

The term dv/dt is the rate of change of voltage versus time. By definition, this parameter is called *slew rate* (SR). The base unit for slew rate is volts per second; however, given the speed of typical devices, slew rate is normally specified in volts per microsecond. Slew rate is very important in that it helps to determine whether a circuit can accurately amplify high-frequency or pulse-type waveforms. In order to create a fast op amp, either the charging current i must be large or the compensation capacitor C must be very small. Since C also plays a role in determining the gain-bandwidth product, there is a lower limit to its size. A typical op amp might use a 30 pF compensation capacitor and the driving stage may effectively produce a charging current of 100 μA.

The resulting slew rate would be

$$SR = \frac{dv}{dt} = \frac{i}{C}$$

$$SR = \frac{100\ \mu A}{30\ pF}$$

$$SR = 3.33\ MV/S$$

$$SR = 3.33\ V/\mu S$$

This means that the output of the op amp can change no faster than 3.33 V over the course of 1 μS. It would take this op amp about 3 μS for its output signal to change a total of 10 V. It can go no faster than this.

The ideal op amp would have an infinite slew rate. While this is a practical impossibility, it is possible to find special high-speed devices which exhibit slew rates in the range of several thousand volts per microsecond. Typical slew rates for a few popular devices are shown in Table 5.1 for comparison.

Table 5.1
Slew rates for some
common devices

Device	Slew Rate
741	.5 V/μS
351	13 V/μS
318	70 V/μS

Slew rate is always output-referred. In this way, the circuit gain need not be taken into account. Slew rate is normally the same whether the signal is positive- or negative-going.

There are a few devices which exhibit an asymmetrical slew rate. One example is the 3900. It has a slew rate of .5 V/μS for positive swings, but shows 20 V/μS for negative swings.

The Effect of Slew Rate on Pulse Signals

An ideal pulse waveform will shift from one level to the other instantaneously, as shown in Figure 5.9. In reality, the rising and falling edges are limited by the slew rate. If this signal is fed into a 741 op amp, the output pulse would be decidedly trapezoidal, as shown in Figure 5.10. The 741 has a slew rate of .5 V/μS. Since the voltage change is 2 V, it takes the 741 four microseconds to traverse from low to high, or from high to low. The resulting waveform is still recognizable as a pulse, though. Gross distortion of the pulse occurs if the pulse width is decreased, as in Figures 5.11 and 5.12. Here, the pulse width is

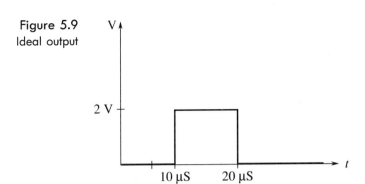

Figure 5.9
Ideal output

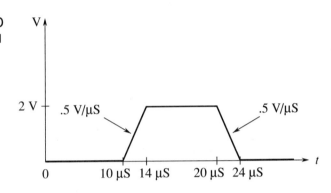

Figure 5.10
Slewed output of the 741

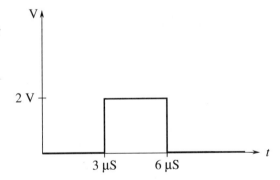

Figure 5.11
Ideal output

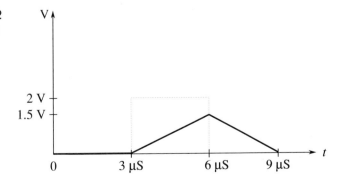

Figure 5.12
Slewed output of the 741

———————   **Figure 5.13**
Slewed output of the 351

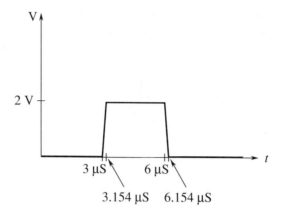

only three microseconds, so the 741 doesn't even have enough time to reach the high level. In three microseconds, the 741 can only change 1.5 volts. By the time the 741 gets to 1.5 volts, the input signal is already swinging low, so the 741 attempts to track it. The result is a triangular waveform of reduced amplitude. This same effect can occur if the amplitude of the pulse is increased. Obviously, pulses that are both fast and large require high slew rate devices. Note that a 351 op amp would produce a nice output in this example. Since its slew rate is 13 V/μS, it requires only .154 μS for the 2 V output swing. Its output waveform is shown in Figure 5.13.

The Effect of Slew Rate on Sinusoidal Signals and Power Bandwidth

Slew rate limiting produces an obvious effect on pulse signals. It can also affect sinusoidal signals. All that is required for slewing to take place is that the signal change faster than the device's slew rate. If the rate of change of the signal is never greater than the slew rate, slewing will never occur.

To find out just how fast a given sine wave changes, we need to find the first derivative with respect to time. The input sine wave has a frequency f, and a peak amplitude K.

$$v(t) = K \sin 2\pi f t$$

$$\frac{dv}{dt} = 2\pi f K \cos 2\pi f t$$

where dv/dt is the rate of change of the signal with respect to time. The maximum rate of change will occur when the sine wave passes through zero (i.e., at $t = 0$).

To find this maximum value, substitute 0 for t, and solve the equation

$$\frac{dv}{dt} = 2\pi f K \tag{5.10}$$

So, the rate of change of the signal is directly proportional to the signal's frequency (f), and its amplitude (K). From this, it is apparent that high-amplitude high-frequency signals require high–slew-rate op amps in order to prevent slewing.

We can rewrite our equation in a more convenient form:

$$\text{required SR} = 2\pi V_p f_{max} \tag{5.11}$$

where V_p is the peak voltage swing required and f_{max} is the highest-frequency sine wave reproduced. Often, it is desirable to know just how "fast" a given op amp is. A further rearranging yields

$$f_{max} = \frac{SR}{2\pi V_p}$$

In this case, f_{max} represents the highest-frequency sine wave which the op amp can reproduce without producing slewing induced **d**istortion (SID). This frequency is commonly referred to as the *power bandwidth*. To be on the conservative side, set V_p to the op amp's clipping level.

Note that slew rate calculations are not dependent on either the circuit gain or small-signal bandwidth. Power bandwidth and small-signal bandwidth (f_2) are **not** the same thing. This is a very important point.

The effects of slewing can be subtle or dramatic. Small amounts of SID are very difficult to see directly on an oscilloscope, and require the use of a distortion analyzer or a spectrum analyzer for verification. Heavy slewing turns a sine wave into a triangular wave. An example is shown in Figure 5.14.

Figure 5.14
Sine wave distorted by heavy slewing

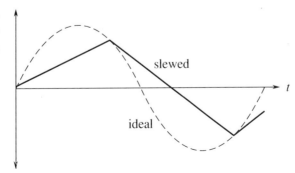

Example 5.7   A 741 is used as part of a motor control system. If the highest reproducible frequency is 3 kHz, and the maximum output level is 12 V peak, does slewing ever occur?

Another way of stating the problem is to ask "Is the 741's power bandwidth at least 3 kHz?"

$$f_{max} = \frac{SR}{2\pi V_p}$$

$$f_{max} = \frac{.5 \text{ V}/\mu S}{2\pi \times 12 \text{ V}}$$

$$f_{max} = \frac{.5 \text{ MV}/S}{\pi \times 24 \text{ V}}$$

$$f_{max} = 6631 \text{ Hz}$$

For this application, the 741 is twice as fast as it needs to be. Note how in the calculation the slew rate is transferred from V/μS into V/S, and how the volts units cancel between denominator and numerator. This leaves units of "1/Seconds," which is another way of saying "Hertz." If the calculation produced a smaller value, say 2 kHz, then slewing is a possibility for certain signals.

Example 5.8   An audio preamplifier needs to reproduce signals as high as 20 kHz. The maximum output swing is 10 V peak. What is the minimum acceptable slew rate for the op amp used?

$$SR = 2\pi V_p f_{max}$$

$$SR = 2\pi \times 10 \text{ V} \times 20 \text{ kHz}$$

$$SR = 1.257 \text{ MV}/S$$

$$SR = 1.257 \text{ V}/\mu S$$

For this design, a 741 would not be fast enough. The aforementioned 351 and 318 would be satisfactory.

A Design Hint

There is a convenient way of graphically determining whether the output of an op amp will be distorted. It involves graphing output levels versus frequency. The two major distortion causes are clipping and slewing. We start with a

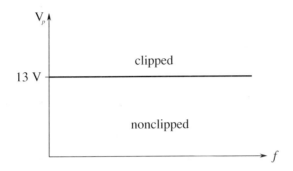

grid measuring frequency on the horizontal axis, and output voltage on the vertical axis. The first step is to plot the output level limit imposed by clipping. Clipping is dependent on the circuit's power supply and is independent of frequency. Therefore, a horizontal line is drawn across the graph at the clipping level (see Figure 5.15). If we assume a standard ± 15 V power supply, this level will be around ± 13 V. The output level cannot swing above this line because clipping will result. Everything below this line represents unclipped signals. The second step is to plot the slewing line. To do this, a point needs to be calculated for f_{max}. In the preceding example a 741 was used and 12 volts produced an f_{max} of 6631 Hz. Plot this point on the graph. Now, since the slew rate is directly proportional to f_{max} and V_p, it follows that doubling f_{max} while halving V_p results in the same slew rate. This new point lets you graphically determine the slope of the slew-limiting line. Plot this new point and connect the two points with a straight line (see Figure 5.16). Everything above this line represents slewed signals, and everything below this line represents nonslewed signals. As long as the desired output signal falls within the lower intersection area of the two lines, the signal will not experience either slewing or clipping. A quick glance at the graph allows you to tell what forms of distortion may affect a given signal.

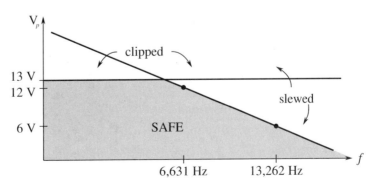

Figure 5.17
Multistage circuit
incorporating the 741,
351, and 318

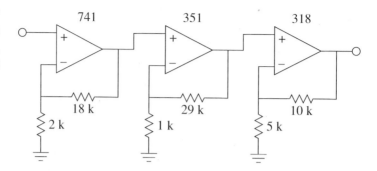

Slew Rate and Multiple Stages

Consider the three-stage circuit shown in Figure 5.17. The slew rates for each device are found in Table 5.1. What is the effective slew rate of the system? You might think that it is set by the slowest device (741 at .5 V/μS), or perhaps by the final device (318 at 70 V/μS). The fact is that the system slew rate could be set by any of the devices, and it depends on the gains of the stages.

The system slew rate will never be faster than the final device. In this example, the slew rate cannot be greater than 70 V/μS. It may be less than this, though. The trick to finding the effective system slew rate is to start at the output of the first stage, and then determine the maximum rate of change for the following stages in sequence. The maximum output rate at stage 1 is .5 V/μS. This is the maximum rate of change going into stage 2. Since stage 2 has a gain of 30, it will attempt to increase this rate to 15 V/μS. This cannot happen, though, because the 351 has a slew rate of only 13 V/μS. Therefore the 351 is the limiting factor at this point. The maximum rate of change out of stage 2 is 13 V/μS. This signal is then applied to stage 3 which has a gain of 3. So, the 318 triples its input signal to 39 V/μS. Since the 318 is capable of changing as fast as 70 V/μS, 39 V/μS becomes the limiting output factor. The system slew rate is 39 V/μS. This is the value used to calculate the system power bandwidth, if needed. The first op amp to slew in this circuit is the 351, even though it is about 26 times faster than the 741 used in stage 1. The reason for this is that it must handle signals 30 times as large. Note that if the final stage had a larger gain, say 10, the 318 would become the limiting factor. The important thing to remember is that the front end stages of a system don't need to be as fast as the final stages since they handle smaller signals.

Noncompensated Devices

As discussed in Chapter 3, all op amps need some form of frequency compensation in order to ensure that their closed-loop response is stable. The most straightforward way to do this is to add a compensation capacitor which forces a 20 dB/decade rolloff to f_{unity}. In this way, no matter what gain you choose, the circuit will be stable. While this is very convenient, it is not the most efficient form of compensation for every circuit. High-gain circuits need less compensation capacitance than do low-gain circuits. The advantage of using a smaller compensation capacitor is that slew rate is increased. Also, available loop gain at higher frequencies is increased. This allows the resulting circuit to have a wider small-signal bandwidth (the effect is as if f_{unity} increased). Therefore, if you are designing a high-gain circuit, you are not producing the maximum slew rate and small-signal bandwidth that you might. The compensation capacitor is large enough to achieve unity gain stability, but your circuit is a high-gain design. The bottom line is that you are trading off slew rate and bandwidth for unity gain stability.

To get around this problem, manufacturers offer noncompensated op amps. No internal capacitor is used. Instead, connections are brought out to the IC package so that you may add your own capacitor. In this way, the op amp can be tailored to your application. There is no set way of determining the values for the external compensation circuit (which may be more complex than a single capacitor). Compensation details are given on manufacturers' data sheets.

One example of a noncompensated op amp is the 301. You can think of a 301 as a 741 without a compensation capacitor. If a 33 pF compensation capacitor is used, the 301 will be unity-gain stable, and produce an f_{unity} of 1 MHz and a slew rate of .5 V/μS. For higher gains, a smaller capacitor can be used. A 10 pF unit will produce an effective f_{unity} of 3 MHz and a slew rate of 1.5 V/μS. Comparative Bode gain plots are shown in Figure 5.18 for the 301.

Figure 5.18
Bode gain plots for
the 301

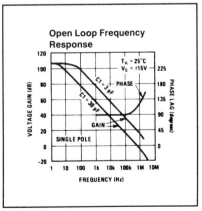

Reprinted with permission of National
Semiconductor Corporation

Feedforward Compensation

Besides the ordinary form of compensation, noncompensated devices like the 301 may utilize other methods which can make them even faster, such as *feedforward compensation*. The concept of feedforward is in direct contrast to feedback. As its name suggests, feedforward involves adding a portion of the input signal to the output, thus bypassing certain sections of the system. Compared to ordinary feedback, feedforward is seldom used as part of the design of an amplifier.[1]

Ordinarily, a compensation capacitor must be large enough to maintain sufficient gain and phase margin for the slowest stage inside an op amp. Quite often, the slowest stage in an op amp is one of the first stages, such as a level shifter. Here is where feedforward comes into play: If the high-frequency content of the input signal can be shunted around this slow stage, the effective bandwidth and speed of the op amp can be increased. This is the key behind the technique of feedforward compensation. Normally, manufacturers will provide details of specific feedforward realizations for their op amps. (It would be very difficult to create successful feedforward designs without detailed knowledge of the internal design of the specific op amp being used.) Not all op amps lend themselves to feedforward techniques.

In summary, noncompensated op amps require a bit more work to configure than fully compensated devices, but offer higher performance. This means faster slew rates and higher upper-break frequencies.

Decompensated Devices

Straddling the worlds of compensated and noncompensated op amps is the *decompensated* device. Decompensated devices are also known as *partially compensated* devices. They include some compensation capacitance, but not enough to make them unity-gain stable. Usually these devices are stable for gains above 3 to 5. Since the majority of applications require gains in this range or above, decompensated devices offer the ease of use of compensated devices and the increased performance of customized noncompensated units. If required, extra capacitance can usually be added to make the circuit unity-gain stable. One example of this type is the 5534. Its Bode gain plot is shown in Figure 5.19. Note how the addition of an extra 22 pF reduces the open-loop gain. This 22 pF is enough to make the device unity-gain stable. It also

---

[1] Feedforward compensation can offer similar advantages, such as a reduction in distortion. Also, feedforward and feedback techniques can be combined in order to achieve complementary increases in performance. For an example, see M. J. Hawksford, "Reduction of Transistor Slope Impedance Dependent Distortion in Large Signal Amplifiers," *Journal of the Audio Engineering Society*, Vol. 36, No. 4, pp. 213–222.

Figure 5.19
Bode gain plot for
the 5534

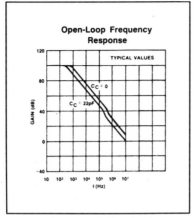

Courtesy of Signetics Company

Figure 5.19
Bode gain plot for
the 5534

has a dramatic effect on the slew rate, as seen in the 5534's spec sheet. Without the 22 pF capacitor, the slew rate is 13 V/μS, but with it, the slew rate drops to 6 V/μS. The performance that is lost in order to achieve unity-gain stability is obvious here.

There is one more interesting item to note about the 5534. A close look at the Bode gain curve shows a hump in the rolloff region. Some other devices, such as the 318, exhibit this hump, too. This is a nice extra. In essence, the manufacturer has been able to achieve a slightly higher open-loop gain than a normal device would allow. This means that your circuits will have higher loop gains, and therefore, the desirable effects of negative feedback will remain active to higher frequencies. This also means that very high-gain circuits will be able to achieve a higher f_2 than is predicted by the gain-bandwidth calculation.

5.5 Offsets

Offsets are undesirable DC levels appearing at the output of a circuit. If op amps were perfect, there would be no such thing as an offset. Even though part matching is very close when ICs are made, the parts will not be identical. One example is that the transistors used for the differential amplifier stage will not have identical characteristics. Because of this, their DC bias points are slightly different. This difference, or unbalance, is amplified by the remaining stages and will eventually produce a DC voltage at the output. Since all op amps are slightly different, we never know what the exact output offset will be. For measurement applications, this offset creates uncertainty in readings. For example, if the circuit output measures 100 mV, the signal might be 99 mV with 1 mV of offset. It might also be 101 mV with −1 mV offset. In other applications, offsets can harm following stages or loads. Dynamic loudspeakers and headphones are two loads which should not be fed DC signals. This will reduce their maximum volume and increase their distortion. In short, offsets are not desirable. Let's see what the causes are, and how we can reduce or eliminate their effect.

Offset Sources and Compensation

For bipolar input sections, the major cause of input current mismatch is the variation of β. Base-emitter junction voltage variation is the major cause of input voltage deviation. For field effect devices, current variation is much less of a problem since the magnitude of input current is very low to begin with. Unfortunately, FETs do suffer from larger input voltage variations due to transconductance curve mismatches.

As mentioned in Chapter 2, the input current into the bases (or gates, in the case of an FET) of the first stage is called I_B, the input bias current. In reality, this is an average of the two input currents, I_{B+} and I_{B-}. The magnitude of their difference is called the *input offset current*, I_{OS} (some manufacturers use the symbol I_{IO}). Note that the actual direction of I_B is normally not specified, but can usually be determined from the manufacturer's circuit diagram. I_B flows into the op amp if the input devices are NPN, and out of the op amp if the input devices are PNP.

$$I_B = \frac{I_{B+} + I_{B-}}{2}$$

$$I_{OS} = |I_{B+} - I_{B-}| \tag{5.12}$$

The voltage difference for the input stage is referred to as the *input offset voltage*, V_{OS} (some manufacturers use the symbol V_{IO}). This is the potential required between the two inputs to null the output, that is, to make sure that the output is 0 V DC. Both I_{OS} and V_{OS} are available on data sheets. The absolute magnitude of these offsets generally gets worse at the temperature extremes. Table 5.2 shows some typical values. Note the low I_B and I_{OS} values for the FET input 351.

Remember, these numbers are absolutes, so when I_{OS} is specified as 10 nA, it means that the actual I_{OS} can be anywhere between −10 nA and +10 nA. I_B, I_{OS}, and V_{OS} combine with other circuit elements to produce an *output offset voltage*. Since this is a linear circuit, superposition may be used to separately calculate their effects. The model in Figure 5.20 will be used. R_i and R_f are the standard feedback components, while R_{off} is called the *offset compensation resistor* (in some cases it may be zero). Since the input signal is grounded, this model is valid for both inverting and noninverting amplifiers.

Table 5.2

Input bias current, input offset current, and input offset voltage for some typical devices.

Device	I_B	I_{OS}	V_{OS}
5534	800 nA	10 nA	.5 mV
351	50 pA	25 pA	5 mV
318	150 nA	30 nA	4 mV
741	80 nA	20 nA	1 mV

V_{OS} is seen as a small input voltage and is multiplied by the circuit's noise gain in order to find its contribution to the output offset. Since offsets are by nature DC, it is important to use the DC noise gain. Consequently, any capacitors found within the feedback loop should be mathematically "opened" for this calculation (for example, when working with the filter circuits presented in Chapter 11).

$$V_{out-offset1} = A_{noise}V_{OS} \tag{5.13}$$

where

$$A_{noise} = 1 + \frac{R_f}{R_i} = \frac{R_i + R_f}{R_i}$$

I_B and I_{OS} pass through input and feedback resistors to produce their output contributions. First, consider the effect of I_{B+}, which creates a voltage across R_{off}. This voltage is then multiplied by the circuit noise gain to yield its portion of the output offset:

$$V_{out-offset2} = I_{B+}R_{off}A_{noise} \tag{5.14}$$

Figure 5.20
Offset model

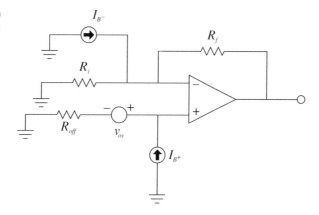

As for the effect of I_{B-}, recall that the inverting input is at virtual ground. This implies that the voltage across R_i must be zero, and therefore, the current through R_i must be zero. Consequently, all of I_{B-} flows through R_f. This creates a relative negative potential at the output.

$$V_{out-offset3} = -I_{B-}R_f \tag{5.15}$$

So the combination of the input bias current effects is

$$V_{out-offset(I_B)} = \left| I_{B+}R_{off}A_{noise} - I_{B-}R_f \right| \tag{5.16}$$

Expanding this produces

$$V_{out-offset(I_B)} = I_{B+}R_{off}\frac{R_i + R_f}{R_i} - I_{B-}R_f$$

$$V_{out-offset(I_B)} = \left(I_{B+}R_{off}\frac{R_i + R_f}{R_iR_f} - I_{B-} \right)R_f$$

By noting the product-sum rule for resistor combination R_i, R_f, this can be further simplified to

$$V_{out-offset(I_B)} = \left(\frac{I_{B+}R_{off}}{R_i \parallel R_f} - I_{B-} \right)R_f \tag{5.17}$$

If R_{off} is set to equal $R_i \parallel R_f$, this reduces to

$$V_{out-offset(I_B)} = (I_{B+} - I_{B-})R_f$$

By definition,

$$I_{OS} = I_{B+} - I_{B-}$$

so we finally arrive at

$$V_{out-offset(I_B)} = I_{OS}R_f \tag{5.18}$$

If it is possible, R_{off} should be set to $R_i \parallel R_f$. This drastically reduces the effect of the input bias current. Note that the value of R_{off} includes the driving source internal resistance. If $R_i \parallel R_f = 2$ k and the driving source resistance is 100 Ω, the required resistance value will be 1.9 k. If setting R_{off} to the optimum value is not possible, you can at least reduce the effect of I_B by using a partial value. Also, note that it is possible to determine the polarity of the offset caused by I_B (Equation 5.17) if the type of device used in the diff amp stage is known. The circuit of Figure 5.20 assumes that NPN devices are being used, hence the currents are drawn as entering the op amp. PNP input devices would produce the opposite polarity.

For a final result, we can combine our components. If $R_{off} = R_i \| R_f$,

$$V_{out\text{-}offset} = V_{OS}A_{noise} + I_{OS}R_f \tag{5.19}$$

and, if $R_{off} \neq R_i \| R_f$,

$$V_{out\text{-}offset} = V_{OS}A_{noise} + |I_{B+}R_{off}A_{noise} - I_{B-}R_f| \tag{5.20}$$

There is one special case involving the selection of R_{off}, and that deals with a voltage follower. Normally for a follower, $R_f = 0$. What if the driving source resistance is perhaps 50 Ω? The calculation would require an R_{off} of 0 Ω, and thus a -50 Ω resistor to compensate for the source resistance. This is, of course, impossible! To compensate for the 50 Ω source, use 50 Ω for R_f. The circuit gain will still be unity, but I_B will now be compensated for. This is shown in Figure 5.21.

Figure 5.21
Offset compensation for a
follower

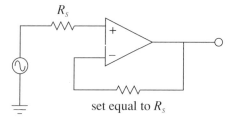

set equal to R_s

Example 5.9   Determine the typical output offset voltage for the circuit of Figure 5.22 if R_{off} is 0 Ω. Then determine an optimum size for R_{off} and calculate the new offset.

Figure 5.22
Circuit for Example 5.9

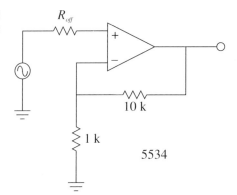

From the data sheet for the 5534, we find $V_{OS} = .5$ mV, $I_{OS} = 10$ nA, and $I_B = 800$ nA. Since this is an approximation, assume $I_{B+} = I_{B-} = I_B$.

$$A_{noise} - 1 + \frac{R_f}{R_i}$$

$$A_{noise} = 1 + \frac{10 \text{ k}}{1 \text{ k}}$$

$$A_{noise} = 11$$

$$V_{out\text{-}offset} = V_{OS}A_{noise} + |I_{B+}R_{off}A_{noise} - I_{B-}R_f|$$

$$V_{out\text{-}offset} = .5 \text{ mV} \times 11 + |800 \text{ nA} \times 0 \times 11 - 800 \text{ nA} \times 10 \text{ k}|$$

$$V_{out\text{-}offset} = 5.5 \text{ mV} + 8 \text{ mV}$$

$$V_{out\text{-}offset} = 13.5 \text{ mV}$$

Remember, this is the magnitude of the offset; it could be anywhere within ± 13.5 mV. It might be worse if this is not a typical device. Now we find the optimum offset compensating resistor:

$$R_{off} = R_i \parallel R_f$$

$$R_{off} = 1 \text{ k} \parallel 10 \text{ k}$$

$$R_{off} = 909 \; \Omega$$

For this case, the offset equation reduces to:

$$V_{out\text{-}offset} = V_{OS}A_{noise} + I_{OS}R_f$$

$$V_{out\text{-}offset} = .5 \text{ mV} \times 11 + 10 \text{ nA} \times 10 \text{ k}$$

$$V_{out\text{-}offset} = 5.5 \text{ mV} + 100 \; \mu\text{V}$$

$$V_{out\text{-}offset} = 5.6 \text{ mV}$$

By adding R_{off}, the output offset voltage is more than halved. This may lead you to think that it is always wise to add R_{off}. Such is not the case. There are two times when you may prefer to leave it out: 1) to optimize noise characteristics, as we shall see shortly, and 2) when using FET input op amps. FET input devices have very small input bias and offset currents to begin with, so their effect is negligible when using typical resistor values.

Example 5.10   The circuit of Figure 5.23 is used as part of a measurement system. Assuming that the DC input signal is 20 mV, how much uncertainty is there in the output voltage typically?

Figure 5.23
Circuit for Example 5.10

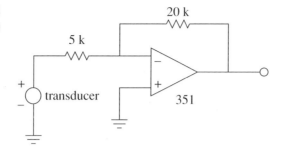

The desired output from the amplifier is

$$V_{out} = A_v V_{in}$$

$$V_{out} = -\frac{R_f}{R_i} V_{in}$$

$$V_{out} = -\frac{20 \text{ k}}{5 \text{ k}} \times 20 \text{ mV}$$

$$V_{out} = -80 \text{ mV}$$

The specs for the 351 are $V_{OS} = 5$ mV, $I_{OS} = 25$ pA, and $I_B = 50$ pA.

$$A_{noise} = 1 + \frac{R_f}{R_i}$$

$$A_{noise} = 1 + \frac{20 \text{ k}}{5 \text{ k}}$$

$$A_{noise} = 5$$

$$V_{out\text{-}offset} = V_{OS} A_{noise} + |I_{B+} R_{off} A_{noise} - I_{B-} R_f|$$

$$V_{out\text{-}offset} = 5 \text{ mV} \times 5 + |50 \text{ pA} \times 0 \times 11 - 50 \text{ pA} \times 20 \text{ k}|$$

$$V_{out\text{-}offset} = 25 \text{ mV} + 1 \text{ } \mu\text{V}$$

$$V_{out\text{-}offset} = 25 \text{ mV}$$

The output can vary by as much as ± 25 mV. Since this is a DC measurement system, the results are devastating. The output can be anywhere from -80 mV $- 25$ mV $= -105$ mV, to -80 mV $+ 25$ mV $= -55$ mV. That is almost a 2:1 spread, and it's caused solely by the op amp. Note that the addition of R_{off} would have little effect here. Since the 351 uses a FET input, its I_B contribution is only 1 μV.

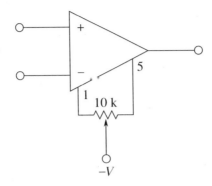

Figure 5.24
A typical nulling
connection

So then, how do you keep output offsets to a minimum? First and foremost, make sure that the chosen op amp has low I_{os} and V_{os} ratings. Second, use the offset compensation resistor, R_{off}. Third, keep the circuit resistances as low as possible. Finally, if the output offset is still too large, it can be reduced by manually nulling the circuit.

Nulling involves summing in a small signal that is of opposite polarity to the existing offset. The new signal will completely cancel the offset and the output will show 0 V DC. This is much easier than it sounds. Most op amps have connections for null circuits. These are specified by the manufacturer and usually consist of a single potentiometer and perhaps one or two resistors. An example of a nulling connection is shown in Figure 5.24. Usually the potentiometer is a multiturn trim type to allow for fine adjustment. To null the circuit, the technician monitors the output with a very sensitive DC voltmeter. The input is grounded (or perhaps tied to ground through a resistor equal to the driving source resistance if it's large). The potentiometer is then adjusted until the meter reads zero.

The drawback to this procedure is that it requires someone (or perhaps some thing) to perform the nulling. Also, the unit will require periodic adjustment to compensate for aging and environmental effects.

5.6 Drift

Drift is a variation in the output offset voltage. Often, it is temperature-induced. Even if a circuit has been manually nulled, an output offset can be produced if the temperature changes. This is because V_{os} and I_{os} are temperature-sensitive. The only way around this pitfall is to keep the circuit in a static environment, which can be very costly. If the drift can be kept within an acceptable range, the added cost of cooling and heating equipment can be removed. As you might expect, the magnitude of the drift depends on the size of the temperature change. It also depends on the I_{os} and V_{os} sensitivities. These items are $\Delta V_{os}/\Delta T$, the change in V_{os} with respect to temperature, and $\Delta I_{os}/\Delta T$, the change in I_{os} with respect to temperature. Drift rates are specified in terms of change per degree centigrade. These parameters are usually specified as worst-case values and can produce either a positive or negative potential.

The development of the drift equation closely follows that of the equation for offsets. The only difference is that offset parameters are replaced by their temperature coefficients and the temperature change. The products of the coefficients and the change in temperature produce an input offset voltage and current:

$$V_{drift} = \frac{\Delta V_{os}}{\Delta T} \Delta T A_{noise} + \frac{\Delta I_{os}}{\Delta T} \Delta T R_f \tag{5.21}$$

As with the offset calculation, the drift result may be either positive or negative. Also, since I_{os} is so small for FET input devices, $\Delta I_{os}/\Delta T$ is often not listed, since it is almost always small enough to ignore. For lowest drift, it is assumed that the op amp uses the offset compensation resistor R_{off}, and that the circuit has been nulled.

Example 5.11   Determine the output drift for the circuit of Figure 5.22 for a target temperature of 80°C. Assume that $R_{off} = 909\ \Omega$ and that the circuit has been nulled at 25°C.

The parameters for the 5534 are $\Delta V_{os}/\Delta T = 5\ \mu V/C°$, $\Delta I_{os}/\Delta T = 200$ pA/C°. The noise gain was already determined to be 11 in Example 5.9. The total temperature change is from 25°C to 80°C, or 55C°.

$$V_{drift} = \frac{\Delta V_{os}}{\Delta T} \Delta T A_{noise} + \frac{\Delta I_{os}}{\Delta T} \Delta T R_f$$

$$V_{drift} = 5\ \mu V/C° \times 55C° \times 11 + 200\ pA/C° \times 55C° \times 10\ k$$

$$V_{drift} = 3.025\ mV + .11\ mV$$

$$V_{drift} = 3.135\ mV$$

Note that for this circuit the V_{os} drift is the major source of error. At 80°C the output of the circuit may have up to $\pm 3.135\ mV$ of DC error.

As with offsets, drift is partially a function of the circuit gain. Therefore high-gain circuits often appear to have excessive drift. In order to compare different amplifiers, *input referred drift* is often used. To find the input referred drift, just divide the output drift by the **signal gain** of the amplifier. Don't use the noise gain! In this way, both inverting and noninverting amplifiers can be compared on an equal footing. For this circuit, the input referred drift is

$$V_{drift(input)} = \frac{V_{drift}}{A_v}$$

$$V_{drift(input)} = \frac{3.135 \text{ mV}}{11}$$

$$V_{drift(input)} = 285 \text{ }\mu\text{V}$$

For many applications, particularly those primarily concerned with AC performance, drift specification is not very important. A communications amplifier, for example, might need to use an output coupling capacitor to block any drift or offset from reaching the output. Drift is usually important for applications involving DC or very low frequencies where coupling capacitors are not practical.

5.7 CMRR and PSRR

CMRR stands for **c**ommon **m**ode **r**ejection **r**atio. It is a measure of how well the two halves of the input differential amplifier stage are matched. A common-mode signal is a signal which is present on both inputs of the diff amp. Ideally, a differential amplifier completely suppresses or rejects common-mode signals. Common-mode signals should not appear at the circuit output. However, because of the nonperfect matching of transistors, some portion of the common-mode signal will make it to the output. Exactly how much this signal is reduced relative to desired signals is measured by the CMRR.

Ideally, CMRR is infinite. A typical value for CMRR would be 100 dB. In other words, if an op amp had both desired (i.e., differential) and common-mode signals at its input that were the same size, the common-mode signal would be 100 dB smaller than the desired signal at the output.

CMRR is particularly important when using the op amp in differential mode (Chapter 4) or when making an instrumentation amplifier (Chapter 6). There are two broad uses for these circuits. First, the amplifier may be receiving a low-level balanced signal over a considerable distance. Good examples of this are a microphone cable in a recording studio, or an instrumentation cable on a factory floor. Interference signals tend to be induced into the cable in phase (common-mode). Since the desired signal is presented out of phase (differential), a high CMRR will effectively remove the interference signal. Second, the op amp may be used as part of a bridge-type measurement system. Here, the desired signal is seen as a small variation between two DC potentials. The op amp must amplify the difference signal, but suppress the DC outputs of the bridge circuit.

Example 5.12 An amplifier has a closed-loop voltage gain of 20 dB and a CMRR of 90 dB. If a common-mode signal is applied to the input at -60 dBV, what is the output?

If the input signal were differential instead of common-mode, the output would be

$$v'_{out} = A'_v + v'_{in}$$

$$v'_{out} = 20 \text{ dB} + -60 \text{ dBV}$$

$$v'_{out} = -40 \text{ dBV}$$

Since this is a common-mode signal, it is reduced by the CMRR:

$$v'_{out} = -40 \text{ dBV} - 90 \text{ dB}$$

$$v'_{out} = -130 \text{ dBV}$$

This signal is so small that it is probably overshadowed by the circuit noise.

One final note concerning CMRR is that it is specified for DC. In truth, CMRR is frequency-dependent. The shape of its curve is reminiscent of the open-loop gain curve. The stated CMRR may remain at its DC level up to perhaps 100 or 1000 Hz, and then fall off as frequency increases. For example, the 318 data sheet states a typical CMRR of 100 dB. By looking at the CMRR graph though, you can see that it starts to roll off noticeably around 10 kHz. By the time it hits 1 MHz, only 40 dB of rejection remains. A more gentle rolloff is exhibited by the 351. At 1 MHz, 60 dB of rejection remains. This means that the op amp cannot suppress high-frequency interference signals as well as it suppresses low-frequency interference.

Similar to CMRR is PSRR, or **p**ower **s**upply **r**ejection **r**atio. Ideally, all ripple, hum, and noise from the power supply will be prevented from reaching the output of the op amp. PSRR is a measure of exactly how well the op amp attains this ideal. Typical values for PSRR are in the 100 dB range. Like CMRR, PSRR is frequency-dependent and shows a rolloff as frequency increases. If an op amp is powered by a 60 Hz source, the ripple frequency from a standard full-wave rectifier will be 120 Hz. At the output, this ripple will be reduced by the PSRR. Higher frequency noise components on the power supply line are not reduced by as much since PSRR rolls off. Normally, PSRR is consistent between power rails. Sometimes, there is a marked performance difference between the positive and negative PSRR. One good example of this is the 351. The positive rail exhibits about a 30 dB improvement over the negative rail. Note that PSRR is virtually nonexistent for the negative rail by the time it reaches 1 MHz.

Example 5.13   The power supply for an op amp has a peak-to-peak ripple voltage of .5 V. If the op amp's PSRR is 86 dB, how much of this ripple is seen at the circuit output?

First, determine the PSRR as an ordinary value.

$$PSRR = log_{10}^{-1} \frac{PSRR'}{20}$$

$$PSRR = log_{10}^{-1} \frac{86 \text{ dB}}{20}$$

$$PSRR = 20{,}000$$

Divide the ripple voltage by the PSRR to find the amount which is seen at the output.

$$v_{out\text{-}ripple} = \frac{v_{ripple}}{PSRR}$$

$$v_{out\text{-}ripple} = \frac{.5 \text{ V}_{pp}}{20{,}000}$$

$$v_{out\text{-}ripple} = 25 \ \mu V_{pp}$$

5.8 Noise

Generally speaking, noise refers to undesired output signals. The background hiss found on audio tapes is a good example of noise. If noise levels get too high, the desired signals are lost. We shall narrow our definition down a bit by considering only the signals that are created by the op amp circuit.

Noise comes from a variety of sources. First of all, all resistors have *thermal*, or *Johnson, noise*. This is due to the random effects that thermal energy produces on the electrons. Thermal noise is also called *white noise*, because it is equally distributed across the frequency spectrum. Semiconductors exhibit other forms of noise. *Shot noise* is caused by the fact that charges move as discrete particles (electrons). It is also white. *Popcorn noise* is dominant at lower frequencies and is caused by manufacturing imperfections. Finally, *flicker noise* has a $1/f$ spectral density. This means that it increases as the frequency drops. It is sometimes referred to as $1/f$ noise.

Any op amp circuit exhibits noise from all of these sources. Since we are not designing the op amps, we don't really need to distinguish the exact sources of the noise; rather, we'd just like to find out how much total noise arrives at the output. This will allow us to determine the *signal-to-noise ratio* (S/N) of the circuit, or how quiet the amplifier is.

If noise performance for a particular design is not paramount, it can be quickly estimated from manufacturers' data sheets. Some manufacturers will specify RMS noise voltages for specified signal bandwidths and source impedances. A typical plot is found in Figure 5.25. To use this plot, simply find the source impedance of your circuit on the horizontal axis, and by using the appropriate signal bandwidth curve, find the noise voltage on the vertical axis. This noise voltage is input-referred. In order to find the output noise voltage, multiply this number by the noise gain of the circuit. The result will not be very exact, but it will put you in the ballpark.

A more exact approach involves the use of two op amp parameters, *input noise voltage density* (v_{ind}), and *input noise current density* (i_{ind}). v_{ind} is usually specified in nanovolts per root Hertz. i_{ind} is specified in picoamps per root Hertz. Refer to the 5534 data sheet for example specifications. (Some manufacturers square these values and give units of volts-squared per Hertz and amps-squared per Hertz. To translate to the more common form, just take the square root of the values given. The data sheet for the 741 is typical of this form.) These two parameters take into account noise from all internal sources. As a result, these parameters are frequency-dependent. Due to the flicker noise component, the curves tend to be rather flat at higher frequencies and then suddenly start to increase at lower frequencies. The point at which

Figure 5.25
Noise voltage for given bandwidth versus source resistance

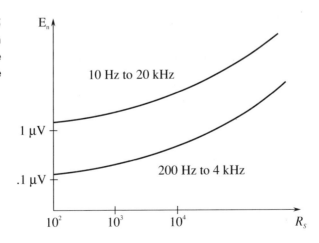

the graphs start to rise is called the *noise corner frequency*. Generally, the lower this frequency is, the better.

In order to find the output noise we will combine the noise from three sources:

1. v_{ind}

2. i_{ind}

3. the thermal noise of the input and feedback resistors

Before we start, there are a few points to note. First, the strength of the noise is dependent on the *noise bandwidth* of the circuit. For general-purpose circuits with 20 dB/decade rolloffs, the noise bandwidth (BW_{noise}) is 1.57 times larger than the small-signal bandwidth. It is larger because some noise still exists in the rolloff region. The noise bandwidth would equal the small-signal bandwidth only if the rolloff rate were infinitely fast. Second, since the noise-bandwidth factor is common to all three sources, instead of calculating its effect three times, we will first combine the partial results of the three sources, and then apply the noise-bandwidth effect. This will make the calculation faster, as we have effectively factored out BW_{noise}. Each of the three sources will be noise-voltage densities, all having units of volts per root Hertz. Since v_{ind} is already in this form, we only need to calculate the thermal and i_{ind} effects before doing the summation. Finally, since noise signals are random, they do not add coherently. In order to find the effective sum we must perform an RMS summation, that is, a *square root of the sum of the squares*. This will result in the input noise voltage. To find the output noise voltage we will then multiply by A_{noise}.

The very first thing that must be done is to determine the *input noise resistance* (R_{noise}). R_{noise} is the combination of the resistance seen from the inverting input to ground and from the noninverting input to ground. To do this, short the voltage source and ground the output. You will end up with a circuit like Figure 5.26. Note that R_i and R_f are effectively in parallel. Therefore,

$$R_{noise} = R_s + R_i \parallel R_f \tag{5.22}$$

You might note that R_s occupies the same position that R_{off} held in the offset calculations. For absolute minimum noise, the offset-compensating resistor is not used. R_{noise} is used to find the thermal noise and the contribution of i_{ind}. As always, Ohm's Law still applies, so as you might expect, $i_{ind}R_{noise}$ produces a noise voltage density with our desired units of volts per root Hertz.

The only source left is the thermal noise. The general equation for thermal noise is

$$e_{th} = \sqrt{4\ K\ T\ BW_{noise}R_{noise}} \tag{5.23}$$

Figure 5.26
Equivalent noise analysis
circuit

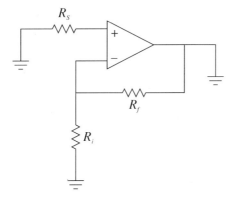

Figure 5.26
Equivalent noise analysis
circuit

where

        e_{th} is the thermal noise

        K is Boltzmann's constant, 1.38×10^{-23} Joules/Kelvin degree

        T is the temperature in Kelvin degrees (Celsius + 273)

        BW_{noise} is the effective noise bandwidth

        R_{noise} is the equivalent noise resistance

Since we are interested in finding the noise density, we can pull out the BW_{noise} factor. Also, when we perform the RMS summation, this quantity will need to be squared. Instead of taking the square root and then squaring it again, we can just leave it as e_{th}^2. These two considerations leave us with the *mean squared thermal noise voltage density*:

$$e_{th}^2 = 4\,K\,T\,R_{noise} \qquad (5.24)$$

Now that we have the components, we can perform the summation:

$$e_{total} = \sqrt{v_{ind}^2 + (i_{ind} \times R_{noise})^2 + e_{th}^2} \qquad (5.25)$$

where e_{total} is the total input noise voltage density. Its units are volts per root Hertz.

At this point we can include the noise bandwidth effect. In order to find BW_{noise}, multiply the small-signal bandwidth by 1.57. If the amplifier is DC-coupled, the small-signal bandwidth is equal to f_2; otherwise it is equal to $f_2 - f_1$. For most applications, setting the bandwidth to f_2 is sufficient.

$$BW_{noise} = 1.57 f_2 \qquad (5.26)$$

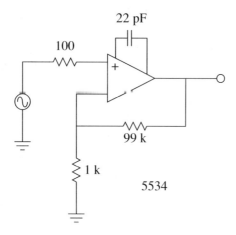

Figure 5.27
Circuit for Example 5.14

For the final step, e_{total} is multiplied by the square root of BW_{noise}. Note that the units for e_{total} are volts per root Hertz. Consequently, we need a *root Hertz* bandwidth. We are performing a mathematical shortcut here. You could square e_{total} in order to get units of volts squared per Hertz, multiply by BW_{noise}, and then take the square root of the result to get back to units of volts, but the first way is quicker. Anyway, we end up with the input-referred RMS noise voltage (e_n):

$$e_n = e_{total}\sqrt{BW_{noise}} \qquad (5.27)$$

Example 5.14   Determine the output noise voltage for the circuit of Figure 5.27. For a nominal output of 1 V RMS, what is the signal-to-noise ratio? Assume T = 300°K (room temperature).

The 5534 data sheet shows the following specs: $v_{ind} = 4\ nV/\sqrt{Hz}$, $i_{ind} = .6\ pA/\sqrt{Hz}$. (These values do rise at lower frequencies, but we shall ignore this effect for now.) Also, f_{unity} is 10 MHz.

$$A_v = 1 + \frac{R_f}{R_i}$$

$$A_v = 1 + \frac{99\ k}{1\ k}$$

$$A_v = 100 = 40\ dB$$

For the noninverting amplifier, $A_v = A_{noise}$, so

$$A_{noise} = 100 = 40\ dB$$

$$R_{noise} = R_s + R_f \parallel R_i$$

$$R_{noise} = 100 + 99 \text{ k} \parallel 1 \text{ k}$$

$$R_{noise} = 1090 \ \Omega$$

$$e_{th}^2 = 4 \ K \ T \ R_{noise}$$

$$e_{th}^2 = 4 \times 1.38 \times 10^{-23} \times 300 \times 1090$$

$$e_{th}^2 = 1.805 \times 10^{-17} \ \text{V}^2/\text{Hz}$$

$$e_{total} = \sqrt{v_{ind}^2 + (i_{ind}R_{noise})^2 + e_{th}^2}$$

$$e_{total} = \sqrt{(4 \text{ nV}/\sqrt{\text{Hz}})^2 + (.6 \text{ pA}/\sqrt{\text{Hz}} \times 1090)^2 + 1.805 \times 10^{-17} \ \text{V}^2/\text{Hz}}$$

$$e_{total} = \sqrt{1.6 \times 10^{-17} + 4.277 \times 10^{-19} + 1.805 \times 10^{-17}}$$

$$e_{total} = 5.87 \text{ nV}/\sqrt{\text{Hz}}$$

Note that the major noise contributors are v_{ind} and e_{th}.

Now to find BW_{noise}. First we must find f_2:

$$f_2 = \frac{f_{unity}}{A_{noise}}$$

$$f_2 = \frac{10 \text{ MHz}}{100}$$

$$f_2 = 100 \text{ kHz}$$

We now can apply Equation 5.26:

$$BW_{noise} = 1.57 f_2$$

$$BW_{noise} = 1.57 \times 100 \text{ kHz}$$

$$BW_{noise} = 157 \text{ kHz}$$

followed by Equation 5.27:

$$e_n = e_{total}\sqrt{BW_{noise}}$$

$$e_n = 5.87 \text{ nV}/\sqrt{\text{Hz}}\sqrt{157 \text{ kHz}}$$

$$e_n = 2.33 \ \mu\text{V RMS}$$

To find the output noise, multiply by the noise gain.

$$e_{n\text{-}out} = e_n A_{noise}$$

$$e_{n\text{-}out} = 2.33 \ \mu\text{V} \times 100$$

$$e_{n\text{-}out} = 233 \ \mu\text{V RMS}$$

For a nominal output signal of 1 V RMS, the signal-to-noise ratio is

$$S/N = \frac{\text{Signal}}{\text{Noise}}$$

$$S/N = \frac{1 \text{ V}}{233 \ \mu V}$$

$$S/N = 4290$$

Normally S/N is given in dB.

$$S/N' = 20 \log_{10} S/N$$

$$S/N' = 20 \log_{10} 4290$$

$$S/N' = 72.6 \text{ dB}$$

As was noted earlier, the noise curves increase at lower frequencies. How do you take care of this effect? Well, first of all, if you are using a wideband design and the noise corner frequency is relatively low (as in Example 5.14), it can be safely ignored. If the low frequency portion takes up a sizable chunk of the signal frequency range, it is possible to split the calculation into two or more segments. One segment would be for the constant part of the curves. Other segments can be made for the lower frequency portions. In these regions, an averaged value would be used for v_{ind} and i_{ind}. This is a rather advanced treatment and we will not pursue it here.

Finally, it is common to use the parameter *input referred noise voltage*. Input referred noise voltage is the output noise voltage divided by the circuit signal gain:

$$e_{in\text{-}ref} = \frac{e_{n\text{-}out}}{A_v} \tag{5.28}$$

This value is the same as e_n for noninverting amplifiers, but varies a bit for inverting amplifiers because A_{noise} does not equal A_v for inverting amplifiers.

Summary

In this chapter we have taken a closer look at op amp characteristics. First of all, we found that the upper frequency limit is a function of the op amp parameter f_{unity}, also known as the gain-bandwidth product, and the circuit's noise gain. The higher the gain is, the lower the upper break frequency will be. Op amps are capable of flat response down to DC. If coupling capacitors are used, the lower break frequency can be found by using standard lead network analysis. When stages are cascaded, the results echo those of cascaded discrete stages. The lowest f_2 is dominant and becomes the system f_2. The highest f_1 is dominant and sets the system f_1. If more than one stage exhibits the dominant critical frequency, the actual critical frequency will be somewhat lower for f_2, and somewhat higher for f_1.

In order to make the op amp unconditionally stable, a compensation capacitor is used to tailor the open-loop frequency response. Besides setting f_{unity}, this capacitor also sets the slew rate. Slew rate is the maximum rate of change of output voltage with respect to time. Slewing slows down the edges of pulse signals and distorts sinusoidal signals. The highest frequency which an amplifier can produce without slewing is called the power bandwidth. In order to optimize f_{unity} and slew rate, some amplifiers are available without the compensation capacitor. The designer then adds just enough capacitance to make the design stable.

Due to slight imperfections between the input transistors, op amps may produce small DC output voltages called offsets. Offsets can be reduced through proper resistor selection. Simple nulling circuits can be used to completely remove the offset. A variable offset due to temperature variation is called drift. The larger the temperature variation, the larger the drift will be. The transistor mismatch also means that common-mode signals will not be completely suppressed. Just how well common-mode signals are suppressed is measured by the common-mode rejection ratio, CMRR. Similar to CMRR is PSRR, the power supply rejection ratio. PSRR measures how well power-supply noise and ripple are suppressed by the op amp. Both PSRR and CMRR are frequency-dependent. Their maximum values are found at DC and then decrease as frequency increases.

Finally, noise is characterized as an undesired random output signal. The noise in op amp circuits can be characterized by three components: the thermal noise of the input and feedback resistors; the op amp's input noise voltage density, v_{ind}; and its input noise current density, i_{ind}. The combination of these elements requires an RMS summation. In order to find the output noise voltage, the input noise voltage is multiplied by the circuit's noise gain. The ratio of the desired output signal and the noise voltage is called, appropriately enough, the signal-to-noise ratio, S/N. Normally, signal-to-noise ratio is specified in decibels.

Self Test Questions

1. Define gain-bandwidth product. What is its use?
2. How do you determine f_2 and f_1 for a multistage circuit?
3. What happens if two or more stages share the same break frequency?

4. What is slew rate?

5. How is power bandwidth determined?

6. How do power bandwidth and small-signal bandwidth differ?

7. What are the advantages and disadvantages of noncompensated op amps?

8. What are decompensated op amps?

9. What causes DC offset voltage?

10. What causes DC drift voltage?

11. What is CMRR?

12. What is PSRR?

13. What parameters describe an op amp's noise performance?

14. What is S/N?

Problem Set

▷ Analysis Problems

1. Determine f_2 for the circuit in Figure 5.3 if a 351 op amp is used.

2. Determine f_2 for the circuit in Figure 5.5 if a 318 op amp is used.

3. What is the minimum acceptable f_{unity} for the op amp in Figure 5.3 if the desired f_2 is 250 kHz?

4. What is the minimum acceptable f_{unity} for the op amp in Figure 5.5 if the desired f_2 is 20 kHz?

5. Determine the power bandwidth for Problem 1. Assume $V_p = 10$ V.

6. Determine the power bandwidth for Problem 2. Assume $V_p = 12$ V.

7. What is the minimum acceptable slew rate for the circuit of Figure 5.3 if the desired power bandwidth is 20 kHz with a V_p of 10 V?

8. What is the minimum acceptable slew rate for the circuit of Figure 5.5 if the desired power bandwidth is 40 kHz with a V_p of 5 V?

9. A circuit has the following specifications: ± 15 V power supply, voltage gain equals 26 dB, desired power bandwidth equals 80 kHz at clipping. Determine the minimum acceptable slew rate for the op amp.

10. Determine the system f_2 in Figure 5.6 if all three devices are 318's.

11. Determine the system slew rate for Figure 5.7. The first device is a 741 and the second unit is a 351.

12. Find the output offset voltage for Figure 5.3.

13. If $R_s = R_f = 100 \, \Omega$ in Figure 5.21, find the output offset voltage using a 318 op amp.

14. Assume that the circuit of Figure 5.22 is nulled at 25°C and that an optimum value for R_{off} is used. Determine the drift at 75°C.

15. Assuming that the 120 Hz power supply ripple in Figure 5.28 is 50 mV, how large is its contribution to the output?

16. Utilizing a 5534 op amp, what is the approximate input noise voltage for a source resistance of 1 k and a bandwidth from 10 Hz to 20 kHz?

17. Assuming that the op amp of Problem 16 is connected like Figure 5.28, what is the approximate output noise voltage? What is the approximate input referred noise voltage?

——— Figure 5.28

18. Assume that $R_s = 0 \, \Omega$, $R_i = 500 \, \Omega$, and $R_f = 10 \, k$ in Figure 5.26. Find the input noise voltage if the op amp is a 351.

19. For a nominal output voltage of 2 V RMS, determine the signal-to-noise ratio for Problem 18.

20. Assume that the input to Figure 5.21 is a 5 V peak 50 kHz square wave, and that the op amp is a 741. Draw the output waveform.

21. Repeat Problem 20 using a 10 V peak 100 kHz sine wave.

▷ Design Problems

22. Determine an optimum value for R_{off} in Figure 5.3 and determine the resulting offset voltage.

23. Determine the optimum value for R_{off} in Figure 5.28. Assuming that the circuit has been nulled at 25°C, find the drift at 60°C.

24. Determine a new value for the 100 k resistor in Figure 5.8 in order to minimize the output offset.

25. Using the optimum resistor found in Problem 24, determine a new value for the input capacitor that will maintain the original f_1.

26. Design a circuit with a gain of 32 dB and an f_2 of at least 100 kHz. You may use any of the op amps studied so far.

27. Design a circuit with a gain of 50 and an f_{max} of at least 50 kHz, given a maximum output swing of 10 V peak. You may use any of the op amps studied so far.

28. Design a circuit with a gain of 12 dB, a small-signal bandwidth of at least 100 kHz, and an f_{max} of at least 100 kHz for a peak output swing of 12 V.

29. Utilizing two or more stages, design a circuit with a gain of 150 and a small-signal bandwidth of at least 600 kHz.

▷— Challenge Problems

30. Determine the system f_2 in Figure 5.17.

31. Determine the input noise voltage for the circuit of Figure 5.28. Assume $R_{off} = 950\ \Omega$.

32. Determine the output noise voltage and the input referred noise voltage for Problem 31.

33. Assuming that driving source resistance in Figure 5.7 is 0, how much offset voltage is produced at the output of the circuit? Assume that both devices are 351's.

34. Assume that you have one each of: 351, 741, 318. Determine the combination which will yield the highest system slew rate in Figure 5.29.

——— Figure 5.29

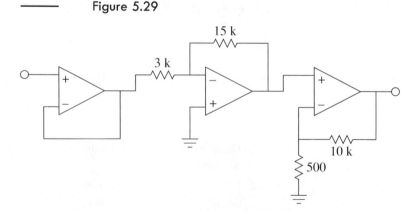

35. Repeat Problem 34 in order to produce the highest system f_2.

36. Assume that the driving source resistance is 0 in Figure 5.7 and that both devices are 351's. Determine the output noise voltage.

37. Derive Equation 5.9 from the text.

⊳ SPICE Problems

38. Use SPICE to create a Bode plot for Problem 10. If a macro model for the 318 is not available, use the 741's instead.

39. Create a time domain representation of the output voltage of Problem 20 using SPICE.

40. Create a time domain representation of the output voltage of Problem 21 using SPICE. Repeat the simulation using an LM318 op amp in place of the 741. What do the results indicate?

41. Simulate the circuit designed in Problem 26 using SPICE. Verify f_2 and A_v through a Bode plot.

42. Using a SPICE Bode plot, verify f_2 and A_v for the circuit designed in Problem 29.

43. Using a SPICE Bode plot, verify f_2 for the circuit designed in Problem 35.

6 Specialized Op Amps

After completing this chapter, you should be able to:

❑ Analyze *instrumentation amplifiers*.

❑ Discuss the advantages of using instrumentation amplifiers versus simple op amp differential amplifiers.

❑ Analyze and detail the advantages of *programmable* op amps, and give possible applications.

❑ Detail the advantages of high-speed and high-power op amps, and give possible applications.

❑ Describe an *OTA*, noting how it differs from a programmable op amp, and give possible applications.

❑ Describe the operation of a *Norton amplifier*, compare how it is used relative to ordinary op amps, and analyze circuitry which makes use of Norton amplifiers.

❑ Describe the operation of *current feedback* op amps, and explain what their primary advantages are.

❑ Describe the need and usage of application-specific integrated circuits.

6.1 Introduction

So far, you have seen how to analyze and design op amp circuits for a variety of general-prupose applications. There are many applications in which the ease of op amp circuit design would be welcome, but in which the average op amp's performance is not suitable. Over the past few years, manufacturers have extended the performance of op amps outside their original low-power low-frequency realm. A variety of special-purpose op amps and op amp derivatives now exist for the designer's convenience. We are going to take a look at a number of the areas where specialized op amps may now be used.

Perhaps the two most noticeable features of extended device performance are increased power handling and speed. At one time, op amps were considered to be suitable for use below 1 MHz. Today it is possible to find high-speed devices designed for applications such as video, where bandwidths are measured in the tens of MHz. As for power, the general-purpose op amp typically operates from a ± 15 V supply and can produce output currents in the neighborhood of 25 mA. New devices can produce output currents measured in amperes, while other devices can produce signals in hundreds of volts. This makes it possible to directly connect low-impedance loads to the op amp, which is very useful.

There are a variety of other useful variations on the basic op amp theme. These include devices optimized for single-supply operation, dedicated voltage followers, devices which allow you to trade off speed for power consumption, and application-specific items like low-noise audio preamplifiers and voltage-controlled amplifiers. This chapter will introduce you to a selection of these devices and take a look at a few of their applications. While these devices may not find as wide an appeal as, for instance, a 351 or 741, they allow you to extend your designs without "reinventing the wheel" each time.

6.2 Instrumentation Amplifiers

There are numerous applications in which a differential signal needs to be amplified. These include low-level bridge measurements, balanced microphone lines, communications equipment, thermocouple amplifiers, and more. The immediate answer to these applications is the differential op amp configuration noted in Chapter 4. There are limitations to that form, unfortunately. For starters, it is practically impossible to achieve matched high-impedance inputs while maintaining high gain and satisfactory offset and noise performance. For that matter, the input impedances are not isolated; indeed, the impedance of one input may very well be a function of the signal present on the other input.

Simply put, this is an unacceptable situation when a precision amplifier is needed, particularly if the source impedance is not very low.

An *instrumentation amplifier* overcomes these problems. Instrumentation amplifiers offer very high impedance, isolated inputs along with high gain, and excellent CMRR performance. Some people like to think of instrumentation amplifiers as a form of "souped-up" differential amplifier. Instrumentation amplifiers can be fashioned from separate op amps. They are also available on a single IC for highest performance.

Instrumentation amplifiers are, in essence, a three-amplifier design. To understand how they work, it is best to start with a differential amplifier based on a single op amp, as seen in Figure 6.1a. One way to increase the input impedances and also maintain input isolation is to place a follower in front of each input. This is shown in Figure 6.1b. The source now drives the very high input impedance followers. The followers exhibit very low output impedance and have no trouble driving the differential stage. In this circuit, op amp 3 is used for common-mode rejection as well as for voltage gain. Note that the gain-bandwidth requirement for op amp 3 is considerably higher than for the input followers. An enhancement to this circuit is shown in Figure 6.2. Here, op amps 1 and 2 are used for signal gain along with their previous duty of input-buffering. The major problem with this configuration is that it requires very close matching of resistors in order to keep the gains the same. Any mismatch in gain between the two inputs will result in a degradation of CMRR. For optimum common-mode rejection, the undesired signal must be identical at the inputs to op amp 3. For high CMRR systems, resistor matching may need to be better than .01%. This is a very expensive requirement if discrete resistors are used. Another way around this would be to allow some adjustment

----

Figure 6.1

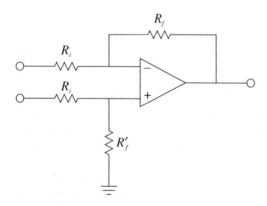

a. Basic differential amplifier

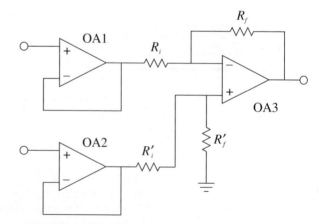

b. Differential amplifier with buffers

of the gain to compensate for gain mismatches, perhaps by using a resistor/potentiometer combination in place of R'_f for op amp 3. For reasons of cost and time, adjustments are frowned upon.

Fortunately, there is a very slight modification to the circuit shown in Figure 6.2 that removes the problem with mismatched gains. This modification joins the R_i values of op amps 1 and 2 into a single resistor, as is shown in Figure 6.3 (page 222). In order to prevent possible confusion with op amp 3, the three resistors used for the input section have been labeled as R_1, R_2, and R_3. To analyze this circuit, we will use the approximation techniques examined earlier.

Figure 6.2
Basic instrumentation
amplifier

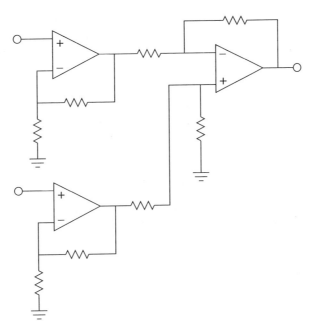

First of all, from preceding work we already know the gain equation for the differential section:

$$v_{out} = \frac{R_f}{R_i}(v_b - v_a) \tag{6.1}$$

Our goal is to find equations for v_b and v_a. Let's examine op amp 1 for v_a. First, due to the approximation that the error voltage of the op amps is 0, the inverting and noninverting inputs of each op amp are the same. We can therefore say that v_x must equal v_{in-}, and that v_y must equal v_{in+}.

$$v_x = v_{in-} \tag{6.2}$$

$$v_y = v_{in+} \tag{6.3}$$

Figure 6.3
Improved instrumentation
amplifier

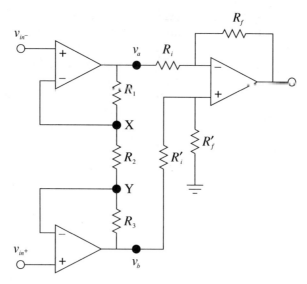

The output voltage v_a must equal v_x plus the drop across R_1.

$$v_a = v_x + v_{R_1} \qquad (6.4)$$

The "positive" potential of v_{R_1} is arbitrary. We are assuming that the current is flowing down through R_1. If the actual numbers show the opposite, v_{R_1} will be negative. The voltage drop v_{R_1} is found by Ohm's Law:

$$v_{R_1} = R_1 i_{R_1} \qquad (6.5)$$

To expand,

$$i_{R_1} = i_{R_2} + i_{op\,amp_1}$$

$i_{op\,amp_1}$ is approximately 0. Therefore,

$$v_{R_1} = R_1 i_{R_2}$$

The current through R_2 is set by its value and the drop across it:

$$i_{R_2} = \frac{v_x - v_y}{R_2} \qquad (6.6)$$

Substituting Equations 6.6 and 6.5 into Equation 6.4 yields

$$v_a = v_x + \frac{R_1(v_x - v_y)}{R_2} \qquad (6.7)$$

Substituting Equations 6.2 and 6.3 into Equation 6.7 yields

$$v_a = v_{in-} + \frac{R_1(v_{in-} - v_{in+})}{R_2}$$

$$v_a = v_{in-} + \frac{R_1}{R_2}(v_{in-} - v_{in+})$$

$$v_a = v_{in-} + v_{in-}\frac{R_1}{R_2} - v_{in+}\frac{R_1}{R_2}$$

$$v_a = v_{in-}\left(1 + \frac{R_1}{R_2}\right) - v_{in+}\frac{R_1}{R_2} \tag{6.8}$$

A close look at Equation 6.8 reveals that it is made up of two terms. The first term is v_{in-} times the noninverting gain of op amp 1, as you may have expected. The second term is v_{in+} times the inverting gain of op amp 1. This output potential is important to note. Even though it may not appear as though op amp 1 will clip a given signal, it might do so if the input to the second op amp is large enough and of the proper polarity.

By a similar derivation, the equation for v_b is found:

$$v_b = v_{in+}\left(1 + \frac{R_3}{R_2}\right) - v_{in-}\frac{R_3}{R_2} \tag{6.9}$$

For gain matching, R_3 is set equal to R_1. R_2 can then be used to control the gain of the input pair, in tandem.

Finally, substituting Equations 6.8 and 6.9 into Equation 6.1 yields

$$v_{out} = \frac{R_f}{R_i}\left(\left[v_{in+}\left(1 + \frac{R_1}{R_2}\right) - v_{in-}\frac{R_1}{R_2}\right] - \left[v_{in-}\left(1 + \frac{R_1}{R_2}\right) - v_{in+}\frac{R_1}{R_2}\right]\right) \tag{6.10}$$

Combining terms produces

$$v_{out} = \frac{R_f}{R_i}\left[(v_{in+} - v_{in-})\left(1 + \frac{R_1}{R_2}\right) + (v_{in+} - v_{in-})\frac{R_1}{R_2}\right]$$

$$v_{out} = (v_{in+} - v_{in-})\left(\frac{R_f}{R_i}\right)\left(1 + 2\frac{R_1}{R_2}\right) \tag{6.11}$$

where the first term is the differential input voltage. The second term is the gain produced by op amp 3, and the third term is the gain produced by op amps 1 and 2. Note that the system common-mode rejection is no longer solely dependent on op amp 3. A fair amount of common-mode rejection is produced by the first section, as evidenced by Equations 6.8 and 6.9. Since the inverting and noninverting gains are almost the same for very high values, high input gains tend to optimize the system CMRR. The remaining common-mode signals can then be dealt with by op amp 3.

Example 6.1   The instrumentation amplifier of Figure 6.4 is used to amplify the output of a balanced microphone. The output of the microphone is 6 mV peak (12 mV differential), and a common-mode hum signal is induced into the lines at 10 mV peak (0 mV differential). If the system has a CMRR of 100 dB, what is the output signal?

First, let's check the outputs of the first section to make sure that no clipping is occurring. We will use superposition and consider the desired signal and hum signal separately.

$$v_a = v_{in-}\left(1 + \frac{R_1}{R_2}\right) - v_{in+}\frac{R_1}{R_2}$$

$$v_a = -6\text{ mV}\left(1 + \frac{20\text{ k}}{400}\right) - 6\text{ mV}\frac{20\text{ k}}{400}$$

$$v_a = -306\text{ mV} - 300\text{ mV}$$

$$v_a = -606\text{ mV}$$

Performing the same calculation on the hum signal yields a contribution of

$$v_a = 10\text{ mV}\left(1 + \frac{20\text{ k}}{400}\right) - 10\text{ mV}\frac{20\text{ k}}{400}$$

$$v_a = 10\text{ mV}$$

Figure 6.4
Instrumentation amplifier
for Example 6.1

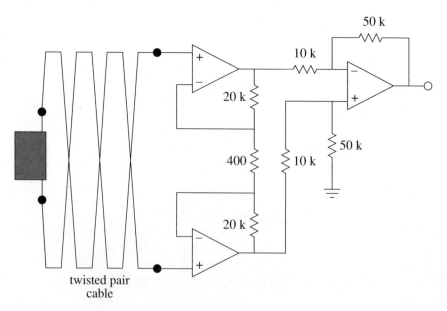

For worst case, these two components' magnitudes add to yield 616 mV, which is far below clipping. The same results are produced for v_b, except that the desired signal is positive.

Now for the output voltage. The second section has a gain of

$$A_v = \frac{R_f}{R_i}$$

$$A_v = \frac{50\,k}{10\,k}$$

$$A_v = 5$$

The desired differential input signal is $v_b - v_a$, so

$$v_{out} = A_v(v_b - v_a)$$

$$v_{out} = 5(606\,mV - -606\,mV)$$

$$v_{out} = 6.06\,V$$

This result can also be produced in one step by using Equation 6.11:

$$v_{out} = (v_{in+} - v_{in-})\frac{R_f}{R_i}\left(1 + 2\frac{R_1}{R_2}\right)$$

$$v_{out} = 12\,mV\,\frac{50\,k}{10\,k}\left(1 + 2\frac{20\,k}{400}\right)$$

$$v_{out} = 6.06\,V$$

Note that the total gain is 505.

Since this amplifier is not perfect, some common-mode signal gets through. It is suppressed by 100 dB over a desired signal. 100 dB translates to a factor of 10^5 in voltage gain. To find the hum signal at the output, multiply the hum by the ordinary signal gain, and then divide it by the CMRR:

$$v_{hum} = v_{out(cm)}$$

$$v_{out(cm)} = v_{in(cm)}\frac{A_v}{CMRR}$$

$$v_{out(cm)} = 10\,mV\,\frac{505}{10^5}$$

$$v_{out(cm)} = 50.5\,\mu V$$

Notice how the hum signal started out just as large as the desired signal, but is now many times smaller. The very high CMRR of the instrumentation amplifier is what makes this possible. In Figure 6.5 (pages 226–228), SPICE

Figure 6.5
SPICE circuits and output
for the instrumentation
amplifier of Figure 6.4

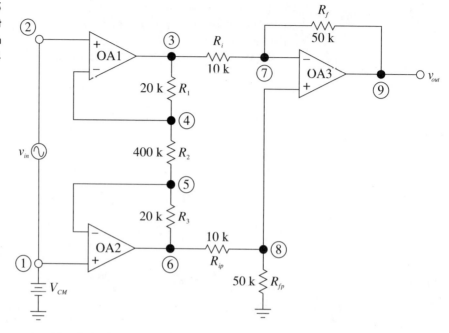

a. Amplifier (complete)

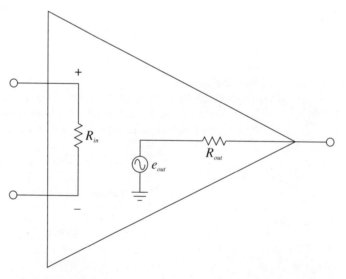

b. Op amp model

———— Figure 6.5   (*Continued*)

```
*****************************AMIGASPICE 5.0*************************************
****      INPUT LISTING                          TEMPERATURE = 27.000 DEG C
*******************************************************************************
******* Start of simple op amp *********
*
.SUBCKT BLORTAMP 1   2    3    4
*                IN+ IN- GND OUT
RIN   1   2   1MEG
ROUT  5   4   75
EOUT  5   3   1   2   50K
.ENDS
*
*****  End of simple op amp ******
*
*****  Instrumentation amplifier description
*
R1    3   4   20K
R2    4   5   400
R3    5   6   20K
Ri    3   7   10K
Rf    7   9   50K
Rip   6   8   10K
Rfp   8   0   50K
VCM   1   0   DC 1
VIN   2   1   SIN ( 0 1M 300HZ )
XOA1  2   4   0   3   BLORTAMP
XOA2  1   5   0   6   BLORTAMP
XOA3  8   7   0   9   BLORTAMP
*
* Analysis directives
*
.TRAN 100US 7MS
.PLOT TRAN V(9)
.WIDTH OUT=80
.END


*****************************AMIGASPICE 5.0*************************************
****     INITIAL TRANSIENT SOLUTION              TEMPERATURE =  27.000 DEG C
*******************************************************************************

NODE    VOLTAGE      NODE    VOLTAGE      NODE    VOLTAGE      NODE    VOLTAGE

(  1)    1.0000      (  2)    1.0000      (  3)    1.0000      (  4)    1.0000
(  5)    1.0000      (  6)    1.0000      (  7)     .8333      (  8)     .8333
(  9)    0.0000      ( 10)    1.0012      ( 11)    1.0012      ( 12)    -.0012

VOLTAGE SOURCE CURRENTS

NAME      CURRENT
VCM      -4.005D-11
VIN      -2.002D-11
```

c. SPICE input and printout

(Continued)

Figure 6.5   (Continued)

```
*************************************AMIGASPICE 5.0*************************************
****          TRANSIENT ANALYSIS                         TEMPERATURE = 27.000 DEG C
**************************************************************************************

TIME       V(9)
                  -1.000D+00    -5.000D-01    0.000D-01    5.000D-01    1.000D+00
           -------------------------------------------------------------------
0.000D-01   1.525D-07  .           .           .       *   .           .
1.000D-04  -9.426D-02  .           .           .     * .   .           .
2.000D-04  -1.849D-01  .           .           .   *     .           .
3.000D-04  -2.691D-01  .           .           . *       .           .
4.000D-04  -3.438D-01  .           .         * .         .           .
5.000D-04  -4.062D-01  .           .       *   .         .           .
6.000D-04  -4.543D-01  .           .   *.      .         .           .
7.000D-04  -4.863D-01  .           . *         .         .           .
8.000D-04  -5.011D-01  .           .*          .         .           .
9.000D-04  -4.981D-01  .           .*          .         .           .
1.000D-03  -4.775D-01  .           .*          .         .           .
1.100D-03  -4.400D-01  .           . *         .         .           .
1.200D-03  -3.868D-01  .           .   *       .         .           .
1.300D-03  -3.200D-01  .           .      *    .         .           .
1.400D-03  -2.418D-01  .           .        *  .         .           .
1.500D-03  -1.551D-01  .           .          *.         .           .
1.600D-03  -6.288D-02  .           .           . *       .           .
1.700D-03   3.158D-02  .           .           .  *      .           .
1.800D-03   1.249D-01  .           .           .    *    .           .
1.900D-03   2.138D-01  .           .           .      *  .           .
2.000D-03   2.952D-01  .           .           .        *.           .
2.100D-03   3.661D-01  .           .           .         . *         .
2.200D-03   4.240D-01  .           .           .         .   *       .
2.300D-03   4.669D-01  .           .           .         .    *      .
2.400D-03   4.932D-01  .           .           .         .     *     .
2.500D-03   5.021D-01  .           .           .         .     *     .
2.600D-03   4.932D-01  .           .           .         .     *     .
2.700D-03   4.668D-01  .           .           .         .    *.     .
2.800D-03   4.239D-01  .           .           .         .   * .     .
2.900D-03   3.660D-01  .           .           .         . *   .     .
3.000D-03   2.951D-01  .           .           .        *.     .     .
3.100D-03   2.137D-01  .           .           .      *  .     .     .
3.200D-03   1.248D-01  .           .           .    *    .     .     .
3.300D-03   3.148D-02  .           .           .  *      .     .     .
3.400D-03  -6.298D-02  .           .           . *       .     .     .
3.500D-03  -1.552D-01  .           .          *.         .     .     .
3.600D-03  -2.419D-01  .           .        *  .         .     .     .
3.700D-03  -3.201D-01  .           .      *    .         .     .     .
3.800D-03  -3.869D-01  .           .   *       .         .     .     .
3.900D-03  -4.400D-01  .           . *         .         .     .     .
4.000D-03  -4.775D-01  .           .*          .         .     .     .
4.100D-03  -4.982D-01  .          *.           .         .     .     .
4.200D-03  -5.011D-01  .          *.           .         .     .     .
4.300D-03  -4.863D-01  .          *.           .         .     .     .
4.400D-03  -4.543D-01  .           .*          .         .     .     .
4.500D-03  -4.062D-01  .           .  *        .         .     .     .
4.600D-03  -3.437D-01  .           .    *      .         .     .     .
4.700D-03  -2.690D-01  .           .       *   .         .     .     .
4.800D-03  -1.848D-01  .           .         * .         .     .     .
4.900D-03  -9.404D-02  .           .           . *       .     .     .
5.000D-03   4.999D-05  .           .           .*        .     .     .
         :         :
7.000D-03  -2.962D-01
           -------------------------------------------------------------------
```

is used to simulate this same amplifier. The simple op amp model presented in Chapter 2 is used here as a subcircuit, and is called three times. In order to clearly see the common-mode rejection, the desired differential input signal is set to a 1 mV sine wave, while the common-mode signal is set to 1 V DC. The initial bias solution shows that op amps 1 and 2 pass both the AC and DC portions of the input, while the final common-mode rejection is left up to op amp 3. This is evidenced by the fact that nodes 7 and 8 both see the same DC potential. At the output, the sine wave has been amplified by approximately 500, as expected. There is no DC offset at the output, indicating rejection of the common-mode DC signal.

Having seen how useful instrumentation amplifiers are, it should come as no surprise to find that manufacturers have produced these devices on a single IC. Two such devices from National Semiconductor are the LH0036 and the LM363. The LH0036 is a hybrid device while the LM363 is monolithic.

The LM363 is billed as a precision-instrumentation amplifier, and allows you to set the gain of the device internally by opening or shorting certain pins. Three preset gains are available: 10, 100, and 1000. The gains can be altered with external components. The LM363 offers other features for more exacting applications.

The LH0036 is suited for more general work and is very similar in layout to Figure 6.3. The equivalent circuit diagram of the LH0036 is shown in Figure 6.6. It features a CMRR of 100 dB, and an input impedance of 300 MΩ. Gain is adjustable from 1 to 1000. Gain is set by placing an appropriate resistor between the *gain set pins*. Generally, this value will be in the range of 100 Ω to 10 k. The gain may be approximated as

$$A_v = 1 + \frac{50\ \mathrm{k}}{R_g}$$

Other pins are available for adjusting the circuit bandwidth and CMRR for optimal results. There is one interesting addition to this circuit: the section labeled *guard drive*. This signal is derived from the outputs of the first section. By using two equal-valued resistors, the differential signals cancel, and thus the guard-drive signal is equal to the common-mode signal. This signal will then be buffered and used to drive the shields of the input wires as shown in Figure 6.7. This is done to enhance the high-frequency performance of the system, particularly of a system with long input leads.

To understand how performance is improved, refer to Figure 6.8 on page 231. Here, the cables are replaced with a simple model. R represents the cable resistance and C represents the cable capacitance. The cable model is little

Figure 6.6
LH0036 equivalent circuit

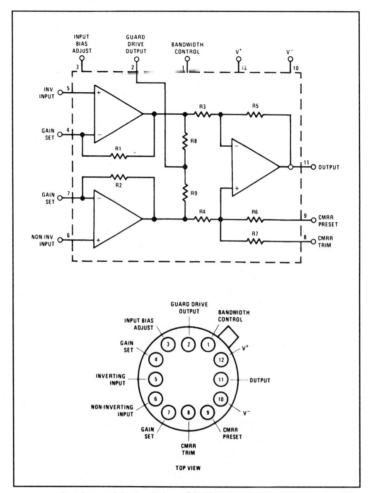

Reprinted with permission of National Semiconductor Corporation

Figure 6.7
Guard drive

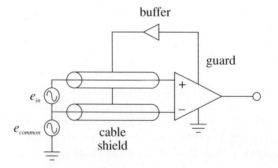

Figure 6.8
Unbalanced lag network
equivalents

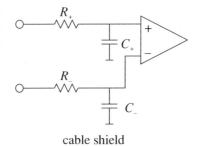

cable shield

more than a lag network. As you know, this causes high-frequency rolloff and phase change. If the two sections are not identical, the rolloffs and phase changes will not be the same in the two lines. These changes will affect the common-mode signal and can lead to a degradation of the system CMRR. In the guard-drive circuit, the cable shields are not connected to ground at the signal source. They are connected only to the guard-buffer output, which is an important point. By driving the shield with a signal equal to the common-mode signal, the voltage developed across C will be zero. Both ends of the capacitor see the same potential; therefore, the drop is nonexistent. This reduces the cable effects considerably.

Example 6.2 The signal produced by a transducer in a factory automation system provides a nominal level of .1 V. For proper use, the signal needs to be amplified up to 1 V. Since this signal must pass through the relatively noisy (electrically speaking) environment of the production floor, a balanced cable with an instrumentation amplifier is appropriate. Using the LH0036, design a circuit to meet these requirements.

For the power supply, a standard ± 15 V unit will suffice. R_g is used to set the desired gain of ten.

$$A_v = 1 + \frac{50\ \text{k}}{R_g}$$

$$R_g = \frac{50\ \text{k}}{A_v - 1}$$

$$R_g = \frac{50\ \text{k}}{10 - 1}$$

$$R_g = 5.556\ \text{k}$$

The completed circuit is shown in Figure 6.9 on page 232. Neither the bandwidth-limiting nor CMRR-adjust pins have been used.

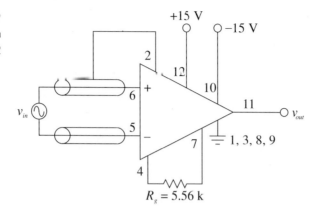

Figure 6.9
Completed instrumentation
amplifier for Example 6.2

$R_g = 5.56 \text{ k}$

6.3 Programmable Op Amps

As noted in Chapter 5, there is always a trade-off between the speed of an op amp and its power consumption. In order to make an op amp fast (i.e., high slew rate and f_{unity}), the charging current for the compensation capacitor needs to be fairly high. Other requirements may also increase the current draw of the device. This situation is not unlike that of an automobile where high fuel mileage and fast acceleration tend to be mutually exclusive. Op amps are now available which can be set for predetermined performance levels. If such a device needs to be very fast, the designer can adjust an external voltage or resistance, and optimize the device for speed. At the other end of the spectrum, such devices can be optimized for low power draw. Because of their ability to change operational parameters, these devices are commonly called *programmable op amps*.

Generally, programmable op amps are just like general-purpose op amps, except that they also include a programming input. The current into this pin is called I_{set}. I_{set} controls a host of parameters including slew rate, f_{unity}, input bias current, standby supply current, and others. A higher value of I_{set} increases f_{unity}, slew rate, standby current, and input bias current. The value of I_{set} can be controlled by a simple transistor current source, or even a single resistor in most cases.

One programmable op amp is the Motorola MC1776. The outline of the MC1776 is shown in Figure 6.10. Two possible ways of adjusting I_{set} are shown in Figures 6.11 and 6.12. A variety of graphs showing the variation of device parameters with respect to I_{set} are shown in Figure 6.13 on pages 234 and 235. As an example, assume that $I_{set} = 10 \ \mu A$ and that we're using standard ± 15 V

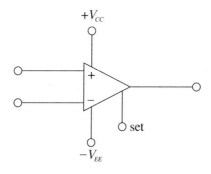

Figure 6.10
Basic connections of the
MC1776

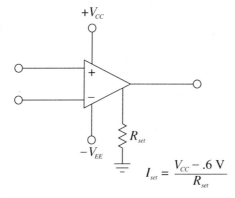

Figure 6.11
Programming via resistor

$$I_{set} = \frac{V_{CC} - .6 \text{ V}}{R_{set}}$$

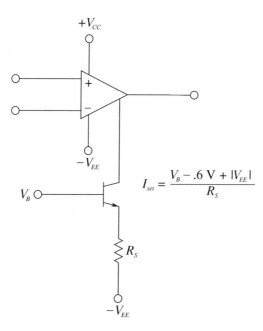

Figure 6.12
Programming via voltage

$$I_{set} = \frac{V_B - .6 \text{ V} + |V_{EE}|}{R_S}$$

Figure 6.13   MC1776 device curves

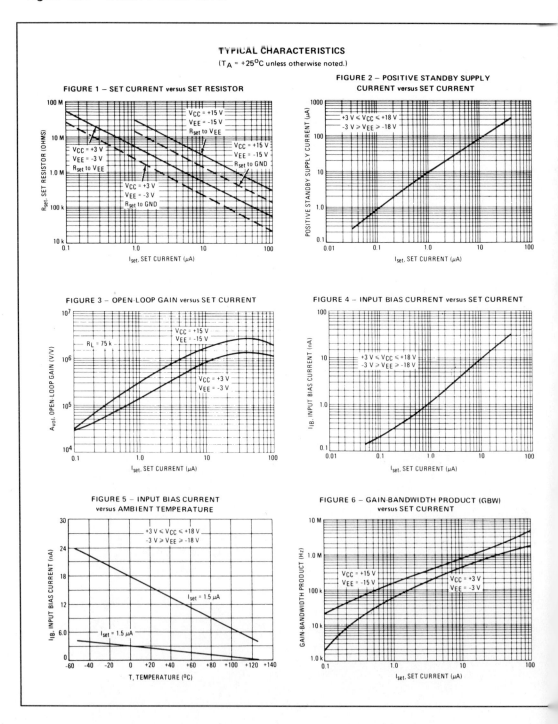

TYPICAL CHARACTERISTICS (continued)

(T$_A$ = +25°C unless otherwise noted.)

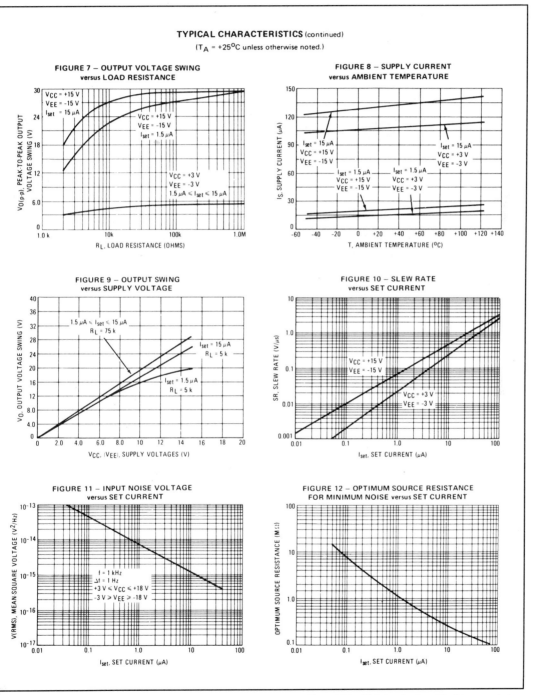

FIGURE 7 – OUTPUT VOLTAGE SWING
versus LOAD RESISTANCE

FIGURE 8 – SUPPLY CURRENT
versus AMBIENT TEMPERATURE

FIGURE 9 – OUTPUT SWING
versus SUPPLY VOLTAGE

FIGURE 10 – SLEW RATE
versus SET CURRENT

FIGURE 11 – INPUT NOISE VOLTAGE
versus SET CURRENT

FIGURE 12 – OPTIMUM SOURCE RESISTANCE
FOR MINIMUM NOISE versus SET CURRENT

supplies. At this current, we find that f_{unity} is 800 kHz, slew rate is .5 V/μS, and the standby current draw is 80 μA. In terms of speed, this is not much different from a 741, but the current draw is decidedly lower. This makes the device very attractive for applications where power draw must be kept to a minimum. Battery-operated systems generally fall into the low power-consumption category. By keeping current drain down, battery life is extended. Perhaps the strongest feature of this class of device is its ability to go into a *power down* mode. When the device is sitting idle, I_{set} can be lowered, and thus the standby current drops considerably. For instance, if I_{set} for the MC1776 is lowered to .1 μA, standby current drops to under 1 μA. By doing this, the device may dissipate only a few microwatts of power during idle periods. When the op amp again needs to be used for amplification, I_{set} can be brought back up to a normal level. This form of real-time adjustable performance can be instituted with a simple controlling voltage and a transistor current source arrangement as in Figure 6.12.

Example 6.3   Determine the power bandwidth for $V_p = 10$ V for the circuit of Figure 6.14. Also find f_2. Power supplies are ± 15 V.

In order to find these quantities, we need to know the slew rate and f_{unity}. These are controlled by I_{set}, so finding I_{set} is the first order of business.

$$I_{set} = \frac{V_{CC} - .6}{R_{set}}$$

$$I_{set} = \frac{15 - .6}{3 \text{ M}\Omega}$$

$$I_{set} = 4.8 \text{ }\mu\text{A}$$

Figure 6.14
Circuit for Example 6.3

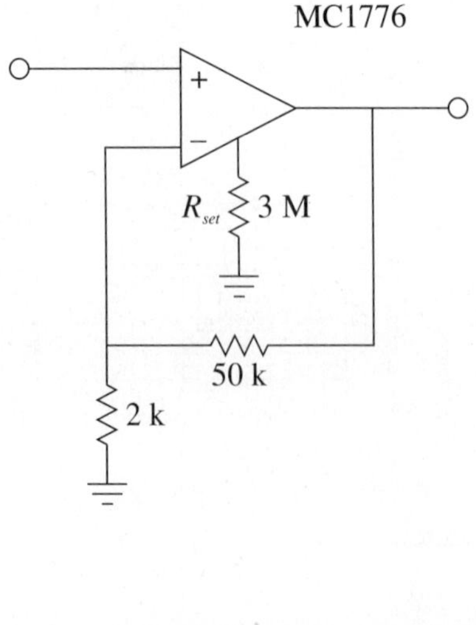

MC1776

From the graph "Gain-Bandwidth Product versus Set Current" in Figure 6.13, we find that 4.8 μA yields an f_{unity} of approximately 500 kHz. The graph "Slew Rate versus Set Current" yields approximately .3 V/μS for the slew rate.

$$A_{noise} = 1 + \frac{R_f}{R_i}$$

$$A_{noise} = 1 + \frac{50\text{ k}}{2\text{ k}}$$

$$A_{noise} = 26$$

$$f_2 = \frac{f_{unity}}{A_{noise}}$$

$$f_2 = \frac{500\text{ kHz}}{26}$$

$$f_2 = 19.2\text{ kHz}$$

$$f_{max} = \frac{SR}{2\pi V_p}$$

$$f_{max} = \frac{.3\text{ V}/\mu\text{S}}{2\pi \times 10\text{ V}}$$

$$f_{max} = 4.77\text{ kHz}$$

As you can see, once the set current and associated parameters are found, analysis proceeds as in any general purpose op amp.

6.4 Op Amps for High Current, Power, and Voltage Applications

General-purpose op amps normally run on ± 15 V power rails and generally produce less than 40 mA of output current, making it impossible to directly connect them to a low-impedance load such as a loudspeaker or motor. Their lack of high-voltage capability prohibits their use in many applications, including many forms of display technology. In short, general-purpose op amps are low-power devices. There are a few ways around these limitations. One way of boosting the output current capability by using a discrete follower was shown in Chapter 4. In recent years, manufacturers have begun producing a variety of op amps with extended power performance.

High Current Devices

Perhaps the earliest demand for higher current capacity op amps came from the audio community. If an op amp could be directly connected to a loudspeaker, a great deal of time and money could be saved for general-purpose audio design work. Instead of a collection of perhaps half a dozen transistors and a handful of required biasing resistors and capacitors, an audio amplifier could be produced with a single op amp and just a few resistors and capacitors. Indeed, some of the first high-output devices were aimed squarely at the audio market. By 1980 it was possible to choose from a number of amplifiers designed to drive loudspeakers with power levels of 1 to 5 W. In an effort to make designs ever easier, the strict op amp form was modified and devices with preset and programmable gains were produced. Due to the increased dissipation requirements, the standard plastic dual inline package was abandoned in favor of multilead TO-220 and TO-3 style cases. More recently, devices have been created which can produce output currents in excess of 10 As. Besides their use in audio/communications, these high-current devices find use in direct motor drive applications, power supply and regulation design, and in other areas.

Let's take a closer look at two representative devices. First, on the low end of the scale is the LM386. This is a low-voltage device ideally suited to battery-powered designs. It operates from a single power supply between 4 and 12 V and is capable of producing .8 watts into a 16 Ω load with 3% THD. The gain is set internally to 20, but can be increased up to 200 with the addition of a few extra components. A minimum configuration layout requires only one external component, as shown in Figure 6.15. In the minimum configuration, the PSRR bypass capacitor is omitted. This produces a PSRR of less than 10 dB. A more respectable PSRR of 50 dB (rolling off below 100 Hz) can be achieved by adding a 50 μF capacitor from the PSRR bypass pin to ground. The small number of required external components is due to the fact that an internal feedback path has already been established by the manufacturer. Normally, this device would not be set up in the forms we have already examined. Set up as in Figure 6.15, the circuit exhibits a 300 kHz bandwidth and an input impedance of 50 k. The LM386 is housed in an 8-pin mini-DIP.

A typical application of the LM386 is shown in Figure 6.16. This is a personal amplifier called The Pocket Rockit. It is designed for musicians, so that they can plug a guitar, keyboard, or other electronic instrument into the unit and hear what they play through a pair of headphones. In essence, it is a personal practice amp. In keeping with its small portable design, The Pocket Rockit runs on a single 9 V radio battery. Note how the single-supply biasing techniques explained in Chapter 4 are used to bias op amps 1a and 1b. The circuit is laid out in three main blocks. The first block serves as a preamplifier and distortion circuit. Since the load for the first stage consists of a pair of diodes, D_2 and D_3, large output signals will be clipped at the diode forward voltage. Normally, clipping is not a desired result in linear circuit design. However, a great number of musicians find this effect to be very rewarding, particularly

Figure 6.15
Simple power amp

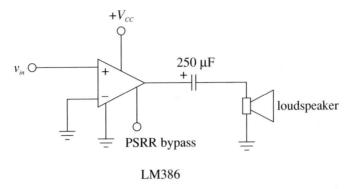

when applied to guitar. (This is one case where subjective valuations can make a designer do rather odd things.) The second stage is a bass/mid/treble control. This is little more than a frequency-selective amplifier, a subject which will be covered in detail in Chapter 11. From the musician's view, it allows control over the tone color, or timbre, of the sound. The final stage utilizes an LM386 to drive a pair of mini-headphones. As you can see, the power amplifier section

Figure 6.16
Pocket Rockit schematic

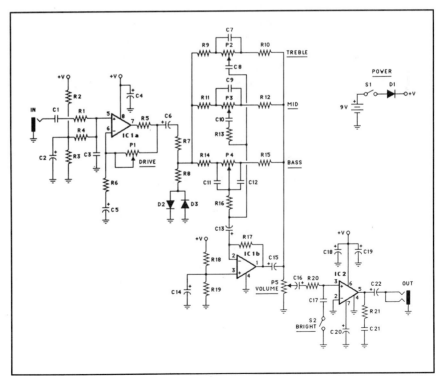

From *Electronic Musician*, Vol. 3, No. 6, June 1987. Reprinted with permission.

is the smallest of the three. It utilizes the default internal gain of 20, and includes C_{20} for optimum PSRR. This is an important detail since this circuit may also be powered from 9 V AC adapters, which are not known for their low noise. The volume control is merely a potentiometer which acts as a voltage divider to the input signal. Coupling capacitors are used throughout to prevent DC bias signals from affecting the load or adjacent stages. They are also used to make sure that DC potentials don't appear across the potentiometers, which can increase adjustment noise.

At the other end of the spectrum is the LM12. The LM12 can produce 150 watts of "sine wave power" into a 4 Ω load. The device may be operated from ± 40 V supplies and delivers up to ± 10 amps to its load. The device comes in a TO-3–type power case with five connections: $\pm V_{supply}$, $\pm$ input, and output. To make the designer's life easier, the LM12 features an array of useful characteristics: thermal limiting, overvoltage shutdown, output-current limiting, and dynamic output-transistor safe-area protection. Unlike many power op amps, the LM12 is unity-gain stable. It shows an f_{unity} of 700 kHz and a slew rate of 9 V/μS. For even higher output currents, it is possible to connect LM12's in parallel. Due to the high-power nature of this device, extra care needs to be exercised in power supply bypassing, ground loops, and reactive loads. Whereas general-purpose op amps may bypass their power supplies with .1 μF to 1 μF capacitors, the LM12 requires 20 μF bypass capacitors. Also, the device may go into oscillation when driving capacitive loads. In this case some form of load isolation/compensation is used. This usually takes the form of a resistor/inductor combination. Generally, all high-power op amps require a bit more attention than general-purpose devices do.

As an example, Figure 6.17 shows a low distortion ($<.01\%$ THD) audio amplifier. Note the large (3900 μF) bypass capacitors and the output clamping diodes. As with discrete power amplifiers, the clamp diodes are used to prevent an overvoltage condition from occurring when driving inductive loads to the clipping level. The stored energy in the inductor can raise the output of the op amp above the power supply rail while the output transistors are in saturation, which will damage the amplifier. The diodes are used to shunt the stored energy to the power supply. The 4 μH/2.2 Ω combination is used to stabilize the system in the event of adversely reactive loads. Obviously, with the sort of power dissipation this device is capable of, appropriate cooling methods, such as heat sinks, must not be overlooked.

Besides driving loudspeakers, the LM12 can also drive motors. If only a single-polarity supply is available, a pair of devices can be hooked up in bridge mode in order to achieve bidirectional load current. Figure 6.18 shows a single-supply bridge configuration driving a servo motor. Either input may be grounded. This system utilizes current drive output (i.e., it is a voltage-to-current transducer). Current output is sometimes preferred with servo motors, as it helps stabilize the system in the presence of the sizable inductance of the servo motor.

Figure 6.17
High-power amplifier

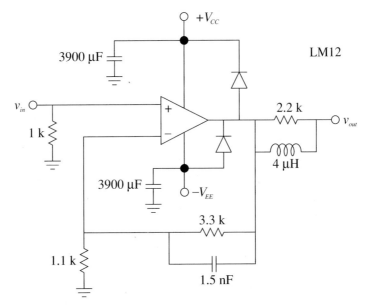

Figure 6.18
Motor drive circuit

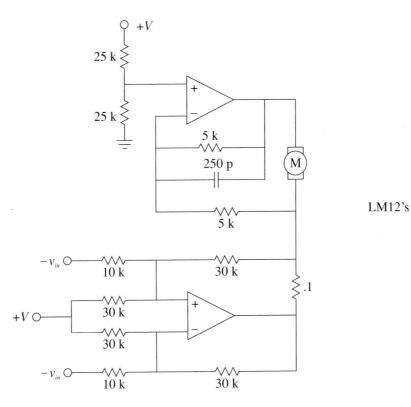

High-Voltage Devices

Sometimes an amplifier needs to deliver output voltages in excess of the 12 to 13 V limits presented by most op amps. In this case, we can turn to a number of high-voltage op amps. These devices do not necessarily also have a higher output-current capability. Basically, they may be treated as general-purpose op amps which can be operated from higher supply voltages. Because of this, their output compliance is increased. Devices are available which can operate from power supplies in excess of ±100 V.

One high-voltage amplifier is the LM144. The LM144 is capable of generating ±30 V output swings when powered by a ±36 V supply. The output current limit is no different from that of the average op amp (20 mA). Like the general-purpose 301, the 144 is noncompensated. Unity gain stability is achieved with a single 27 pF capacitor. A 3 pF capacitor will allow the amplifier to be stable for gains greater than 10, and will produce a 30 V/μS slew rate.

6.5 High-Speed Amplifiers

There are many applications where amplification in the 1 MHz to 100 MHz region is desired. In this range fall the broadcast spectrums of FM radio, VHF television, citizen's band radio, and video processing in general. The bulk of general-purpose op amps exhibit f_{unity}'s in the 1 MHz to 10 MHz range. Also, slew rates tend to be below 20 V/μS. These characteristics make them wholly unsuitable for higher-frequency applications. Even standard "fast" op amps like the LM318 and LF357 offer f_{unity}'s of no more than 20 MHz with 50 V/μS slew rates.

In contrast to general-purpose devices, "ultra fast" op amps exhibit gain-bandwidth products in excess of 50 MHz and slew rates over 100 V/μS. Devices are available with slew rates greater than 1000 V/μS and others boast gain-bandwidth products approaching 1 GHz. In order to optimize performance, these devices are normally noncompensated so the largest possible bandwidth can be produced for any given gain.

Of course, when working with such fast devices, circuit layout is far more critical than normal: A very good circuit design may be crippled by a careless layout. First of all, PC-board ground planes are advised. A ground plane offers isolation along with low resistance and inductance pathways. These parasitic and stray impedance effects, which are often ignored at lower frequencies, can reduce circuit performance at high frequencies. Shielding around input and output paths may also be required for some applications, but care must be exercised since the resulting capacitance may also produce high-frequency attenuation. In general, the circuit layout should minimize stray capacitance between the output and feedback point, from the feedback-summing node to ground, and between the circuit input and output. One common technique is to extend a ground plane in order to isolate the input and output traces. The traces should be kept short to minimize coupling effects, and thus, compact layouts are generally preferred. IC sockets are generally avoided due to the increased lead inductance and capacitance they present. Finally, when breadboarding high speed circuits, the popular multi-row socket boards used for general-purpose work are normally unacceptable due to high lead inductance and capacitance effects.

As with power op amps, some high-speed op amps have been optimized for specific applications and are no longer interchangeable with the generic op amp model. For example, specific gains can be selected by grounding or opening certain pins on the integrated circuit.

The LH0032 is one example of an ultra-fast op amp. It is a noncompensated device and boasts a 500 V/μS slew rate with a 70 MHz bandwidth. Due to its open-loop Bode plot, the LH0032 is stable for gains greater than 50 and requires no compensation. Unity gain compensation is possible with a pair of external capacitors. As with any high-frequency device, power-supply bypass capacitors are required.

Creating a 40 dB amplifier with the LH0032 is straightforward. At gains this high, compensation capacitance is not required. Figure 6.19 shows this connection. A circuit such as this could be used to amplify signals in the AM broadcast band.

Lower gains (such as 10) require some form of compensation. This may involve a single capacitor, a pair of capacitors, or perhaps a resistor/capacitor combination. As for most noncompensated devices, there is no blanket formula for determining the required capacitance; specifics for each op amp are given by the manufacturer.

Another application which requires high speed is fiber-optic communication. On page 244, Figure 6.20 shows an LH0032 used as a fiber-optic receiver. The light pulses from the fiber-optic cable are turned into current pulses by the photo diode. These current pulses are then passed into the op amp, which

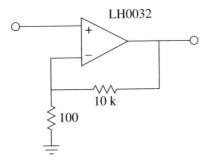

Figure 6.19
40 dB amplifier (no compensation capacitor required)

Figure 6.20
Photo diode amplifier
(power supply bypass not
shown)

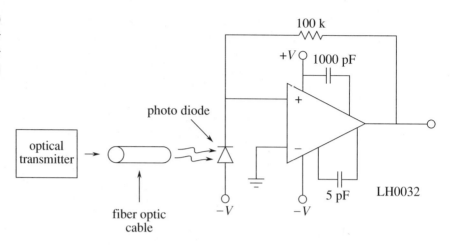

Figure 6.20
Photo diode amplifier
(power supply bypass not
shown)

is connected as a current-to-voltage converter. A relatively large value for R_f is required in order to achieve a reasonable output level. This circuit can handle input data rates up to 3.5 MHz. The 5 pF and 1000 pF capacitors are used for compensation. Yet another high-speed device is found in the next section.

6.6 Voltage Followers and Buffers

Unity-gain noninverting buffers (voltage followers) are used in wide variety of applications. Any time a signal source needs to be isolated, a buffer is called for. As you have seen, connecting op amps in a follower configuration is a very straightforward exercise. But this is not always the best choice. By optimizing the amplifier design for buffer operation, the manufacturer can increase performance, or in some cases, reduce the case size. This is possible because followers need very few connections: input, output, and power supply. Also, since the devices must, by nature, be unity-gain stable, an external compensation connection is normally not desired.

One buffer amplifier optimized for high performance is the LH0063. This device is designed with video applications, high speed drivers, and analog-to-digital converters in mind. Besides the standard power-supply bypass capacitors and perhaps a line-termination resistor, no other parts are needed. Figure 6.21 shows a minimum configuration for driving 50 Ω coaxial cable. This buffer is capable of supplying ±250 mA to its load, so direct 50 Ω connection poses no problem. Slew rate is typically 2400 V/μS with a 50 Ω load, and can be as high as 6000 V/μS when used with higher load impedances. Circuit bandwidth is 200 MHz.

Figure 6.21
High speed buffer (power
supply bypass and offset
adjust not shown)

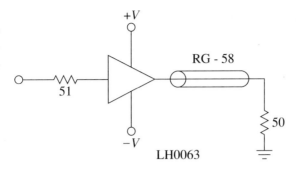

Example 6.4 Determine the approximate bandwidth required for a high-resolution video-display amplifier. The display has a resolution of 1024 pixels wide by 1024 pixels high, with a refresh rate of 60 frames per second.[1] This specification is typical of a workstation or a very high-quality personal computer.

In its simplest form (monochrome, or black and white), each pixel is either on or off. Expressed as a controlling voltage, this means either a high or a low. Sixty times each second, an updated grid of 1024 by 1024 dots must be drawn on the display monitor. From this, we can determine the pixel, or dot, rate:

$$\text{pixel rate} = \text{height} \times \text{width} \times \text{refresh}$$
$$\text{pixel rate} = 1024 \times 1024 \times 60$$
$$\text{pixel rate} = 62{,}914{,}560 \text{ pixels per second}$$

This indicates that each pixel requires 1/62,914,560 seconds, or about 15.9 nS, to reproduce. For proper pulse shape, the risetime should be no more than 30% of the pulse width (preferably less). From the risetime/bandwidth relationship established in Chapter 1,

$$f_2 = \frac{.35}{T_r}$$
$$f_2 = \frac{.35}{.3 \times 15.9 \text{ nS}}$$
$$f_2 = \frac{.35}{4.77 \text{ nS}}$$
$$f_2 = 73.4 \text{ MHz}$$

---

[1] The term *pixel* is a contraction of *picture element*. The display is considered to be nothing more than a large grid of dots. The term *refresh rate* refers to how quickly the display is updated or redrawn.

Theoretically, a 73.4 MHz bandwidth is required. Once overhead (such as vertical retrace) is added, a practical circuit will require a bandwidth on the order of 100 MHz.

6.7 Operational Transconductance Amplifier

The operational transconductance amplifier, or OTA as it is normally abbreviated, is primarily used as a controlled-gain block. What this means is that an external controlling signal, either a current or a voltage, will be used to set a key parameter of the circuit, such as closed-loop gain or f_2. This is not the same as the programmable op amps discussed earlier.

OTAs exhibit high output impedances. They are intended to be used as current sources. Therefore, the output voltage is a function of the output current and the load resistance. The output current is set by the differential input signal and the OTA's transconductance. Transconductance is, in turn, con-

Figure 6.22 LM3080 equivalent circuit

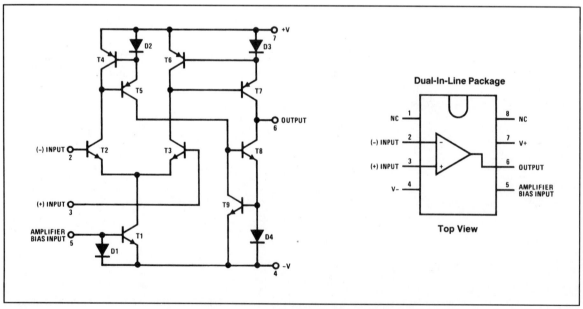

trolled by a setup current, I_{abc}. What this means is that the output signal is a function of I_{abc}. If I_{abc} is increased, transconductance rises, and with it, output voltage. Normally, OTAs are operated without feedback. One major consequence of this is that the differential input signal must remain relatively small in order to avoid distortion. Typically, the input signal remains below 100 mV peak to peak.

One popular OTA is the LM3080. The schematic of the LM3080 is shown in Figure 6.22. The differential amplifier stage comprised of T_2 and T_3 is biased by T_1. The current set by T_1 establishes the transconductance of this stage. T_1 is part of a current mirror made with D_1. T_1 mirrors the current flowing through D_1. The current flowing through D_1 is I_{abc}, and is set externally. Thus, the differential amplifier is directly controlled by the external current.

An improved version of the LM3080 is the LM13600. The LM13600 features two OTAs in a single 16-pin DIP. Since OTAs are modeled as output current sources, the LM13600 includes simple emitter follower buffers for each OTA. Connection to the buffers is optional. Also included are input-linearizing diodes which allow higher input levels, and thus, superior signal-to-noise performance.

Example 6.5

One example of OTA use is shown in Figure 6.23. This circuit is a simple voltage-controlled amplifier (VCA). Let's walk through the circuit to see how the component values were derived. The gain of the system is set by $V_{control}$. The 22 k resistor is used to turn $V_{control}$ into I_{abc}. The maximum value for I_{abc} is specified as 1 mA. By referring to Figure 6.22, an expression for the maximum

Figure 6.23
Voltage-controlled amplifier

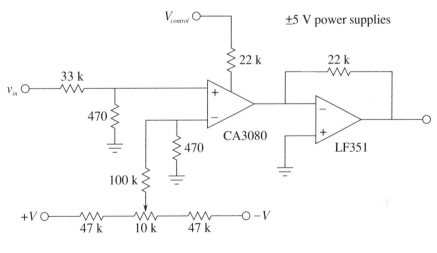

offset adjust

I_{abc} may be determined from Ohm's Law:

$$I_{abc} = \frac{V_+ - V_- - V_{D_1}}{R_{control}}$$

$$I_{abc} = \frac{5 - (-5) - .7}{22 \text{ k}}$$

$$I_{abc} = .423 \text{ mA}$$

This places us at about half the allowed maximum, giving a twofold safety margin. The LM3080 transconductance graph indicates that we may expect a maximum value of about 10 millisiemens with our maximum I_{abc}. Note how the input signal must be attenuated in order to prevent overload. For reasonable distortion levels, the input to the LM3080 should be below 50 mV peak to peak. This is accomplished by the 33-k/470-Ω voltage divider. This divider produces an input impedance of approximately 33 k, and a loss factor of

$$\text{loss} = \frac{470}{33 \text{ k} + 470}$$

$$\text{loss} = .014$$

This means that the circuit can amplify peak-to-peak signals as large as

$$50 \text{ mV} = v_{in(max)}(\text{loss})$$

$$v_{in(max)} = \frac{50 \text{ mV}}{\text{loss}}$$

$$v_{in(max)} = 3.56 \text{ V}$$

The current out of the LM3080 drives an LF351 configured as a current-to-voltage converter. The transresistance of this portion is 22 k. Using the maximum 50 mV input, the maximum output signal will be

$$i_{out\text{-}3080} = V_{in\text{-}3080} \, gm$$

$$i_{out\text{-}3080} = 50 \text{ mV} \times 10 \text{ mS}$$

$$i_{out\text{-}3080} = .5 \text{ mA}$$

This current will drive the LF351 to a maximum output of

$$v_{out} = i_{out\text{-}3080} R_f$$

$$v_{out} = .5 \text{ mA} \times 22 \text{ k}$$

$$v_{out} = 11 \text{ V}$$

With the signals shown, this circuit acts as a simple amplitude modulator. The $V_{control}$ signal is impressed upon the input signal. These are the audio and carrier signals, respectively. This setup can also be used as part of an AGC (automatic gain control) loop. Also, note the adjustable DC signal applied to the inverting input. This is used to null the amplifier, which is an important point. Any DC offset is also affected by the gain change, in effect, causing the control signal to "leak" into the output. This can produce many undesirable side effects, including clicking and popping noises if the circuit is used as an audio VCA. By nulling the output offset, these undesirable effects are eliminated. In order to keep the inputs balanced, a 470 Ω resistor is added to the inverting input. Now, each input sees approximately 470 Ω to ground.

When designing a VCA as in Example 6.5, we usually start with known maximum input and output voltages. The feedback resistor of the current-to-voltage converter is chosen to yield the desired maximum output potential when the OTA is producing its maximum output current. The OTA's maximum input potential times its transconductance will yield the OTA's maximum output current. Once this current has been established, the value of $R_{control}$ may be determined given the range of $V_{control}$. The maximum allowable OTA input potential is determined from the OTA distortion graph, and, in association with the desired maximum input voltage, dictates the ratio of the input voltage divider.

OTAs can also be used to create voltage-controlled filters (VCF), where the cutoff frequency is a function of $V_{control}$. Discussion of filters in general is found in Chapter 11. OTAs lend themselves to several unique applications. The downside is that they have limited dynamic range: The range of signal amplitudes which they can handle without excessive distortion or noise is not as great as many other devices.

6.8 Norton Amplifiers

It is possible to create an input-differencing function without using a differential amplifier. One alternative is to use a current-mirror arrangement to form a *current-differencing amplifier*. Since the input function deals with a difference of current instead of voltage, amplifiers of this type are often referred to as *Norton amplifiers*. Norton amplifiers have the distinct advantages of low cost and the ability to operate from a single-polarity power supply. Perhaps the most popular Norton amplifier is the LM3900. The LM3900 is a quad device, which means that four amplifiers are combined in a single package.

Figure 6.24

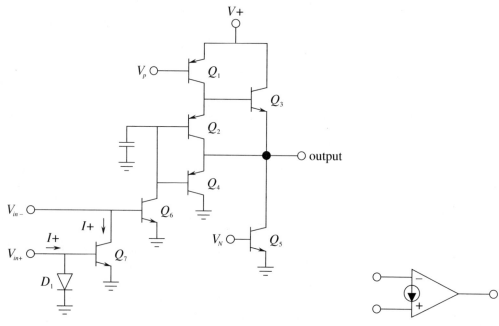

a. LM3900 Norton schematic

b. Schematic symbol

The internal circuitry of one Norton amplifier is shown in Figure 6.24a. Norton amplifiers also use a slightly modified schematic symbol, shown in Figure 6.24b, to distinguish them from ordinary op amps. The amplifier is comprised of two main sections: a current-differencing input portion consisting of Q_6, Q_7, and D_1, and a gain stage comprised of Q_1 through Q_5. The current-differencing portion is of greatest interest to us, and is based upon the current-mirror concept. D_1 and Q_7 make up a current mirror which is fed by the noninverting input's source current, I_+. This current flows down through D_1, and is mirrored into Q_7. Therefore, the collector current of Q_7 is equal to I_+. This current is effectively subtracted from the current which enters the inverting input. The differential current, $I_- - I_+$, is the net input current which feeds Q_6, and is amplified by the following stage. Note how the schematic symbol of Figure 6.24b echoes the current mirror subtraction process.

Due to its unique input configuration, the equivalent model of the Norton amplifier is rather different from the standard op amp. This is illustrated in Figure 6.25. This clearly shows that the Norton amplifier is a current-sensing device. Also, note that the input impedance for the noninverting input is little more than the dynamic resistance of D_+. This resistance is dependent on the

Figure 6.25
Norton equivalent circuit

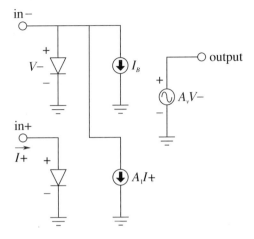

input current, and may be found using the standard diode resistance equation

$$Z_{in+} = \frac{.026}{I_+}$$ (6.12)

Another important point is that both the inverting and noninverting inputs are locked at approximately one diode drop above ground. This can be beneficial for a current-summing node, but it does require that some form of input resistor be used if a voltage input is expected. A side benefit of this is that given a large enough input resistor, there is virtually no limit to the input common-mode voltage range, since the potential will drop across the input resistor. For proper AC operation, an appropriate biasing current must be used. A typical inverting amplifier is shown in Figure 6.26. Note the input and output coupling capacitors, as well as R_B, which serves to convert the bias voltage into a bias current. A generalized model of this circuit is shown in

Figure 6.26
Typical inverting amplifier

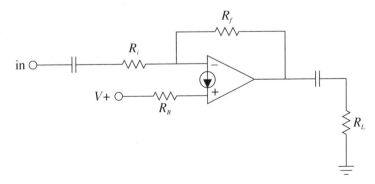

Figure 6.27. From this, we will be able to determine a voltage gain expression, and a technique for determining a proper value for R_B. We'll start by setting up an expression for the output voltage for the bias equivalent:

$$V_{out} = V_{D_-} + R_f(I_+ + I_B)$$

and

$$I_+ = \frac{V_{supply} - V_{D_+}}{R_B}$$

Figure 6.27
Generalized model of the
amplifier in Figure 6.26

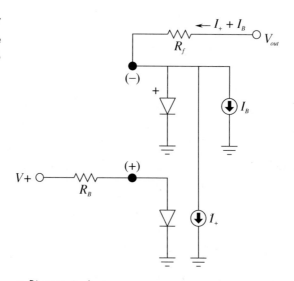

a. Bias equivalent

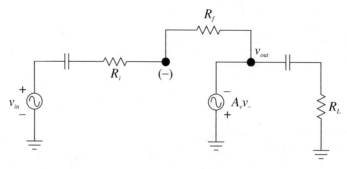

b. AC equivalent

Combining the previous two equations yields

$$V_{out} = V_{D_-} + R_f\left(\frac{V_{supply} - V_{D_+}}{R_B} + I_B\right) \tag{6.13}$$

Normally, $V_{supply} \gg V_D$, and I_B is relatively small. Given these factors, Equation 6.13 may be approximated as

$$V_{out} = V_{supply}\frac{R_f}{R_B} \tag{6.14}$$

Therefore, for a midpoint bias,

$$R_B = 2R_f \tag{6.15}$$

For the AC equivalent circuit, the gain equation may be derived in the same fashion as it was in Chapter 4 for a standard inverting op amp configuration:

$$A_v = -\frac{R_f}{R_i} \tag{6.16}$$

As in the standard op amp inverting amplifer, the input impedance may be approximated as R_i.

Example 6.6 Using the LM3900, design an amplifier which operates from a single $+9$ V battery, has an inverting voltage gain of 20, an input impedance of at least 50 k, and a lower break frequency of no more than 100 Hz.

First, R_i can be established from the input impedance spec at 50 k. From there, both R_f and R_B can be determined.

$$A_v = -\frac{R_f}{R_i}$$

$$R_f = -A_v R_i$$

$$R_f = -(-20) \times 50 \text{ k}$$

$$R_f = 1 \text{ M}\Omega$$

$$R_B = 2R_f$$

$$R_B = 2 \times 1 \text{ M}$$

$$R_B = 2 \text{ M}$$

Figure 6.28
Completed design

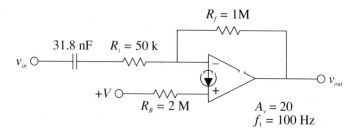

The input coupling capacitor will be used to set the lower break frequency in association with R_i. This is basically unchanged from the single-supply bias calculations presented in Chapter 4.

$$f_c = \frac{1}{2\pi RC}$$

$$C = \frac{1}{2\pi R f_c}$$

$$C = \frac{1}{2\pi 50 \text{ k } 100 \text{ Hz}}$$

$$C = 31.8 \text{ nF}$$

The completed circuit is shown in Figure 6.28. While this circuit does not detail the output coupling capacitor and load resistance, it would be handled in a similar vein. One possible problem with this circuit is that any power-supply ripple present on V_{supply} will be passed into the amplifier via R_B, and appear at the output. In order to minimize this problem, a power-supply decoupling circuit such as that presented in Chapter 4 can be used. Figure 6.29 shows a typical inverting amplifier with an added decoupling circuit. Note

Figure 6.29
Optimized bias circuitry

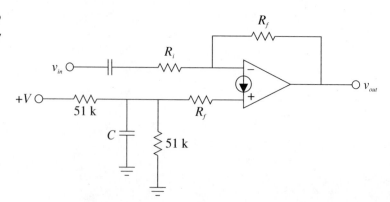

that R_B is not set to twice R_f, but is instead set equal to it. This is because R_B is being driven by only one-half of the supply voltage due to the lower-resistance decoupling divider.

Norton amplifiers may be used in a variety of other configurations as well, including noninverting and differential amplifiers. A simple noninverting amplifier is shown in Figure 6.30. Unlike the ordinary op amp version, the Norton amplifier requires an input resistor. Remembering that the input impedance of the noninverting input may be quite low (Equation 6.12), we can derive equations for both circuit input impedance and voltage gain.

The input impedance is the series combination of R_i and the impedance looking into the noninverting input, Z_{in+}.

$$Z_{in} = R_i + Z_{in+} \tag{6.17}$$

In order to find voltage gain, note that due to the virtual ground at the inverting input, the drop across R_f is equivalent to v_{out}.

$$v_{out} = v_{R_f} = R_f I_- \tag{6.18}$$

v_{in} drops across the system input impedance. Given Equation 6.17, we can say that

$$v_{in} = I_+(R_i + Z_{in+}) \tag{6.19}$$

Since, by definition, $A_v = v_{out}/v_{in}$, we may say

$$A_v = \frac{R_f I_-}{I_+(R_i + Z_{in+})}$$

The two currents are identical in the AC equivalent circuit, so this simplifies to

$$A_v = \frac{R_f}{R_i + Z_{in+}} \tag{6.20}$$

Figure 6.30
Noninverting amplifier

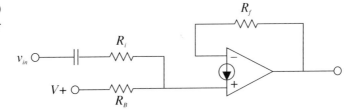

In summary, there are a few things which you should remember about Norton amplifiers:

1. The inputs are current-sensing, and thus require some form of input-current limiting resistance.

2. As with an ordinary op amp, voltage gain is set by the two feedback resistors.

3. A DC bias potential must be fed into the noninverting input via a current limiting resistor, R_B. This resistor is normally twice the size of R_f.

6.9 Current Feedback Amplifiers

A relatively recent device in the world of linear integrated circuits is the current feedback amplifier. Its prime advantage is that it does not suffer from the strict gain/bandwidth trade-off typical of ordinary op amps, which means that it is possible to increase its gain without incurring an equal decrease in bandwidth. This has obvious advantages for high-speed applications. Also, these devices tend to be free of slewing effects.

An equivalent model of a noninverting current feedback amplifier is shown in Figure 6.31. This device uses a transimpedance source (a current-controlled voltage source) to translate the feedback current into an output voltage. Instead of relying on negative feedback to keep the inverting and noninverting terminals at the same potential, the current feedback amplifier utilizes a unity-gain buffer placed between the two inputs. In this manner, the potential which drives the noninverting input must also appear at the inverting input. Also, the buffer exhibits a low output impedance into or out of which the feedback current can flow. From this diagram, we can derive the general voltage-gain expression.

By inspection,

$$v_{in} = v_+ = v_- \tag{6.21}$$

$$i_1 = \frac{v_{in}}{R_1} \tag{6.22}$$

$$i_2 = \frac{v_{out} - v_{in}}{R_2} \tag{6.23}$$

$$i_f = i_1 - i_2 \tag{6.24}$$

$$v_{out} = A_{ol} i_f \tag{6.25}$$

Figure 6.31
Current feedback amplifier
equivalent

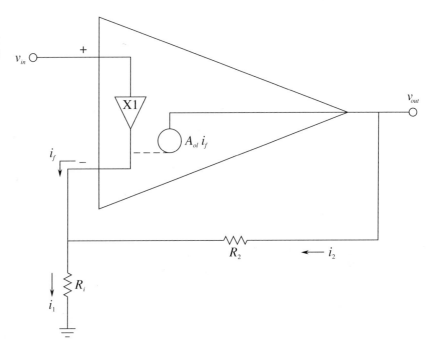

Combining Equations 6.22, 6.23, and 6.24 with Equation 6.25 yields

$$v_{out} = A_{ol}\left(\frac{v_{in}}{R_1} - \frac{v_{out} - v_{in}}{R_2}\right) \tag{6.26}$$

Simplification follows:

$$\frac{v_{out}}{A_{ol}} = \frac{v_{in}}{R_1} - \frac{v_{out} - v_{in}}{R_2}$$

$$\frac{v_{out}}{A_{ol}} = \frac{v_{in}R_2 - v_{out}R_1 + v_{in}R_1}{R_1 R_2}$$

$$\frac{v_{out}R_1 R_2}{A_{ol}} = v_{in}R_2 - v_{out}R_1 + v_{in}R_1$$

$$\frac{v_{out}R_1 R_2}{A_{ol}} + v_{out}R_1 = v_{in}(R_2 + R_1)$$

$$v_{out}\left(\frac{R_1 R_2}{A_{ol}} + R_1\right) = v_{in}(R_2 + R_1) \tag{6.27}$$

We now solve Equation 6.27 in terms of v_{out}/v_{in}, which is A_v.

$$\frac{v_{out}}{v_{in}} = \frac{R_1 + R_2}{\dfrac{R_1 R_2}{A_{ol}} + R_1}$$

$$A_v = \frac{R_1 + R_2}{\dfrac{R_1 R_2}{A_{ol}} + R_1}$$

$$A_v = \frac{\dfrac{R_2}{R_1} + 1}{\dfrac{R_2}{A_{ol}} + 1} \tag{6.28}$$

If, for convenience, we define $(R_2/R_1) + 1$ as G, Equation 6.28 becomes

$$A_v = \frac{G}{\dfrac{R_2}{A_{ol}} + 1} \tag{6.29}$$

If A_{ol} is sufficiently large, this can be approximated as

$$A_v = G$$

$$A_v = \frac{R_2}{R_1} + 1 \tag{6.30}$$

Equation 6.30 is identical to the approximate gain equation for an ordinary noninverting op amp stage. A closer look at the exact equation (6.29) reveals an important difference from the exact series-parallel expression examined in Chapter 3:

$$A_v = \frac{A_{ol}}{1 + \beta A_{ol}}$$

As you may recall, β is the feedback factor, and determines the closed-loop gain. To a first approximation, $\beta A_{ol} \gg 1$, and thus, the series-parallel gain is ideally $1/\beta$. As frequency is raised, however, the value of A_{ol} drops, and soon this approximation no longer holds. It is very important to note that **the point at which the idealization ceases to be accurate is a function of β**. For higher closed-loop gains, β is smaller, and thus, the βA_{ol} product reaches unity at a lower frequency. It is this very interdependency which causes the gain/band-

────────

Figure 6.32

Pinout of the Apex WA01

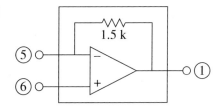

width trade-off. A close look at Equation 6.29 shows no such interdependency. In this case, the denominator is not influenced by the feedback factor. In its place is the static resistor value R_2. Normally, R_2 is set by the manufacturer of the device, and is usually around 1 k to 2 k in size. Closed-loop gain is set by altering the value of R_1. This means that the closed-loop upper-break frequency is not dependent on the closed-loop gain. This is the ideal case. In reality, secondary effects will cause some decrease of f_2 with increasing A_v. This effect is noticeable at medium to high gains, but is still considerably less than what an ordinary op amp would produce. For example, a jump in gain from 10 to 100 may drop f_2 from 150 MHz to only 50 MHz, whereas an ordinary device would drop to 15 MHz.

The WA01 from Apex is one example of a current feedback amplifier. It boasts a slew rate of 5000 V/μS, a bandwidth of 100 MHz, and an output-current capability of 400 mA. This makes it ideal for applications such as high-speed line drivers or video-display drivers. The equivalent circuit is shown in Figure 6.32. Note the 1.5 k internal resistor for R_2. Figure 6.33 shows the inverting and noninverting configurations of the WA01. From the WA01's data sheet in Appendix A, note the very small variation in f_2 as the gain is changed from 5 to 30.

──────── Figure 6.33

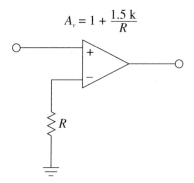

a. Noninverting amplifier

b. Inverting amplifier

Example 6.7   Using the WA01, design a noninverting 100 MHz amplifier with a voltage gain of 20 dB.

By design, the WA01 will meet our bandwidth requirement as long as the desired gain is not excessive. A voltage gain of 20 dB translates to an ordinary gain of 10, which is not excessive. Referring to Figure 6.33a, we find the gain equation and circuit schematic.

$$A_v = 1 + \frac{1.5 \text{ k}}{R}$$

$$R = \frac{1.5 \text{ k}}{A_v - 1}$$

$$R = \frac{1.5 \text{ k}}{10 - 1}$$

$$R = 167 \text{ }\Omega$$

As with any high-speed amplifier, care must be taken when laying out this circuit. Failure to do so may seriously degrade amplifier performance.

6.10 Other Specialized Devices

Op amps have been further refined into other specialized devices. With them, a wide range of specialized linear integrated circuits have emerged. Simple amplifiers are available which are designed to work primarily from single-polarity power supplies. Also, a large array of specialty amplifiers exist for such applications as low-noise phono and tape preamplifiers. For very low offset applications, Chopper Auto Zero or Commutating Auto Zero (CAZ) amplifiers are available. CAZ amplifiers actually monitor their DC offset, and actively correct it.

Moving away from the strict confines of op amps, linear integrated circuits are available for most popular functions, making the designer's job even easier. Some of these items will be explored in later chapters. Other devices are for very specific applications and will not be covered here. A quick sampling will give you an idea of the level of integration now in wide use: Some of the applications now covered by single linear integrated circuits include TV stereo decoders, radio control receivers, FM stereo demodulators, dynamic noise reduction, LED bar graph meters, AM radio systems, and ultrasonic transceivers.

Some manufacturers develop ICs solely for specific market segments. As an example, in the world of music, Precision Monolithics/Solid State Micro-Technology and Curtis Electromusic make very specialized ICs for synthesizer manufacturers. These include very-high-precision voltage-controlled amplifiers, voltage-controlled filters, and voltage-controlled oscillators. Of course, nothing prevents a designer from using one of these integrated circuits in a nonmusical product.

6.11 Extended Topic: ASICs

For specific applications under large-scale production, it is often feasible for designers to create their own custom ICs. This is the final word for specialized designs. For lower production volumes, the design cost of a fully custom IC must be borne by fewer units, and thus, may be prohibitively expensive. At the same time, physical size and manufacturing constraints may well preclude the use of an ordinary discrete approach. The answer to this problem is often found in the **a**pplication **s**pecific **i**ntegrated **c**ircuit, or ASIC.

In essence, an ASIC is a partly designed or partly fabricated integrated circuit. The circuit is "finished" by the designer of the application circuit. By using an ASIC, a designer may not have to be concerned with the manufacturing intricacies of the IC's internal components. Instead, the designer is free to concentrate on the particular application. This greatly reduces the overall design time and cost. Effectively, the basic setup costs for the ASIC are shared among those who use a given ASIC core, or starting point. While the per-part cost of an ASIC may be higher than that of a fully custom design, the reduction in setup costs makes the ASIC more economical when fewer units are being manufactured. ASICs are popular in digital design, but are also used for analog design. For highly integrated applications, a combination of analog and digital circuitry may be used. Eventually, if production volumes increase to an appropriate level, the ASIC may be abandoned in favor of a fully custom design. The custom device may be given the exact same pin configuration as the ASIC, so that a simple part replacement is all that is required for future circuit production.

There are two main approaches to configuring an ASIC core: *standard cells*, or *array-based*. The choice of methodology will depend on the application. For designs which utilize only a small percentage of analog functions (say, less than one-third of the total application circuit), the standard cell scheme appears to have favor, particularly for higher production volumes. The array-based scheme is generally preferred when a larger percentage of functions are analog in nature, and when the performance of the analog circuitry is of prime importance.

Figure 6.34   A tiled analog array

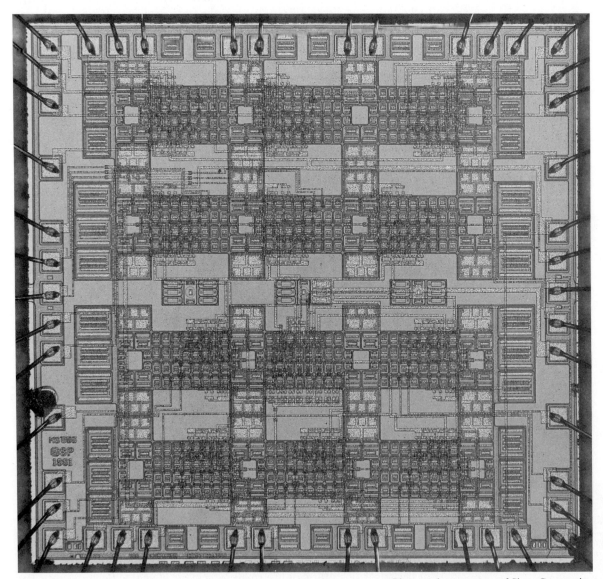

Photo and art courtesy of Sipex Corporation

a. Integrated circuit close-up photograph

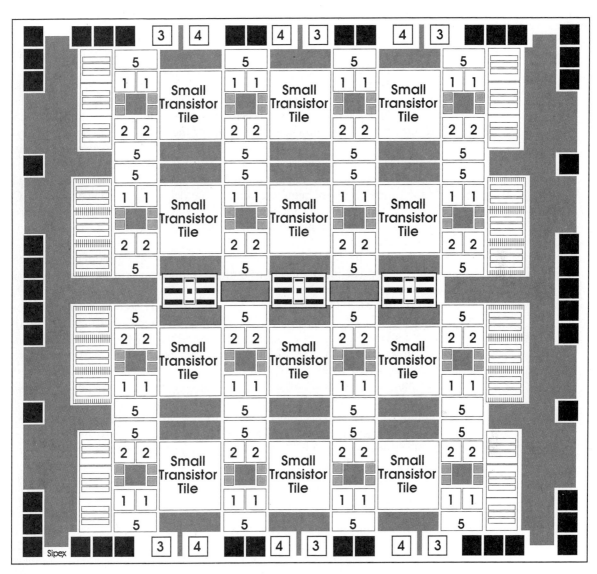

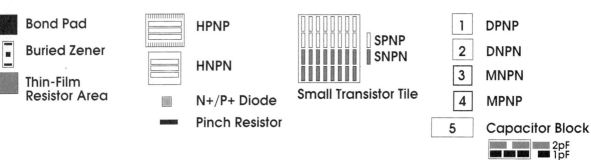

b. Functional tile layout

Standard cell ASICs derive their name from the fact that the IC is formed of equal-height cells. The cells, in turn, will contain the necessary transistors used in the design. The width of the cell can be adjusted if more devices are needed. By making the cells a standard height, interconnections, signal paths, and power supply tie-ins are dealt with more efficiently. The ASIC manufacturer will maintain a library of possible cells and cell variants. For example, a given cell may be used for a low-level logic gate, or the layout of computer memory. The actual shape of the cell can vary. The cells are defined as required by the application, and are connected to form the final ASIC. Since the ASIC manufacturer maintains the cell library, the manufacturer has intimate knowledge of the processing steps required to make the defined cells. Effectively, the IC is partly designed before the application designer even starts to work. Thus, production of the standard-cell ASIC is less expensive than starting a design from scratch.

The array-based ASIC, in contrast, is basically a collection of separate components. The interconnections of these components are determined by the application designer in order to create the finished IC. The ASIC is, in fact, partly fabricated. Only the required interconnections need to be specified. In this respect, using the array-based ASIC is not very different from designing with conventional discrete components. For larger functional blocks (such as op amps), a preformed set of interconnections may be used. These items are usually referred to as *macros*. Designing an ASIC with macros is akin to designing a circuit with large-scale functions such as op amps. For example, an instrumentation amplifier requires three op amps. If the application circuit required the use of an instrumentation amplifier, this function could be achieved by connecting three op amp macros. Since the circuit is partly fabricated, the application designer only defines the interconnection masks for the IC, rather than the series of masks required to define individual devices. The array-based ASIC is therefore a relatively inexpensive solution for lower production volumes.

Such ease of use does have drawbacks. A given array may not contain all of the elements which an application circuit might require. This may force a redesign of the circuit. On the other hand, the larger arrays may hold thousands of devices, or the equivalent of several dozen op amps. While a more complex design may find all of the requisite components in a larger array, a certain amount of waste may occur, resulting in a less-than-optimal cost. It is advisable to be aware of the array elements and choices before ever beginning the application design. In this manner, the most efficient use of a given ASIC core may be found. Also, certain design techniques are better utilized in integrated circuits. For example, internal-component matching is very good, but exacting tolerances are not easy to achieve. A smart designer will exploit the inherent strengths of the integrated circuit in favor of more traditional discrete design techniques.

An example of an ASIC is shown in Figure 6.34 (pages 262–263).

Summary

In this chapter we have examined operational amplifiers which exhibit extended performance, and those which have been tailored to more specific applications. Requirements for precision differential amplification are met by the instrumentation amplifier. Instrumentation amplifiers may be formed from three separate op amps, or, they may be purchased as single hybrid or monolithic ICs. Instrumentation amplifiers offer isolated high-impedance inputs, and excellent common-mode rejection characteristics.

Programmable op amps allow the designer to set desired performance characteristics. In this way, an optimum mix of parameters such as f_{unity} and slew rate versus power consumption is achieved. Programmable op amps can also be set to a very low power consumption standby level. This is ideal for battery-powered circuits. Generally, programming is performed by either a resistor (for static applications) or via an external current or voltage (for dynamic applications).

The output-drive capabilities of the standard op amp have been pushed to high levels of current and voltage. Power op amps may be directly connected to low-impedance loads such as servo motors or loudspeakers. Moderate power devices exist which have been designed for line drivers and audio applications. Due to the higher dissipation requirements of these applications, power op amps are often packaged in TO-220 and TO-3 type cases.

Along with higher-power devices, still other devices show increased bandwidth and slewing performance. These fast devices are particularly useful in video applications. Perhaps the fastest amplifiers are those which rely on current feedback and utilize a transimpedance output stage. These amplifiers are a significant departure from the ordinary op amp. They do not suffer from strict gain-bandwidth limitations, and can achieve very wide bandwidth with moderate gain.

Operational transconductance amplifiers, or OTAs, can be used as building blocks for larger circuits such as voltage-controlled amplifiers or filters. Norton amplifiers rely on a current mirror to perform a current-differencing operation. They are relatively inexpensive and operate directly from single-polarity power supplies with little support circuitry. Since they are current-sensing, input-limiting resistors are required.

Self Test Questions

1. What are the advantages of using an instrumentation amplifier versus a simple op amp differential amplifier?

2. How might an instrumentation amplifier be constructed from general-purpose op amps?

3. What are the advantages of using programmable op amps?

4. What are the results of altering the programming current in a programmable op amp?

5. Give at least two applications for a high-power op amp.

6. Give at least two applications for a high-speed op amp.

7. What is an OTA?

8. How is an OTA different from a programmable op amp?

9. Give an application which might use an OTA.

10. Give at least three applications which could benefit from specialty linear integrated circuits.

11. Describe how a Norton amplifier achieves input differencing.

12. List a few of the major design differences which must be considered when working with Norton amplifiers versus ordinary op amps.

13. Explain why a current-feedback amplifier does not suffer from the same gain-bandwidth limitations that ordinary op amps do.

14. What is an ASIC?

Problem Set

▷ Analysis Problems

1. An instrumentation amplifier has a differential input signal of 5 mV and a common-mode hum input of 2 mV. If the amplifier has a differential gain of 32 dB and a CMRR of 85 dB, what are the output levels of the desired signal and the hum signal?

2. Determine v_{out} in Figure 6.35 if $v_{in+} = +20$ mV DC and $v_{in-} = -10$ mV DC.

3. Repeat Problem 2 for a differential input signal of 10 mV peak to peak.

4. Repeat Problem 2 for a differential input signal of 20 mV peak to peak and a common-mode signal of 5 mV peak to peak. Assume the system CMRR is 75 dB.

5. Determine the programming current in Figure 6.11 if $R_{set} = 1$ M. Assume standard ± 15 V supplies.

——— Figure 6.35

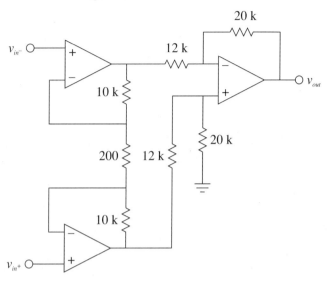

6. Using an MC1776 programmable amplifier, determine the following parameters if $I_{set} = 5 \ \mu A$: slew rate, f_{unity}, input noise voltage, input bias current, standby supply current, and open-loop gain.

7. Determine the voltage gain, f_2, and power bandwidth (assume $V_p = 10$ V) for the circuit of Figure 6.36.

——— Figure 6.36

8. The circuit of Figure 6.37 uses a light-dependent resistor (LDR) to enable or disable the amplifier. Under full light, this LDR exhibits a resistance of 1 k, while under no light conditions, the resistance is 50 M. What are the standby current and f_{unity} values under the conditions of full light and no light?

————— Figure 6.37

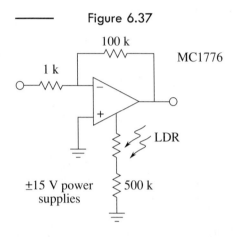

9. Determine the required capacitance to set PSRR to at least 20 dB at 100 Hz for an LM386 power amp.

10. What is the power bandwidth for a 5 V peak signal in Figure 6.21?

11. Determine the device dissipation for an LM386 delivering .5 W into an 8 Ω load. Assume a 12 V power supply.

12. Determine the input resistance, output resistance, and transconductance for an LM3080 OTA with $I_{abc} = 100$ μA.

13. Determine the value of I_{abc} in Figure 6.23 if $V_{control} = .5$ V DC.

14. Recalculate the value in R_i for the circuit in Figure 6.28 if a voltage gain of 45 is desired.

15. Recalculate the input capacitance value for the circuit in Figure 6.28 if a lower break frequency of 15 Hz is desired.

▷ Design Problems

16. Design an instrumentation amplifier with a gain of 20 dB using the LH0036.

17. Utilizing the LM3900, design an inverting amplifier with a gain of 12 dB, an input impedance of at least 100 k, and a lower break frequency no greater than 25 Hz.

18. Utilizing the LM3900, design a noninverting amplifier with a gain of 24 dB, an input impedance of at least 40 k, and a lower break frequency no greater than 50 Hz.

19. Utilizing the WA01, design a noninverting amplifier with a gain of 32 dB.

20. Utilizing the WA01, design an inverting amplifier with a gain of 16 dB.

▷ Challenge Problems

21. Using Figure 6.35 as a guide, design an instrumentation amplifier with a differential gain of 40 dB. The system bandwidth should be at least 50 kHz. Indicate which op amps you intend to use.

22. Determine the gain, f_2, power bandwidth ($V_p = 10$ V), and standby current for the circuit of Figure 6.38 if $V_B = 0$ V, and if $V_B = -15$ V.

—————— Figure 6.38

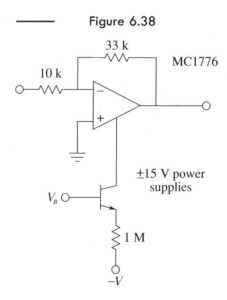

23. Using the LM386, design a power amplifier with a voltage gain of 32 dB.

24. Determine v_{out} in Figure 6.23 if $V_{control} = 1$ V DC and $v_{in} = 100$ mV sine wave.

25. Sketch v_{out} for Figure 6.23 if $V_{control} = 1 \sin 2\pi1000t$, and $v_{in}(t) = .5 \sin 2\pi300,000t$.

26. Alter Figure 6.23 such that a 100 mV input signal will yield a 1 V output signal when $V_{control}$ is $+2$ V.

27. Derive Equation 6.9 from the text.

▷ SPICE Problems

28. Use SPICE to verify the design produced in Problem 21.

29. Use SPICE to simulate the output of the circuit shown in Figure 6.28. Use a $+15$ V DC power supply with $v_{in}(t) = .01 \sin 2\pi 100t$.

30. Use SPICE to verify the gain of the design produced in Problem 18.

31. Device mismatching can adversely affect the CMRR of an instrumentation amplifier. Rerun the simulation of the instrumentation amplifier (Figure 6.5) with a $\pm 5\%$ tolerance applied to the value of R'_i (Rip). Based on these simulations, what do you think would happen if all of the circuit resistors had this same tolerance?

7 Nonlinear Circuits

After completing this chapter, you should be able to:

❑ Understand the advantages and disadvantages of active rectifiers versus those of passive rectifiers.

❑ Analyze the operation of various half- and full-wave precision rectifiers.

❑ Analyze the operation of peak detectors, limiters, and clampers.

❑ Utilize function generation circuits to alter the amplitude characteristics of an input signal.

❑ Detail the usefulness of a Schmitt trigger, and compare its performance to an ordinary open-loop comparator.

❑ Utilize dedicated comparator ICs, such as the LM311.

❑ Analyze log and antilog amplifiers and multipliers.

7.1 Introduction

Besides their use as amplification devices, op amps may be used for a variety of other purposes. In this chapter, we will look into a number of circuits which fall under the general classification of nonlinear applications. The term *nonlinear* is used because the input/output transfer characteristic of the circuit is no longer a straight line. Indeed, the characteristic may take on a wide variety of shapes, depending on the application. It might seem odd to be studying nonlinear circuits when op amps are generally considered to be linear devices. Rest assured that nonlinear elements are often essential tools in a larger system. The general usage of nonlinear elements is that of wave shaping. Wave shaping is the process of reforming signals in a desired fashion. You might think of it as signal sculpting.

Wave shaping runs the gamut from simple half-wave rectification up through transfer-function generation. While many of the applications presented in this chapter can be realized without op amps, the inclusion of the op amp lends the circuit a higher level of precision and stability. This is not to say that the op amp versions are always superior to a passive counterpart. Due to bandwidth and slew rate limitations, the circuits presented perform best at lower frequencies—primarily in the audio range and below. Also, there are definite power-handling and signal-level limitations. In spite of these factors, the circuits do see quite a bit of use in the appropriate areas. Many of the applications fall into the categories of instrumentation and measurement.

Besides the wave-shaping functions, the other uses noted here are comparators and logarithmic amplifiers. General-purpose comparators made from simple op amps were explained in Chapter 2. In this chapter, specialized devices are examined, and their superiority to the basic op amp comparator is noted. Logarithmic amplifiers have the unique characteristic of compressing an input signal. That is, they reduce the range of signal variation. The mirror image of the log amplifier is the antilog amplifier. It increases the range of input signal variation.

7.2 Precision Rectifiers

Imagine for a moment that you would like to half-wave rectify the output of an oscillator. Probably the first strategy that pops into your head is to use a diode, as in Figure 7.1. As shown, the diode passes positive half-waves and blocks negative half-waves. But what happens if the input signal is only .5 V peak? Rectification never occurs, because the diode requires .6 to .7 V to turn on. Even if a germanium device is used with a forward drop of .3 V, a sizable portion of the signal will be lost. Not only that, but the circuit of Figure 7.1

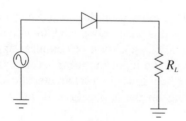

Figure 7.1
Passive rectifier

Figure 7.2
Precision half-wave
rectifier

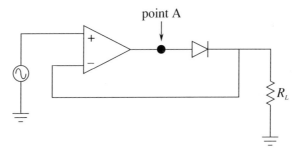

point A

exhibits vastly different impedances to the driving source. Even if the signal is large enough to avoid the forward drop difficulty, the source impedance must be relatively low. At first glance it seems as though it is impossible to rectify a small AC signal with any hope of accuracy.

One of the items noted about negative feedback (Chapter 3) was the fact that it tended to compensate for errors. Negative feedback tends to reduce errors by an amount equal to the loop gain. This being the case, it should be possible to reduce the diode's forward drop by a very large factor by placing it inside a feedback loop. This arrangement is shown in Figure 7.2, and is called a *precision half-wave rectifier*. To a first approximation, when the input is positive, the diode is forward-biased. In essence, the circuit reduces to a simple voltage follower with a high input impedance and a voltage gain of one, so the output looks just like the input. On the other hand, when the input is negative, the diode is reverse-biased, opening up the feedback loop. No signal current is allowed to the load, so the output voltage is zero. Thanks to the op amp though, the driving source still sees a high impedance. The output waveform consists of just the positive portions of the input signal, as shown in Figure 7.3. Due to the effect of negative feedback, even small signals can be

Figure 7.3
Output signal

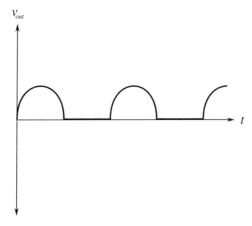

V_{out}

t

Figure 7.4
Transfer characteristic

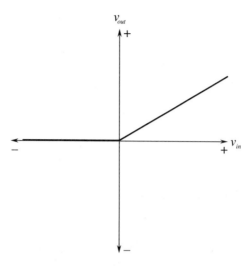

properly rectified. The resulting transfer characteristic is presented in Figure 7.4. A perfect one-to-one input/output curve is seen for positive input signals, while negative input signals produce an output potential of zero.

In order to create the output waveform, the op amp creates an entirely different waveform. For positive portions of the input, the op amp must produce a signal which is approximately .6 to .7 V greater. This extra signal effectively compensates for the diode's forward drop. Since the feedback signal is derived after the diode, the compensation is as close as the available loop gain allows. At low frequencies where the loop gain is high, the compensation is almost exact, producing a near perfect copy of positive signals. When the input signal swings negative, the op amp tries to sink current in response. As it does, the diode becomes reverse-biased, and current flow is halted. At this point the op amp's noninverting input will be seeing a large negative potential relative to the inverting input. The resulting negative error signal forces the op amp's output to go to negative saturation. Since the diode remains reverse-biased, the circuit output stays at 0 V. The op amp is no longer able to drive the load. This condition will persist until the input signal goes positive again, at which point the error signal becomes positive, forward-biasing the diode, and allowing load current to flow. The op amp and circuit output waveforms are shown in Figure 7.5.

One item to note about Figure 7.5 is the amount of time it takes for the op amp to swing in and out of negative saturation. This time is determined by the device's slew rate. Along with the decrease of loop gain at higher frequencies, slew rate determines how accurate the rectification will be. Suppose that the op amp is in negative saturation and that a quick positive input pulse occurs. In order to track this, the op amp must first climb out of negative

Figure 7.5
Output of op amp

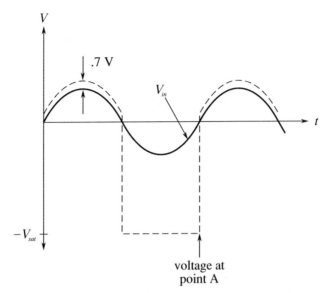

saturation. Using a 741 op amp with 15 V supplies, it will take about 26 μS to go from negative saturation (−13 V) to zero. If the aforementioned pulse in only 20 μS wide, the circuit doesn't have enough time to produce the pulse. The input pulse will have gone negative again, before the op amp has a chance to "climb out of its hole." If the positive pulse were a bit longer, say 50 μS, the op amp would be able to track a portion of it. The result would be a distorted signal as shown in Figure 7.6. In order to accurately rectify fast-moving signals, op amps with high f_{unity} and slew rate are required. If only

Figure 7.6
High-frequency errors

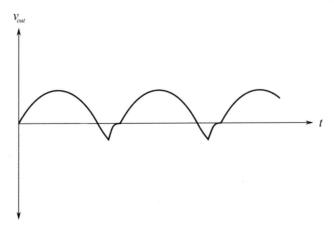

Figure 7.7
Rectifier with gain

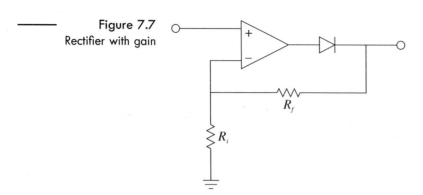

slow signals are to be rectified, it is possible to configure the circuit with moderate gain as a cost-saving measure. This is shown in Figure 7.7. Finally, for negative half-wave output, the only modification required is the reversal of the diode.

A SPICE simulation of the circuit shown in Figure 7.2 is presented in Figure 7.8. The circuit is shown redrawn in 7.8a with the nodes labeled. Figure 7.8b shows the input file listing. This example utilizes the 741 op amp model examined earlier. The output waveform (using Probe) is shown in Figure 7.8c on page 278. Note the accuracy of the rectification. The output of the op amp is also shown, so that the effects of negative feedback illustrated in Figure 7.5 are clearly visible. Since this circuit utilizes an accurate op amp model, it is very instructive to rerun the simulation for higher input frequencies. In this way, the inherent speed limitations of the op amp are shown, and effects such as those presented in Figure 7.6 may be noted.

Figure 7.8
SPICE simulation of the precision rectifier of Figure 7.2

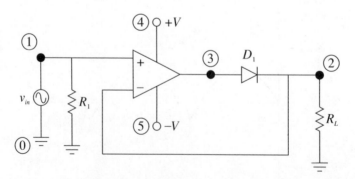

a. SPICE schematic

Figure 7.8
(Continued)

```
Precision Half-wave Rectifier
*******Start of UA741 op amp*******
.SUBCKT UA 741  1    2    3    4    5    6
*                IN+  IN- GND  V+  V-  OUT
Q1   11   1   13 UA741QA
Q2   12   2   14 UA741QB
RC1   4  11   5.305165E+03
RC2   4  12   5.305165E+03
C1   11  12   5.459553E-12
RE1  13  10   2.151297E+03
RE2  14  10   2.151297E+03
IEE  10   5   1.666000E-05
CE   10   3   3.000000E-12
RE   10   3   1.200480E+07
GCM   3  21   10 3 5.960753E-09
GA   21   3   12 11 1.884655E-04
R2   21   3   1.000000E+05
C2   21  22   3.000000E-11
GB   22   3   21 3 2.357851E+02
RO2  22   3   4.500000E+01
D1   22  31   UA741DA
D2   31  22   UA741DA
EC   31   3   6 3 1.0
RO1  22   6   3.000000E+01
D3    6  24   UA741DB
VC    4  24   2.80323E+00
D4   25   6   UA741DB
VE   25   5   2.80323E+00
.ENDS
.MODEL UA741DA D (IS=9.762287E-11)
.MODEL UA741DB D (IS=8.000000E-16)
.MODEL UA741QA NPN (IS=8.000000E-16 BF=9.166667E+01)
.MODEL UA741QB NPN (IS=8.309478E-16 BF=1.178571E+02)
*****End of UA741 op amp*****
*
.MODEL DIN914 D (IS=100E-15 RS=16 CJO=2PF TT=12NS BV=100 IBV=4E-10)
********Main circuit description********

R1    0  1   10K
RL    0  2   10K
D1    3  2   DIN914
VCC   4  0   DC 15
VEE   5  0   DC -15
XOA1  1  2   0 4 5 3  UA741
VIN   1  0   SIN (0 2 300Hz)

********Analysis Options********
*Calculate transient analysis in 100 uS steps up to .01 S.
.TRAN 100US .01S
*Plot the time domain response at the output (node 2).
.PLOT TRAN V(2) V(3)
*Use .Probe with PSpice only
.Probe
.END
```

b. SPICE input file

(Continued)

Figure 7.8
(Continued)

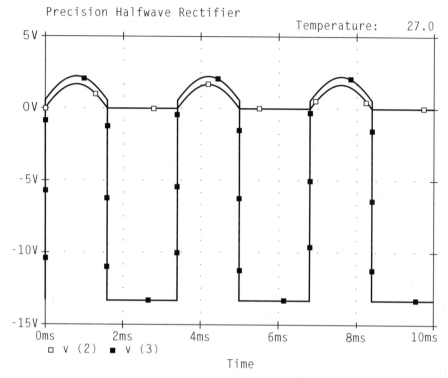

c. PSpice-Probe output

Peak Detectors

One variation on the basic half-wave rectifier is the *peak detector*. This circuit will produce an output which is equal to the peak value of the input signal. This can be configured for either positive or negative peaks. The output of a peak detector can be used for instrumentation or measurement applications. It can be thought of as an analog pulse stretcher.

A simple positive peak detector is shown in Figure 7.9. Here is how it works: The first portion of the circuit is a precision positive half-wave rectifier. When its output is rising, the capacitor, C, is being charged. This voltage is presented to the second op amp, which serves as a buffer for the final load. Since the output impedance of the first op amp is low, the charge-time constant is very fast, and thus the signal across C is very close to the input signal. When the input signal starts to swing back toward ground, the output of the first op amp starts to drop along with it. Due to the capacitor voltage the diode ends up in reverse bias, thus opening the drive to C. C starts to discharge, but the discharge-time constant will be much longer than the charge-time constant.

Figure 7.9
Peak detector

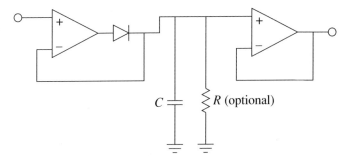

The discharge resistance is a function of R, the impedance looking into the noninverting input of op amp 2, and the impedance looking into the inverting input of op amp 1, all in parallel. Normally, FET input devices are used, so from a practical standpoint, R sets the discharge rate.

The capacitor will continue to discharge toward zero until the input signal rises enough to overtake it again. If the discharge-time constant is much longer than the input period, the circuit output will be a DC value equal to the peak value of the input. If the discharge-time constant is somewhat shorter, it lengthens the pulse time. It also has the effect of producing the overall contour, or envelope, of complex signals, so it is sometimes called an *envelope detector*. Possible output signals are shown in Figure 7.10. For very long discharge times, large capacitors must be used. Larger capacitors will, of course, produce a lengthening of the charge time (i.e., the risetime will suffer). Large capacitors can also degrade slewing performance. The rate at which C is charged is limited because a given op amp can only produce a finite current. This is no different from the case presented with compensation capacitors in Chapter 5. As an

Figure 7.10
Effect of τ on pulse shape

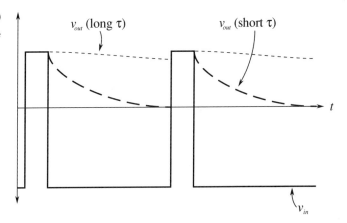

example, if C is 10 μF, and the maximum output current of the op amp is 25 mA,

$$i = C \frac{dv}{dt}$$

$$\frac{dv}{dt} = \frac{i}{C}$$

$$\frac{dv}{dt} = \frac{25 \text{ mA}}{10 \text{ } \mu\text{F}}$$

$$\frac{dv}{dt} = 2500 \text{ V/S} = 2.5 \text{ mV/}\mu\text{S}$$

This is a very slow slew rate! If FET input devices are used, the effective discharge resistance can be very high, thus lowering the requirement for C. For typical applications, C will be many times smaller than the value used here. For long discharge times, high-quality capacitors must be used as the internal leakage will place the upper limit on discharge resistance.

Example 7.1   A positive-peak detector is used along with a simple comparator in Figure 7.11 to monitor input levels and warn of possible overload. Explain how it works and determine the point at which the LED lights.

 The basic problem when trying to visually monitor a signal for overloads is that the overloading peak may come and go faster than the human eye can detect. For example, the signal might be sent to a comparator which could light an LED when a preset threshold is exceeded. When the input signal falls,

Figure 7.11
Detector for Example 7.1

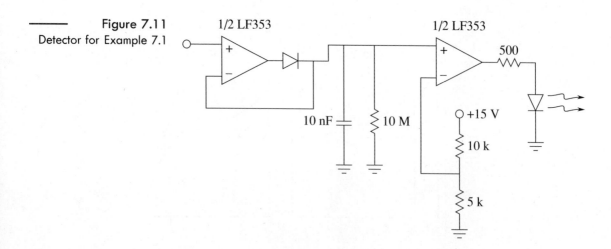

1/2 LF353   1/2 LF353

10 nF   10 M   +15 V   500   10 k   5 k

the comparator and LED will go into the off state. Even though the LED does light at the peak, it remains for such a short time that humans can't detect it. This sort of result is quite possible in the communications industry, where the output of a radio station's microphone or turntable will produce very dynamic waves with a great many peaks. These peaks can cause havoc in other pieces of equipment down the line. The LED needs to remain on for longer periods.

The circuit of Figure 7.11 uses a peak detector to stretch out the positive pulses. These stretched pulses are then fed to a comparator which drives an LED. Since the input pulses are expanded, the LED will remain on for longer periods.

The discharge-time constant is set by R and C. Since FET input devices are used, their impedance is high enough to ignore.

$$T = RC$$

$$T = 10 \text{ M} \times 10 \text{ nF}$$

$$T = .1 \text{ S}$$

The 10 nF capacitor is small enough to maintain a reasonable slew rate. You may wish to verify this as an exercise. The comparator trip point is set by the 10 k/5 k voltage divider at 5 V. When the input signal rises above 5 V, the comparator output goes high. Assuming that the LED forward drop is about 2.5 V, the 500 Ω resistor limits the output current to

$$I_{LED} = \frac{V_{sat} - V_{LED}}{500}$$

$$I_{LED} = \frac{13 \text{ V} - 2.5 \text{ V}}{500}$$

$$I_{LED} = \frac{10.5 \text{ V}}{500}$$

$$I_{LED} = 21 \text{ mA}$$

The LF353 should be able to deliver this current.

In summary then, the input pulses are stretched by the peak detector. If any of the resulting pulses are greater than 5 V, the comparator trips, and lights the LED. An example input/output wave is shown in Figure 7.12 on page 282. The one problem with this remedy is that only positive peaks are

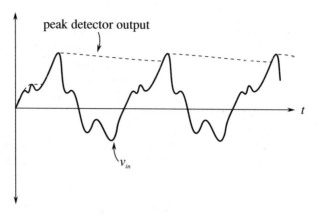

a. Output of detector portion

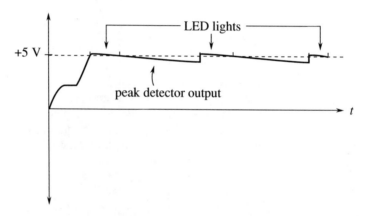

b. Comparator/LED action

detected. If large negative peaks exists, they will not cause the LED to light. It is possible to use a similar circuit to detect negative peaks and use that output to drive a common LED along with the positive peak detector. Another way to accomplish this is to utilize a full-wave rectifier/detector.

Precision Full-Wave Rectifiers

The input/output characteristic of a full-wave rectifier is shown in Figure 7.13. No matter what the input polarity is, the output is always positive. For this reason, this circuit is often referred to as an *absolute value* circuit. The design of a precision full-wave rectifier is a little more involved than the single-polarity types. One way of achieving this design is to combine the outputs of negative

Figure 7.13

Transfer characteristic for
full-wave rectification

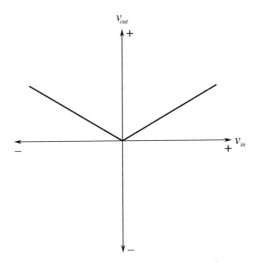

Figure 7.13

Transfer characteristic for
full-wave rectification

and positive half-wave circuits with a differential amplifier. Another way is
shown in Figure 7.14.

The precision rectifier of the circuit in Figure 7.14 is convenient in that it
only requires two op amps, and that all resistors (save one) are the same value.
This circuit is comprised of two parts: an inverting half-wave rectifier, and a
weighted summing amplifier. The rectifier portion is redrawn in Figure 7.15
(page 284). Let's start the analysis with this portion.

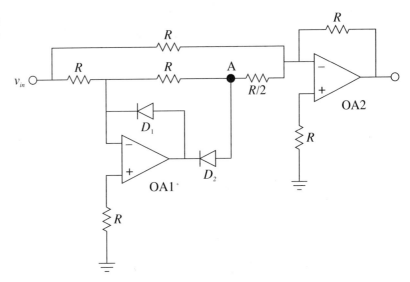

Figure 7.14

Precision full-wave rectifier

Figure 7.15
Inverting half-wave
rectifier

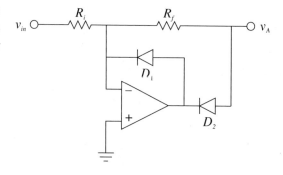

First, note that the circuit is based on an inverting voltage amplifier, with the diodes D_1 and D_2 added. For positive input signals, the input current will attempt to flow through R_f to create an inverted output signal with a gain of R_f/R_i. (Normally, gain is set to unity.) Since the op amp's inverting input is more strongly positive than the noninverting input, the op amp tries to sink output current. This forces D_2 on, completing the feedback loop, while also forcing D_1 off. Since D_2 is inside the feedback loop, its forward drop is compensated for. Thus, positive input signals are amplified and inverted as in a normal inverting amplifier.

If the input signal is negative, the op amp will try to source current. This turns D_1 on, creating a path for current flow. Since the inverting input is at virtual ground, the output voltage of the op amp is limited to the .6 to .7 V drop of D_1. Thus, the op amp does not saturate; rather, it delivers the current required to satisfy the source demand. The op amp's output polarity also forces D_2 off, leaving the circuit output at an approximate ground. Therefore, for negative input signals, the circuit output is zero.

The combination of the positive and negative input swings creates an inverted, half-wave–rectified output signal, as shown in Figure 7.16. This circuit

Figure 7.16
Output of half-wave
rectifier

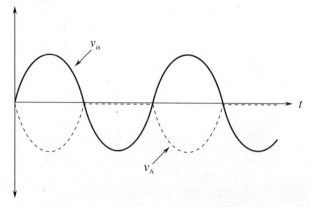

Figure 7.17   Combination of signals produces output

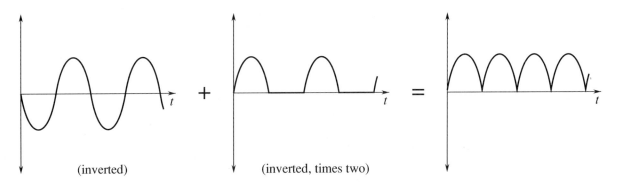

(inverted)   (inverted, times two)

can be used on its own as a half-wave rectifier if need be. Its major drawback is a somewhat limited input impedance. On the plus side, since the circuit is nonsaturating, it may prove to be faster than the half-wave rectifier first discussed.

The voltage at point A in Figure 7.14 is the output of the half-wave rectifier and is shown in Figure 7.16. This is one of two signals applied to the summer configured around op amp 2. The other input to the summer is the main circuit's input signal. This signal is given a gain of unity, while the half-wave signal is given a gain of 2. These two signals will combine as shown in Figure 7.17 to create a positive full-wave output. Mathematically,

for the first 180°: $v_{out} = -K \sin \omega t + 2K \sin \omega t$

$$v_{out} = K \sin \omega t$$

for the second 180°: $v_{out} = K \sin \omega t + 0$

$$v_{out} = K \sin \omega t$$

In order to produce a negative full-wave rectifier, simply reverse the polarity of D_1 and D_2.

An example application of a rectifier based on an op amp is shown in Figure 7.18 (page 286). This circuit is used in the Phase Linear model A-15 audio power amplifier to sense signal levels and warn the user of possible overload. Op amp Z102 is configured as an inverting half-wave rectifier with unity gain. It is functionally similar to the circuit found in Figure 7.15, and produces only the inverted positive input half-wave, as shown in Figure 7.16. A noninverting unity-gain rectifier is configured around op amp Z103. This portion will reproduce the negative-going peaks of the input signal. Together, these two elements drive a lag network comprised of R128, R134, and C101. This lag

Figure 7.18
Phase Linear A-15 power
amplifier

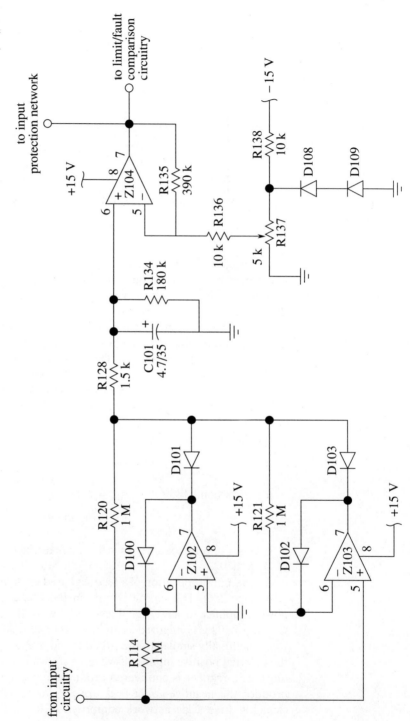

network will tend to negatively peak-detect the composite signal, producing a level which is proportional to the input signal strength. This level, which grows negatively as the input volume increases, is applied to the input of op amp Z104, after which it will be compared to a signal derived from the output of the amplifier. If these signals are not properly matched, the amplifier is most likely clipping. This condition will cause the limit/fault LED to light.

7.3 Wave Shaping

Active Clampers

Clampers are used to add a specific amount of DC to a signal. Generally, the amount of DC will be equal to the peak value of the signal. Clampers are "intelligent" in that they can adjust the amount of DC if the peak value of the input signal changes. Ideally, clampers will not change the shape of the input signal, rather, the output signal is simply a vertically shifted version of the input, as shown in Figure 7.19. One common application of the clamper is in television work. Here, a clamper is referred to as a *DC restorer*. After the video signal has been amplified by AC coupled gain stages, the DC restorer returns the video to its normal orientation. Clampers are commonly made with passive components, just as rectifiers are. Like rectifiers, simple clampers produce errors caused by the diode's forward voltage drop. Active clampers remove this error.

Figure 7.19   Effect of clampers on input signal

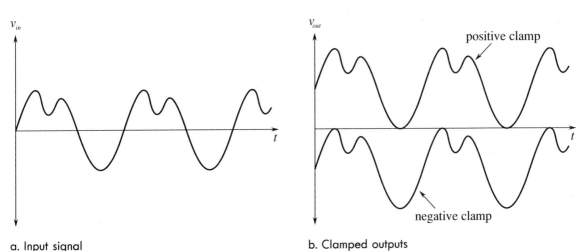

a. Input signal

b. Clamped outputs

Figure 7.20

Single clamper model

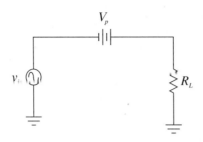

Conceptually, a clamper needs to sense the amplitude of the input signal and create a DC signal of equal value. This DC signal is then added to the input as shown in Figure 7.20. If the peak value is known, and does not change, it is possible to create this function with a summing amplifier. If the signal level is dynamic, a different course needs to be taken. The trick is in getting the DC source to properly track input level changes. Obviously, a simple DC supply is not appropriate; however, a charged capacitor will fit the bill nicely— as long as its discharge-time constant is much longer than the period of the input wave. All that needs to be done is to have the capacitor charge to the peak level of the input. When the capacitor voltage is added to the input signal, the appropriate DC shift will result.

Figure 7.21 shows an active positive clamper. On the first negative-going peak of the input, the source will attempt to pull current through the capacitor in the direction shown. This forces the inverting input of the op amp to go slightly negative, thus creating a positive op amp output. The result of this action will be to forward-bias the diode and supply charging current to the capacitor. The capacitor will charge to the negative peak value of the input. The output resistance of the op amp is very low, so charging is relatively fast, limited only by the maximum output current of the op amp. When the source

Figure 7.21

Active positive clamper

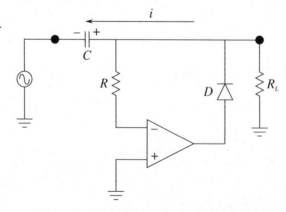

Figure 7.22
Clamped output with
offset

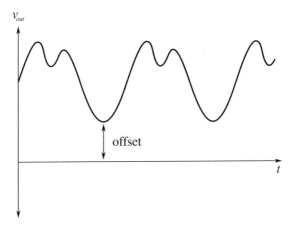

changes direction, the op amp will produce a negative output, thus turning off the diode and effectively removing the op amp from the circuit. Since the discharge time for C is much longer than the input period, its potential stays at roughly V_{p-}. It now acts like a voltage source. These two sources are added, so we can see that the output must be

$$v_{out} = v_{in} + |V_{p-}| \qquad (7.1)$$

The op amp will remain in saturation until the next negative peak, at which point the capacitor will be recharged. During the charging period, the feedback loop is closed, and thus, the diode's forward drop is compensated for by the op amp. In other words, the op amp's output will be approximately .6 to .7 V above the inverting input's potential. The discharge time of the circuit is set by the load resistor, R_L. If particularly long time constants are required, a buffer stage may be used, along with an FET input clamping op amp. The resistor R is used to prevent possible damage to the op amp from capacitor discharge. This value is normally in the low kilohm region. In order to make a negative clamper, just reverse the polarity of the diode.

Sometimes, it is desirable to clamp a signal and add a fixed offset as well. An example output waveform of this type is shown in Figure 7.22. This function is relatively easy to add to the basic clamper. In order to include the offset, all that needs to be done is to change the op amp's reference point. In the basic clamper, the noninverting input is tied to ground. Consequently, this establishes the point at which the charge/discharge cycle starts. If this reference is altered, the charge/discharge point is altered, too. To create a positive offset, a DC signal equal to the offset is applied to the noninverting input, as shown in Figure 7.23 on the next page. A similar arrangement may be used for negative clampers.

Figure 7.23
Active positive clamper
with offset

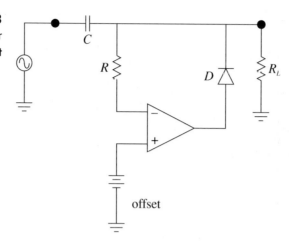

Example 7.2   An active clamper is shown in Figure 7.24. An LF353 op amp is used. (The LF353 is a dual-package version of the LF351.) Determine the capacitor's voltage and verify that the time constant is appropriate for the input waveform. Also, sketch the output waveform and determine the maximum differential input voltage for the first op amp.

Since this is a positive clamper, the capacitor's voltage is the sum of the offset potential and the negative peak potential of the input waveform. V_{offset} is set by the 10 k/1.5 k divider:

$$V_{offset} = V_{CC} \frac{R_2}{R_1 + R_2}$$

$$V_{offset} = 15 \text{ V} \frac{1.5 \text{ k}}{10 \text{ k} + 1.5 \text{ k}}$$

$$V_{offset} = 1.96 \text{ V}$$

$$V_c = V_{offset} + |V_{p-}|$$
$$V_c = 1.96 \text{ V} + |-1 \text{ V}|$$
$$V_c = 2.96 \text{ V}$$

This is the amount of DC added to the input signal. The output waveform will look just like the input, except that it will be shifted up by 2.96 V. This is shown in Figure 7.25.

The discharge time constant should be much larger than the period of the input wave. The input wave has a frequency of 1 kHz, and thus, a period of

Figure 7.24 Active clamper and waveform for Example 7.2

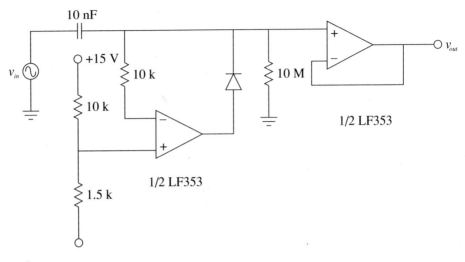

a. Circuit

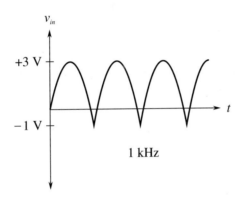

b. Input wave form

Figure 7.25
Output of clamper for
Example 7.2

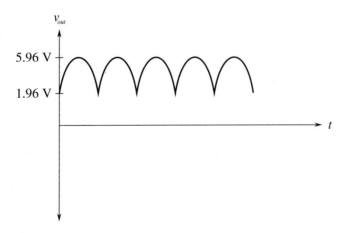

1 mS. The capacitor discharges through the load resistor. Since FET input op amps are used, their effect on the discharge rate is minimal.

$$T = R_L C$$

$$T = 10 \text{ M} \times 10 \text{ nF}$$

$$T = .1 \text{ S}$$

The discharge rate is 100 times longer than the input period, so the capacitor droop will be satisfactory.

The maximum differential input signal is the worst-case difference between the inverting and noninverting input potentials. The noninverting input is tied directly to 1.96 V. The inverting input sees the output waveform. The maximum value of the output is 5.96 V. So, the difference is

$$v_{in\text{-}diff} = v_{in+} - v_{in\text{-}diff}$$

$$v_{in\text{-}diff} = 1.96 \text{ V} - 5.96 \text{ V}$$

$$v_{in\text{-}diff} = -4 \text{ V}$$

The LF353 op amp will have no problem with a differential input of this size.

Active Limiters

A *limiter* is a circuit which places a maximum restriction on its output level. The output of a limiter can never be above a specific preset level. In one sense, all active circuits are limiters, in that they will all eventually clip the signal by going into saturation. A true limiter, though, prohibits the signal at levels considerably lower than saturation. The exact level is fairly easy to set. Limiters can be used to protect following stages from excessive input levels. They can also be seen as a type of wave shaper. One form of limiter was shown in Chapter 6, in the Pocket Rockit's schematic (Figure 6.16). In that circuit, a limiter was used to purposely clip the music signal for artistic effect. The limitation of that form is that the limit potential is locked at ±.7 V by the parallel signal diodes. For a more general form, an arbitrary limit potential is desired. Instead of using parallel signal diodes, a series combination of zener diodes will prove useful.

An example limiter is shown in Figure 7.26. It is based upon the inverting-voltage amplifier form. As long as the output signal is below the zener potential, the output equals the input times the voltage gain. If the output voltage tries to rise above the zener potential, one of the diodes will go into zener conduction,

Figure 7.26
Active limiter

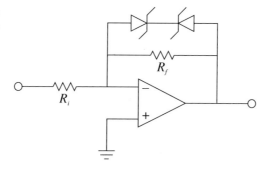

and the other diode will go into forward bias. Once this happens, the low dynamic resistance of the diodes will disallow any further increase in output potential. The output will not be allowed to move outside the zener potential (plus the .7 V turn-on for the second diode):

$$\left|v_{out}\right| \leq V_Z + .7 \text{ V} \tag{7.2}$$

Or we might reword this as

$$V_{limit} = \pm(V_Z + .7 \text{ V}) \tag{7.3}$$

This effect is shown graphically in Figure 7.27, which illustrates the limiter's transfer characteristic.

Figure 7.27
Transfer characteristic
of limiter

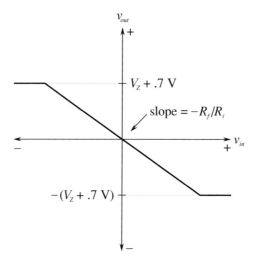

Example 7.3   If the input to the circuit of Figure 7.28 is a 4 V peak triangle wave, sketch v_{out}.

Figure 7.28
Limiter for Example 7.3

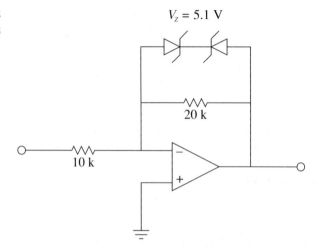

$V_z = 5.1$ V

20 k

10 k

Without the zener limit diodes, this amplifier would simply multiply the input by a gain of 2, and invert it.

$$v_{out} = A_v v_{in}$$

$$v_{out} = -2 \times 4 \text{ V} \quad \text{(peak)}$$

$$v_{out} = -8 \text{ V} \quad \text{(peak)}$$

So, an inverted 8 V peak triangle would be the output. With the inclusion of the diodes, the maximum output is

$$V_{limit} = \pm(V_z + .7 \text{ V})$$

$$V_{limit} = \pm(5.1 \text{ V} + .7 \text{ V})$$

$$V_{limit} = \pm 5.8 \text{ V}$$

Consequently, the output wave is clipped at 5.8 V, as shown in Figure 7.29 at the top of the next page.

Figure 7.29
Input/output signals of
limiter for Example 7.3

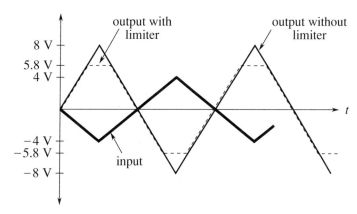

By using different zeners, it is also possible to produce asymmetrical limiting. For example, positive signals might be limited to 10 V, while negative signals could be limited to −5 V.

7.4 Function Generation

Function generation circuits are used to create arbitrary transfer characteristics. These can be used for a variety of purposes, including transducer linearization and sine shaping. Indeed, many modern laboratory signal generators do not directly create a sine wave; rather, they generate a triangle wave and pass it through a function circuit, which will then produce the desired sine wave. In essence, a function circuit gives different gains to different parts of the input signal. It impresses its shape on the input waveform. If a "straight-sided" waveform such as a ramp or triangle is fed into a function circuit, the resulting output waveform will bear a striking resemblance to the circuit's transfer curve. (This effect can be put to good use when simulating function circuits with SPICE.)

Basically, there are two ways in which to create a function circuit with an op amp. The first way is an extension of the zener limiter circuit. The second form relies on a biased diode network. In both cases, the base circuit form is that of an inverting-voltage amplifier. Also, both techniques allow the resulting transfer curve's slope to increase or decrease at specified break points and rates.

Figure 7.30
Simple function generator

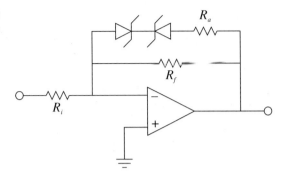

A simple function circuit is shown in Figure 7.30. Note that it is very similar to the limiter of Figure 7.26. The only difference is the inclusion of R_a. Below the zener potential, the circuit gain is still set by R_f. When the zener potential is eventually reached, the diodes can no longer force the output to an unchanging value since R_a is in series with them. Instead of an ideal short across R_f as in the limiter, R_f is effectively paralleled by R_a. In other words, once the zener potential is reached, the gain drops to $(R_f \parallel R_a)/R_i$. This is shown in Figure 7.31. To be a bit more precise, the gain change is not quite as abrupt as this approximation. In reality, the transfer curve is smoother and somewhat delayed, as indicated by the dashed line in the figure.

In order to achieve an increasing slope, the zener/resistor combination is placed in parallel with R_i. When the zener potential is reached, R_a will effectively parallel R_i, reducing gain. This has the unfortunate side effect of reducing the

Figure 7.31
Transfer characteristic
(decreasing slope)

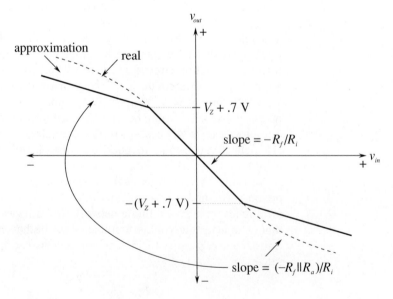

Figure 7.32

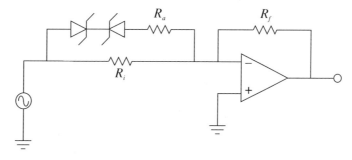

a. Simple function generator (increasing gain)

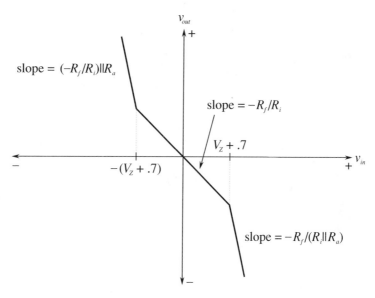

b. Transfer characteristic (increasing slope)

input impedance, so it may be necessary to include an input buffer. An example of such a circuit and the resulting transfer curve are shown above in Figure 7.32.

If more than two slopes are required, it is possible to parallel multiple zener/ resistor combinations. This technique can be used to create a piecewise linear approximation of a desired transfer characteristic. A two-section circuit is shown in Figure 7.33 on the next page. Note how the resistors are constantly placed in parallel, thus reducing gain. If this multiple-section technique is used with a circuit such as the one in Figure 7.32, we will increase the gain.

Figure 7.33

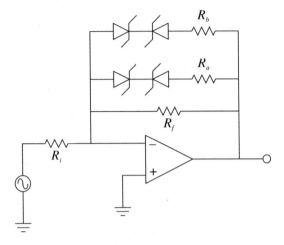

a. Multiple section function generator

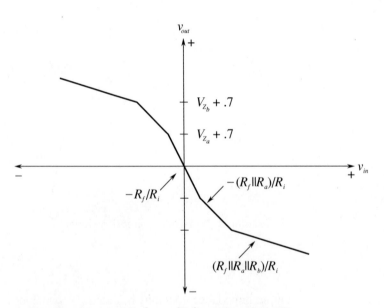

b. Transfer characteristic for multiple section generator

Example 7.4   Sketch the transfer curve for the circuit in Figure 7.34.

The first step is to note the breakpoints in the curve. Since this circuit uses a decreasing gain scheme with diodes across R_f, the output break points are set by the zener potentials. Also note that the break points are symmetrical

Figure 7.34
Function generator for
Example 7.4

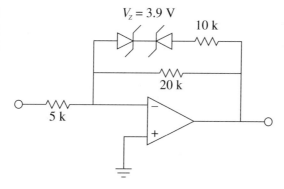

about zero since identical diodes are used.

$$V_{break} = \pm(V_Z + .7 \text{ V})$$

$$V_{break} = \pm(3.9 \text{ V} + .7 \text{ V})$$

$$V_{break} = \pm 4.6 \text{ V}$$

The next item to determine is the base voltage gain. This is the gain without the zener/resistor effect.

$$A_v = -\frac{R_f}{R_i}$$

$$A_v = -\frac{20 \text{ k}}{5 \text{ k}}$$

$$A_v = -4$$

The second level gain occurs when the diodes are on, placing R_a in parallel with R_f.

$$A_{v2} = -\frac{R_f \| R_a}{R_{i_{\,}}}$$

$$A_{v2} = -\frac{20 \text{ k} \| 10 \text{ k}}{5 \text{ k}}$$

$$A_{v2} = -1.33$$

The resulting plot is shown in Figure 7.35 (page 300). In order to find the breakpoint for the input signal (in this case only the output break is found), you need to divide the output break by the slope of the transfer curve. For

Figure 7.35
Transfer characteristic of
the circuit in Figure 7.34

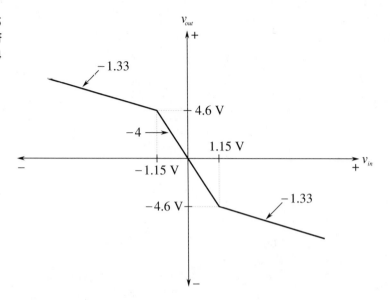

an output of 4.6 V, the input signal will be 4 times less at 1.15 V. If multiple
sections are used, the changes due to each section are added in order to find
the corresponding input signal. If the circuit is an increasing-gain type (as in
Figure 7.32), then the input break points are known, and the output points
are found by multiplying by the slope. Both of these items are illustrated in
the following example.

Example 7.5   Draw the transfer curve for the circuit of Figure 7.36.

Figure 7.36
Multiple section circuit for
Example 7.5

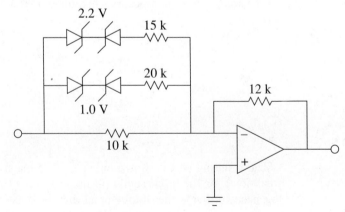

All diode pairs are identical, so symmetry is maintained. By inspection, the input breakpoints are:

$$V_{break1-in} = 1.0 \text{ V} + .7 \text{ V} = 1.7 \text{ V}$$

$$V_{break2-in} = 2.2 \text{ V} + .7 \text{ V} = 2.9 \text{ V}$$

The base gain is

$$A_v = -\frac{R_f}{R_i}$$

$$A_v = -\frac{12 \text{ k}}{10 \text{ k}}$$

$$A_v = -1.2$$

The second level gain is

$$A_{v2} = -\frac{R_f}{R_i \parallel R_a}$$

$$A_{v2} = -\frac{12 \text{ k}}{10 \text{ k} \parallel 20 \text{ k}}$$

$$A_{v2} = -1.8$$

The third level gain is

$$A_{v3} = -\frac{R_f}{R_i \parallel R_a \parallel R_b}$$

$$A_{v3} = -\frac{12 \text{ k}}{10 \text{ k} \parallel 20 \text{ k} \parallel 15 \text{ k}}$$

$$A_{v3} = -2.6$$

The corresponding output breaks are:

$$V_{break1-out} = A_v V_{break1-in}$$

$$V_{break1-out} = -1.2 \times 1.7$$

$$V_{break1-out} = -2.04 \text{ V}$$

$$V_{break2-out} = V_{break1-out} + A_{v2}(V_{break2-in} - V_{break1-in})$$

$$V_{break2-out} = -2.04 \text{ V} + -1.8(2.9 \text{ V} - 1.7 \text{ V})$$

$$V_{break2-out} = -4.2 \text{ V}$$

The resulting curve is shown in Figure 7.37 on the next page.

Figure 7.37
Characteristic of multiple-
section circuit of
Figure 7.36

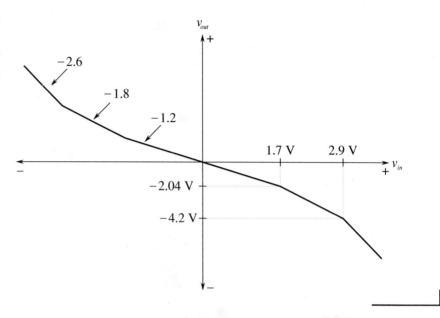

──────   **Example 7.6**   A temperature transducer's response characteristic is plotted in Figure 7.38. Unfortunately, the response is not consistent across a large temperature range. At very high or low temperatures, the device becomes more sensitive. Design a circuit which will compensate for this error, so that the output will remain at a sensitivity of 1 V/10C° over a large temperature range.

──────   **Figure 7.38**
Temperature transducer's
response for Example 7.6

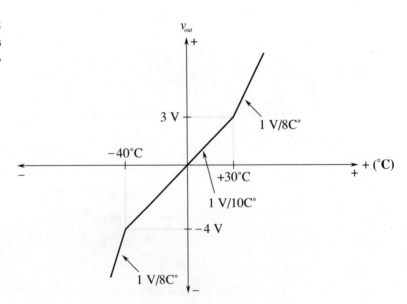

Figure 7.39
Desired transfer curve for
Example 7.6

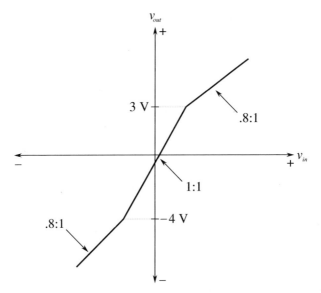

Figure 7.39

Desired transfer curve for Example 7.6

First of all, in order to compensate for an increasing curve, a mirror-image decreasing curve is called for. The circuit will need only a single diode section; however, it will not be symmetrical. The positive break occurs at 30°C (3 V), and the negative break occurs at -40°C (-4 V). In this range the gain will remain at unity. Outside of this range, the gain must fall. In order to bring a ratio of 1 V/8C° back to 1 V/10C°, it must be multiplied by its reciprocal—a gain of 8/10. The desired transfer curve is shown in Figure 7.39. If R_f and R_i are arbitrarily chosen to be 10 k, the resulting combination of $R_a \parallel R_f$ must be 8 k to achieve this gain. R_a can then be found:

$$R_a \parallel R_f = 8 \text{ k}$$

$$\frac{1}{R_a} + \frac{1}{10 \text{ k}} = \frac{1}{8 \text{ k}}$$

$$R_a = 40 \text{ k}$$

Since the break points have already been determined to be 3 V and -4 V, respectively, all that needs to be done is to compensate for the forward drop of the other diode.

$$V_{Z^+} = 3 \text{ V} - .7 \text{ V}$$

$$V_{Z^+} = 2.3 \text{ V}$$

$$V_{Z^-} = \left| -(4 \text{ V} - .7 \text{ V}) \right|$$

$$V_{Z^-} = \left| -3.3 \text{ V} \right|$$

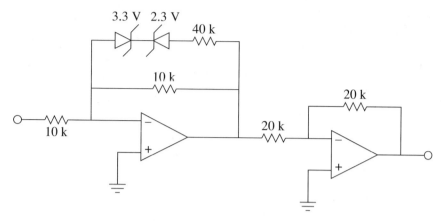

Since this circuit inverts the signal, it may be wise to include an inverting buffer to compensate. The resulting circuit is shown in Figure 7.40.

──────

One of the problems with the zener form is that the transfer curves change slope slowly. Because of this, fine control over the shape is not very easy to come by. Consequently, a good deal of trial and error is sometimes needed during the design sequence. Also, there is a limited range of zener diode values available. A form which is somewhat more exacting utilizes a biased-diode network.

An increasing gain-function circuit which utilizes a biased-diode approach is shown in Figure 7.41. Its corresponding transfer curve is shown in Figure 7.42 on page 306. For each breakpoint, one diode and a pair of resistors are required. Even if the breakpoints are symmetrical for positive and negative inputs, one set of components will be required for each polarity. The basic concept is the same as that of the zener approach: By turning on sections with higher and higher input voltages, resistors will be placed in parallel with the base feedback resistors, thus changing the gain. As with the zener approach, a decreasing gain characteristic is formed by placing the network across R_f, instead of R_i. Let's take a look at how a single section reacts.

Each breakpoint requires a blocking diode, a gain resistor, and a bias resistor. Together with diode D_1, resistors R_2 and R_3 make up a single section for negative input voltages. Under low signal conditions, D_1 is off and blocks current flow. Effectively, the section is an open circuit. If the input potential goes strongly negative enough, D_1 will turn on, thus placing R_2 in parallel with R_1 and increasing the gain. The slope (gain) of the transfer curve is

$$\text{Slope} = -\frac{R_f}{R_1 \parallel R_2}$$

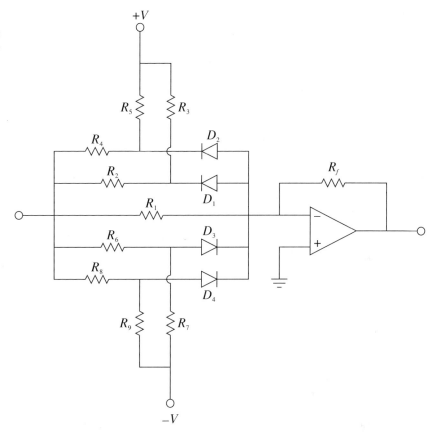

Figure 7.41
Biased diode function
generator

The breakpoint occurs when the cathode of D_1 goes to approximately $-.7$ V (assuming silicon). At this point, the drop across R_3 must be

$$V_{R_3} = V_{CC} + .7 \text{ V}$$

and the drop across R_2 must be

$$V_{R_2} = V_{in} - .7 \text{ V}$$

Just before D_1 conducts, the current through R_2 must equal the current through R_3.

$$\frac{V_{R_3}}{R_3} = \frac{V_{R_2}}{R_2}$$

$$\frac{V_{CC} + .7 \text{ V}}{R_3} = \frac{V_{in} - .7 \text{V}}{R_2}$$

Figure 7.42
Transfer characteristic of
biased diode function
generator

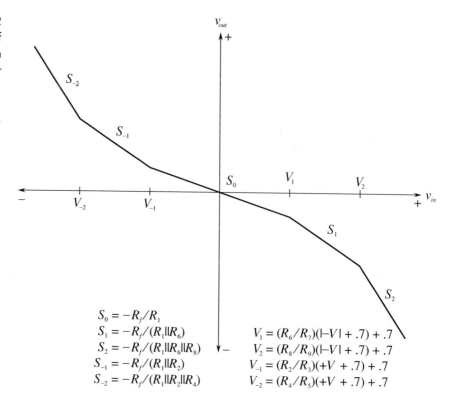

Figure 7.42
Transfer characteristic of
biased diode function
generator

$$S_0 = -R_f/R_1$$
$$S_1 = -R_f/(R_1\|R_6)$$
$$S_2 = -R_f/(R_1\|R_6\|R_8)$$
$$S_{-1} = -R_f/(R_1\|R_2)$$
$$S_{-2} = -R_f/(R_1\|R_2\|R_4)$$

$$V_1 = (R_6/R_7)(|-V| + .7) + .7$$
$$V_2 = (R_8/R_9)(|-V| + .7) + .7$$
$$V_{-1} = (R_2/R_3)(+V + .7) + .7$$
$$V_{-2} = (R_4/R_5)(+V + .7) + .7$$

At this instant, V_{in} equals the breakpoint. Solving for V_{in} gives us

$$V_{in} = V_{breakpoint} = \frac{R_2}{R_3}(V_{CC} + .7 \text{ V}) + .7 \text{ V}$$

or to a rough approximation,

$$V_{breakpoint} \approx \frac{R_2}{R_3} V_{CC}$$

The approximation error can be minimized by using germanium diodes. Note that since the breakpoints are set by a resistor ratio, very tight control of the curve is possible. When designing a function circuit, the required slopes dictate the values of the gain resistors. For ease of use the bias potentials are set to the positive and negative supply rails. Given these elements and the desired breakpoint voltages, the necessary bias resistor values can be calculated.

Figure 7.43
False turn-off spike

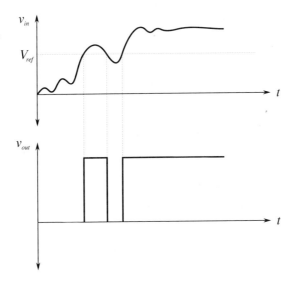

7.5 Comparators

The simple open-loop op amp comparator was discussed in Chapter 2. While this circuit is functional, it is not the final word on comparators. It suffers from two faults: 1) it is not particularly fast, 2) it does not use *hysteresis*. Hysteresis provides a margin of safety and "cleans up" switching transitions. Providing a comparator with hysteresis means that its reference depends on its output state. As an example, for a positive-going transition, the reference might be 2 V, but for a negative transition, the reference might be 1 V. This effect can dramatically improve performance when used with noisy inputs.

Take a look at the noisy signal in Figure 7.43. If this signal is fed into a simple comparator, the noise will produce a false turn-off spike. If this same signal is fed into a comparator with hysteresis, as in Figure 7.44 (page 308), a clean transition results. In order to change from low to high, the signal must exceed the upper reference. In order to change from high to low, the signal must drop below the lower reference. This is a very useful feature.

A comparator with hysteresis is shown in Figure 7.45 (page 308). The output voltage of this circuit will be either $+V_{sat}$ or $-V_{sat}$. Let's assume that the device is at $-V_{sat}$. To change to $+V_{sat}$, the inverting input must go lower than the noninverting input. The noninverting input is derived from the R_1/R_2 voltage divider. The divider is driven by the output of the op amp, in this case, $-V_{sat}$. Therefore, to change state, V_{in} must be

$$V_{in} = V_{lower\ thres} = -V_{sat} \frac{R_2}{R_1 + R_2} \tag{7.4}$$

Figure 7.44
Clean transition using
hysteresis

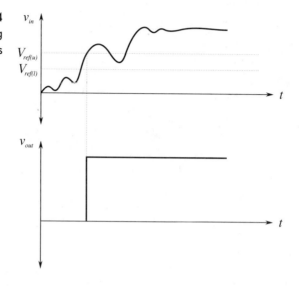

Figure 7.45
Comparator with
hysteresis

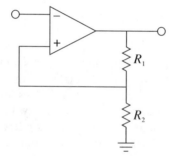

Similarly, to go from positive to negative,

$$V_{in} = V_{upper\ thres} = +V_{sat} \frac{R_2}{R_1 + R_2} \tag{7.5}$$

The upper and lower trip points are commonly referred to as the upper and lower thresholds. The adjustment of R_1 and R_2 creates an "error band" around zero (ground). Note that when the trip point is reached, the comparator output state changes, thus changing the reference. This reinforces the initial change. In effect, the comparator is now using positive feedback (note how the feedback signal is tied to the noninverting terminal). This circuit is sometimes referred to as a *Schmitt trigger*.

A noninverting version of the Schmitt trigger is shown in Figure 7.46. Note that in order to change state, the noninverting terminal's voltage will be ap-

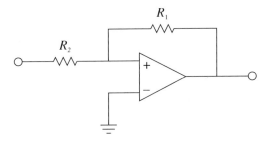

Figure 7.46
Noninverting comparator
with hysteresis

proximately zero. If the circuit is in the low state, the voltage across R_1 will equal $-V_{sat}$ at the time of transition. At this point, the voltage across R_2 will equal V_{in}. Since the current into the op amp is negligible, the current through R_1 equals that through R_2.

$$\frac{V_{R_1}}{R_1} = \frac{V_{R_2}}{R_2}$$

$$\frac{-(-V_{sat})}{R_1} = \frac{V_{in}}{R_2}$$

Since V_{in} equals the threshold voltage at transition,

$$V_{in} = V_{upper\ thres} = V_{sat}\frac{R_2}{R_1} \tag{7.6}$$

Similarly, for the opposite transition,

$$V_{in} = V_{lower\ thres} = -V_{sat}\frac{R_2}{R_1} \tag{7.7}$$

Example 7.7 Sketch the output waveform for the circuit of Figure 7.47 if the input signal is a 5 V peak sine wave.

Figure 7.47
Comparator for
Example 7.7

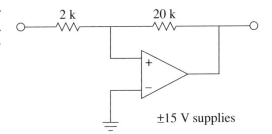

2 k   20 k

±15 V supplies

Figure 7.48
Comparator waveforms

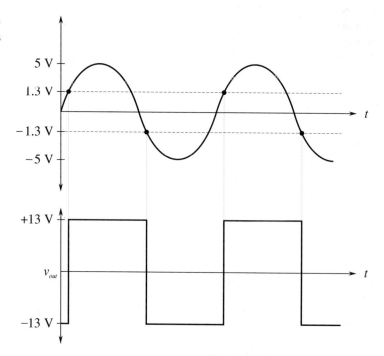

Figure 7.48
Comparator waveforms

First, determine the upper and lower threshold voltages.

$$V_{upper\ thres} = V_{sat}\frac{R_2}{R_1}$$

$$V_{upper\ thres} = 13\ \text{V}\ \frac{2\ \text{k}}{20\ \text{k}}$$

$$V_{upper\ thres} = 1.3\ \text{V}$$

$$V_{lower\ thres} = -V_{sat}\frac{R_2}{R_1}$$

$$V_{lower\ thres} = -13\ \text{V}\ \frac{2\ \text{k}}{20\ \text{k}}$$

$$V_{lower\ thres} = -1.3\ \text{V}$$

The output will go to $+13$ V when the input exceeds $+1.3$ V, and will go to -13 V when the input drops to -1.3 V. The input/output waveform sketches are shown in Figure 7.48.

While the use of positive feedback and hysteresis is a step forward, switching speed is still dependent on the speed of the op amp. Also, the output levels are approximately equal to the power supply rails, so interfacing to other circuitry (such as TTL logic) requires extra circuitry. To cure these problems, specialized comparator circuits have evolved. Generally, comparator ICs can be broken down into a few major categories: general purpose, high speed, and low power/low cost. A typical general-purpose device is the LM311. The LM360 is a high-speed device with differential outputs. Examples of the low-power variety include the LM393 dual comparator and the LM339 quad comparator. As was the case with ordinary op amps, there is a definite trade-off between comparator speed and power consumption. As you might guess, the high-speed LM360 suffers from the highest power consumption, while the more miserly LM393 and LM339 exhibit considerably slower switching speeds. High-speed devices often exhibit high input bias and offset currents as well.

The general-purpose LM311 is one of the more popular comparators in use today. An FET input version, the LF311, is also available. Generally, all of the op amp–based comparator circuits already mentioned can be adapted for use with the LM311. The LM311 is far more flexible than the average op amp comparator, though. An outline and data sheet for the LM311 are shown in Figures 7.49a and b.

First of all, note that the LM311 is reasonably fast, producing a response time of approximately 200 nS. This places it squarely in the midperformance range, 10 or so times faster than a low-power comparator, but at least 10 times slower than high-speed devices. The voltage gain is relatively high, at 200,000 typically. Input offset voltage is moderate at 2 mV typically, and 7.5 mV maximum, at room temperature. The input offset and bias currents are 50 nA and

Figure 7.49

Outline and data sheet for the LM311 comparator

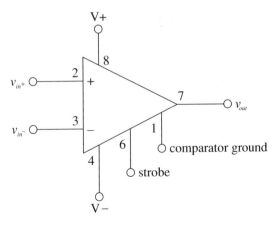

a. The LM311

(*Continued*)

Figure 7.49 (Continued)

Absolute Maximum Ratings for the LM111/LM211

If Military/Aerospace specified devices are required, contact the National Semiconductor Sales Office/Distributors for availability and specifications. (Note 7)

Total Supply Voltage (V_{84})	36V
Output to Negative Supply Voltage (V_{74})	50V
Ground to Negative Supply Voltage (V_{14})	30V
Differential Input Voltage	±30V
Input Voltage (Note 1)	±15V
Power Dissipation (Note 2)	500 mW
Output Short Circuit Duration	10 sec

Operating Temperature Range LM111	−55°C to 125°C
LM211	−25°C to 85°C
Storage Temperature Range	−65°C to 150°C
Lead Temperature (Soldering, 10 sec)	260°C
Voltage at Strobe Pin	V+ −5V

Soldering Information
Dual-In-Line Package
Soldering (10 seconds) . . . . 260°C
Small Outline Package
Vapor Phase (60 seconds) . . . . 215°C
Infrared (15 seconds) . . . . 220°C

See AN-450 "Surface Mounting Methods and Their Effect on Product Reliability" for other methods of soldering surface mount devices.

ESD rating to be determined.

Electrical Characteristics for the LM111 and LM211 (Note 3)

Parameter	Conditions	Min	Typ	Max	Units
Input Offset Voltage (Note 4)	$T_A = 25°C$, $R_S \le 50k$		0.7	3.0	mV
Input Offset Current (Note 4)	$T_A = 25°C$		4.0	10	nA
Input Bias Current	$T_A = 25°C$		60	100	nA
Voltage Gain	$T_A = 25°C$	40	200		V/mV
Response Time (Note 5)	$T_A = 25°C$		200		ns
Saturation Voltage	$V_{IN} \le -5$ mV, $I_{OUT} = 50$ mA $T_A = 25°C$		0.75	1.5	V
Strobe ON Current (Note 6)	$T_A = 25°C$	2.0	3.0	5.0	mA
Output Leakage Current	$V_{IN} \ge 5$ mV, $V_{OUT} = 35V$ $T_A = 25°C$, $I_{STROBE} = 3$ mA		0.2	10	nA
Input Offset Voltage (Note 4)	$R_S \le 50$ k			4.0	mV
Input Offset Current (Note 4)				20	nA
Input Bias Current				150	nA
Input Voltage Range	$V^+ = 15V$, $V^- = -15V$, Pin 7 Pull-Up May Go To 5V	−14.5	13.8,-14.7	13.0	V
Saturation Voltage	$V^+ \ge 4.5V$, $V^- = 0$ $V_{IN} \le -6$ mV, $i_{SINK} \le 8$ mA		0.23	0.4	V
Output Leakage Current	$V_{IN} \ge 5$ mV, $V_{OUT} = 35V$		0.1	0.5	μA
Positive Supply Current	$T_A = 25°C$		5.1	6.0	mA
Negative Supply Current	$T_A = 25°C$		4.1	5.0	mA

Note 1: This rating applies for ±15 supplies. The positive input voltage limit is 30V above the negative supply. The negative input voltage limit is equal to the negative supply voltage or 30V below the positive supply, whichever is less.

Note 2: The maximum junction temperature of the LM111 is 150°C, while that of the LM211 is 110°C. For operating at elevated temperatures, devices in the TO-5 package must be derated based on a thermal resistance of 150°C/W, junction to ambient, or 45°C/W, junction to case. The thermal resistance of the dual-in-line package is 110°C/W, junction to ambient.

Note 3: These specifications apply for $V_S = \pm 15V$ and Ground pin at ground, and $-55°C \le T_A \le +125°C$, unless otherwise stated. With the LM211, however, all temperature specifications are limited to $-25°C \le T_A \le +85°C$. The offset voltage, offset current and bias current specifications apply for any supply voltage from a single 5V supply up to ±15V supplies.

Note 4: The offset voltages and offset currents given are the maximum values required to drive the output within a volt of either supply with a 1 mA load. Thus, these parameters define an error band and take into account the worst-case effects of voltage gain and input impedance.

Note 5: The response time specified (see definitions) is for a 100 mV input step with 5 mV overdrive.

Note 6: Do not short the strobe pin to ground; it should be current driven at 3 to 5 mA.

Note 7: Refer to RETS111X for the LM111H, LM111J and LM111J-8 military specifications.

b. LM311 data sheet

250 nA worst case, respectively. Devices with improved offset and bias performance are available, but these values are acceptable for most applications. (For example, the FET input LF311 shows input offset and bias currents of about 1000 times less, with only a slight increase in worst-case input offset voltage.) The saturation voltage indicates exactly how low a "low state" potential really is. For a typical TTL-type load, the LM311 low output will be no more than .4 V. Finally, note the current draw from the power supplies. Worst-cast values are 7.5 mA and 5.0 mA, from the positive and negative supplies, respectively. By comparison, low-power comparators are normally in the 1 mA range, while high-speed devices can range up toward 20 to 30 mA.

The LM311 can be configured to drive TTL or MOS logic circuits, and loads referenced to ground, the positive supply or the negative supply. Finally, it can directly drive relays or lamps with its 50 mA output current capability.

The operation of the LM311 is as follows:

1. If $v_{in+} > v_{in-}$ the output goes to an open-collector condition. Therefore, a pull-up resistor is required to establish the high output potential. The pull-up does not have to go back to the same supply as the LM311. This aides in interfacing with various logic levels. For a TTL interface, the pull-up resistor will be tied back to the $+5$ V logic supply.

2. If $v_{in-} > v_{in+}$ the output will be shorted through to the "comparator ground" pin. Normally this pin goes to ground, indicating a low logic level of 0 V. It can be tied to other potentials if need be.

3. The strobe pin affects the overall operation of the device. Normally, it is left open. If the strobe pin is connected to ground through a current-limiting resistor, the output will go to the open-collector state regardless of the input levels. Normally, an LM311 is strobed with a logic pulse which turns on a switching transistor.

An LM311-based comparator circuit is shown in Figure 7.50 on page 314. It is powered by ± 15 V supplies so that the inputs are compatible with general op amp circuits. The output uses a $+5$ V pull-up source so that the output logic is TTL compatible. A small-signal transistor such as a 2N2222 is used to strobe the LM311. A logic low on the base of the transistor turns the transistor off, thus leaving the LM311 in normal operation mode. A high on the base will turn the transistor on, thus placing the LM311's output at a logic high.

An example of commercial use of the LM311 can be found in the Digital Sampling Musical Instrument application of Chapter 12. In this application, the comparator serves as part of an *analog-to-digital converter*.

Our final type of comparator circuit is the *window comparator*. The window comparator is used to determine if a particular signal is within an allowable range of levels. This circuit features two different threshold inputs, an *upper*

Figure 7.50
LM311 with strobe

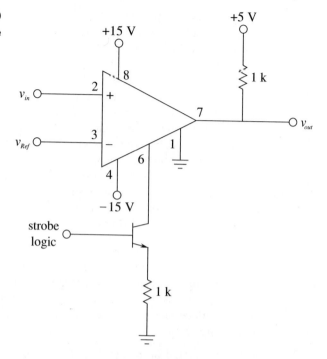

Figure 7.50
LM311 with strobe

threshold and a *lower threshold*. Do not confuse these items with the similarly-named levels associated with Schmitt triggers. A block diagram of a window comparator is shown in Figure 7.51. This circuit is comprised of two separate comparators with common inputs and outputs. As long as the input signal is between the upper and lower thresholds, the output of both comparators will

Figure 7.51
Window comparator

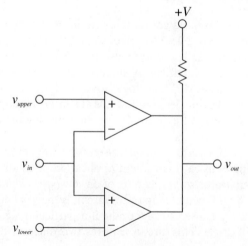

be high, thus producing a logic high at the circuit output. If the input signal is greater than the upper threshold or less than the lower threshold (i.e., outside the allowed window), then one of the comparators will short its output to comparator ground, producing a logic low. The other comparator will go to an open-collector condition. The net result is that the circuit output will be a logic low. Since the window comparator requires two separate comparators, a dual-comparator package such as the LM319 can prove to be convenient.

7.6 Log and Antilog Amplifiers

It is possible to design circuits with logarithmic response. The uses for this are manyfold. Recalling your logarithm fundamentals, remember that processes such as multiplication and division are performed by addition and subtraction of logs. Also, powers and roots are performed by multiplication and division. Bearing this in mind, if an input signal is processed by a logarithmic circuit, multiplied by a gain, A, and then processed by an antilog circuit, effectively the signal will have been raised to the Ath power. This brings to light a great many possibilities. For example, in order to take the square root of a value, the signal would be passed through a log circuit, divided by two (perhaps with something as simple as a voltage divider), and then passed through an antilog circuit. One possible application is in true RMS detection circuits. Other applications for log/antilog amplifiers include signal compression and process control. Signals are often compressed in order to decrease their dynamic range (the difference between the highest and lowest level signals). In telecommunications systems, this compression may be required in order to achieve reasonable voice or data transmission with limited resources. Seeing their possible uses, our question then, is how do we design log/antilog circuits?

Figure 7.52 shows a basic log circuit. The input voltage is turned into an input current by R_i. This current feeds the transistor. Note that the output voltage appears across the base-emitter junction, V_{BE}. The transistor is being

Figure 7.52
Basic log circuit

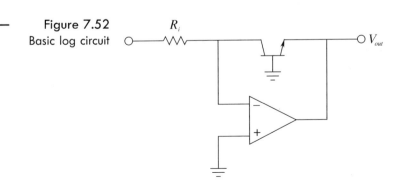

used as a current-to-voltage converter. The voltage/current characteristic of a transistor is logarithmic, thus the circuit produces a log response. In order to find an output equation, we start with the basic Shockley equation for PN junctions:

$$I_c = I_s(\varepsilon^{qV_{BE}/KT} - 1) \qquad (7.8)$$

where I_s is the reverse saturation current, ε is log base, q is the charge on one electron 1.6×10^{-9} coulombs, K is Boltzmann's constant 1.38×10^{-23} Joules per degree Kelvin, and T is the absolute temperature in degrees Kelvin.

Using 300°K (approximately room temperature) and substituting these constants into Equation 7.8 produces

$$I_c = I_s(\varepsilon^{38.6V_{BE}} - 1) \qquad (7.9)$$

Normally, the exponent term is much larger than 1, so this can be approximated as

$$I_c = I_s\varepsilon^{38.6V_{BE}} \qquad (7.10)$$

Using the inverse log relationship and solving for V_{BE}, this is reduced to

$$V_{BE} = .0259 \ln \frac{I_c}{I_s} \qquad (7.11)$$

Earlier it was noted that the current I_c is a function of the input voltage and R_i. Also, note that $V_{out} = -V_{BE}$. Substituting these elements into Equation 7.11 yields

$$V_{out} = -.0259 \ln \frac{V_{in}}{R_iI_s} \qquad (7.12)$$

We now have an amplifier which takes the log of the input voltage, and also multiplies the result by a constant. It is very important that the antilog circuit multiply by the reciprocal of this constant, or errors will be introduced. If the input voltage (or current) is plotted against the output voltage, the result will be a straight line when plotted on a semilog graph, as shown in Figure 7.53. The rapid transition at approximately .6 V is due to the fast turn-on of the transistor's base-emitter junction. If output voltages greater than .6 V are required, amplifying stages will have to be added.

There are several properties to note about this circuit. First, the range of input signals is small. Large input currents will force the transistor into less-than-ideal logarithmic operation. Also, the input signal must be unipolar. Finally, the circuit is rather sensitive to temperature changes. Try using a

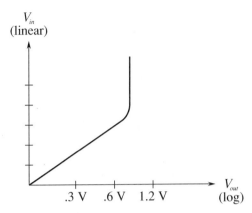

Figure 7.53
Input/output characteristic
of log amplifier

slightly different value for T in the above derivation, and you will see a sizable
change in the resulting constant.

A basic antilog amplifier is shown in Figure 7.54. Note that the transistor
is used to turn the input voltage into an input current, with a log function.
This current then feeds R_f, which produces the output voltage. The derivation
of the input/output equation is similar to the log circuit's:

$$V_{out} = -R_f I_c$$

Recalling Equation 7.10,

$$I_c = I_s \varepsilon^{38.6 V_{BE}}$$

$$V_{out} = -R_f I_s \varepsilon^{38.6 V_{BE}} \tag{7.13}$$

The same comments about stability and signal limits apply to both the
antilog amplifier and the log amplifier. Also, there is one interesting observation
worth remembering: in a general sense, the response of the op amp system
echoes the characteristics of the elements used in the feedback network. When
only resistors (which have a linear relation between voltage and current) are
used, the resulting amplifier exhibits a linear response. If a logarithmic PN
junction is used, the result is an amplifier with a log or antilog response.

Figure 7.54
Basic antilog circuit

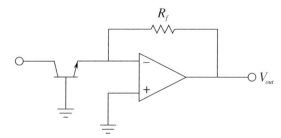

Example 7.8   For the circuit of Figure 7.52, assume that $R_i = 50$ k, $I_s = 30$ nA, and $T = 300°$K. Determine the output voltage for inputs of 1 V, .5 V, and 2 V.

For $V_{in} = 1$ V:

$$V_{out} = -.0259 \ln \frac{V_{in}}{R_i I_s}$$

$$V_{out} = -.0259 \ln \frac{1 \text{ V}}{50 \text{ k } 30 \text{ nA}}$$

$$V_{out} = -.0259 \ln 666.6$$

$$V_{out} = -.1684 \text{ V}$$

For $V_{in} = .5$ V:

$$V_{out} = -.0259 \ln \frac{.5 \text{ V}}{50 \text{ k } 30 \text{ nA}}$$

$$V_{out} = -.0259 \ln 333.3$$

$$V_{out} = -.1504 \text{ V}$$

For $V_{in} = 2$ V:

$$V_{out} = -.0259 \ln \frac{2 \text{ V}}{50 \text{ k } 30 \text{ nA}}$$

$$V_{out} = -.0259 \ln 1333$$

$$V_{out} = -.1864 \text{ V}$$

Notice that for each doubling of the input signal, the output signal went up a constant 18 mV. Such is the nature of a log amplifier.

As noted, the basic log/antilog forms presented have their share of problems. It is possible to create more elaborate and accurate designs, but generally, circuit design of this type is not for the faint-hearted. The device-matching and temperature-tracking considerations are not minor.

Some manufacturers supply relatively simple-to-use log/antilog circuits in IC form. This takes much of the "grind" out of log circuit design. An example device is the 4127 from Burr-Brown. An equivalent circuit for the 4127 is shown in Figure 7.55. This device is a hybrid IC. The circuit uses matched transistors to increase the circuit accuracy. A recovery amplifier and a zener-regulated current source round out the package. The 4127 accepts bipolar input signals and has an input current dynamic range of $10^6:1$. With the addition of just a few external resistors, the device can be configured to act as either a log or an antilog amplifier. Other circuits are also possible. Further details on the design and application of higher-quality log amplifiers (such as the 4127) can be found in the Extended Topic section at the end of this chapter.

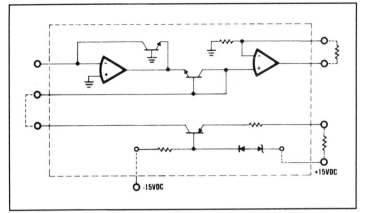

Reprinted, in whole or in part, with the permission of Burr-Brown Corporation

Four-Quadrant Multipliers

A *four-quadrant multiplier* is a device with two inputs and a single output. The output potential is the product of the two inputs along with a scaling factor, K.

$$v_{out} = Kv_x v_y \tag{7.14}$$

Typically, K is .1 in order to minimize the possibility of output overload. The schematic symbol for the multiplier is shown in Figure 7.56. It is called a four-quadrant device since both inputs and the output may be positive or negative. An example device is the Motorola MC1494. The 1494 utilizes mono-lithic construction and can be powered by standard ± 15 V supplies. While multipliers are not really nonlinear in and of themselves, they can be used in a variety of out-of-the-ordinary applications, and in areas where log amps might be used.

Multipliers have many uses including squaring, dividing, balanced modulation/demodulation, frequency modulation, amplitude modulation, and automatic gain control. The most basic operation, multiplication, involves using one input as the signal input, and the other input as the gain control potential. Unlike the simple VCA circuit discussed earlier, this gain control potential is allowed to swing both positive and negative. A negative polarity will produce

Figure 7.56
Schematic symbol for multiplier

Figure 7.57
Multiplication circuit

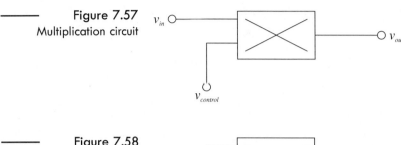

Figure 7.58
Squaring circuit

an inverted output. This basic connection is shown in Figure 7.57. This same circuit can be used as a balanced modulator. This is very useful for creating dual sideband signals for communications work. Multiplier ICs may have external connections for scale factor and offset adjust potentiometers. Also, they can be modeled as current sources, so an external op amp connected as a current-to-voltage converter may be required.

If the two inputs are tied together and fed from a single input, the result will be a squaring circuit. Squaring circuits can be very useful for RMS calculations and for frequency doubling. An example is shown in Figure 7.58.

The multiplier can also be used for division or square root functions. A divider circuit is shown in Figure 7.59. Here's how it works: First, note that the output of the multiplier, v_m, is a function of v_x and the output of the op amp, v_{out}.

$$v_m = K v_x v_{out} \tag{7.15}$$

Figure 7.59
Divider circuit

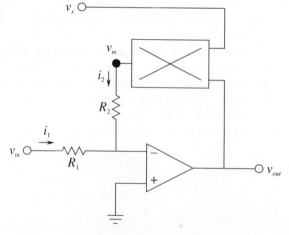

v_m produces a current through R_2. Note that the bottom end of R_2 is at a virtual ground, so that all of v_m drops across R_2.

$$i_2 = \frac{v_m}{R_2}$$

As we have already seen,

$$i_1 = \frac{v_{in}}{R_1}$$

Now, since the current into the op amp is assumed to be zero, i_1 and i_2 must be equal and opposite. It is apparent that the arbitrary direction of i_2 is negative. Therefore, if R_1 is set equal to R_2,

$$v_m = -v_{in}$$

Using Equation 7.15,

$$-v_{in} = Kv_x v_{out}$$

$$v_{out} = -\frac{v_{in}}{Kv_x} \qquad (7.16)$$

Note that v_x sets the magnitude of the division. There are definite size restrictions on v_x. If it is too small, output saturation will result. In a similar vein, if v_x is also tied back to the op amp's output, a square-root function will result. Picking up the derivation at the next-to-last step,

$$-v_{in} = Kv_{out} v_{out}$$

$$v_{out} = \sqrt{\frac{-v_{in}}{K}} \qquad (7.17)$$

Example 7.9 Determine the output voltage in Figure 7.58 if $K = .1$ and the input signal is $2 \sin 2\pi 60t$.

$$v_{out} = Kv_x v_y$$

Since both inputs are tied together, this reduces to

$$v_{out} = Kv_{in}^2$$

$$v_{out} = .1(2 \sin 2\pi 60t)^2 \qquad (7.18)$$

Substituting the following basic trig identity

$$(\sin \omega)^2 = .5 - .5 \cos 2\omega$$

into Equation 7.18,

$$v_{out} = .1(1 - 1 \cos 2\pi 120t)$$

$$v_{out} = .1 - .1 \cos 2\pi 120t$$

This result means that the output signal is .1 V peak, and is riding on a .1 V DC offset. It also indicates a phase shift of $-90°$ and a doubling of the input frequency. This last attribute makes this connection very useful. We have seen many approaches to increasing a signal's amplitude, but this circuit allows us to increase a signal's frequency. The waveforms are shown in Figure 7.60.

Figure 7.60
Waveforms of the squaring circuit for Example 7.9

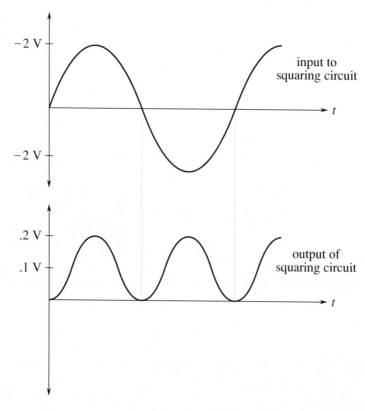

7.7 Extended Topic: A More Exacting Log Amplifier

The basic log amplifier discussed earlier suffers from two major problems: repeatability and temperature sensitivity. Let's take a closer look at Equation 7.12. First, if you work backward through the derivation, you will notice that the constant .0259 is actually KT/q using a temperature of 300°K. We can rewrite this equation as

$$V_{out} = -\frac{KT}{q} \ln \frac{V_{in}}{R_i I_s} \tag{7.19}$$

Note that the output voltage is directly proportional to the circuit temperature, T. Normally, this is not desired. The second item of interest is I_s. This current can vary considerably between devices, and is also temperature-sensitive, approximately doubling for each 10C° rise. For these reasons, commercial log amps, such as the 4127 mentioned earlier, are preferred. This does not mean that it is impossible to create stable log amps from the basic op amp building blocks. To the contrary, a closer look at practical circuit solutions will point out why ICs such as the 4127 are sucessful. We shall treat the two problems separately.

One way of removing the effect of I_s from our log amp is to subtract an equal effect. The circuit of Figure 7.61 does exactly this. This circuit utilizes two log amps. While each is drawn with an input resistor and input voltage

Figure 7.61   A high-quality log amplifier

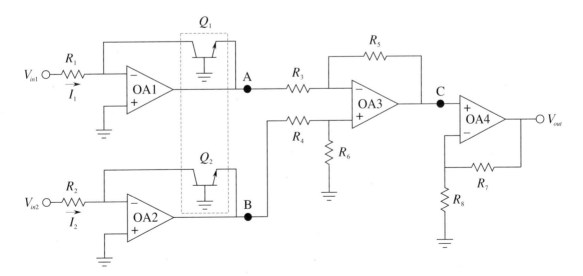

source, removal of R_1 and R_2 would allow current sensing inputs. Each log amp's transistor is part of a *matched pair*. These two transistors are fabricated on the same silicon wafer and exhibit nearly identical characteristics (in our case, the most notable is I_s). Based on Equation 7.19 we find

$$V_A = -\frac{KT}{q} \ln \frac{V_{in1}}{R_1 I_{s1}}$$

$$V_B = -\frac{KT}{q} \ln \frac{V_{in2}}{R_2 I_{s2}}$$

where, typically, $R_1 = R_2$. These two signals are fed into a differential amplifier comprised of op amp 3 and resistors R_3 through R_6. It is normally set for a gain of unity. The output of the differential amplifier (point C) is

$$V_C = V_B - V_A$$

$$V_C = -\frac{KT}{q} \ln \frac{V_{in2}}{R_2 I_{s2}} - \left(-\frac{KT}{q} \ln \frac{V_{in1}}{R_1 I_{s1}} \right)$$

$$V_C = \frac{KT}{q} \left(\ln \frac{V_{in1}}{R_1 I_{s1}} - \ln \frac{V_{in2}}{R_2 I_{s2}} \right) \tag{7.20}$$

Using the basic identity that subtracting logs is the same as dividing their arguments, Equation 7.20 becomes

$$V_C = \frac{KT}{q} \ln \frac{\dfrac{V_{in1}}{R_1 I_{s1}}}{\dfrac{V_{in2}}{R_2 I_{s2}}} \tag{7.21}$$

Since R_1 is normally set equal to R_2, and I_{s1} and I_{s2} are identical because Q_1 and Q_2 are matched devices, Equation 7.21 simplifies to

$$V_C = \frac{KT}{q} \ln \frac{V_{in1}}{V_{in2}} \tag{7.22}$$

As you can see, the effect of I_s has been removed. V_C is a function of the ratio of the two inputs. Therefore, this circuit is called a *log ratio amplifier*. The only remaining effect is that of temperature variation. Op amp 4 is used to compensate for this. This stage is little more than a standard SP noninverting amplifier. What makes it unique is that R_8 is a temperature-sensitive resistor. This component has a positive temperature coefficient of resistance, meaning that as temperature rises, so does its resistance. Since the gain of this stage is $1 + R_7/R_8$, a temperature rise causes a decrease in gain. Combining this with

Equation 7.22 produces

$$V_{out} = \left(1 + \frac{R_7}{R_8}\right)\frac{KT}{q}\ln\frac{V_{in1}}{V_{in2}} \tag{7.23}$$

If the temperature coefficient of R_8 is chosen correctly, the first two temperature-dependent terms of Equation 7.23 will cancel, leaving a temperature-stable circuit. This coefficient is approximately $1/300°K$, or $.33\%$ per Centigrade degree in the vicinity of room temperature.

Our log ratio circuit is still not complete. While the major stability problems have been eliminated, other problems do exist. It is important to note that the transistor used in the feedback loop loads the op amp just as the ordinary feedback resistor R_f would. The difference lies in the fact that the effective resistance which the op amp sees is the dynamic base-emitter resistance, r'_e. This resistance varies with the current passing through the transistor, and has been found to equal 26 mV/I_E at room temperature. For higher input currents this resistance can be very small, and might lead to an overload condition. A current of 1 mA, for example, would produce an effective load of only 26 Ω. This problem can be alleviated by inserting a large resistance value in the feedback loop. This resistor, labelled R_E, is shown in Figure 7.62. An appropriate value for R_E can be found by realizing that at saturation, virtually all of the output potential will be dropped across R_E, save V_{BE}. The current through R_E is the maximum expected input current plus load current. Using Ohm's Law,

$$R_E = \frac{V_{sat} - V_{BE}}{I_{max}} \tag{7.24}$$

Figure 7.62 also shows a compensation capacitor, C_c. This capacitor is used to roll off high-frequency gain in order to supress possible high-frequency oscillations. An optimum value for C_c is not easy to determine since the feedback

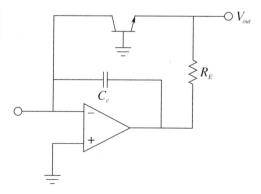

Figure 7.62
Compensation components

Figure 7.63
Light transmission
measurement

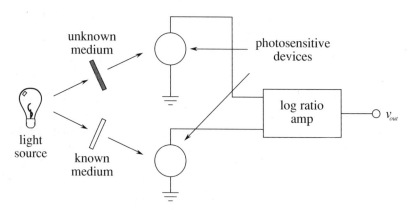

element's resistance changes with the input level. It can be found empirically in the laboratory. A typical value will be in the vicinity of 100 pF.[1]

One possible application for the log ratio circuit is found in Figure 7.63. This system is used to measure the light transmission of a given material. Since variations in the light source will affect a direct measurement, the measurement is instead made *relative to a known material*. In this example, light is passed through a known medium (such as a vacuum) while it is simultaneously passed through the medium under test. On the far side of both materials are light-sensitive devices such as photodioides, phototransistors, or photomultiplier tubes. These devices produce currents proportional to the amount of light striking them. In this system, these currents are fed into the log ratio circuit, which will then produce an output voltage proportional to the light transmission abilities of the new material. This setup eliminates the problem of light-source fluctuation because each input will see an equal percentage change in light intensity. Effectively, this is a common-mode signal which is suppressed by the differential amplifier section. This system also eliminates the difficulty of generating calibrated light-intensity readings for comparisons. By its very nature, this system performs relative readings.

Summary

In this chapter, you have examined a variety of nonlinear op amp circuits. They are deemed nonlinear because their input/output characteristics are not a straight line.

---

[1] A detailed derivation for C_c can be found in *Nonlinear Circuits Handbook*, published by Analog Devices, 2nd edition, pp. 174–178.

Precision rectifiers use negative feedback to compensate for the forward drop of diodes. By doing so, very small signals can be successfully rectified. Both half-wave and full-wave versions are available. An extension of the precision rectifier is the precision peak detector. This circuit uses a capacitor to effectively lengthen the duration of peaks. It can also be used as an envelope detector.

Other active diode circuits include clampers and limiters. Clampers adjust the DC level of an input signal so that the output is either always positive or always negative. The furthest peaks will just hit 0 V. Clampers can also be used with a DC offset for greater control of the output signal. Limiters force their output to be no greater than a preset value. They can be thought of as programmable signal clippers.

Function-generation circuits are used to create piecewise linear approximations of transfer functions. These can be used to compensate for the nonlinearities of transducers and measurement devices. There are two basic approaches for realizing a design: 1) using zener diodes, 2) using a biased-signal diode network. The zener scheme is a bit easier to set up, but does not offer the performance of the biased diode form.

In order to improve the performance of comparators, positive feedback can be used. The resulting circuits are called Schmitt triggers, and have much greater noise immunity than do simple open-loop op amp comparators. Specialized comparators are also available. These devices are generally faster than simple op amps, and offer more flexibility for logic family interfacing.

Log and antilog amplifiers utilize a transistor in their feedback loops. The result is a form of signal compression for the log amp, and an inverse signal expansion for the antilog amp. Specialized log/antilog ICs are available which take much of the tedium out of amplifier design and testing.

Finally, a four-quadrant multiplier produces an output which is proportional to the product of its two inputs. These devices can be used for a number of applications including balanced modulation, frequency doubling, and more.

In short, while op amps are linear devices, they can be used with other components to form nonlinear circuits.

Self Test Questions

1. What are the advantages of active rectifiers versus passive rectifiers?
2. What are the disadvantages of active rectifiers versus passive rectifiers?
3. What is a peak detector?
4. What is a limiter?

5. What is the function of a clamper?

6. What are the differences between active and passive clampers?

7. What is a transfer function generator circuit, and what is its use?

8. Explain how a function circuit might be used to linearize an input transducer.

9. What is a Schmitt trigger?

10. What is the advantage of a Schmitt-type comparator versus an ordinary open-loop op amp comparator?

11. What are the advantages of dedicated comparators such as the LM311 versus ordinary op amp comparators?

12. How are log and antilog amplifiers formed?

13. What is the effect of passing a signal through a log or antilog amplifier?

14. What is a four-quadrant multiplier?

15. How might a multiplier be used?

16. What is the difference between a multiplier and a VCA?

Problem Set

▷ Analysis Problems

1. Sketch the output of Figure 7.2 if the input is a 100 Hz 2 V peak triangle wave.

2. Repeat Problem 1 for a square wave input.

3. Sketch the output of Figure 7.7 if $R_f = 20$ k and $R_i = 5$ k. Assume that v_{in} is a .5 V peak triangle wave.

4. Repeat Problem 3, but with the diode reversed.

5. Sketch the output of Figure 7.9 if $C = 50$ nF and $R = 5$ M. Assume that the input is a 1 V peak pulse with 10% duty cycle. The input frequency is 1 kHz.

6. Repeat Problem 5 with the diode reversed.

7. Sketch the output of Figure 7.14 if $R = 25$ k and $v_{in} = 2 \sin 2\pi 500t$.

8. Repeat Problem 7 with v_{in} equal to a 3 V peak square wave.

9. Sketch the output of Figure 7.15 if the diodes are reversed. Assume that v_{in} is a 100 mV peak-to-peak sine wave at 220 Hz.

10. In Figure 7.23, assume that $R = 10$ k, $R_L = 1$ M, $C = 100$ nF, and $V_{offset} = 2$ V. Sketch the output for a 10 kHz 1 V peak sine wave input.

11. Sketch the output of Figure 7.26 if $R_i = 10$ k, $R_f = 40$ k and the zener potential is 3.9 V. The input signal is a 2 V peak sine wave.

12. Sketch the transfer curve for the circuit of Problem 11.

13. For Figure 7.30, sketch the transfer curve if $R_i = 5$ k, $R_f = 33$ k, $R_a = 20$ k and $V_Z = 5.7$ V.

14. Sketch the output of the circuit in Problem 13 for an input signal equal to a 2 V peak triangle wave.

15. Repeat Problem 14 with a square-wave input.

16. If $R_i = 4$ k, $R_a = 5$ k, $R_f = 10$ k, and $V_Z = 2.2$ V in Figure 7.32, determine the minimum and maximum input impedance.

17. Draw the transfer curve for the circuit of Problem 16.

18. Sketch the output voltage for the circuit of Problem 16 if the input signal is a 3 V peak triangle wave.

19. Sketch the transfer curve for the circuit of Figure 7.33 if $R_i = 1$ k, $R_f = 10$ k, $R_a = 20$ k, $R_b = 18$ k, $V_{Z_a} = 3.9$ V, and $V_{Z_b} = 5.7$ V.

20. Sketch the output of the circuit in Problem 19 if the input is a 1 V peak triangle wave.

21. If $R_1 = 10$ k and $R_2 = 33$ k in Figure 7.45, determine the upper and lower thresholds if the power supplies are ± 15 V.

22. Determine the upper and lower thresholds for Figure 7.46 if $R_1 = 4.7$ k and $R_2 = 2.2$ k, with ± 12 V power supplies.

23. Sketch the output of Figure 7.50 if $v_{in} = 2 \sin 2\pi 660t$, $V_{ref} = 1$ V DC, and $V_{strobe} = 5$ V DC.

24. Repeat Problem 23 for $V_{strobe} = 0$ V DC.

25. Determine the output of Figure 7.52 if $V_{in} = .1$ V, $R_i = 100$ k, and $I_s = 60$ nA.

26. Determine the output of Figure 7.54 if $V_{in} = 300$ mV, $R_f = 20$ k, and $I_s = 40$ nA.

27. Sketch the output signal of Figure 7.57 if $v_{in} = 2 \sin 2\pi 1000t$, K $= .1$, and $V_{control} = 1$ V DC, -2 V DC, and 5 V DC.

28. Sketch the output of Figure 7.58 if $v_{in} = 1 \sin 2\pi 500t$ and K $= .1$.

29. Sketch the output of Figure 7.59 if $v_{in} = 5 \sin 2\pi 2000t$, K $= .1$, $v_x = 4$ V DC, and $R_1 = R_2 = 10$ k.

⊳ **Design Problems**

30. Determine values of R and C in Figure 7.9 so that stage 1 slewing is at least 1 V/μS, along with a time constant of 1 mS. Assume that op amp 1 can produce at least 20 mA of current.

31. In Figure 7.21, assume that $C = 100$ nF. Determine an appropriate value for R_L if the input signal is at least 2 kHz. Use a time constant factor of 100 for your calculations.

32. Determine new values for R_a and R_f in Problem 13 such that the slopes are -5 and -3.

33. Determine new resistor values for the circuit of Figure 7.32 such that the slopes are -10 and -20. The input impedance should be at least 3 k.

34. Determine the resistor values required in Figure 7.36 to produce slopes of -5, -8, and -12, if the input impedance must be at least 1 k.

35. Determine the resistor values required in Figure 7.41 for $S_0 = -1$, $S_1 = S_{-1} = -5$, $S_2 = S_{-2} = -12$. Also, $V_1 = |V_{-1}| = 3$ V, and $V_2 = |V_{-2}| = 4$ V. Use $R_f = 20$ k.

36. Sketch the transfer curve for Problem 35.

37. Determine a value for R_i in Figure 7.52 such that a 3 V input will produce an output of .2 V. Assume $I_s = 60$ nA.

38. Determine a value for R_f in Figure 7.54 such that a .2 V input will produce a 3 V output. Assume $I_s = 40$ nA.

39. If the multiplier of Figure 7.59 can produce a maximum current of 10 mA, what should the minimum sizes of R_1 and R_2 be? (Assume ± 15 V supplies.)

⊳ **Challenge Problems**

40. Assuming that 5% resistors are used in Figure 7.14, determine the worst-case mismatch between the two halves of the rectified signal.

41. Design a circuit which will light an LED if the input signal is beyond ± 5 V peak. Make sure that you include some form of pulse-stretching element so that the LED remains visible for short duration peaks.

42. A pressure transducer produces an output of 500 mV per atmosphere up to 5 atmospheres. From 5 to 10 atmospheres, the output is 450 mV per atmosphere. Above 10 atmospheres, the output falls off to 400 mV per atmosphere. Using the zener form, design a circuit to linearize this response.

43. Repeat the problem above using the biased-diode form.

44. Design a circuit to produce the transfer characteristic shown in Figure 7.64.

—————— **Figure 7.64**

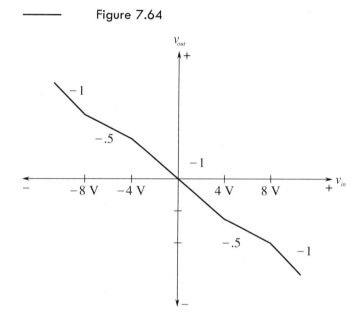

45. Sketch the output waveform of Figure 7.46 if $R_1 = 22$ k, $R_2 = 4.7$ k, and $v_{in} = 2 + 8 \sin 2\pi 120t$. Assume that the power supplies are ± 15 V.

46. Determine the output of Figure 7.56 if $K = .1$, $v_x = 1 \sin 2\pi 440t$, and $v_y = 3 \sin 2\pi 200{,}000t$.

⇥ **SPICE Problems**

47. In preceding work it was noted that the use of an inappropriate device model can produce SPICE simulations which are utterly inaccurate. An easy way to see this is to simulate the circuit of Figure 7.2 using an accurate model (such as the 741 simulation presented in this chapter), and a simple model, such as the controlled-voltage–source version presented in Chapter 2. Run two simulations of this circuit for each of these models. One simulation should use a lower input frequency, such as 1 kHz, while the second simulation should use a higher frequency where the differences in the models will be very apparent, such as 500 kHz.

48. Simulate the circuit designed in Problem 30 using a square wave input. Perform the simulation for several different input frequencies between 10 kHz and 10 Hz. Do the resulting waveforms exhibit the proper shapes?

49. Perform SPICE simulations for the circuit of Figure 7.14. Use R = 12 k. For the input waveform, use both sine and square waves, each 1 V peak at 200 Hz.

50. Component tolerances can directly affect the rectification accuracy of the full-wave circuit shown in Figure 7.14. This effect is easiest to see using a square wave input. If the positive and negative portions of the input signal see identical gains, the output of the circuit will be a DC level. Any inaccuracy or mismatch will produce a small square wave riding on this DC level. Simulate this effect using the SPICE Monte Carlo analysis option. If Monte Carlo analysis is not available on your system, you can still see the effect by manually altering the resistor values within a preset tolerance band for each of a series of simulation runs.

51. Use SPICE to verify the results of the limiter examined in Example 7.3.

52. Use SPICE to verify the response of the circuit shown in Figure 7.36. For the stimulus, use a low-frequency triangle wave of 4 V peak amplitude. Does the resulting waveform exhibit the same hard "corners" shown on the transfer curve of Figure 7.37?

8 Voltage Regulation

Chapter Objectives

After completing this chapter, you should be able to:

❑ Understand the need and usefulness of voltage regulators.

❑ Detail the difference between load and line regulation.

❑ Detail the need for a stable reference voltage.

❑ Describe the operation of a basic linear regulator.

❑ Describe the operation of a basic switching regulator.

❑ Outline the differences between linear and switching regulators, giving the advantages and disadvantages of each.

❑ Detail the function of a pass transistor in both linear and switching regulators.

❑ Utilize many popular linear IC regulators, such as the 78XX and 723.

❑ Analyze system heat sink requirements.

8.1 Introduction

In this chapter, the need for voltage regulation is examined. The different schemes for achieving voltage regulation, and the typical ICs used, are presented. By the end of this chapter, you should be able to use standard voltage regulator devices in your work, and understand the advantages and disadvantages of the various types.

Generally, voltage regulators are used to keep power supply potentials constant in spite of changes in load current or source voltage. Without regulation, some circuits may be damaged by the fluctuations present in the power supply voltage. Even if devices are not damaged, the fluctuations may degrade circuit performance.

Two general forms are presented, the *linear regulator* and the *switching regulator*. Both forms have distinct advantages and drawbacks. The trick is to figure out which form works best in a given situation. Due to their wide use in modern power-supply design, many regulators have achieved high levels of internal sophistication and robust performance. Often, the inclusion of voltage-regulator ICs is a very straightforward—almost cookbook—affair.

The chapter concludes with a discussion of heat sink theory and application. Since regulators are called upon to dissipate a bit of power, attention to thermal considerations is an important part of the design process.

8.2 The Need For Regulation

Modern electronic circuits require stable power-supply voltages. Without stable potentials, circuit performance may degrade, or if variations are large enough, the circuit may cease to function altogether and various components may be destroyed. There are many reasons why a power supply may fluctuate. No matter what the cause, it is the job of the regulator to absorb the fluctuations, and thereby maintain constant operating potentials.

A basic power-supply circuit is shown in Figure 8.1. First, a transformer is used to isolate the circuit from the AC power source. It is also used to step down (or step up) the AC source potential to a more manageable level. A rectifier then converts the scaled AC signal to pulsating DC, in the form of either half-wave, or more typically, full-wave rectification. The variations in the pulsating DC signal are then filtered out in order to produce a (hopefully) stable DC potential, which feeds the load. The filter can take the form of a simple capacitor, or possibly a more complex network comprising both capacitors and inductors.

Figure 8.1 Basic power supply

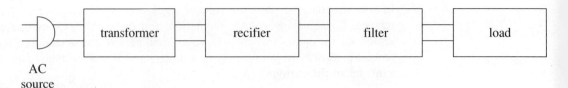

AC
source

There are two main causes of power-supply output variation. First, if the AC source signal changes, a proportional change will be seen at the output. If, for example, a *brown out*[1] occurs, the nominal 120 V AC source used in the U.S. may drop to say, 100 V AC. This represents a decrease of one sixth, or about 16.7%. This same change will be reflected by the transformer, so the output of the transformer will be about 16.7% low as well. This reduction carries right through the rectifier and filter to the load. In some circuits, this will not present a major problem. For example, this may mean that an op amp will be running off a 12 V supply instead of a 15 V supply. On the other hand, a TTL logic circuit may not operate correctly if only 4 V instead of 5 V are applied. Of course, if an overvoltage occurred, both the op amp and the logic circuit could be damaged.

How well a circuit handles variations in the AC line signal is denoted by the parameter *line regulation*.

$$\text{Line Regulation} = \frac{V_{max} - V_{min}}{V_{min}} \qquad (8.1)$$

where V_{max} is the load voltage produced at the maximum AC line potential and V_{min} is the load voltage for the lowest AC line potential. Normally, this value is expressed as a percentage. Ideally, a power supply will always produce the same output potential, and therefore, the perfect line regulation figure would be 0%.

The other major source of supply variation is variation of load current. Indeed, load current variations can be far, far greater than the usual variations seen in the AC line. This normally has the effect of increasing or decreasing the amount of AC *ripple* seen in the output voltage. Ideally, there would be no ripple in the output. Ripple is caused when heavy load currents effectively reduce the discharge-time constant of the filter. The result is that the filter gives up its stored energy more quickly, and cannot successfully "fill in the gaps" of the pulsating DC signal fed to it. The ripple signal is effectively an AC signal which rides on the DC output. This rapid variation of supply potential can find its way into an audio or signal processing path, and create a great deal of interference. Along with this variation, the effective DC value of the supply can drop as the load current is increased. The figure of merit for stability in spite of load changes is called *load regulation*.

$$\text{Load Regulation} = \frac{V_{max} - V_{min}}{V_{min}} \qquad (8.2)$$

---

[1] A reduction in the standard supplied potential may be introduced by a power utility as a way of dealing with particularly heavy load conditions, such as on a particularly hot summer's day when numerous air conditioners are running.

──────
Figure 8.2
Two forms of regulator
control elements: series
and parallel

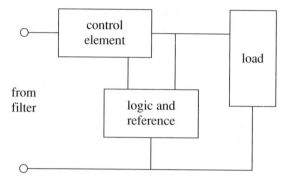

a. Series regulator

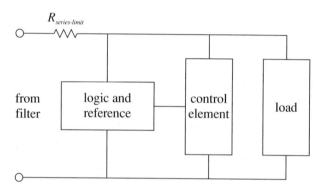

b. Shunt regulator

where V_{max} is the largest load voltage produced, and V_{min} is the minimum load voltage produced. These points usually occur at the minimum and maximum load currents, respectively. Again, this value is normally expressed as a percentage and is ideally 0%.

In order to maintain a constant output voltage, the power-supply regulator needs to sense its output and then compensate for any irregularities. This implies a number of things: First, some form of reference is needed for stable comparison to the output signal. Second, some form of comparator or amplifier is required in order to make use of this comparison. Finally, some form of control element is needed to absorb the difference between the desired output and the input to the regulator circuit. This control element can either appear in series or in parallel with the load, as seen in Figure 8.2. The first form is shown in Figure 8.2a and is normally referred to as a *series mode regulator*. The control element allows current to pass through to the load, but drops a

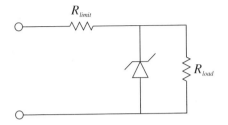

Figure 8.3
Simple zener shunt
regulator

specific amount of voltage. The voltage which appears across the control element is the difference between the filter's output and the desired load voltage.

Figure 8.2b shows a *shunt mode regulator*. Here, the control element is in parallel with the load, and draws enough current to keep the output level constant. For many applications, shunt mode regulators are not as efficient as series mode regulators. One example of a simple shunt mode regulator is the simple resistor/zener diode arrangement shown in Figure 8.3. Note that under no-load conditions (when the load impedance is infinite), considerable current flows in the regulating circuit.

8.3 Linear Regulators

If a regulator's control element operates in its linear region, the regulator is said to be a *linear regulator*. Most linear regulator ICs are series mode types. The main advantages of linear regulators are their ease of use and accuracy of control. Their main disadvantage is low efficiency.

A basic linear regulator is shown in Figure 8.4 (page 338). The series control element is the transistor Q_1. This component is most often called a *pass transistor*, since it allows current to pass through to the load. The pass transistor is driven by the op amp. The pass transistor's job is to amplify the output current of the op amp. The op amp constantly monitors its two inputs. The noninverting input sees the zener voltage of device D_1. R_1 is used to properly bias D_1. The op amp's inverting input sees the voltage drop across R_3. Note that R_2 and R_3 make up a voltage divider with V_{load} as the divider's source. Remembering the basics of negative feedback, the op amp will produce enough current to keep its two inputs at approximately the same level, thus keeping V_{error} at zero. In other words, the voltage across R_3 should be equal to the zener potential. As long as the op amp has a high enough output-current capability, this equality will be maintained. Note that the V_{BE} drop produced by Q_1 is compensated for since it is within the feedback loop. The difference between the filtered input signal and the final V_{load} is dropped across the collector/emitter of Q_1.

Figure 8.4
Basic op amp linear
regulator

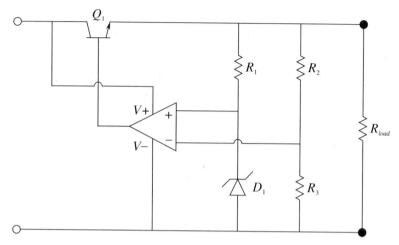

In essence, the circuit of Figure 8.4 is a noninverting amplifier. V_Z is the input potential, and R_2 and R_3 take the place of resistors R_f and R_i, respectively. V_{load} is found by using a variation of the basic gain formula:

$$V_{load} = V_Z \frac{R_2 + R_3}{R_3} \tag{8.3}$$

By choosing the appropriate ratios of resistors along with a suitable zener, a wide range of output potentials may be achieved. Note that the op amp is fed from the nonregulated input. Any ripple which exists on this line will hardly affect the op amp's function. The ripple will be reduced by the op amp's PSRR. The zener bias resistor R_1 is chosen so that it sets up a current which will guarantee zener conduction. This is usually in the low milliamp range.

Example 8.1   Determine the output of Figure 8.4 if $R_1 = 5$ k, $R_2 = 20$ k, $R_3 = 10$ k, and $V_Z = 3.9$ V. Assume that the filtered input is 20 V DC with no more than 3 V peak-to-peak ripple.

First, note that the input signal varies between a minimum of 18.5 V to a maximum of 21.5 V. (20 V $\pm$ 1.5 V peak) As long as the circuit doesn't try to produce more than the minimum input, and as long as all current and power dissipation limits are obeyed, everything should function correctly.

$$V_{load} = V_Z \frac{R_2 + R_3}{R_3}$$

$$V_{load} = 3.9 \text{ V} \times \frac{20 \text{ k} + 10 \text{ k}}{10 \text{ k}}$$

$$V_{load} = 11.7 \text{ V}$$

Due to the V_{BE} drop, the op amp needs to produce about .7 V more than this. Also, note that the zener current can now be found.

$$I_Z = \frac{V_{load} - V_Z}{R_1}$$

$$I_Z = \frac{11.7 \text{ V} - 3.9 \text{ V}}{5 \text{ k}}$$

$$I_Z = 1.56 \text{ mA}$$

This is a reasonable value.

Example 8.2 Determine the power dissipation for Q_1 and the efficiency of the circuit in Example 8.1 if the effective load resistance is 20 Ω.
First, I_{load} must be determined.

$$I_{load} = \frac{V_{load}}{R_{load}}$$

$$I_{load} = \frac{11.7 \text{ V}}{20}$$

$$I_{load} = .585 \text{ A}$$

The dissipation of Q_1 is the product of the current passing through it and the voltage across it. The current through Q_1 is the load current. The voltage across Q_1 is the difference between the filter's output and the load voltage. Since the filter's output contains a relatively small AC signal riding on DC, the average will equal the DC value, in this case, 20 V.

$$V_{CE} = V_C - V_E$$
$$V_{CE} = 20 \text{ V} - 11.7 \text{ V}$$
$$V_{CE} = 8.3 \text{ V}$$

$$P_D = I_C V_{CE}$$
$$P_D = .585 \text{ A} \times 8.3 \text{ V}$$
$$P_D = 4.86 \text{ W}$$

So, the pass transistor must dissipate 4.86 W, and tolerate a current of .585 A. The maximum collector-emitter voltage will occur at the peak input of 21.5 V. Therefore, the maximum differential is 9.8 V.

Finally, note that a .585 A output would require a minimum β of

$$\beta = \frac{I_C}{I_B}$$

$$\beta = \frac{585 \text{ mA}}{20 \text{ mA}}$$

$$\beta = 29.25$$

This assumes that the op amp can produce 20 mA, and also ignores the small zener and voltage divider currents at the output.

As far as efficiency is concerned, the input power and load power need to be calculated. To find the load power:

$$P_{load} = I_{load} V_{load}$$

$$P_{load} = .585 \text{ A} \times 11.7 \text{ V}$$

$$P_{load} = 6.844 \text{ W}$$

Ignoring the current requirements of the op amp, zener, and R_2/R_3 divider, the supplied current equals .585 A. The average input voltage is 20 V.

$$P_{in} = I_{in} V_{in}$$

$$P_{in} = .585 \text{ A} \times 20 \text{ V}$$

$$P_{in} = 11.7 \text{ W}$$

Efficiency, η, is defined as the ratio of useful output to required input, so,

$$\eta = \frac{P_{load}}{P_{in}}$$

$$\eta = \frac{6.844 \text{ W}}{11.7 \text{ W}}$$

$$\eta = .585 \text{ or } 58.5\%$$

Therefore, 41.5% of the input power is wasted.

In order to minimize the waste and maximize the efficiency, the input/output differential voltage needs to be as small as possible. For proper operation of the op amp and pass transistor, this usually means a differential lower than 2 to 3 V is impossible. Consequently, this form of regulation is very inefficient for lower power-supply output levels.

SPICE Analysis

A SPICE simulation of an op amp–based linear regulator, such as the one in Figure 8.4, is shown in Figure 8.5. The input signal is 20 V average, with a 120 Hz 2 V peak sine wave riding on it (representing ripple). A 741 op amp is used as the comparison element along with a generic transistor for the pass device. The reference is a 5.2 V zener diode. Given the circuit values, a manual calculation shows

$$V_{out} = V_Z\left(1 + \frac{R_f}{R_i}\right)$$

$$V_{out} = 5.2 \text{ V} \times \left(1 + \frac{14 \text{ k}}{11 \text{ k}}\right)$$

$$V_{out} = 11.8 \text{ V}$$

The SPICE output file (not shown) indicates a constant output at approximately 12.2 V. The discrepancy can be attributed to the non-ideal nature of the zener diode (i.e., its internal resistance). Indeed, the initial solution indicates that the zener potential (node 5) is nearly 5.4 V. A more accurate specification of the zener diode (in particular, the BV and IBV parameters) would result in

Figure 8.5
SPICE simulation of a linear regulator based on op amps

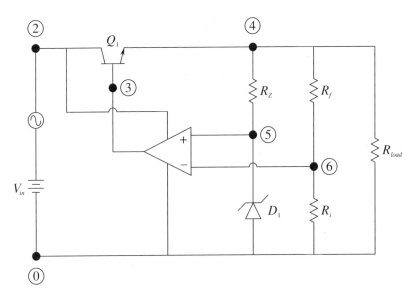

a. SPICE schematic

(Continued)

———

Figure 8.5   *(Continued)*

```
Op amp regulator circuit

*******Start of UA741 op amp*********
.SUBCKT UA741 1   2   3  4  5   6
*               IN+ IN- GND V+ V- OUT
Q1  11  1   13 UA741QA
Q2  12  2   14 UA741QB
RC1  4 11   5.305165E+03
RC2  4 12   5.305165E+03
C1  11 12   5.459553E-12
RE1 13 10   2.151297E+03
RE2 14 10   2.151297E+03
IEE 10  5   1.666000E-05
CE  10  3   3.000000E-12
RE  10  3   1.200480E+07
GCM  3 21   10 3 5.960753E-09
GA  21  3   12 11 1.884655E-04
R2  21  3   1.000000E+05
C2  21 22   3.000000E-11
GB  22  3   21 3 2.357851E+02
RO2 22  3   4.500000E+01
D1  22 31   UA741DA
D2  31 22   UA741DA
EC  31  3   6 3 1.0
RO1 22  6   3.000000E+01
D3   6 24   UA741DB
VC   4 24   2.803238E+00
D4  25  6   UA741DB
VE  25  5   2.803238E+00
.ENDS
.MODEL UA741DA D (IS=9.762287E-11)
.MODEL UA741DB D (IS=8.000000E-16)
.MODEL UA741QA NPN (IS=8.000000E-16 BF=9.166667E+01)
.MODEL UA741QB NPN (IS=8.309478E-16 BF=1.178571E+02)
*****End of UA741 op amp******
.MODEL D1N752 D (IS=0.5UA RS=6 BV=5.2 IBV=110UA)
.MODEL QPASS NPN (IS=1.0E-14 BF=75 CJE=50PF CJC=30PF)

********Main circuit description********
RZ    4  5   2K
RL    0  4   100
RF    6  4   14K
RI    6  0   11K
D1    0  5   D1N752
Q1    2  3   4  QPASS
XOA1  5  6   0 2 0 3  UA741
*Source is 20 VDC with 2 VAC ripple. In reality, the ripple will not be
*perfectly sinusoidal, but this is a reasonable approximation.
VIN   2  0   SIN (20  2 120HZ)

********Analysis options*********
*Calculate transient analysis in 400 uS steps up to 20MS.
.TRAN 400US 20MS
*Plot the time domain response at the output (node 4), and the input (node 2).
.PLOT TRAN V(4) V(2)
*
.END
```

b. SPICE input file

Figure 8.5
(Continued)

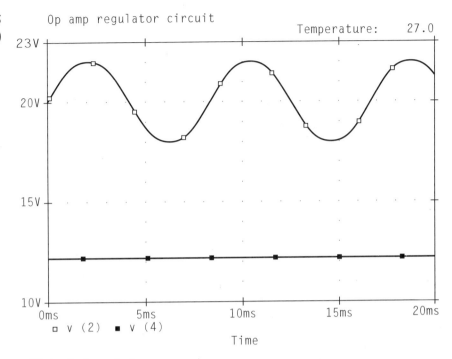

c. PSpice-Probe output

a much closer prediction. Another way of looking at this is to say that the 5.2 V zener specification is too far below the actual zener bias point for utmost accuracy (that is, the manual calculation is somewhat sloppy, and is, therefore, the source of the error). This minor discrepancy aside, the time-domain plot does show the stability of the output signal in spite of the sizable input variations.

Three-Terminal Devices

In an effort to make the designer's job ever easier, manufacturers provide circuits such as the one shown in Figure 8.4 in a single package. If you notice, the circuit really needs only three pins with which to connect to the outside world: input from the filter, ground, and output. These devices are commonly known as *three-pin regulators*—hardly a creative nametag, but descriptive at least. Several different devices are available for varying demands for current and power dissipation.

The LM78XX and LM79XX are members of a typical "3-pin" device family. The LM78XX series is for positive outputs, while the LM79XX series produces negative outputs. The "XX" indicates the rated load voltage. For example, the

LM7805 is a +5 V regulator, while the LM7912 is a −12 V unit. The most popular sizes are the 5, 12, and 15 V units, although many others are also available. A similar series is the LM340-XX/LM360-XX family. Again, the "XX" denotes the rated output voltage. The LM340's are the positive members of the family.

A data sheet for the LM78XX series is shown in Figure 8.6. This regulator is manufactured in two variants, the LM78XX CK (TO-3 case), and the LM78XX CT (TO-220 case). The TO-3 case version offers somewhat higher power-dissipation capability. Output currents in excess of 1 A are available. As a side note, for lighter loads with lower current demands, the LM78LXX may be used. This regulator offers a 100 mA output and is available in TO-92, TO-39, and mini-DIP packages.

The practical power-dissipation limit of the LM78XX series, like any power device, is highly dependent on the type of heat sink used. The device graphs show that it can be as high as 20 W with the TO-3 case. The typical output voltage is within approximately five percent of the nominal value. This indicates the inherent accuracy of the IC, and is not a measure of its regulation abilities. Figures for both load and line regulation are given in the data sheet. For a variance in the line voltage of over 2:1, we can see that the output voltage varies by no more than one percent of the nominal output. Worst-case load regulation is just as good, showing a one-percent deviation for a load current change from 5 mA up to 1.5 A. The regulator can also be seen to draw very little standby current (only 8.5 mA over a wide temperature range).

The output of the LM78XX is fairly clean. The LM7815 typically shows an output noise voltage of 90 μV and an average ripple rejection of 70 dB. This means that ripple presented to the input of the regulator is reduced by 70 dB at the output. Since 70 dB translates to a factor of over 3000, this means that an input ripple signal of one volt will be reduced to less than one-third of a millivolt at the output. It is worthwhile to note that this figure is frequency-dependent, as shown by the Ripple Rejection graph. Fortunately, the maximum value occurs in the desired range of 60 to 120 Hz.

The last major item in the LM78XX data sheet is an extremely important one: Input Voltage Required To Maintain Line Regulation. The value shown is approximately 2.5 to 2.7 V greater than the nominal rating of the regulator. Without this headroom, the regulator will cease to function properly. While it is desirable to keep the input/output differential voltage low in order to decrease device power dissipation and maximize efficiency, a value that is too low will cripple the regulator. As seen in the Peak Output Current graph, the highest load currents are achieved when the differential is in the 5 to 10 V range.

In all cases, the devices offer thermal shutdown and output short-circuit protection. The maximum input voltage is limited to 35 V. Besides the LM78XX series, special high-current types with outputs in the 3 A and higher range are also available.

Figure 8.6   LM78XX data sheet

Absolute Maximum Ratings

If Military/Aerospace specified devices are required, contact the National Semiconductor Sales Office/ Distributors for availability and specifications.

Input Voltage (V_O = 5V, 12V and 15V)	35V
Internal Power Dissipation (Note 1)	Internally Limited
Operating Temperature Range (T_A)	0°C to +70°C

Maximum Junction Temperature	
(K Package)	150°C
(T Package)	150°C
Storage Temperature Range	−65°C to +150°C
Lead Temperature (Soldering, 10 sec.)	
TO-3 Package K	300°C
TO-220 Package T	230°C

Electrical Characteristics LM78XXC (Note 2) 0°C ≤ Tj ≤ 125°C unless otherwise noted.

Symbol	Parameter	Conditions	5V / 10V Min	Typ	Max	12V / 19V Min	Typ	Max	15V / 23V Min	Typ	Max	Units
V_O	Output Voltage	Tj = 25°C, 5 mA ≤ I_O ≤ 1A	4.8	5	5.2	11.5	12	12.5	14.4	15	15.6	V
		P_D ≤ 15W, 5 mA ≤ I_O ≤ 1A	4.75		5.25	11.4		12.6	14.25		15.75	V
		V_{MIN} ≤ V_{IN} ≤ V_{MAX}	(7.5 ≤ V_{IN} ≤ 20)			(14.5 ≤ V_{IN} ≤ 27)			(17.5 ≤ V_{IN} ≤ 30)			V
ΔV_O	Line Regulation	I_O = 500 mA, Tj = 25°C		3	50		4	120		4	150	mV
		ΔV_{IN}	(7 ≤ V_{IN} ≤ 25)			(14.5 ≤ V_{IN} ≤ 30)			(17.5 ≤ V_{IN} ≤ 30)			V
		0°C ≤ Tj ≤ +125°C			50			120			150	mV
		ΔV_{IN}	(8 ≤ V_{IN} ≤ 20)			(15 ≤ V_{IN} ≤ 27)			(18.5 ≤ V_{IN} ≤ 30)			V
		I_O ≤ 1A, Tj = 25°C			50			120			150	mV
		ΔV_{IN}	(7.5 ≤ V_{IN} ≤ 20)			(14.6 ≤ V_{IN} ≤ 27)			(17.7 ≤ V_{IN} ≤ 30)			V
		0°C ≤ Tj ≤ +125°C			25			60			75	mV
		ΔV_{IN}	(8 ≤ V_{IN} ≤ 12)			(16 ≤ V_{IN} ≤ 22)			(20 ≤ V_{IN} ≤ 26)			V
ΔV_O	Load Regulation	Tj = 25°C, 5 mA ≤ I_O ≤ 1.5A		10	50		12	120		12	150	mV
		250 mA ≤ I_O ≤ 750 mA			25			60			75	mV
		5 mA ≤ I_O ≤ 1A, 0°C ≤ Tj ≤ +125°C			50			120			150	mV
I_Q	Quiescent Current	I_O ≤ 1A, Tj = 25°C			8			8			8	mA
		0°C ≤ Tj ≤ +125°C			8.5			8.5			8.5	mA
ΔI_Q	Quiescent Current Change	5 mA ≤ I_O ≤ 1A			0.5			0.5			0.5	mA
		Tj = 25°C, I_O ≤ 1A			1.0			1.0			1.0	mA
		V_{MIN} ≤ V_{IN} ≤ V_{MAX}	(7.5 ≤ V_{IN} ≤ 20)			(14.8 ≤ V_{IN} ≤ 27)			(17.9 ≤ V_{IN} ≤ 30)			V
		I_O ≤ 500 mA, 0°C ≤ Tj ≤ +125°C			1.0			1.0			1.0	mA
		V_{MIN} ≤ V_{IN} ≤ V_{MAX}	(7 ≤ V_{IN} ≤ 25)			(14.5 ≤ V_{IN} ≤ 30)			(17.5 ≤ V_{IN} ≤ 30)			V
V_N	Output Noise Voltage	T_A = 25°C, 10 Hz ≤ f ≤ 100 kHz		40			75			90		μV
$\frac{\Delta V_{IN}}{\Delta V_{OUT}}$	Ripple Rejection	I_O ≤ 1A, Tj = 25°C or	62	80		55	72		54	70		dB
		f = 120 Hz { I_O ≤ 500 mA	62			55			54			dB
		0°C ≤ Tj ≤ +125°C										
		V_{MIN} ≤ V_{IN} ≤ V_{MAX}	(8 ≤ V_{IN} ≤ 18)			(15 ≤ V_{IN} ≤ 25)			(18.5 ≤ V_{IN} ≤ 28.5)			V
R_O	Dropout Voltage	Tj = 25°C, I_{OUT} = 1A		2.0			2.0			2.0		V
	Output Resistance	f = 1 kHz		8			18			19		mΩ
	Short-Circuit Current	Tj = 25°C		2.1			1.5			1.2		A
	Peak Output Current	Tj = 25°C		2.4			2.4			2.4		A
	Average TC of V_{OUT}	0°C ≤ Tj ≤ +125°C, I_O = 5 mA		0.6			1.5			1.8		mV/°C
V_{IN}	Input Voltage Required to Maintain Line Regulation	Tj = 25°C, I_O ≤ 1A	7.5			14.6			17.7			V

Note 1: Thermal resistance of the TO-3 package (K, KC) is typically 4°C/W junction to case and 35°C/W case to ambient. Thermal resistance of the TO-220 package (T) is typically 4°C/W junction to case and 50°C/W case to ambient.

Note 2: All characteristics are measured with capacitor across the input of 0.22 μF, and a capacitor across the output of 0.1 μF. All characteristics except noise voltage and ripple rejection ratio are measured using pulse techniques (t_w ≤ 10 ms, duty cycle ≤ 5%). Output voltage changes due to changes in internal temperature must be taken into account separately.

(Continued)

Figure 8.6 (*Continued*)

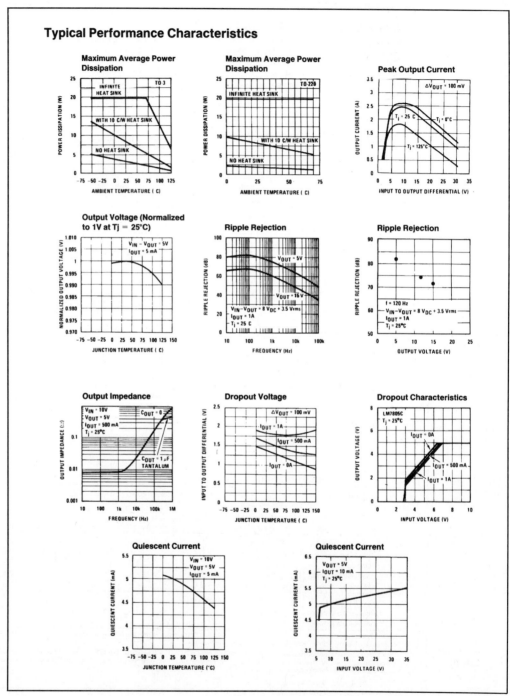

Figure 8.7   Dual supply for op amp circuits

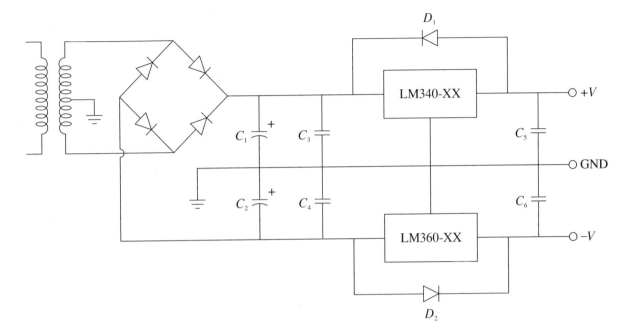

As an example, a bipolar power supply is shown in Figure 8.7. A center-tapped transformer is used to generate the required positive/negative polarities. C_1 and C_2 serve as the filter capacitors. The positive peak across each capacitor should not exceed 35 V, or the regulators may be damaged. The minimum voltage across the capacitors should be at least three volts greater than the desired output level. The exact size of the capacitor depends on the amount of load current expected. In a similar manner, the magnitude of the load current will determine which case style will be used. Capacitors C_3 and C_4 are .22 μF units and are only required if the regulators are located several inches or more from the filter capacitors. If their use is required, C_3 and C_4 must be located very close to the regulator. C_5 and C_6 are used to improve the transient response of the regulator. They are not used for filtering or for circuit stabilization. Through common misuse, C_5 and C_6 are often called "stability capacitors," although this is not their function. Finally, diodes D_1 and D_2 protect the regulators from output overvoltage conditions, such as those experienced with inductive loads.

As you can see, designing moderate fixed voltage-regulated power supplies with this template can be a very straightforward exercise. Once the basic power supply is configured, all that needs to be added are the regulators and a few capacitors and diodes. Quite simply, the regulator is "tacked on" to a basic capacitor-filtered unregulated supply. For a simple single-polarity supply, the additional components can be reduced to a single regulator IC, ignoring the extra capacitors and diode.

Current Boosting

For very high output currents, it is possible to utilize external pass transistors to augment the basic linear regulator ICs. An example is shown in Figure 8.8. When the input current rises to a certain level, the drop across R_1 will be large enough to turn on transistor Q_1. When Q_1 starts to conduct, a current will flow through R_3. As this current rises, the potential across R_3 will increase to the point where the power transistor Q_2 will turn on. Q_2 will handle any further increases in load current. Normally, R_1 is set so that the regulator is running at better than one-half of its maximum rating before switchover occurs. A typical value would be 22 Ω. For example, if switchover occurs at 1 A, the regulator will supply all of the load's demand up to 1 A. If the load requires more than 1 A, the regulator will supply 1 A, and the power transistor will supply the rest. Circuits of this type often need a minimum load current to function correctly. This "bleed-off" can be achieved by adding a single resistor in parallel with the load (across the output terminals).

Another possibility for increasing output current is accomplished by paralleling devices and adding small ballast resistors. An example of this is shown in Figure 8.9. The ballast resistors are used to create local feedback. This reduces current-hogging and forces the regulators to equally share the load current. (This is the same technique which is commonly used in high-power amplifiers so that multiple output transistors may be connected in parallel.) The size of the ballast resistors is rather low, usually .5 Ω or less.

Figure 8.8
Pass transmitters for
increased I_{load}

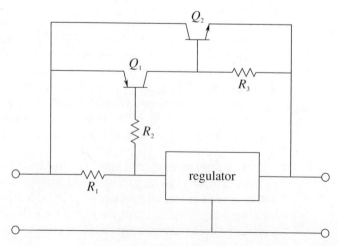

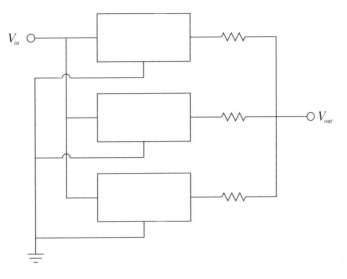

Programmable and Tracking Regulators

Along with the simple 3-pin fixed regulators, a number of adjustable or programmable devices are available. Some devices also include features such as programmable current-limiting. Also, it is possible to configure multiple regulators so that they *track*, or follow, each other.

One popular adjustable regulator is the LM317. This device is functionally similar to the 340 series discussed in the previous section. In essence, it has an internal reference of 1.25 V. By using an external voltage divider, a wide range of output potentials are available. The LM317 will produce a maximum current of 1.5 A, and a maximum voltage of 37 V. A basic connection diagram is shown in Figure 8.10 (page 350). The R_1/R_2 divider sets the output voltage according to the formula

$$V_{out} = 1.25 \text{ V} \left(1 + \frac{R_2}{R_1} \right) + I_{adj} R_2$$

I_{adj} is the current flowing through the bottom *adjust pin*. I_{adj} is about 100 μA, and being this small, can be ignored to a first approximation. R_1 is set to 240 Ω. D_1 serves as a protection device for output overvoltages as seen in the preceding section. The 10 μF capacitor is used to increase the ripple rejection of the device. The addition of this capacitor will increase ripple rejection by at least 10 dB. Diode D_2 is used to prevent possible destructive discharges from the 10 μF capacitor. In practice, R_2 is set to a fixed-size resistor for static supplies, or is a potentiometer for user-adjustable supplies.

Figure 8.10
Basic LM317 regulator

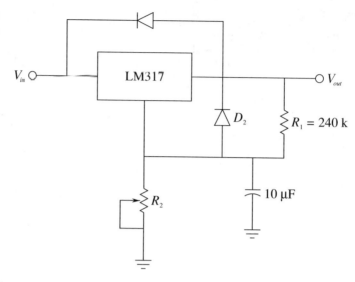

Example 8.3   Determine a value for R_2 such that the output is adjustable from a minimum of 1.25 V to a maximum of 15 V.

For the minimum value, R_2 should be 0 k. Ignoring the effect of I_{adj}, the maximum value is found by

$$V_{out} = 1.25 \text{ V} \left(1 + \frac{R_2}{R_1}\right)$$

$$R_2 = R_1 \times \left(\frac{V_{out}}{1.25 \text{ V}} - 1\right)$$

$$R_2 = 240 \times \left(\frac{15 \text{ V}}{1.25 \text{ V}} - 1\right)$$

$$R_2 = 2.64 \text{ k}$$

Normally, such a value is not available for stock potentiometers. A 2.5 k unit may be readily available, so a slight drop in R_1 may be in order for compensation. If a minimum value greater than 1.25 V is needed, the pot may be placed in series with a fixed resistor. Finally, precise preset values may be obtained through the use of a rotary switch and a bank of fixed resistors, as shown in Figure 8.11. As a matter of fact, these techniques can also be applied to the fixed 3-pin devices presented in the previous section. For example, an LM7805 can be used as a 15 V regulator just by adding an external divider network as shown in Figure 8.12. You can think of the LM7805 as an LM317

Figure 8.11   Adjustable regulation

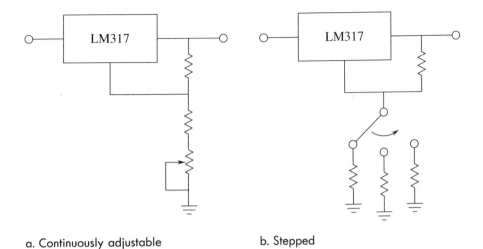

Figure 8.11   Adjustable regulation

a. Continuously adjustable   b. Stepped

with a 5 V reference. The exact values are not critical: what is important is
that the ratio of the two resistors is 2:1.

One of the major problems with adjustable linear regulators is a "bottom
end" limitation. It is easier for a regulator of this type to generate a high load
current at a high load voltage than it is to generate a high load current at a
low load voltage. The reason for this is that at low output voltages, the internal
pass transistor sees a very high differential voltage. This results in very high
power dissipation. If the device gets too hot, the thermal protection circuits
may activate. From a user's standpoint, this means that the power supply may
be able to produce a 15 V 1 A output, but be unable to produce a 5 V 1 A
output.

Another popular regulator IC is the LM723. This device is of a more modular
design and allows for preset current limiting. The equivalent circuit of the 723

Figure 8.12
Varying V_{out} with a fixed
regulator

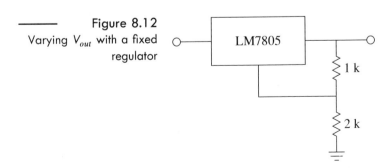

Figure 8.13   LM723 equivalent circuit

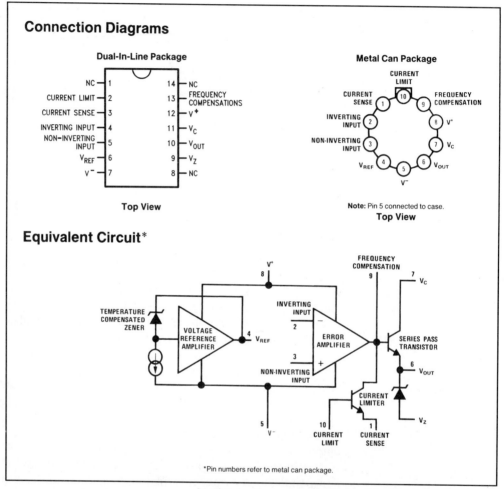

Connection Diagrams

Dual-In-Line Package

NC	1	14 NC
CURRENT LIMIT	2	13 FREQUENCY COMPENSATIONS
CURRENT SENSE	3	12 V^+
INVERTING INPUT	4	11 V_C
NON−INVERTING INPUT	5	10 V_{OUT}
V_{REF}	6	9 V_Z
V^-	7	8 NC

Top View

Metal Can Package

Note: Pin 5 connected to case.
Top View

Equivalent Circuit*

*Pin numbers refer to metal can package.

Reprinted with permission of National Semiconductor Corporation

is shown in Figure 8.13. By itself, the LM723 is only capable of a 150 mA output. With the use of external pass transistors, an LM723-based regulator can produce several amps of load current. The internal reference is approximately 7.15 V. Because of this, two basic configurations exist, one for outputs below 7 V and one for outputs above 7 V. These configurations and their output formulas are shown in Figure 8.14. By combining Figure 8.13 with Figure 8.14 and redrawing the results as in Figures 8.15a and 8.15b (page 355), you can see a strong resemblance to the basic op amp regulator encountered

Figure 8.14   LM723 "hook-ups"

TABLE I. Resistor Values (kΩ) for Standard Output Voltage

Positive Output Voltage	Applicable Figures	Fixed Output ±5%		Output Adjustable ±10% (Note 5)			Negative Output Voltage	Applicable Figures	Fixed Output ±5%		5% Output Adjustable ±10%		
	(Note 4)	R1	R2	R1	P1	R2			R1	R2	R1	P1	R2
+3.0	1, 5, 6, 9, 12 (4)	4.12	3.01	1.8	0.5	1.2	+100	7	3.57	102	2.2	10	91
+3.6	1, 5, 6, 9, 12 (4)	3.57	3.65	1.5	0.5	1.5	+250	7	3.57	255	2.2	10	240
+5.0	1, 5, 6, 9, 12 (4)	2.15	4.99	0.75	0.5	2.2	−6 (Note 6)	3, (10)	3.57	2.43	1.2	0.5	0.75
+6.0	1, 5, 6, 9, 12 (4)	1.15	6.04	0.5	0.5	2.7	−9	3, 10	3.48	5.36	1.2	0.5	2.0
+9.0	2, 4, (5, 6, 9, 12)	1.87	7.15	0.75	1.0	2.7	−12	3, 10	3.57	8.45	1.2	0.5	3.3
+12	2, 4, (5, 6, 9, 12)	4.87	7.15	2.0	1.0	3.0	−15	3, 10	3.65	11.5	1.2	0.5	4.3
+15	2, 4, (5, 6, 9, 12)	7.87	7.15	3.3	1.0	3.0	−28	3, 10	3.57	24.3	1.2	0.5	10
+28	2, 4, (5, 6, 9, 12)	21.0	7.15	5.6	1.0	2.0	−45	8	3.57	41.2	2.2	10	33
+45	7	3.57	48.7	2.2	10	39	−100	8	3.57	97.6	2.2	10	91
+75	7	3.57	78.7	2.2	10	68	−250	8	3.57	249	2.2	10	240

TABLE II. Formulae for Intermediate Output Voltages

Outputs from +2 to +7 volts (Figures 1, 5, 6, 9, 12, [4])	Outputs from +4 to +250 volts (Figure 7)	Current Limiting
$V_{OUT} = \left(V_{REF} \times \dfrac{R2}{R1 + R2} \right)$	$V_{OUT} = \left(\dfrac{V_{REF}}{2} \times \dfrac{R2 - R1}{R1} \right); R3 = R4$	$I_{LIMIT} = \dfrac{V_{SENSE}}{R_{SC}}$

Outputs from +7 to +37 volts (Figures 2, 4, [5, 6, 9, 12])	Outputs from −6 to −250 volts (Figures 3, 8, 10)	Foldback Current Limiting
$V_{OUT} = \left(V_{REF} \times \dfrac{R1 + R2}{R2} \right)$	$V_{OUT} = \left(\dfrac{V_{REF}}{2} \times \dfrac{R1 + R2}{R1} \right); R3 = R4$	$I_{KNEE} = \left(\dfrac{V_{OUT}\,R3}{R_{SC}\,R4} + \dfrac{V_{SENSE}\,(R3 + R4)}{R_{SC}\,R4} \right)$ $I_{SHORT\ CKT} = \left(\dfrac{V_{SENSE}}{R_{SC}} \times \dfrac{R3 + R4}{R4} \right)$

Typical Applications

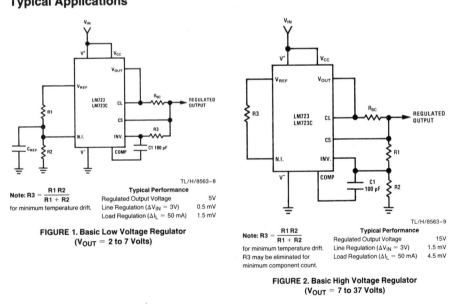

Note: $R3 = \dfrac{R1\,R2}{R1 + R2}$
for minimum temperature drift.

Typical Performance
Regulated Output Voltage   5V
Line Regulation (ΔV_{IN} = 3V)   0.5 mV
Load Regulation (ΔI_L = 50 mA)   1.5 mV

TL/H/8563–8

FIGURE 1. Basic Low Voltage Regulator
(V_{OUT} = 2 to 7 Volts)

Note: $R3 = \dfrac{R1\,R2}{R1 + R2}$
for minimum temperature drift.
R3 may be eliminated for
minimum component count.

Typical Performance
Regulated Output Voltage   15V
Line Regulation (ΔV_{IN} = 3V)   1.5 mV
Load Regulation (ΔI_L = 50 mA)   4.5 mV

TL/H/8563–9

FIGURE 2. Basic High Voltage Regulator
(V_{OUT} = 7 to 37 Volts)

(Continued)

Figure 4.10 (Continued)

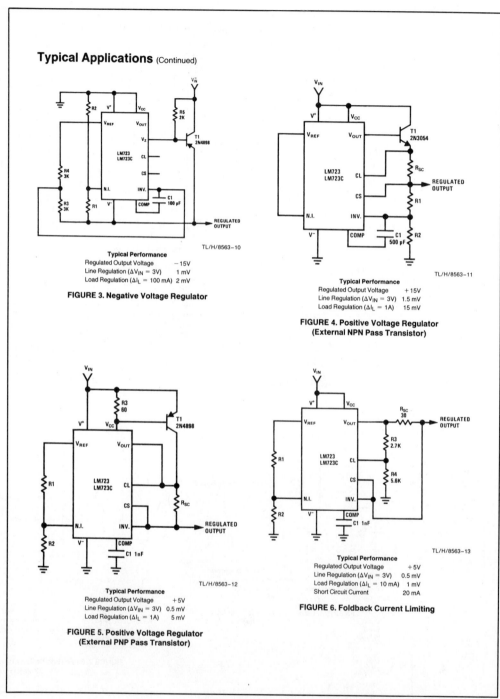

Typical Applications (Continued)

TL/H/8563–10

Typical Performance

Regulated Output Voltage   −15V
Line Regulation (ΔV_{IN} = 3V)   1 mV
Load Regulation (ΔI_L = 100 mA) 2 mV

FIGURE 3. Negative Voltage Regulator

TL/H/8563–11

Typical Performance

Regulated Output Voltage   +15V
Line Regulation (ΔV_{IN} = 3V) 1.5 mV
Load Regulation (ΔI_L = 1A)   15 mV

FIGURE 4. Positive Voltage Regulator
(External NPN Pass Transistor)

TL/H/8563–12

Typical Performance

Regulated Output Voltage   +5V
Line Regulation (ΔV_{IN} = 3V) 0.5 mV
Load Regulation (ΔI_L = 1A)   5 mV

FIGURE 5. Positive Voltage Regulator
(External PNP Pass Transistor)

TL/H/8563–13

Typical Performance

Regulated Output Voltage   +5V
Line Regulation (ΔV_{IN} = 3V) 0.5 mV
Load Regulation (ΔI_L = 10 mA) 1 mV
Short Circuit Current   20 mA

FIGURE 6. Foldback Current Limiting

Reprinted with permission of National Semiconductor Corporation

Figure 8.15   Two basic configurations of an LM723 regulator

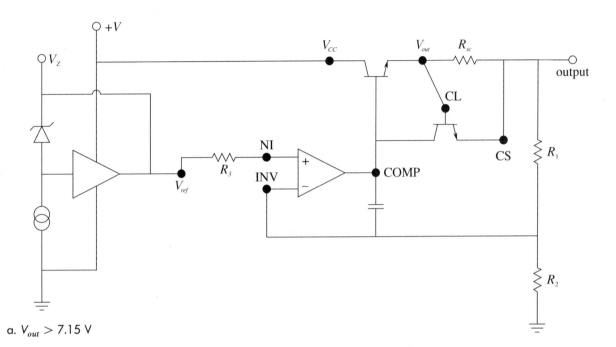

a. $V_{out} > 7.15$ V

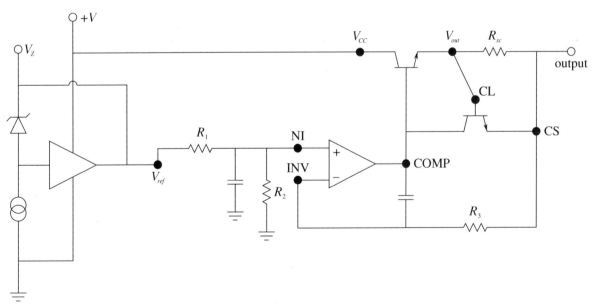

b. $V_{out} < 7.15$ V

earlier. Figure 8.15a is used for outputs greater than the 7.15 V reference. As such, resistors R_1 and R_2 are used to set the gain of the internal amplifier (the amount by which the reference will be multiplied). Resistor R_3 simply serves as DC bias compensation for R_1 and R_2. For outputs less than the reference, R_1 and R_2 serve as a voltage divider which effectively reduces the reference, while the amplifier operates with a gain of unity. Again, R_3 serves as bias compensation. In short, the output is once again determined by the reference voltage in conjunction with a pair of resistors. In both cases, the output current limit is set by

$$I_{limit} = \frac{V_{sense}}{R_{sc}}$$

where V_{sense} is approximately .65 V at room temperature. (For other temperatures, a V_{sense} graph is given in the LM723 data sheet.) This equation can be found on the manufacturer's data sheet, but also can be readily derived by inspecting Figure 8.15. Resistor R_{sc} is placed in series with the load, and thus, the current through it is the load current (ignoring the small current required by the adjustment resistors). The current-limit transistor is connected so that the voltage across R_{sc} is applied to this transistor's base-emitter junction. The collector of the current-limit transistor is connected to the base of the output-pass transistor. If the output current rises to the point where the voltage across R_{sc} exceeds approximately .65 V, the current-limit transistor will turn on, thus shunting output drive current away from the output-pass transistor. As you can see, the formula for I_{limit} is little more than the direct application of Ohm's Law. This scheme is similar to the method examined in Chapter 2 to safely limit an op amp's output current.

Example 8.4 Design a +12 V regulator using the LM723, with a current limit of 50 mA.
The basic form is the circuit labelled "Basic High Voltage Regulator" in Figure 8.14. The appropriate output equation is

$$V_{out} = V_{ref} \frac{R_1 + R_2}{R_2}$$

Choosing an arbitrary value for R_2 of 10 k, and then solving for R_1,

$$R_1 = \frac{V_{out}}{V_{ref}} R_2 - R_2$$

$$R_1 = \frac{12 \text{ V}}{7.15 \text{ V}} 10 \text{ k} - 10 \text{ k}$$

$$R_1 = 6.78 \text{ k}$$

Figure 8.16
Completed 12 V circuit for
Example 8.4

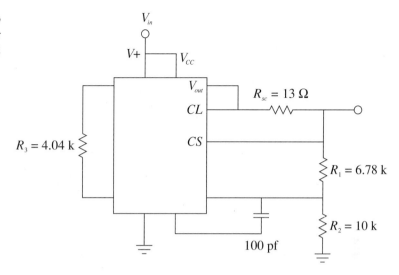

Figure 8.16
Completed 12 V circuit for
Example 8.4

For the current-sense resistor,

$$I_{limit} = \frac{V_{sense}}{R_{sc}}$$

$$R_{sc} = \frac{V_{sense}}{I_{limit}}$$

$$R_{sc} = \frac{.65 \text{ V}}{50 \text{ mA}}$$

$$R_{sc} = 13 \ \Omega$$

Finally, for minimum temperature drift, R_3 is included, and set to $R_1 \| R_2$:

$$R_3 = R_1 \| R_2$$
$$R_3 = 10 \text{ k} \| 6.78 \text{ k}$$
$$R_3 = 4.04 \text{ k}$$

The completed circuit is shown in Figure 8.16.

Example 8.5 Using the LM723, design a continuously-adjustable 2 V to 5 V supply with a current limit of 1.0 A.

Figure 8.17
Circuit for Example 8.5:
2 V–5 V regulator, 1 amp

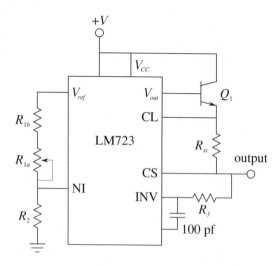

The basic form is the circuit labelled "Basic Low Voltage Regulator" in Figure 8.14. The appropriate output equation is

$$V_{out} = V_{ref} \frac{R_2}{R_1 + R_2}$$

We need to make a few modifications to the basic form in order to accommodate the high output current and the output voltage adjustment. One possibility is shown in Figure 8.17. In order to produce a 1 A load current, an external pass transistor will be used. To obtain the desired voltage adjustment, resistor R_1 is replaced with a series potentiometer/resistor combination (R_{1a}, R_{1b}). In this fashion, the minimum value for R_1 will be R_{1b}, while the maximum value will be $R_{1a} + R_{1b}$. There are several ways in which we can approach the calculation of these three resistors. Perhaps the easiest is to pick a value for R_2, and then determine values for R_{1a} and R_{1b}. Though this is fairly straightforward, it is not very practical because you will most likely wind up with an odd size for the potentiometer. A better, though admittedly more involved, approach revolves around the selection of a reasonable pot value, such as 10 k. Given the desired output potentials, the other two resistors can be found.

First of all, note that these three resistors are nothing more than a voltage divider. The output voltage equation may be rewritten as

$$\frac{R_1 + R_2}{R_2} = \frac{V_{ref}}{V_{out}}$$

For the maximum case, we have

$$\frac{R_{1b} + R_2}{R_2} = \frac{V_{ref}}{V_{out}}$$

$$\frac{R_{1b} + R_2}{R_2} = \frac{7.15 \text{ V}}{5 \text{ V}}$$

$$\frac{R_{1b} + R_2}{R_2} = 1.43$$

If we consider R_2 to be unity, we may say that the ratio of $R_2 + R_{1b}$ to R_2 is 1.43:1, or, that the ratio of R_{1b} to R_2 is .43:1.

For the minimum case, we have

$$\frac{R_{1a} + R_{1b} + R_2}{R_2} = \frac{V_{ref}}{V_{out}}$$

$$\frac{R_{1a} + R_{1b} + R_2}{R_2} = \frac{7.15 \text{ V}}{2 \text{ V}}$$

$$\frac{R_{1a} + R_{1b} + R_2}{R_2} = 3.575$$

We may say that the ratio of the three resistors to R_2 is 3.575:1, or, that the ratio of $R_{1a} + R_{1b}$ to R_2 is 2.575:1. Since we already know that the ratio of R_{1b} to R_2 is .43:1, the ratio of R_{1a} to R_2 must be the difference, or 2.145:1. Since we chose 10 k for R_{1a},

$$R_2 = \frac{R_{1a}}{2.145}$$

$$R_2 = \frac{10 \text{ k}}{2.145}$$

$$R_2 = 4.66 \text{ k}$$

Similarly,

$$R_{1b} = R_2(.43)$$

$$R_{1b} = 4.66 \text{ k}(.43)$$

$$R_{1b} = 2 \text{ k}$$

For the current sense resistor,

$$I_{limit} = \frac{V_{sense}}{R_{sc}}$$

$$R_{sc} = \frac{V_{sense}}{I_{limit}}$$

$$R_{sc} = \frac{.65 \text{ V}}{1.0 \text{ A}}$$

$$R_{sc} = .65 \ \Omega$$

For minimum temperature drift, R_3 is included, and set to $R_1 \parallel R_2$. Since R_1 is adjustable, a midpoint value will be used.

$$R_3 = R_1 \parallel R_2$$

$$R_3 = 7\,\text{k} \parallel 4.66\,\text{k}$$

$$R_3 = 2.8 \ \text{k}$$

The final calculation involves the external pass transistor. Since this design uses a 1 A output, this device must be able to handle this current continuously. Also, a minimum β specification is required. Since the LM723 will be driving the pass transistor, the LM723 needs to produce only base drive current. With a maximum output of 150 mA, this translates to a minimum β of

$$\beta_{min} = \frac{I_c}{I_b}$$

$$\beta_{min} = \frac{1 \text{ A}}{150 \text{ mA}}$$

$$\beta_{min} = 6.67$$

This value should pose no problem for a power transistor.

Besides the applications we have just looked at, the LM723 can also be used to make negative regulators, switching regulators, and other types as well. One useful variation on the basic theme is the use of *foldback current limiting*, as seen in the last circuit shown in Figure 8.14. Unlike the ordinary form of current limiting, foldback limiting actually produces a decrease in output current once the limit point is reached. Figure 8.18a shows the effect of ordinary limiting. Once the limit point is reached, further demands by the load will be

Figure 8.18
Current limiting

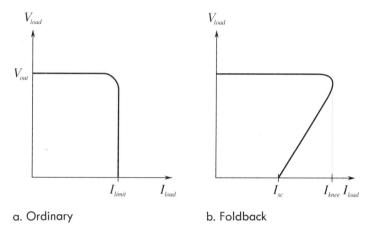

a. Ordinary   b. Foldback

ignored. The problem with this arrangement is that under short-circuit conditions, the pass transistor will be under heavy stress. Since the load is shorted, $V_{load} = 0$, and therefore, a large potential will drop across the pass transistor. Since this device is already handling the full current draw, the resulting power dissipation can be very high. Foldback limiting gets around this problem by lowering the output current as the pass transistor's voltage increases. In Figure 8.18b, note how the current limit curve does not drop straight down to the limit point, but rather, as the load demand increases, the current drops back to I_{sc}. By limiting the current in this fashion, a much lower power dissipation is achieved.

Our last device family of interest in this section is the LM325/LM326 dual-tracking regulator. These devices are of particular interest to the op amp technician and designer. The LM325 is a ± 15 V bipolar regulator with output current up to 100 mA. This is a tracking device, meaning that the positive and negative outputs are closely aligned. (The LM125 upgraded version exhibits outputs balanced to within 1%.) Like the LM723, the LM325 includes provisions for output-current limiting as well as current boosting. The LM326 is similar to the LM325, except that its output is set up for ± 12 V. A pinout diagram and example circuit outlines are given in Figure 8.19. These devices are ideal for powering general-purpose op amp circuits.

The circuit labelled "Basic Regulator" in Figure 8.19 is a minimum-parts configuration dual regulator. All that is required ahead of this circuit is a standard transformer, rectifier, and filter capacitor arrangement, such as that found in Figure 8.7. In order to prevent damage to the LM325/LM326, the maximum output (unloaded) of the filter capacitors should be no more than 30 V. This circuit is certainly simpler than the LM340-based regulator of Figure 8.7. The downside is that its power dissipation and maximum currents are considerably lower than its LM340 counterpart. Still, a 100 mA output capability is sufficient to drive a large number of op amps and other small-signal devices. Current limiting is not shown on this circuit, but can be added through

Figure 8.19   LM325 pinout and examples

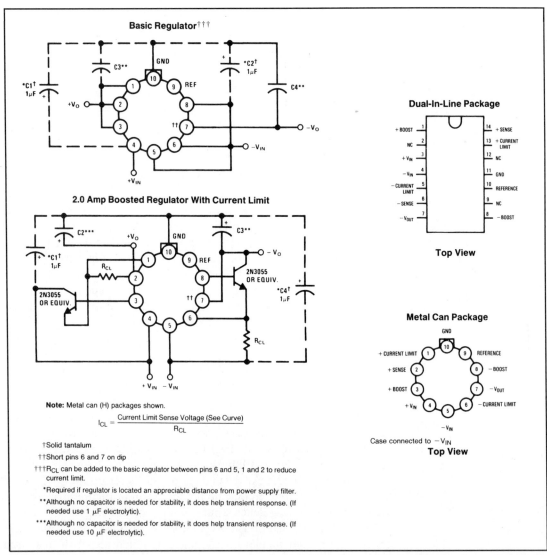

Reprinted with permission of National Semiconductor Corporation

the inclusion of a pair of current-sensing resistors (using essentially the same scheme as the LM723), one each for the positive and negative outputs.

For circuits demanding higher output currents and power dissipations, the circuit labelled "2.0 Amp Boosted Regulator" in Figure 8.19 may be considered. Note that this circuit is only marginally more complex than the basic circuit of Figure 8.19. To each output, a current sense resistor and external pass transistor have been added. Since the LM325/LM326 only needs to produce base drive current for the external pass transistor, output currents of several amps are possible using this configuration. The actual rating depends on the current and power capabilities of the external pass transistors. In this case, the popular 2N3055 NPN power devices yield a 2 A load current. (Note that this indicates a transistor minimum β of 2 A/100 mA, or 20. This is a reasonable value for a medium- to high-power transistor.)

There are many more linear regulator ICs available to the designer than have been presented here. Many of these devices are rather specialized, and all units seem to have their own special set of operational formulas and graphs. There are, however, a few common threads among them all. First, due to the relative internal complexity, manufacturers often give very specific application guidelines for their particular ICs. The resulting design sequence is rather like following a cookbook and makes the designer's life much easier. Second, as mentioned at the outset, all linear regulators tend to be rather inefficient. This inefficiency is inherent in the design and implementation of the linear regulation circuits, and cannot be avoided. At best, the inefficiency can be minimized for a given application. In order to achieve high efficiency, a different topology must be considered. One alternative is the switching regulator.

8.4 Switching Regulators

The major cause of the inefficiency of a linear regulator is that its pass transistor operates in the linear region. This means that it constantly sees both a high current and a high (or at least moderate) voltage. The result is a sizable power dissipation. Switching regulators rely on the efficiency of the transistor switch. When the transistor is off, no current flows, and thus, no power is dissipated. When the transistor is on (in saturation), a high current flows, but the voltage across the transistor ($V_{ce(sat)}$) is very small. This results in a rather modest power dissipation. The only time that both the current and voltage are relatively large at the same time is during the switching interval. This time period is relatively short compared to cycle time, so again, the power dissipation is rather small.

Switching the pass transistor on and off can be very efficient, but unfortunately, the resulting pulses of current are not appropriate for most loads. Some way of smoothing out the pulses into a constant DC level is needed. One way to do this is through an inductor/capacitor arrangement. This concept is essential to the switching regulator.

While the switching regulator does offer an increased efficiency over the linear regulator, it is not without its detractions. First of all, switching regulators tend to be more complex than linear types. This complexity may outstrip the efficiency advantage, particularly for low-power designs. Some of the newest "single chip" switching regulators make switching power supply design almost

Figure 8.20
Basic switching regulator

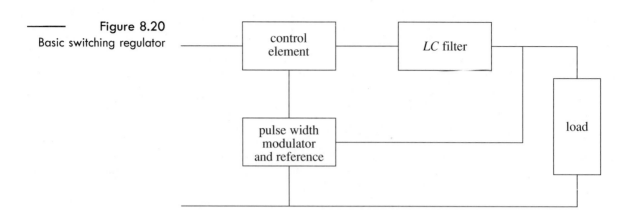

as straightforward as linear design, so this problem is not quite as great as it once was. The other major problem is that the switching process can create a great deal of radiated electrical noise. Without proper shielding and similar precautions, the induced noise signals can produce grave interference in nearby analog or digital circuits. In spite of these obstacles, switching power supplies have seen great acceptance in a variety of areas, including powering personal computers.

A basic switching regulator outline is shown in Figure 8.20. Once again, a control device is used to adjust the incoming signal. Unlike the linear regulator, the signal out of the control element is a pulse train. In order to produce a smooth DC output, the pulse train is passed through a capacitor/inductor network. The heart of the circuit is the pulse-width modulator. This circuit creates a variable-duty cycle pulse waveform which alternately turns the control element on and off. The duty cycle of the pulse is proportional to the load-current demand. Low load impedances require large currents, and thus, will create pulse trains with long "on" times. The peak value of the pulses will tend to be much greater than the average load-current demand; however, if this current pulse is averaged over one cycle, the result will equal the load-current demand. Effectively, the *LC* network serves to integrate the current pulses. This is shown graphically in Figure 8.21. Since the lower load impedance is receiving a proportionally larger current, the load voltage remains constant. This is shown graphically in Figure 8.22.

Due to the constant charging and discharging of the *L* (inductor) and *C* (capacitor) components, switching regulators tend to respond to load changes somewhat more slowly than linear regulators do. Also, if the frequency of the switching pulse is high, the required sizes for *L* and *C* may be reduced, resulting in a smaller circuit and possible cost reductions. Many switching regulators run in the 20 kHz to 100 kHz region. The practical upper limit for switching speed is determined by the speed of the control element. Devices with fast switching times will prove to be more efficient. Power FETs are very attractive for this application because of their inherent speed and low drive requirements.

Figure 8.21
Current waveforms

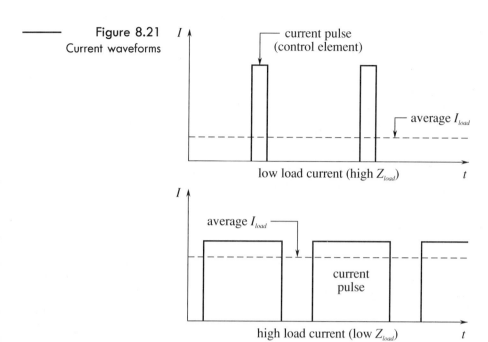

Figure 8.22
Voltage waveforms

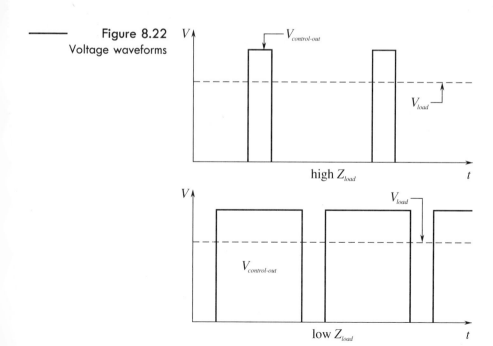

Figure 8.23
Basic stepdown switching
regulator

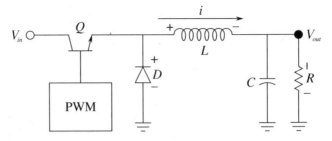

a. Q on: source supplies current and charges inductor

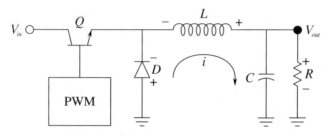

b. Q off: inductor supplies current

Their negative temperature coefficient of transconductance also helps reduce thermal-runaway problems. For the highest powers, bipolar devices are the only present choice. Finally, it is possible to configure many switchers for *step down*, *step up*, or *inverter* operation. As an example, we shall investigate the step down, or *buck*, configuration.

A block diagram of a step down switching regulator is shown in Figure 8.23. Here is how it works: When the transistor, Q, is on, current flows through Q, L, and the load. The inductor current rises at a rate equal to the inductor voltage divided by the inductance. The inductor voltage is equal to the input voltage minus the load voltage and the transistor's saturation potential. While L is charging (i.e., $I_L < I_{load}$), C provides load current. When Q turns off, the inductor's magnetic field starts to collapse, causing an effective polarity reversal. In other words, the inductor is now acting as a source for load current. At this point, diode D is forward-biased, effectively removing the left portion of the circuit. L also supplies current to capacitor C during this time period. Eventually, the inductor current will drop below the value required by the load, and C will start to discharge, making up the difference. Before L completely discharges, the transistor switch will turn back on, repeating the cycle. Since the inductor is never fully discharged, this is called *continuous operation*. (*Discontinuous operation* is also possible, although we will not pursue it here.) One point worth noting is that for proper continuation of this cycle, some load current draw must always be present. This minimum level can be achieved through the use of a paralleled bleed resistor.

Figure 8.24
LM3578 equivalent circuit

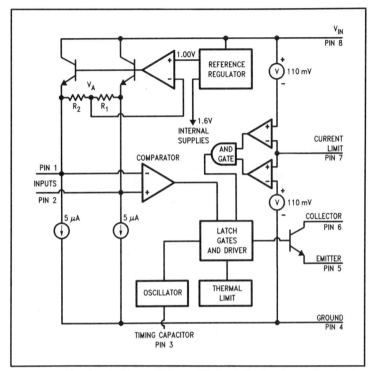

Reprinted with permission of National Semiconductor Corporation

The values for C and L are dependent on the input and output voltages, desired output current, the switching frequency, and the particulars of the switching circuit used. Manufacturers generally give look-up tables and charts for appropriate values and/or formulas. Earlier switching ICs contained the necessary base components and required only a moderate amount of external circuitry. The newest ICs may be configured with no more than a half-dozen external passive parts.

One example of a switch-mode regulator is the LM3578. A functional diagram of the inner workings of this IC is shown in Figure 8.24. The AND gates, comparator, oscillator, and associated latches combine to form the pulse-width modulator. Notice that this circuit includes an internal reference and medium-power switching transistor. Thermal shutdown and current limiting are available. The circuit will operate at inputs up to 40 V, and can produce output currents up to .75 A. The maximum switching frequency is 100 kHz. Step up, step down, and inverter configurations are all possible with this device. To make the design sequence as fast as possible, the manufacturer has included a design chart. This is shown in Figure 8.25 on page 368. As with virtually all highly-specialized ICs, specific design equations only apply to particular

Figure 8.25   LM3578 design chart

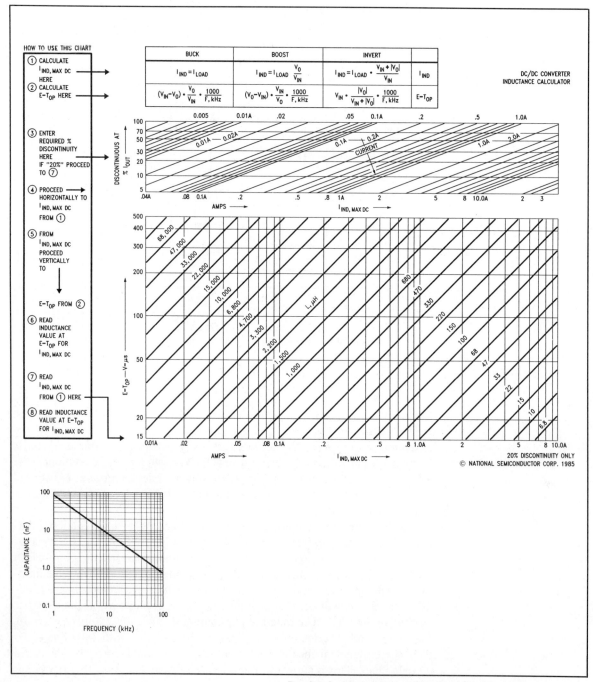

Reprinted with permission of National Semiconductor Corporation

Figure 8.26
Step down regulator using
an LM3578

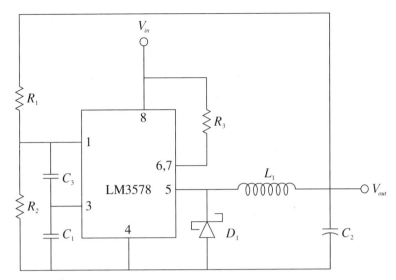

devices, and probably cannot be used with ICs which produce similar functions. Consequently, it is recommended that you do not memorize these formulas!

A step down regulator using the LM3578 is shown in Figure 8.26. C_1 is the frequency-selection capacitor, and can be found from the manufacturer's chart. C_3 is necessary for continuous operation and is generally in the vicinity of 10 to 30 pF. D_1 should be a Schottky-type rectifier. R_1 and R_2 set the step down ratio (they are functionally the same as R_f and R_i in the previous work). R_3 sets the current limit. Finally, L_1 and C_2 are used for the final output filtering. The relevant equations from the manufacturer's data sheets are:

$$V_{out} = \frac{R_1}{R_2} + 1 \qquad \text{(in volts)}$$

$$R_3 = \frac{.11 \text{ V}}{I_{sw(max)}}$$

$$C_2 \geq V_{out} \frac{V_{in} - V_{out}}{8f^2 V_{in} V_{ripple} L_1}$$

$$L_1 = V_{out} \frac{V_{in} - V_{out}}{\Delta I_{out} V_{in} f}$$

$$\Delta I_{out} = 2 \times I_{out} \times \text{discontinuity factor}$$

where $I_{sw(max)}$ is the maximum current through the switching element, the discontinuity factor is (typically) .2, and V_{ripple} is in peak-to-peak volts. Some values can be found using the look-up chart method shown in Figure 8.25.

Example 8.6   Using the LM3578, design a step down regulator which delivers 12 V from a 20 V source, with 200 mA of load current. Use an oscillator frequency of 50 kHz, and a discontinuity factor of .2 (20%). Ripple should be no more than 40 mV. First, determine the R_1 and R_2 values. R_2 is arbitrarily chosen at 10 k.

$$V_{out} = \frac{R_1}{R_2} + 1 \quad \text{(in volts)}$$

$$R_1 = R_2(V_{out} - 1 \text{ V})$$

$$R_1 = 10 \text{ k}(12 \text{ V} - 1 \text{ V})$$

$$R_1 = 110 \text{ k}$$

$$R_3 = \frac{.11 \text{ V}}{I_{sw(max)}}$$

$$R_3 = \frac{.11 \text{ V}}{.75 \text{ A}}$$

$$R_3 = .15 \text{ }\Omega$$

Note that R_3 will always be .15 Ω for this circuit form.

From the oscillator graph, C_1 is estimated at 1700 pF, and C_3 is set to 20 pF, as suggested by the manufacturer.

$$\Delta I_{out} = 2 \times I_{out} \times \text{discontinuity factor}$$

$$\Delta I_{out} = 2 \times 200 \text{ mA} \times .2$$

$$\Delta I_{out} = 80 \text{ mA}$$

$$L_1 = V_{out}\frac{V_{in} - V_{out}}{\Delta I_{out} V_{in} f}$$

$$L_1 = 12 \text{ V} \times \frac{20 \text{ V} - 12 \text{ V}}{80 \text{ mA} \times 20 \text{ V} \times 50 \text{ kHz}}$$

$$L_1 = 1.2 \text{ mH}$$

$$C_2 \geq V_{out}\frac{V_{in} - V_{out}}{8f^2 V_{in} V_{ripple} L_1}$$

$$C_2 \geq 12 \text{ V} \frac{20 \text{ V} - 12 \text{ V}}{8 \times 50 \text{ kHz}^2 \times 20 \text{ V} \times 40 \text{ mV} \times 1.2 \text{ mH}}$$

$$C_2 \geq 5 \text{ }\mu\text{F}$$

To be on the conservative side, C_2 is usually set a bit higher, so a standard 33 μF or 47 μF might be used.

Figure 8.27
Basic inverting switcher

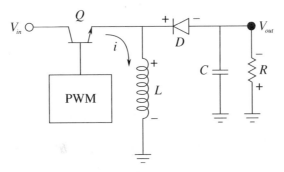

a. Q on: inductor is charged by source

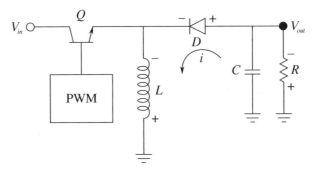

b. Q off: inductor supplies current

A basic inverting switcher is shown in Figure 8.27. This circuit produces a negative output potential from a positive input. It works as follows: When the transistor switch is closed, the inductor L is charged. Diode D is in reverse bias since its anode is negative. When the switch opens, the inductor's collapsing field causes it to appear as a source of opposite polarity. The diode is forward-biased since its cathode is now forced to be lower than its anode. The inductor is now free to deliver current to the load. Eventually, the inductor will discharge to the point where the transistor switch will turn back on. While the inductor is charging, capacitor C will supply the load current. The process repeats, thus maintaining a constant output potential.

A basic step up switcher is shown in Figure 8.28 (page 372). This variation is used when a potential greater than the input is desired, such as deriving a 15 V supply from an existing 5 V source. Here is how the circuit works: When the transistor switch is closed, the inductor L charges. During this time period, the capacitor C is supplying load current. Since the output potential will be much higher than the saturation voltage of the transistor switch, the diode D will be in reverse bias. When the transistor turns off, the magnetic field of the inductor collapses, causing the inductor to appear as a source. This potential is added to the driving source potential since these two elements are in series. This combined voltage is what the load sees; hence, the load voltage is greater

——————   Figure 8.28
Basic step up switching
regulator

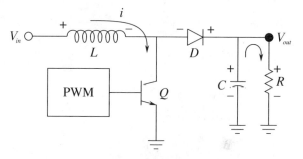

a. Q on: source charges inductor

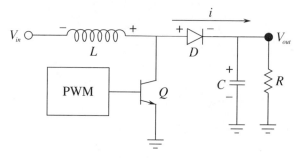

b. Q off: inductor supplies current

than the driving source. Eventually, the inductor will discharge to the point where the transistor switches back on, thus reverse-biasing the diode and recharging L. The capacitor will continue to supply load current during this time. This process will continue in this fashion, producing the desired output voltage.

For the step up and inverter forms, other sets of equations are used for the LM3578. While the design sequence is certainly not quite as straightforward as in the linear regulator circuits, it is definitely not a major undertaking, either. If 750 mA is not sufficient, an external pass transistor can be added to the LM3578. Other switching regulator ICs are available from different manufacturers. Each unit operates on the same basic principle, but the realization of the design may take considerably different routes. Specific device data sheets **must** be consulted for each model.

8.5 Heat Sink Usage

Whenever appreciable amounts of power are dissipated by semiconductor devices, some form of cooling element must be considered. Power-supply regulation circuits are no exception. The pass transistors used in both linear and switching regulators can be forced to dissipate large amounts of power. The result of this is the production of heat. Generally, the lifespan of semiconductors drops as the operation temperature rises. Most silicon-based devices exhibit maximum allowable junction temperatures in the 150°C range. While power transistors utilize heavier metal cases, they are generally not suitable for high dissipation applications by themselves. In order to increase the thermal efficiency of the device, an external heat sink is used. Heat sinks are normally made of aluminum and appear as a series of fins. The fins produce a large surface, which enhances the process of heat convection. In other words, the heat sink can transfer heat to the surrounding atmosphere faster than the power transistor can. By bolting the transistor to the heat sink, the device will be able to dissipate more power at a given operating temperature. Some typical heat sinks are shown in Figure 8.29 on page 374.

Physical Requirements

Heat sinks are designed to work with specific device-case styles. The most common case styles for regulators are the TO-220 "power tab," and the TO-3 "can." Heat sinks are available for these specific styles, including the requisite mounting hardware and insulation spacers. Some of the lower-power regulators utilize TO-5 "mini can" or DIP-type cases. Heat sinks are available for these package types too, but are not quite as common.

There are several general rules which should be followed when using heat sinks:

1. Always use some form of heat sink grease or thermally-conductive pad between the heat sink and the device. This will increase the thermal transfer between the two parts. Note that excessive quantities of heat sink grease will actually **decrease** performance.

2. Mount fins in the vertical plane for optimum natural convective cooling.

3. Do not overcrowd or obstruct devices which use heat sinks.

4. Do not block air flow around heat sinks—particularly directly above and below items which rely on natural convection.

5. If thermal demands are particularly high, consider using forced convection (e.g., fans).

Thermal Resistance

In order to specify a particular heat sink for a given application, a more technical explanation is in order. What we are going to do is create a *thermal circuit equivalent*. In this model, the concept of *thermal resistance* is used. Thermal

Figure 8.29   Typical heat sinks

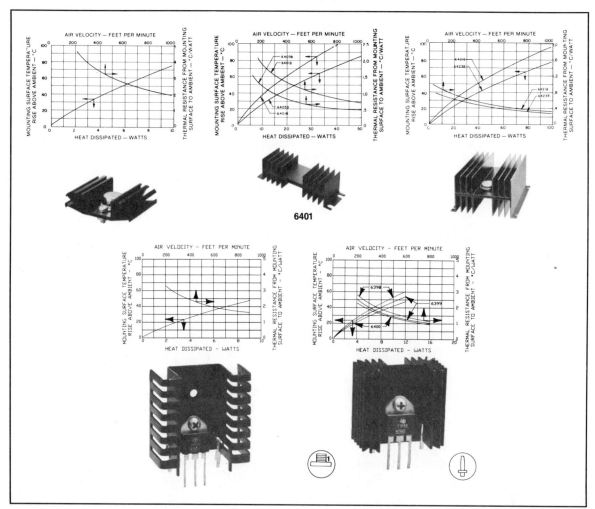

Courtesy of Thermalloy, Inc.

resistance refers to how easy it is to transfer heat energy from one mechanical part to another. The symbol for thermal resistance is θ, and the units are Centigrade degrees per watt. In this model, temperature is analogous to voltage, and thermal power dissipation is analogous to current. A useful equation is

$$P_D = \frac{\Delta T}{\theta_{total}}$$ (8.4)

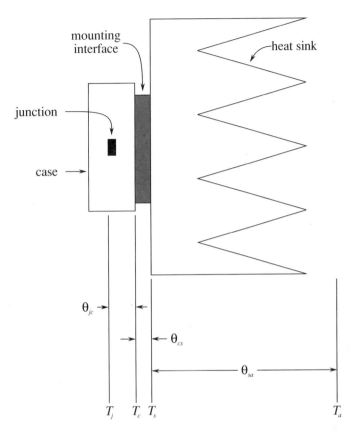

Figure 8.30
Device and heat sink

where P_D is the power dissipated by the semiconductor device in watts, ΔT is the temperature differential, and θ_{total} is the sum of the thermal resistances. Basically, this is a thermal version of Ohm's Law.

In order to construct our model, let's take a closer look at the power-device/heat-sink combination. This is shown in Figure 8.30. T_j is the semiconductor junction temperature. This heat energy source heats the device case to T_c. The thermal resistance between the two entities is θ_{jc}. The case, in turn, heats the heat sink via the interconnection. This thermal resistance is θ_{cs}, and the resulting temperature is T_s. Finally, the heat sink passes the thermal energy to the surrounding air which is at temperature T_a. The thermal resistance of the heat sink is θ_{sa}. The equivalent thermal model is shown in Figure 8.31 on page 376. (While this discussion does not have perfect correspondence with normal circuit analysis, it does illustrate the main points.)

In this model, ground represents absolute zero. Since the circuit is at an ambient temperature T_a, a voltage source of T_a is connected to ground and the heat sink. The three thermal resistances are in series, and are driven by a current source which is set by the present power dissipation of the device. Note

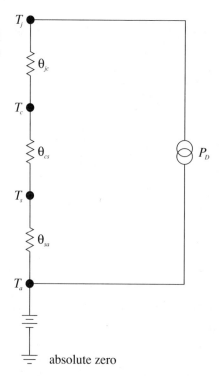

Figure 8.31
Equivalent thermal model
of Figure 8.30

absolute zero

that if the power dissipation is high, the resulting "voltage drops" across the thermal resistances are high. Since voltage is analogous to temperature in this model, this indicates that a high temperature is created. Since there is a maximum limit to T_j, higher power dissipations require lower thermal resistances. Since θ_{jc} is set by the device manufacturer, you have no control over that element. θ_{cs} is a function of the case style and the insulation material used, so you do have some control (but not a lot) over that. On the other hand, as the person who specifies the heat sink, you have a great deal of control over θ_{sa}. Values for θ_{sa} are given by heat sink manufacturers. A useful variation of Equation 8.4 is

$$P_D = \frac{T_j - T_a}{\theta_{jc} + \theta_{cs} + \theta_{sa}} \tag{8.5}$$

Normally, power dissipation, junction and ambient temperatures, θ_{jc} and θ_{cs} are known. The idea is to determine an appropriate heat sink. Both T_j and θ_{jc} are given by the semiconductor device manufacturer. T_a is the ambient temperature and can be determined experimentally. Due to localized warming, it tends to be higher than the actual "room temperature." θ_{cs} can be determined from standard graphs, such as those found in Figure 8.32.

Figure 8.32 θ_{cs} for TO-3 and TO-220

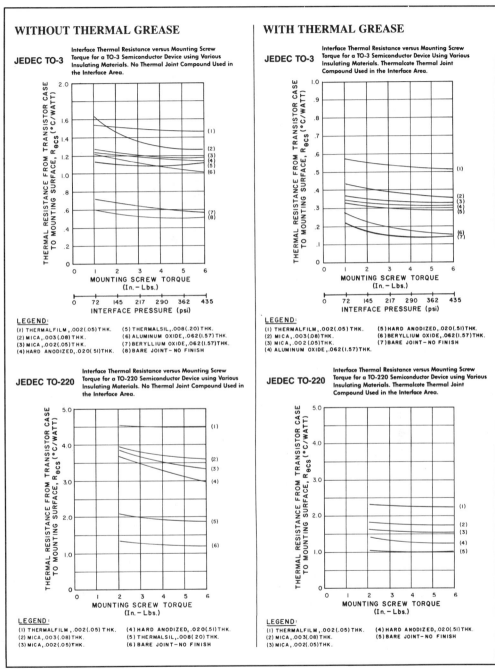

Courtesy of Thermalloy, Inc.

Example 8.7   Determine the appropriate heat-sink rating for a power device rated as follows: $T_{j(max)} = 150°C$, TO-220 case style, $\theta_{jc} = 3.0C°/W$. The device will be dissipating a maximum of 6 W in an ambient temperature of 40°C. Assume that the heat sink will be mounted with heat sink grease and a .002 mica insulator.

    First, find θ_{cs} from the TO-220 graph. Curve 3 is used. The approximate (conservative) value is 1.6C°/W.

$$P_D = \frac{T_j - T_a}{\theta_{jc} + \theta_{cs} + \theta_{sa}}$$

$$\theta_{sa} = \frac{T_j - T_a}{P_D} - \theta_{jc} - \theta_{cs}$$

$$\theta_{sa} = \frac{150°C - 40°C}{6 \text{ W}} - 3.0C°/W - 1.6C°/W$$

$$\theta_{sa} = 13.73C°/W$$

This is the maximum acceptable value for the heat sink's thermal resistance. Note that the use of heat sink grease gives us an extra 2 C°/W or so. Also, note the generally lower values of θ_{sa} for the TO-3 case in comparison to the TO-220. This is one reason why TO-3 cases are used for higher-power devices. This case also makes it easier for the manufacturer to reduce θ_{jc}.

8.6 Extended Topic: Primary Switching Regulators

The switching regulators examined earlier are referred to as *secondary switchers* since the switching elements are found on the secondary side of the power transformer. In contrast to this is the *primary* or *forward switcher*. The switching circuitry in these designs is placed prior to the primary of the power transformer. This positioning offers a distinct advantage over the secondary switcher. Since the power transformer and secondary rectifier will be handling much higher frequencies, they can be made much smaller. The result is a physically smaller and lighter design.

    One possible configuration of a primary switcher is shown in Figure 8.33. This is known as a *push-pull* configuration. The two power transistors are alternately pulsed on and off. That is, when one device is conducting, the other is off. By doing this, opposite polarity pulses are fed into the primary of the transformer, creating a high-frequency alternating current. (As is the case with secondary switchers, primary switchers operate at frequencies well above the nominal 60-Hz power line.) This high-frequency waveform is then stepped up

Figure 8.33   Push-pull primary switching regulator

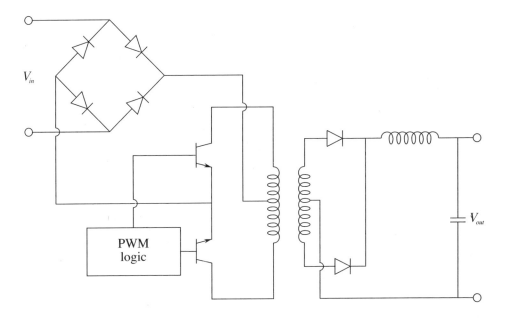

or down to the secondary, where it is again rectified and then filtered, producing a DC output signal.

While the transformer and secondary rectifier/filter may be reduced in size, it is important to note that the main input rectifier and switching transistors are not isolated from the AC power source (as is the case in other power supply designs). These devices must handle high input potentials. For an ordinary 120 V AC line, that translates to over 170 V peak. Also, the power transistors will see an off state potential of twice V_{in}, or over 340 V in this case. Some form of in-rush current limiting will also need to be added.

A somewhat more sophisticated approach is taken in Figure 8.34 (page 380). This circuit is known as a *full bridge* switcher. In this configuration, diagonal pairs of devices (i.e., Q_1/Q_4 and Q_2/Q_3) are simultaneously conducting. By eliminating the center-tapped primary, each device sees a maximum potential of V_{in}, or one-half that of the push-pull switcher. The obvious disadvantage is the need for four power devices instead of just two.

Primary switchers do offer size and weight advantages over secondary switchers and linear regulators. They also maintain the high efficiency characteristics of the secondary switcher. They do tend to be somewhat more complex though, and their application is therefore best suited to cases where circuit size, weight, and efficiency are paramount.

Figure 8.34   Full bridge primary switching regulator

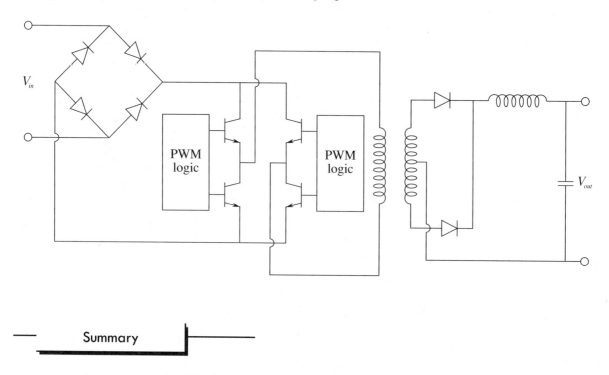

Summary

In this chapter you have examined the basic operation of voltage regulators. Their purpose is simple: to provide constant, nonvarying output voltages despite changes in either the AC source or in the load-current demand. Voltage regulation circuits are an integral part of just about every piece of modern electronic equipment. Due to their wide use, a number of specialized voltage regulator ICs are available from a variety of manufacturers.

Voltage regulation can be achieved through two main methods. These methods are linear regulation and switching regulation. In both cases, a portion of the output voltage is compared to a stable internal reference. The result of this comparison is used to drive a control element, usually a power transistor. If the output voltage is too low, the control element allows more current to flow to the load from the rectified AC source. Conversely, if the output is too high, the control element constricts the current flow. In the case of the linear regulator, the control element is always in the active, or linear, state. Because of this, the linear regulator tends to dissipate quite a bit of power, and as a result, is rather inefficient. On the plus side, the linear regulator is able to quickly react to load variations, and thus exhibits good transient response.

In contrast to the linear regulator, the control device in the switching regulator is either fully on or fully off. As a result, its power dissipation tends to be reduced. For best performance, fast control devices are needed. The device is driven by a pulse-width modulator. The output of this modulator is a rectangular pulse whose duty cycle is proportional to the load-current demand. Since the control device produces current pulses instead of a constant current, some means of smoothing the pulses is necessary. This function is performed by an *LC* filter. The main advantage of the switching regulator is its high efficiency. On the downside, switching regulators are somewhat more difficult to design, do not respond as fast to transient load conditions, and tend to radiate high-frequency interference.

No matter what type of regulator is used, power dissipation can be rather large in the control device, so heat sinking is generally advisable. Heat sinks allow for a more efficient transferral of heat energy to the surrounding atmosphere than the control device exhibits on its own.

Self Test Questions

1. What is the function of a voltage regulator?

2. What is the difference between load regulation and line regulation?

3. Why do regulators need a reference voltage?

4. What is the functional difference between a linear regulator and a switching regulator?

5. What are the main advantages of using linear regulators versus switching regulators?

6. What are the main advantages of using switching regulators versus linear regulators?

7. What is the function of a pass transistor?

8. Describe two ways in which to increase the output current of an IC-based regulator.

9. How can fixed 3-pin regulators be used to regulate at other than their rated voltage?

10. What is the purpose of the output inductor and capacitor in the switching regulator?

11. Explain the correlation between the output current demand and the pulse-width modulator used in switching regulators.

12. What is the purpose of a heat sink?

13. What is meant by the term *thermal resistance*?

14. What are the thermal resistance elements which control heat flow in a typical power-device/heat-sink connection?

15. What are the general rules which should be considered when using heat sinks?

Problem Set

1. If the average input voltage to the circuit of Problem 9 is 22 V, determine the maximum device dissipation for a 900 mA output.

2. If the average input voltage is 25 V for the circuit of Problem 11, determine the maximum output current for each output voltage. Use the TO-220 case style ($P_D = 15$ W, $I_{limit} = 1.5$ A).

3. Draw a block diagram of a complete ± 12 V regulated power supply using LM78XX and LM79XX series parts.

4. Determine the maximum allowable thermal resistance for a heat sink given the following: Ambient temperature = 50°C, maximum operating temperature = 150°C, TO-3 case style with thermal grease and Thermalfilm isolator, power dissipation is 30 W, and the device's thermal resistance is 1.1 C°/W, junction to case.

5. A pass transistor has the following specifications: maximum junction temperature = 125°C, TO-220 case, junction to case thermal resistance = 1.5 C°/W. Determine the maximum power dissipation allowed if this device is connected to a 20 C°/W heat sink with thermal grease, using a .003 mica insulator. The ambient temperature is 35°C.

6. The thermal resistance of the LM723 is 25 C°/W, junction-to-case. Its maximum operating temperature is 150°C. For a maximum dissipation of 500 mW and an ambient temperature of 30°C, determine the maximum allowable thermal resistance for the heat-sink/insulator-interface combination.

⊳ Design Problems

7. Using Figure 8.4, design a 15 V regulator using a 3.3 V zener. The zener bias current should be 2 mA, and the output should be capable of 500 mA.

8. Using Figure 8.4 as a guide, design a variable power supply regulator for a 5 to 15 V range using a 3.9 V zener. $I_{zener} = 3$ mA.

9. Design a $+12$ V regulator using the LM317. The output current capability should be at least 900 mA.

10. Design a $+3$ to $+15$ V regulator using the LM317. The output should be continuously variable.

11. Using the LM317, configure a regulator to produce either $+5$ V, $+12$ V, or $+15$ V.

12. Design a $+12$ V regulator using the LM7805.

13. Design a $+9$ V regulator using the LM723. Use a current limit of 100 mA.

14. Design a $+5$ V regulator with 100 mA current limiting using the LM723.

15. Configure a ± 12 V regulator with 70 mA current limiting. Use the LM326.

16. Reconfigure the circuit of Problem 15 for ± 15 V.

17. Using the LM3578, design a 5 V 400 mA regulator. The input voltage is 15 V. Use a discontinuity factor of .2, and an oscillator frequency of 75 kHz. No more than 10 mV of ripple is allowed.

18. Repeat Problem 17 for a 9 V output.

⊳ Challenge Problems

19. Based on the LM723 adjustable regulator example, design a regulator which will produce a continuously variable output from 5 V to 12 V.

20. The LM317 has a maximum operating temperature of 125°C. The TO-220 case version shows a thermal resistance of 4 C°/W, junction to case. It also shows 50 C°/W, junction to ambient (no heat sink used). Assuming an ambient temperature of 50°C, what is the maximum allowable power dissipation for each setup? Assume that the first version uses a 15 C°/W heat sink with a 2 C°/W case to heat sink interconnection.

21. Forced air cooling of a heat sink device can significantly aid in removing heat energy. As a rule of thumb, forced air cooling at a velocity of 1000 feet per minute will effectively increase the efficiency of a heat sink by a factor of 5. Assuming such a system is applied to the circuit of Problem 20, calculate the new power dissipation.

22. An LM317 (TO-3) is used for a 5 V 1 A power supply. The average voltage into the regulator is 12 V. Assume a maximum operating temperature of 125°C, and an ambient temperature of 25°C. First, determine whether a heat sink is required. If it is, determine the maximum acceptable thermal resistance for the heat sink/insulator combination. For the LM317, thermal resistance $= 2.3$ C°/W, junction to case, and 35 C°/W, junction to ambient.

⊳ SPICE Problems

23. Using SPICE, plot the time domain response of the circuit of Figure 8.12, assuming an input of 22 V with 3 V peak ripple. How does the simulation change if the ripple is increased to 8 V peak?

24. Verify the output waveform for the circuit of Figure 8.16 using SPICE. Use various loads in order to test the current limit operation. The source is 18 V DC, with 2 V peak ripple.

25. Verify the adjustment range for the regulator designed in Example 8.5 in the text using SPICE. Use a load of 200 Ω, and a source equal to 10 V, with 1 V peak ripple.

26. Use several different loads with SPICE in order to test the current limit portion of the regulator designed in Example 8.5 in the text.

9 Oscillators and Frequency Generators

After completing this chapter, you should be able to:

❑ Explain the differences between positive and negative feedback.

❑ Define the *Barkhausen criterion*, and relate it to individual circuits.

❑ Detail the need for level-detecting circuitry in a practical oscillator.

❑ Analyze the operation of Wien bridge and phase shift op amp oscillators.

❑ Analyze the operation of comparator-based op amp oscillators.

❑ Detail which factors contribute to the accuracy of a given oscillator.

❑ Explain the operation of a VCO.

❑ Explain the operation of a PLL, and define the terms *capture range* and *lock range*.

9.1 Introduction

Oscillators are signal sources. Many times, it is necessary to generate waveforms with a known wave shape, frequency, and amplitude. A laboratory signal generator perhaps first comes to mind, but there are many other applications. Signal sources are needed to create and receive radio and television signals, to time events, and to create electronic music, among other uses. Oscillators can produce very low frequencies (a fraction of a cycle per second) to very high frequencies (microwaves, >1 GHz). Oscillators employing op amps are generally used in the range below 1 MHz. Specialized linear circuits can be used at much higher frequencies. The output wave shape may be sinusoidal, triangular, pulse, or some other shape. Oscillators can generally be broken into two broad categories: fixed frequency or variable. For many fixed-frequency oscillators, absolute accuracy and freedom from drift are of prime importance. For variable oscillators, ease of tuning and repeatability are usually important. Also, a variable-frequency oscillator might not be directly controlled by human hands; rather, the oscillator may be tuned by another circuit. A VCO, or *voltage-controlled oscillator*, is one example of this. Depending on the application, other factors such as total harmonic distortion or rise time may be important.

No matter what the application or how the oscillator design is realized, the oscillator circuit will normally employ positive feedback. Unlike negative feedback, positive feedback is regenerative—it reinforces change. Generally speaking, without some form of positive feedback, oscillators could not be built. In this chapter we are going to look at positive feedback and the requirements for oscillation. A variety of small oscillators based on op amps will be examined. Finally, more powerful integrated circuits will be discussed.

Positive Feedback and the Barkhausen Criterion

In earlier work, we examined the concept of negative feedback. Here, a portion of the output signal is sent back to the input, and summed out of phase with the input signal. The difference between the two signals is what is amplified. The result is a stability in the circuit response since the large open-loop gain effectively forces the difference signal to be very small. Something quite different happens if the feedback signal is summed in phase with the input signal, as shown in Figure 9.1. In this case, the combined signal looks just like the output signal. As long as the open-loop gain of the amplifier is larger than the feedback factor, the signal can be constantly regenerated. This means that the signal source can be removed. In effect, the output of the circuit is used to create its own input. As long as power is maintained to the circuit, the output signal will continue practically forever. This self-perpetuating state is called *oscillation*.

Figure 9.1
Positive feedback

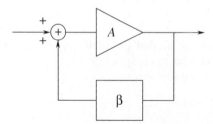

Oscillation will cease if the product of open-loop gain and feedback factor falls below unity, or if the feedback signal is not returned perfectly in phase (0° or some integer multiple of 360°). This combination of factors is called the *Barkhausen oscillation criterion*. We may state this as follows:

> In order to maintain self-oscillation, the closed-loop gain must be unity or greater, and the loop phase must be $N360°$, where $N = 0, 1, 2, 3 \ldots$.

Note that when we examined linear amplifiers, we looked at this from the opposite end. Normally, you don't want amplifiers to oscillate, and thus you try to guarantee that the Barkhausen criterion is never met by setting appropriate gain and phase margins.

A good example of positive feedback is the "squeal" sometimes heard from improperly adjusted public address systems. Basically, the microphone is constantly picking up the ambient room noise, which is then amplified and fed to the loudspeakers. If the amplifier gain is high enough, or if the acoustic loss is low enough (i.e., the loudspeaker is physically close to the microphone), the signal which the microphone picks up from the loudspeaker can be greater than the ambient noise. The result is that the signal constantly grows in proper phase to maintain oscillation, producing the familiar squealing sound. In order to stop the squeal, either the gain or the phase must be disrupted. Moving the microphone may change the relative phase, but it is usually easier to just reduce the volume a little. It is particularly interesting to listen to a system which is on the verge of oscillation. Either the gain or the phase is just not quite perfect, and the result is a rather irritating ringing sound as the oscillation dies out after each word or phrase.

There are some practical considerations to be aware of when designing oscillators. First of all, it is not necessary to provide a "start-up" signal source as seen in Figure 9.1. Normally, there is enough energy in either the input noise level or possibly in a turn-on transient to get the oscillator started. Both the turn-on transient and the noise signal are broad spectrum signals, so the desired oscillation frequency is contained within either of them. The oscillation signal will start to increase as time progresses due to the closed-loop gain being greater than one. Eventually, the signal will reach a point where further level increases are impossible due to amplifier clipping. For a more controlled low-distortion oscillator, it is desirable to have the gain start to roll off before clipping occurs. In other words, the closed-loop gain should fall back to exactly 1. Finally, to minimize frequency drift over time, the feedback network should be selective. Frequencies either above or below the target frequency should see greater attenuation than the target frequency. Generally, the more selective (i.e., higher Q) this network is, the more stable and accurate the oscillation frequency will be. One simple solution is to use an *RLC* tank circuit in the feedback network. Another possibility is to use a piezo-electric crystal. A block diagram of the practical oscillator is shown in Figure 9.2 (page 388).

Figure 9.2
Practical oscillator

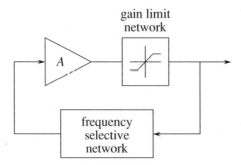

A Basic Oscillator

A real-world circuit which embodies all of the elements is shown in Figure 9.3. This circuit is not particularly efficient or cost-effective, but it does illustrate the important points. Remember, in order to maintain oscillation the closed-loop gain of the oscillator circuit must be greater than one, and the loop phase must be a multiple of 360°. To provide gain, a pair of inverting amplifiers is used. Note that op amp 2 serves to buffer the output signal. Since each stage produces a 180° shift, the shift for the pair is 360°. The product of the gains has to be larger than the loss produced by the frequency selecting network. This network is made up of R_3, L, and C. Since the LC combination produces an impedance peak at the resonant frequency, f_o, a minimum loss will occur. Also, at resonance, the circuit is basically resistive, so no phase change occurs.

Figure 9.3 A basic oscillator

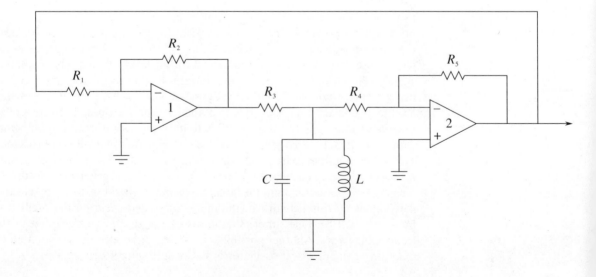

Consequently, this circuit should oscillate at the f_o set by L and C. This circuit can be easily tested in lab. For example, if you drop the gain of one of the op amp stages, there will not be enough system gain to overcome the tank circuit's loss, and thus oscillation will cease. You can also verify the phase requirement by replacing one of the inverting amplifiers with a noninverting amplifier of equal gain. The resulting loop phase of 180° will halt oscillation. This circuit does not include any form of automatic gain adjustment so the output signal may be clipped. If properly chosen, the slew rate of the op amp can be used as the limit factor. (A 741 will work acceptably for f_o in the low kilohertz range.) While this circuit does work, and points out the specifics, it is certainly not a top choice for an oscillator design based on op amps.

Wien Bridge Oscillators

A relatively straightforward design useful for general-purpose work is the *Wien bridge oscillator*. This oscillator is far simpler than the generalized design shown in Figure 9.3, and offers very good performance. The frequency-selecting network is a simple lead/lag circuit, such as that shown in Figure 9.4. This circuit is a frequency-sensitive voltage divider. It combines the response of both the simple lead and lag networks. Normally, both resistors are set to the same value. The same may be said of the two capacitors. At very low frequencies, the capacitive reactance is essentially infinite, and thus, the upper series capacitor is an open. Because of this, the output voltage is zero. Likewise, at very high frequencies the capacitive reactance approaches zero, and the lower shunt capacitor effectively shorts the output to ground. Again, the output voltage is zero. At some middle frequency the output voltage will be at a peak. This will be the preferred, or selected frequency, and will become the oscillation frequency as long as the proper phase relation is held. We need to determine the phase change at this point as well as the voltage divider ratio. These items are needed in order to guarantee that the Barkhausen conditions are met.

Figure 9.4
Lead/lag network

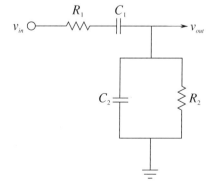

First, note that

$$\beta = \frac{Z_2}{Z_1 + Z_2}$$

where $Z_1 = R_1 - jX_{C_1}$ and $Z_2 = R_2 \parallel -jX_{C_2}$.

$$Z_2 = \frac{-jX_{C_2}R_2}{-jX_{C_2} + R_2}$$

$$Z_2 = \frac{-jX_{C_2}R_2}{-jX_{C_2}\left(1 + \frac{R_2}{-jX_{C_2}}\right)}$$

$$Z_2 = \frac{R_2}{1 + \frac{R_2}{-jX_{C_2}}}$$

Recalling that $X_C = 1/\omega C$, we find that

$$Z_1 = R_1 - \frac{j}{\omega C_1}$$

$$Z_2 = \frac{R_2}{1 + j\omega R_2 C_2}$$

So,

$$\beta = \frac{\dfrac{R_2}{1 + j\omega R_2 C_2}}{\dfrac{R_2}{1 + j\omega R_2 C_2} + R_1 - \dfrac{j}{\omega C_1}}$$

$$\beta = \frac{R_2}{R_2 + R_1 - \dfrac{j}{\omega C_1} + j\omega R_1 R_2 C_2 + \dfrac{R_2 C_2}{C_1}}$$

$$\beta = \frac{R_2}{R_2\left(1 + \dfrac{C_2}{C_1}\right) + R_1 + j\left(\omega R_1 R_2 C_2 - \dfrac{1}{\omega C_1}\right)} \tag{9.1}$$

We can determine the desired frequency from the imaginary portion of Equation 9.1.

$$jwR_1R_2C_2 = \frac{1}{\omega C_1}$$

$$\omega^2 = \frac{1}{R_1R_2C_1C_2}$$

$$\omega = \frac{1}{\sqrt{R_1R_2C_1C_2}} \tag{9.2}$$

Normally, $C_1 = C_2$ and $R_1 = R_2$, so Equation 9.2 reduces to

$$\omega = \frac{1}{RC}$$

or

$$f_o = \frac{1}{2\pi RC} \tag{9.3}$$

To find the magnitude of the feedback factor, and thus the required forward gain of the op amp, we need to examine the real portion of Equation 9.1:

$$\beta = \frac{R_2}{R_2\left(1 + \dfrac{C_2}{C_1}\right) + R_1}$$

Assuming that equal components are used, this reduces to

$$\beta = \frac{R}{3R}$$

or simply

$$\beta = \frac{1}{3}$$

The end result is that the forward gain of the amplifier must have a gain of slightly over 3 and a phase of $0°$ in order to maintain oscillation. It also means that the frequency of oscillation is fairly easy to set, and can even be adjusted if potentiometers are used to replace the two resistors.

The final circuit is shown in Figure 9.5 on page 392. This circuit uses a combination of negative feedback and positive feedback to achieve oscillation. The positive feedback loop utilizes R_t and C. The negative feedback loop utilizes R_a and R_b. R_b must be approximately twice the size of R_a. If it is smaller, the $A\beta$ product will be less than unity and oscillation can not be maintained. If the gain is significantly larger, excessive distortion may result. Indeed, some form of gain reduction at higher output voltages is desired for this circuit. One possibility is to replace R_a with a lamp. As the signal amplitude increases across

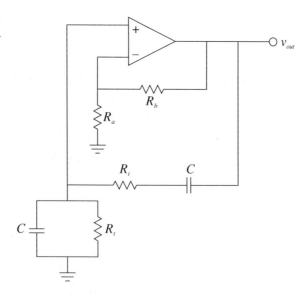

Figure 9.5
Wien bridge oscillator

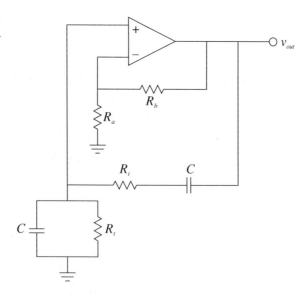

the lamp, its resistance increases, thus decreasing gain. At a certain point the lamp's resistance will be just enough to produce an $A\beta$ product of exactly 1.

Another technique is shown in Figure 9.6. Here an opposite approach is taken. Resistor R_b is first broken into two parts. The smaller part, R_{b2}, is shunted by a pair of signal diodes. For lower amplitudes, the diodes are off and do not affect the circuit operation. At higher amplitudes, the diodes start to turn on, and thus start to short R_{b2}. If correctly implemented, this action

Figure 9.6
Wien bridge oscillator
with gain adjustment

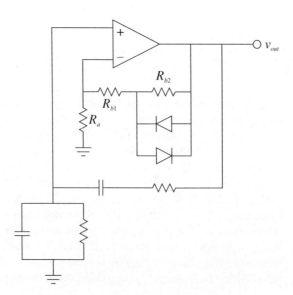

Figure 9.7
Wien bridge oscillator,
redrawn

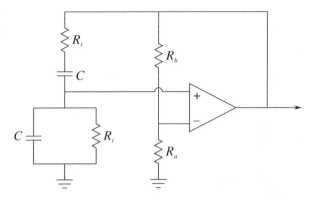

is not instantaneous, and does not produce clipping. It simply serves to reduce the gain at higher amplitudes.

Another way of drawing the Wien bridge oscillator is shown in Figure 9.7. This form clearly shows the Wien bridge configuration. Note that the output of the bridge is the differential input voltage (i.e., error voltage). In operation, the bridge is balanced, and thus, the error voltage is zero.

Example 9.1   Determine the frequency of oscillation for the circuit of Figure 9.8.

$$f_o = \frac{1}{2\pi RC}$$

$$f_o = \frac{1}{2\pi \times 50\text{ k} \times .01\,\mu\text{F}}$$

$$f_o = 318 \text{ Hz}$$

Figure 9.8
Oscillator for Example 9.1

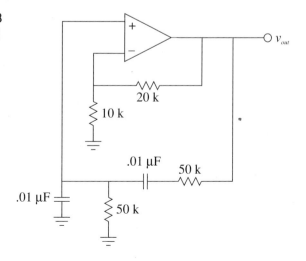

For other frequencies, either R or C can be altered as required. Also, note that the forward gain works out to exactly 3 thus perfectly compensating the positive feedback factor of $\frac{1}{3}$. In reality, component tolerances make this circuit impractical. To overcome this difficulty, a small resistor/diode combination can be placed in series with the 20 k, as shown in Figure 9.6. A typical resistor value would be about one-fourth to one-half the value of R_b, or about 5 k to 10 k in this example. R_b would be decreased slightly as well (or, R_a might be increased).

The ultimate accuracy of f_o depends on the tolerances of R and C. If 10% parts are used in production, a variance of about 20% is possible. Also, at higher frequencies, the op amp circuit will produce a moderate phase shift of its own. Thus the assumption of a perfect noninverting amplifier is no longer valid, and some error in the output frequency will result. With extreme values in the positive feedback network, it is also possible that some shift of the output frequency may occur due to the capacitive and resistive loading effects of the op amp. Normally, this type of loading is not a problem since the op amp's input resistance is very high, and its input capacitance is quite low.

Example 9.2   Figure 9.9 shows an adjustable oscillator. Three sets of capacitors are used to change the frequency range, while a dual-gang potentiometer is used to adjust the frequency within a given range. Determine the maximum and minimum frequency of oscillation within each range.

First, note that the capacitors are spaced by decades. This means that the resulting frequency ranges will also change by factors of ten. The .1 μF capacitor will produce the lowest range, the .01 μF will produce a range ten times higher, and the .001 μF range will be ten times higher still. Thus, we only need to calculate the range produced by the .1 μF capacitor.

The maximum frequency of oscillation within a given range will occur with the lowest possible resistance. The minimum resistance is seen when the 10 k pot is fully shorted, the result being 1.1 k. Conversely, the minimum frequency will occur with the largest resistance. When the pot is fully in the circuit, the resulting sum is 11.1 k. Note that a dual-gang potentiometer means that both units are connected to a common shaft. Thus, both pots track in tandem.

For f_{min} with .1 μF:

$$f_o = \frac{1}{2\pi RC}$$

$$f_o = \frac{1}{2\pi \times 11.1 \text{ k} \times .1 \ \mu F}$$

$$f_o = 143.4 \text{ Hz}$$

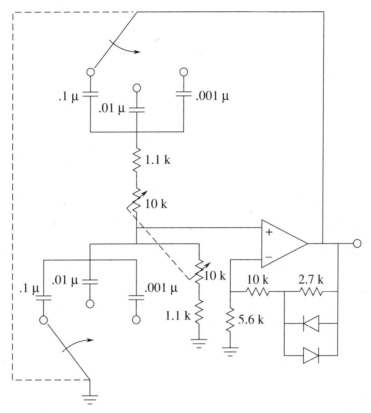

Figure 9.9
Adjustable oscillator

For f_{max} with .1 μF:

$$f_o = \frac{1}{2\pi RC}$$

$$f_o = \frac{1}{2\pi \times 1.1 \text{ k} \times .1 \ \mu\text{F}}$$

$$f_o = 1.447 \text{ kHz}$$

For the .01 μF capacitor, the ranges will be 1.434 kHz to 14.47 kHz, and for the .001 μF capacitor, the ranges will be 14.34 kHz to 144.7 kHz. Note that each range picks up where the previous one left off. In this way, there are no "gaps," or unobtainable frequencies. For stable oscillation, this circuit must have a gain of 3.

For low level outputs, the diodes will not be active, and the forward gain will be

$$A_v = 1 + \frac{R_J}{R_i}$$

$$A_v = 1 + \frac{10\text{ k} + 2.7\text{ k}}{5.6\text{ k}}$$

$$A_v = 3.27$$

As the signal rises, the diodes begin to turn on, thus shunting the 2.7 k resistor, and dropping the gain back to exactly 3.

Phase Shift Oscillators

Considering the Barkhausen criterion, it should be possible to create an oscillator by using a simple phase shift network in the feedback path. For example, if the circuit uses an inverting amplifier ($-180°$ shift), a feedback network with an additional 180° shift should create oscillation. The only other requirement is that the gain of the inverting amplifier be greater than the loss produced by the feedback network. This is illustrated in Figure 9.10. The feedback network can be as simple as three cascaded lead networks. The lead networks will produce a combined phase shift of 180° at only one frequency. This will become the frequency of oscillation. In general, the feedback network will look something like the circuit of Figure 9.11. This form of RC layout is usually referred to as a *ladder network*.

There are many ways in which R and C can be set in order to create the desired 180° shift. Perhaps the most obvious scheme is to set each stage for a

Figure 9.10
Block diagram of phase shift oscillator

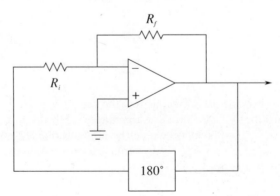

Figure 9.11
Phase shift network

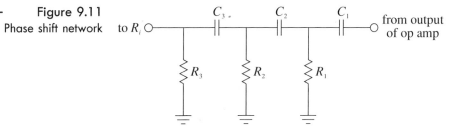

Figure 9.11
Phase shift network

60° shift. The components are determined by finding a combination which produces a 60° shift at the desired frequency. In order to avoid loading effects, each stage must be set to higher and higher impedances. For example, R_2 might be set to 10 times the value of R_1, and R_3 set to 10 times the value of R_2. The capacitors will see a corresponding decrease. Since the tangent of the phase shift yields the ratio of X_C to R, at our desired 60° we find

$$\tan 60° = \frac{X_C}{R}$$

$$1.732 = \frac{X_C}{R}$$

$$X_C = 1.732R$$

Using this in the general reactance formula produces

$$f_o = \frac{1}{2\pi 1.732RC} \tag{9.4}$$

Likewise, a look at the magnitude shows the approximate loss per stage:

$$\beta = \frac{R}{\sqrt{R^2 + X_C^2}}$$

$$\beta = \frac{R}{\sqrt{R^2 + (1.732R)^2}}$$

$$\beta = .5$$

Since there are three stages, the total loss for the feedback network will be .125. Therefore, the inverting amplifier needs a gain of 8 in order to set the $A\beta$ product to unity. Remember, these results are approximate, and will depend on minimum interstage loading. A more exacting analysis will follow shortly.

——— **Example 9.3** Determine the frequency of oscillation in Figure 9.12.

$$f_o = \frac{1}{2\pi 1.732 RC}$$

$$f_o = \frac{1}{2\pi \times 1.732 \times 1\,k \times .1\,\mu F}$$

$$f_o = 919\ Hz$$

_____

Figure 9.12 graphically reveals the major problem with the "60° per stage" concept. In order to prevent loading, the final resistors must be very high. In this case a feedback resistor of 8 MΩ is required. It is possible to simplify the circuit somewhat by omitting the 1 MΩ and connecting the 100 kΩ directly to the op amp as shown in Figure 9.13. This saves one part and does allow the feedback resistor to be dropped in value, but the resulting component spread is still not ideal. If we accept the loading effect, we can simplify things a bit by making each resistor and capacitor the same size, as shown in Figure 9.14. With these values, we can be assured that the resulting frequency will no longer be 919 Hz since Equation 9.4 is no longer valid. Also, it is quite likely that the loss produced by the network will no longer be .125. We need to determine the general input/output ratio of the ladder network, and from this, find the gain and frequency relations for a net phase shift of $-180°$. It will be easiest if we start at the output of the ladder (v_3), and work toward the input (v_0). We will continually strive to keep the equations in terms of the input and output quantities. Since all resistors and capacitors are equal in this variation,

——— **Figure 9.12**
Phase shift oscillator
(minimum loading form)

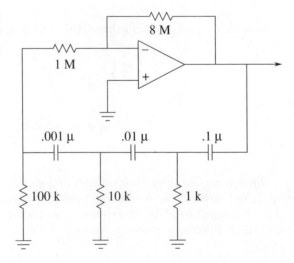

Figure 9.13
Improved phase shift
oscillator

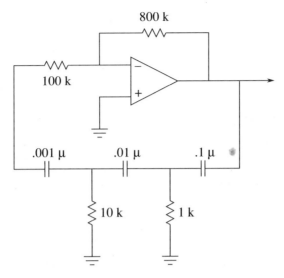

we will be able to simplify our equations readily. By inspection,

$$v_3 = Ri_3$$

which may also be written as

$$i_3 = \frac{v_3}{R} \tag{9.5}$$

v_2 is the sum of the drop across the final capacitor and v_3:

$$v_2 = v_3 + i_3 X_C \tag{9.6}$$

Substituting Equation 9.5 yields

$$v_2 = v_3 + \frac{v_3 X_C}{R}$$

$$v_2 = v_3\left(1 + \frac{X_C}{R}\right) \tag{9.7}$$

Figure 9.14
Phase shift analysis

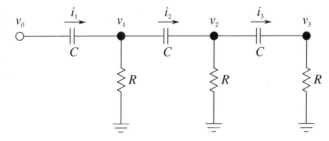

v_1 is the sum of the drop across the middle capacitor and v_2:

$$v_1 = v_2 + i_2 X_C \tag{9.8}$$

i_2 is the sum of two branch currents, i_3, and the current through the middle resistor:

$$i_2 = \frac{v_2}{R} + i_3 \tag{9.9}$$

Using Equation 9.5 this becomes

$$i_2 = \frac{v_2}{R} + \frac{v_3}{R} \tag{9.10}$$

Substituting Equation 9.10 into Equation 9.8 yields

$$v_1 = v_2 + \frac{X_C v_2}{R} + \frac{X_C v_3}{R}$$

$$v_1 = v_2\left(1 + \frac{X_C}{R}\right) + \frac{X_C v_3}{R} \tag{9.11}$$

Rewriting the v_2 factor in terms of v_3 using Equation 9.7 produces

$$v_1 = v_3\left(1 + \frac{X_C}{R}\right)\left(1 + \frac{X_C}{R}\right) + \frac{X_C v_3}{R}$$

$$v_1 = v_3\left(1 + \frac{2X_C}{R} + \frac{X_C^2}{R^2} + \frac{X_C}{R}\right)$$

$$v_1 = v_3\left(1 + \frac{3X_C}{R} + \frac{X_C^2}{R^2}\right) \tag{9.12}$$

v_0 is the sum of the drop across the first capacitor and v_1:

$$v_0 = v_1 + i_1 X_C \tag{9.13}$$

i_1 is reduced as in the preceding fashion. Note the use of Equation 9.10 for i_2:

$$i_1 = \frac{v_1}{R} + i_2$$

$$i_1 = \frac{v_1}{R} + \frac{v_2}{R} + \frac{v_3}{R} \tag{9.14}$$

Substituting Equation 9.14 into Equation 9.13 yields

$$v_0 = v_1 + \frac{X_c v_1}{R} + \frac{X_c v_2}{R} + \frac{X_c v_3}{R}$$

$$v_0 = v_1\left(1 + \frac{X_c}{R}\right) + \frac{X_c v_2}{R} + \frac{X_c v_3}{R} \tag{9.15}$$

Equation 9.15 is now expanded by using Equation 9.7 in place of v_2:

$$v_0 = v_1\left(1 + \frac{X_c}{R}\right) + v_3\left(1 + \frac{X_c}{R}\right)\frac{X_c}{R} + \frac{X_c v_3}{R} \tag{9.16}$$

After multiplying out the middle factors and combining terms, Equation 9.16 becomes

$$v_0 = v_1\left(1 + \frac{X_c}{R}\right) + v_3\left(\frac{2X_c}{R} + \frac{X_c^2}{R^2}\right) \tag{9.17}$$

The v_1 term of Equation 9.17 is now replaced with Equation 9.12:

$$v_0 = v_3\left(1 + \frac{3X_c}{R} + \frac{X_c^2}{R^2}\right)\left(1 + \frac{X_c}{R}\right) + v_3\left(\frac{2X_c}{R} + \frac{X_c^2}{R^2}\right) \tag{9.18}$$

$$v_0 = v_3\left(1 + \frac{4X_c}{R} + \frac{4X_c^2}{R^2} + \frac{X_c^3}{R^3}\right) + v_3\left(\frac{2X_c}{R} + \frac{X_c^2}{R^2}\right)$$

$$v_0 = v_3\left(1 + \frac{6X_c}{R} + \frac{5X_c^2}{R^2} + \frac{X_c^3}{R^3}\right) \tag{9.19}$$

At this point, we are nearly finished with the general equation. All that is left is to substitute $1/j\omega C$ in place of X_C. Remember, $j^2 = -1$.

$$v_0 = v_3\left(1 + \frac{6}{j\omega CR} - \frac{5}{\omega^2 C^2 R^2} - \frac{1}{j\omega^3 C^3 R^3}\right) \tag{9.20}$$

This equation contains both real and imaginary terms. For this equation to be satisfied, the imaginary components ($6/j\omega CR$ and $1/j\omega^3 C^3 R^3$) must sum to zero, and the real components must similarly sum to zero. (Since there are only two terms for each, their magnitudes must be equal.) We can use these facts to find both the gain and the frequency.

$$\frac{6}{j\omega CR} = \frac{1}{j\omega^3 C^3 R^3}$$

$$\frac{1}{j\omega CR} = \frac{1}{6j\omega^3 C^3 R^3}$$

$$1 = \frac{1}{6\omega^2 C^2 R^2}$$

$$\omega^2 = \frac{1}{6C^2 R^2} \tag{9.21}$$

$$\omega = \frac{1}{\sqrt{6}CR} \tag{9.22}$$

or

$$f_o = \frac{1}{2\pi\sqrt{6}CR} \tag{9.23}$$

For the gain, we solve Equation 9.20 in terms of the voltage ratio and zero the imaginary terms, since the result must be real.

$$\frac{v_0}{v_3} = 1 - \frac{5}{\omega^2 C^2 R^2} \tag{9.24}$$

Substituting Equation 9.21 into Equation 9.24 yields

$$\frac{v_0}{v_3} = 1 - \frac{5}{\dfrac{1}{6C^2 R^2} C^2 R^2} \tag{9.25}$$

$$\frac{v_0}{v_3} = 1 - 5 \times 6$$

$$\frac{v_0}{v_3} = -29 \tag{9.26}$$

The gain of the ladder network is v_3/v_0 (the reciprocal of Equation 9.26), or

$$\beta = \frac{1}{-29} \tag{9.27}$$

The loss produced will be $\frac{1}{29}$. This has the disadvantage of requiring a forward gain of 29 instead of 8 (as in the previous form). This disadvantage is minor compared to the advantage of reasonable component values.

Figure 9.15
Equal component phase
shift oscillator

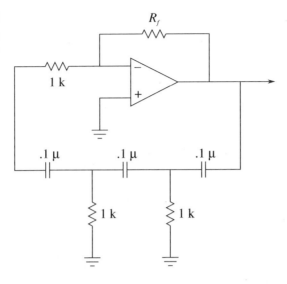

Example 9.4   Determine a value for R_f in Figure 9.15 in order to maintain oscillation. Also determine the oscillation frequency.

Equation 9.27 shows that the inverting amplifier must have a gain of 29.

$$A_v = -\frac{R_f}{R_i}$$

$$R_f = -R_i A_v$$

$$R_f = -1\,\text{k} \times -29$$

$$R_f = 29\,\text{k}$$

Of course, the higher standard value will be used. Also, in order to control the gain at higher levels, a diode/resistor combination (as used in the Wien bridge circuits) should be placed in series with R_f. Without a gain-limiting circuit, excessive distortion may occur.

$$f_o = \frac{1}{2\pi\sqrt{6}RC}$$

$$f_o = \frac{1}{2\pi\sqrt{6} \times 1\,\text{k} \times .1\,\mu\text{F}}$$

$$f_o = 650\,\text{Hz}$$

In order to verify Equations 9.3 and 9.23, the gain and phase responses of the feedback networks of Figures 9.13 and 9.15 are found in Figure 9.16, along

Figure 9.16
SPICE simulation of phase
shift networks

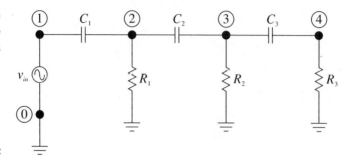

a. SPICE schematic

```
Equal value 3 stage RC network for phase shift oscillator
**********Main circuit description**********
C1      1  2  .1U
R1      2  0  1K
C2      2  3  .1U
R2      3  0  1K
C3      3  4  .1U
R3      4  0  1K
VIN     1  0  AC 1

********Analysis options*******

*Calculate 60 points per decade from 400 Hz to 1 KHz.
.AC DEC 60 400 1K
*
*Plot the output amplitude and phase versus frequency (node 4).
.PLOT AC V(4)
.PLOT AC VP(4)
.END
*
Staggered value 3 stage RC network for phase shift oscillator
**********Main circuit description*********
C1      1  2  .1U
R1      2  0  1K
C2      2  3  .01U
R2      3  0  10K
C3      3  4  .001U
R3      4  0  100K
VIN     1  0  AC 1

********Analysis options************
*
*Calculate 60 points per decade from 600 Hz to 1.2 KHz.
.AC DEC 60 600 1.2K
*
*Plot the output amplitude and phase versus frequency (node 4).
.PLOT AC V(4)
.PLOT AC VP(4)
.END
```

b. SPICE input files

Figure 9.16
(*Continued*)

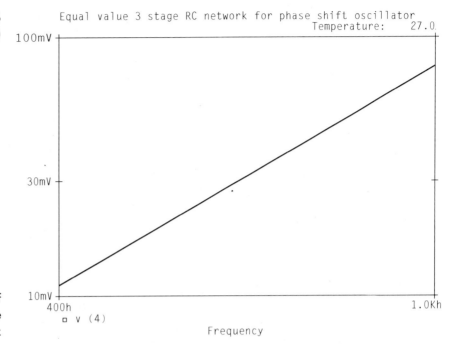

c. Amplitude response of
equal value three-stage
network

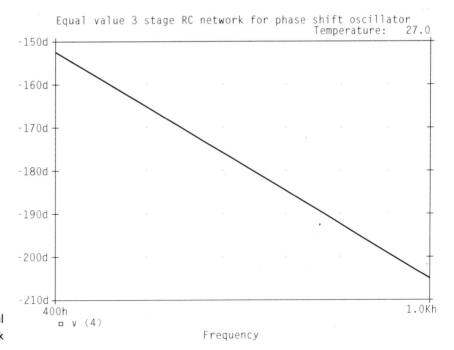

d. Phase response of equal
value three-stage network

(*Continued*)

Figure 9.16
(*Continued*)

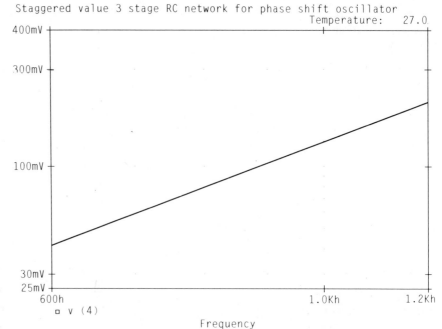

e. Amplitude response
of staggered value
three-stage network

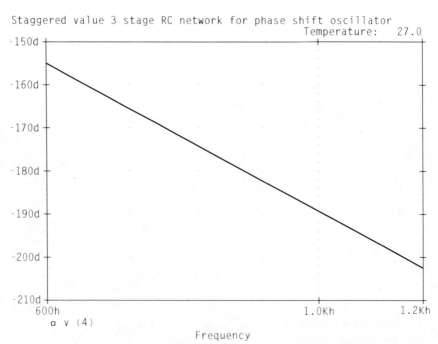

f. Phase response of
staggered value three-
stage network

with the SPICE input files required to generate them. These graphs were obtained using Probe™. Note that the phase for both circuits hits 180° at the predicted frequencies. Likewise, the gain response agrees with the derivations for attenuation at the oscillation frequency. It can be very instructive to analyze these circuits for the gain and phase response at each stage as well.

Square/Triangle Function Generators

Besides generating sine waves, op amp circuits can be employed to generate other wave shapes such as ramps, triangle waves, or pulses. Generally speaking, square- and pulse-type waveforms can be derived from other sources through the use of a comparator. For example, a square wave can be derived from a sine wave by passing it through a comparator, such as those seen in Chapter 7. Linear waveforms such as triangles and ramps can be derived from the charge/discharge action of a capacitor.

As you may recall from basic circuit theory, the voltage across a capacitor will rise linearly if it is driven by a constant current source. One way of achieving this linear rise is with the circuit of Figure 9.17. In essence, this circuit is an inverting amplifier with a capacitor taking the place of R_f. The input resistor, R, turns the applied input voltage into a current. Since the current into the op amp itself is negligible, this current flows directly into capacitor C. As in a normal inverting amplifier, the output voltage is equal to the voltage across the feedback element, but inverted. The relationship between the capacitor current and voltage is

$$\frac{dv}{dt} = \frac{i}{C} \tag{9.28}$$

$$v(t) = \frac{1}{C} \int i \, dt$$

$$v_{out} = -\frac{1}{C} \int i \, dt \tag{9.29}$$

Figure 9.17
Ramp generator

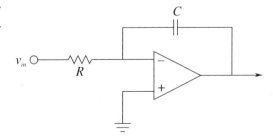

Figure 9.18
Ramp generator
waveforms

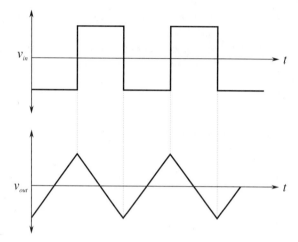

As expected, a fast rise will be caused by either a small capacitor or a large current. (As a side note, this circuit is called an *integrator*, and will be examined in greater detail in the next chapter.)

By choosing appropriate values for R and C, the v_{out} ramp can be set at a desired rate. The polarity of the ramp's slope is determined by the direction of the input current; a positive source will produce a negative-going ramp and vice versa. If the polarity of the input changes at a certain rate, the output ramp will change direction in tandem. The net effect is a triangle wave. A simple way to generate the alternating input polarity is to drive R with a square wave. As the square wave changes from positive to negative, the ramp changes direction. This is shown in Figure 9.18. So, we are now able to generate a triangle wave. The only problem is that a square wave source is needed. How do we produce the square source? As mentioned earlier, a square wave can be derived by passing an AC signal through a comparator. Logically then, we should be able to pass the output triangle wave into a comparator in order to create the needed square wave. The resulting circuit is shown in Figure 9.19. A comparator with hysteresis is used to turn the triangle into a square wave. The square wave then drives the ramp circuit. The circuit produces two simultaneous outputs: a square wave which swings to ± saturation, and a triangle wave which swings to the upper and lower comparator thresholds. This is illustrated in Figure 9.20. The thresholds can be determined from the equations presented in Chapter 7. In order to determine the output frequency, the volt/second rate of the ramp is determined from Equation 9.28. Since we know that the peak-to-peak swing of the triangle is $V_{upper\ thres} - V_{lower\ thres}$, the period of the wave can be found. The output frequency is the reciprocal of the period.

Figure 9.19   Triangle/square generator

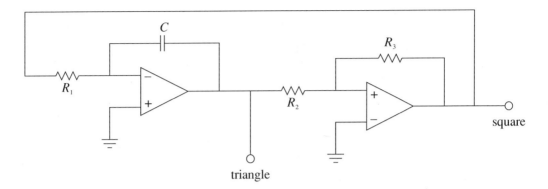

Figure 9.20
Output waveforms of
triangle/square generator

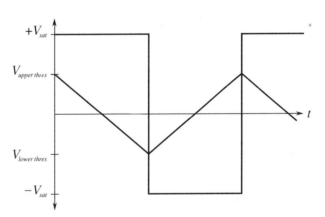

Example 9.5   Determine the output frequency and amplitudes for the circuit of Figure 9.21
(page 410). Use $V_{sat} = \pm 13$ V.

First, note that the comparator always swings between $+V_{sat}$ and $-V_{sat}$.
Now, determine the upper and lower thresholds for the comparator.

$$V_{upper\ thres} = V_{sat} \frac{R_2}{R_1}$$

$$V_{upper\ thres} = 13 \text{ V} \frac{10 \text{ k}}{20 \text{ k}}$$

$$V_{upper\ thres} = 6.5 \text{ V}$$

The lower threshold will be -6.5 V. We now know that the triangle wave
output will be 13 V peak to peak. From this we can determine the output

Figure 9.21   Signal generator for Example 9.5

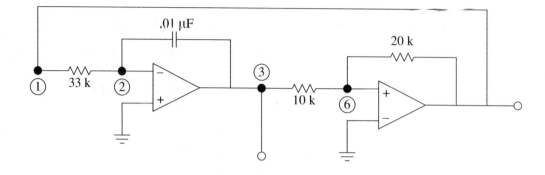

period. Since the ramp generator is driven by a square wave with an amplitude of V_{sat}, Equation 9.28 may be rewritten as

$$\frac{dv}{dt} = \frac{V_{sat}}{RC}$$

$$\frac{dv}{dt} = \frac{13 \text{ V}}{33 \text{ k} \times .01 \text{ } \mu\text{F}}$$

$$\frac{dv}{dt} = 39,394 \text{ V/S}$$

The time required to produce the 13 V peak-to-peak swing is

$$T = \frac{13 \text{ V}}{39,394 \text{ V/S}}$$

$$T = .33 \text{ mS}$$

This represents one-half cycle of the output wave. To go from $+6.5$ V to -6.5 V and back will require .66 mS. Therefore the output frequency is

$$f = \frac{1}{T}$$

$$f = \frac{1}{.66 \text{ mS}}$$

$$f = 1.52 \text{ kHz}$$

This frequency can be adjusted by changing either the 33 k resistor or the .01 μF capacitor. While changing the comparator's resistors can alter the thresholds, and thus alter the frequency, this is generally not recommended since a change in output amplitude will occur as well. By combining steps, the above process may be reduced to a single equation:

$$f = \frac{1}{\dfrac{2V_{pp}}{V_{sat}} RC}$$
(9.30)

where V_{pp} is the difference between $V_{upper\ thres}$ and $V_{lower\ thres}$. Note that if R_1 is 4 times larger than R_2 in the comparator, Equation 9.30 reduces to

$$f = \frac{1}{RC}$$

and the peak triangle wave amplitude is one-fourth of V_{sat}.

Generally, circuits such as this are used for lower-frequency work. For clean square waves, very fast op amps are required. Finally, for lower impedance loads, the outputs should be buffered with voltage followers. The SPICE simulation for this signal generator is shown in Figure 9.22 on pages 412 and 413. The square and triangle outputs are plotted together so that the switching action can be seen. The output plot is delayed several milliseconds in order to guarantee a plot of the steady-state output.

If an accurate triangle wave is not needed, and only a square wave is required, the circuit of Figure 9.19 can be reduced to a single op amp stage. This is shown in Figure 9.23 on page 414. This circuit is, in essence, a comparator. Resistors R_1 and R_2 form the positive feedback portion and set the effective comparator trip point, or threshold. The measurement signal is the voltage across the capacitor. The potentials of interest are shown in Figure 9.24 (page 414). If the output is at positive saturation, the noninverting input will see a percentage of this, depending on the voltage divider produced by R_1 and R_2. This potential is $V_{upper\ thres}$. Since the output is at positive saturation, the capacitor C will be charging toward it. Since it is charging through resistor R, the waveform is an exponential type. Once the capacitor voltage reaches $V_{upper\ thres}$, the noninverting input will no longer be greater than the inverting input, and the device will change to the negative state. At this point, C will reverse its course, and move toward negative saturation. At the lower threshold, the op amp will again change state, and the process repeats. In order to determine the frequency of oscillation, we need to find how long it takes the capacitor to charge between the two threshold points. Normally the circuit will be powered from equal magnitude supplies, and therefore $+V_{sat} = -V_{sat}$

—————

Figure 9.22   SPICE simulation of the signal generator for Example 9.5

```
*********************************AMIGASPICE 5.0*************************************
Triangle/square wave synthesizer
****    INPUT LISTING                TEMPERATURE=27.000 DEG C
**********************************************************************************
*******Start of UA741 op amp*********
.SUBCKT UA741 1    2    3    4    5    6
*                 IN+ IN- GND  V+  V- OUT
Q1   11   1   13 UA741QA
Q2   12   2   14 UA741QB
RC1   4  11   5.305165E+03
RC2   4  12   5.305165E+03
C1   11  12   5.459553E-12
RE1  13  10   2.151297E+03
RE2  14  10   2.151297E+03
IEE  10   5   1.666000E-05
CE   10   3   3.000000E-12
RE   10   3   1.200480E+07
GCM   3  21   10 3   5.960753E-09
GA   21   3   12 11 1.884655E-04
R2   21   3   1.000000E+05
C2   21  22   3.000000E-11
GB   22   3   21 3 2.357851E+02
RO2  22   3   4.500000E+01
D1   22  31   UA741DA
D2   31  22   UA741DA
EC   31   3   6 3 1.0
RO1  22   6   3.000000E+01
D3    6  24   UA741DB
VC    4  24   2.803238E+00
D4   25   6   UA741DB
VE   25   5   2.803238E+00
.ENDS
.MODEL UA741DA D (IS=9.762287E-11)
.MODEL UA741DB D (IS=8.000000E-16)
.MODEL UA741QA NPN (IS=8.000000E-16 BF=9.166667E+01)
.MODEL UA741QB NPN (IS=8.309478E-16 BF=1.178571E+02)
*****End of UA741 op amp******
*Main circuit description
Ri1    1   2   33K
Ri2    3   6   10K
Ct     2   3   .01UF
Rf     6   1   20K
VCC    4   0   DC 15
VEE    5   0   DC -15
X1     0   2   0 4 5 3 UA741
X2     6   0   0 4 5 1 UA741
.TRAN 20US 6500US 5000US
.PLOT TRAN V(1) V(3)
.END
```

a. SPICE input file

Figure 9.22 *(Continued)*

```
***************************** AMIGASPICE 5.0 ***************************
Triangle / square wave synthesizer.
****        TRANSIENT ANALYSIS          TEMPERATURE -   27.000 DEG C
***********************************************************************

LEGEND    *: V(1)   +: V(3)

TIME      V(1)

(*)----------------- -2.000D+01       -1.000D+01       0.000D-01        1.000D+01        2.000D+01
                     - - - - - - - - - - - - - - - - - - - - - - - - - - - - - - - - - - - - - - -
(+)----------------- -1.000D+01       -5.000D+00       0.000D-01        5.000D+00        1.000D+01
                     - - - - - - - - - - - - - - - - - - - - - - - - - - - - - - - - - - - - - - -
 5.000D-03  1.300D+01  .                         .              +  .                   .   *     .
 5.020D-03  1.300D+01  .                         .            +    .                   .   *     .
 5.040D-03  1.300D+01  .                         .         +       .                   .   *     .
 5.060D-03  1.300D+01  .                         .       +         .                   .   *     .
 5.080D-03  1.300D+01  .                       +  .                .                   .   *     .
 5.100D-03  1.300D+01  .                     +    .                .                   .   *     .
 5.120D-03  1.300D+01  .                  +       .                .                   .   *     .
 5.140D-03  1.299D+01  .               +          .                .                   .   *     .
 5.160D-03  3.221D+00  .         +                .        *       .                   .         .
 5.180D-03 -7.728D+00  .         +                .  *             .                   .         .
 5.200D-03 -1.300D+01  .            *+            .                .                   .         .
 5.220D-03 -1.300D+01  .            *    +        .                .                   .         .
 5.240D-03 -1.300D+01  .            *      .+     .                .                   .         .
 5.260D-03 -1.300D+01  .            *         +   .                .                   .         .
 5.280D-03 -1.300D+01  .            *            +.                .                   .         .
 5.300D-03 -1.300D+01  .            *             .   +            .                   .         .
 5.320D-03 -1.300D+01  .            *             .      +         .                   .         .
 5.340D-03 -1.300D+01  .            *             .         +      .                   .         .
 5.360D-03 -1.300D+01  .            *             .           +    .                   .         .
 5.380D-03 -1.300D+01  .            *             .              + .                   .         .
 5.400D-03 -1.300D+01  .            *             .                .  +                .         .
 5.420D-03 -1.300D+01  .            *             .                .     +             .         .
 5.440D-03 -1.300D+01  .            *             .                .        +          .         .
 5.460D-03 -1.300D+01  .            *             .                .           +       .         .
 5.480D-03 -1.300D+01  .            *             .                .             +.     .         .
 5.500D-03 -1.300D+01  .            *             .                .                +   .         .
 5.520D-03 -1.300D+01  .            *             .                .                   + .         .
 5.540D-03 -6.287D+00  .                          .      *         .                   .  +      .
 5.560D-03  5.256D+00  .                          .                .          *        .    +    .
 5.580D-03  1.300D+01  .                          .                .                   .  X      .
 5.600D-03  1.300D+01  .                          .                .                 +.  *      .
 5.620D-03  1.300D+01  .                          .                .              +.   *      .
 5.640D-03  1.300D+01  .                          .                .           +      *      .
 5.660D-03  1.300D+01  .                          .                .        +         *      .
 5.680D-03  1.300D+01  .                          .                .     +            *      .
 5.700D-03  1.300D+01  .                          .                .  +               *      .
 5.720D-03  1.300D+01  .                          .                +                  *      .
 5.740D-03  1.300D+01  .                          .             .+                    *      .
 5.760D-03  1.300D+01  .                          .          +                        *      .
 5.780D-03  1.300D+01  .                          .       +                           *      .
 5.800D-03  1.300D+01  .                          .     +                             *      .
 5.820D-03  1.300D+01  .                          .  +                                *      .
 5.840D-03  1.300D+01  .                       +  .                                   *      .
 5.860D-03  1.300D+01  .                    +  .                                      *      .
 5.880D-03  1.300D+01  .                 +  .                                         *      .
 5.900D-03  1.300D+01  .              +     .                                         *      .
 5.920D-03  8.885D+00  .         +          .                                    *    .
 5.940D-03 -2.069D+00  .         +          .                *                        .
 5.960D-03 -1.284D+01  .            +*      .                                         .
 5.980D-03 -1.300D+01  .            *   +   .                                         .
 6.000D-03 -1.300D+01  .            *      +.                                         .
 6.020D-03 -1.300D+01  .            *       .  +                                      .
 6.040D-03 -1.300D+01  .            *       .      +                                  .
 6.060D-03 -1.300D+01  .            *       .          +                              .
 6.080D-03 -1.300D+01  .            *       .              +                          .
 6.100D-03 -1.300D+01  .            *       .                 +                       .
 6.120D-03 -1.300D+01  .            *       .                  +  .                    .
 6.140D-03 -1.300D+01  .            *       .                   .  +                   .
 6.160D-03 -1.300D+01  .            *       .                   .     +                .
 6.180D-03 -1.300D+01  .            *       .                   .        +             .
 6.200D-03 -1.300D+01  .            *       .                   .           +          .
 6.220D-03 -1.300D+01  .            *       .                   .              +       .
 6.240D-03 -1.300D+01  .            *       .                   .                 +    .
 6.260D-03 -1.300D+01  .            *       .                   .                   .+  .
 6.280D-03 -1.300D+01  .            *       .                   .                   .   +  .
 6.300D-03 -1.230D+01  .         *          .                   .                   .     + .
 6.320D-03 -8.358D-01  .                    .                *  .                   .   .    +
 6.340D-03  1.077D+01  .                    .                   .                 *  .  .    +
 6.360D-03  1.300D+01  .                    .                   .                   . + *  .
 6.380D-03  1.300D+01  .                    .                   .                 .+  *  .
 6.400D-03  1.300D+01  .                    .                   .              +   *  .
 6.420D-03  1.300D+01  .                    .                   .           +      *  .
 6.440D-03  1.300D+01  .                    .                   .        +         *  .
 6.460D-03  1.300D+01  .                    .                   .     +            *  .
 6.480D-03  1.300D+01  .                    .                   .  +               *  .
                     - - - - - - - - - - - - - - - - - - - - - - - - - - - - - - - - - - - - - - -
```

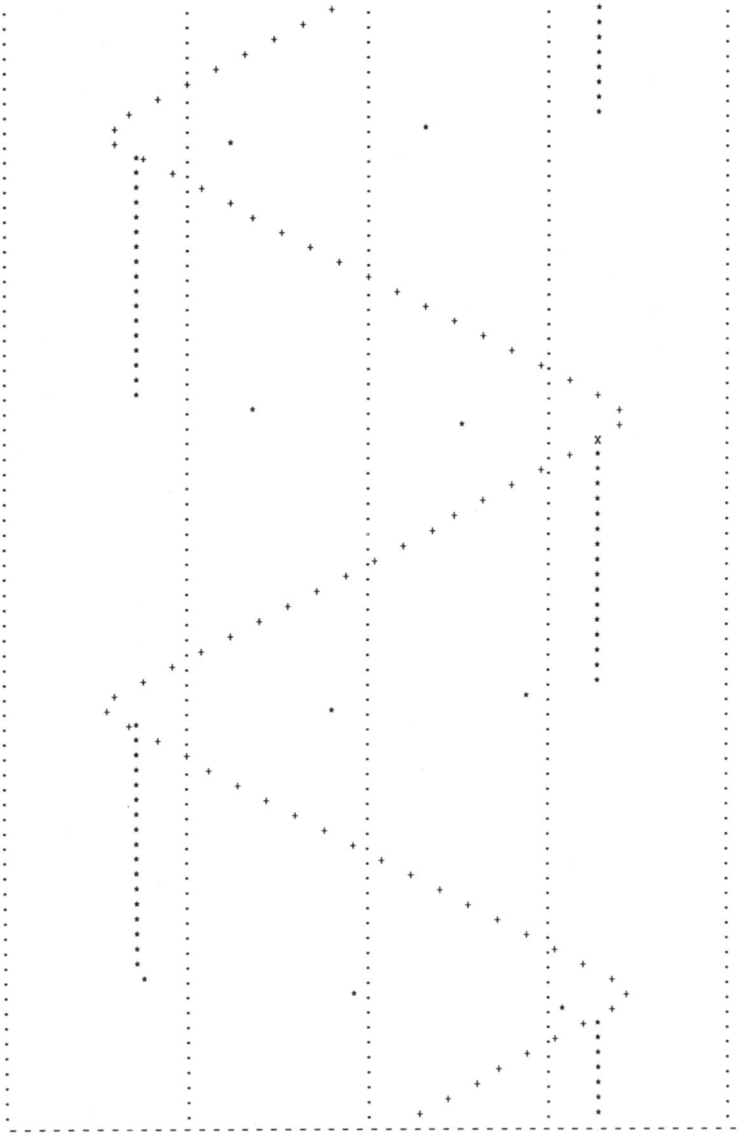

b. SPICE output showing square and triangle outputs

Figure 9.23
Simple square wave
generator

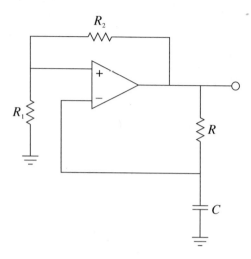

Figure 9.24
Waveforms of a simple
square wave generator

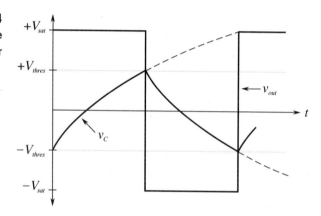

and $V_{upper\ thres} = V_{lower\ thres}$ By inspection,

$$V_{thres} = V_{sat} \frac{R_1}{R_1 + R_2} \tag{9.31}$$

The capacitor voltage is

$$v_C(t) = V_k(1 - \varepsilon^{-t/RC}) \tag{9.32}$$

where V_k is the total potential applied to the capacitor. Since the capacitor will start at one threshold and attempt to charge to the opposite saturation limit, this is

$$V_k = V_{sat} + V_{thres} \tag{9.33}$$

Combining Equations 9.31, 9.32, and 9.33 yields

$$v_C(t) = (V_{sat} + V_{thres})(1 - \varepsilon^{-t/RC}) \qquad (9.34)$$

At the point where the comparator changes state,

$$V_C = 2V_{thres} \qquad (9.35)$$

Combining Equations 9.34 and 9.35 produces

$$2V_{thres} = (V_{sat} + V_{thres})(1 - \varepsilon^{-t/RC})$$

$$\frac{2V_{thres}}{V_{sat} + V_{thres}} = 1 - \varepsilon^{-t/RC}$$

$$\frac{2V_{sat}\dfrac{R_1}{R_1 + R_2}}{V_{sat}\left(1 + \dfrac{R_1}{R_1 + R_2}\right)} = 1 - \varepsilon^{-t/RC}$$

$$\frac{2R_1}{2R_1 + R_2} = 1 - \varepsilon^{-t/RC}$$

$$\frac{R_2}{2R_1 + R_2} = \varepsilon^{-t/RC}$$

$$\ln\left(\frac{R_2}{2R_1 + R_2}\right) = \frac{-t}{RC}$$

$$t = RC \ln\left(\frac{2R_1 + R_2}{R_2}\right)$$

This represents the charge time of the capacitor. Since one period requires two such traverses, we may say

$$T = 2RC \ln\left(\frac{2R_1 + R_2}{R_2}\right)$$

or

$$f_o = \frac{1}{2RC \ln\left(\dfrac{2R_1 + R_2}{R_2}\right)} \qquad (9.36)$$

We can transform Equation 9.36 into "nicer" forms by choosing values for R_1 and R_2 such that the log term turns into a convenient number, such as 1 or .5. If we set $R_1 = .859R_2$, the log term is unity, and Equation 9.36 becomes

$$f_o = \frac{1}{2RC}$$

Example 9.6   Design a 2 kHz square wave generator using the circuit of Figure 9.23. For convenience, set $R_1 = .859R_2$. If R_1 is arbitrarily set to 10 k, then

$$R_1 = .859R_2$$

$$R_2 = \frac{R_1}{.859}$$

$$R_2 = \frac{10\text{ k}}{.859}$$

$$R_2 = 11.64\text{ k}$$

In order to set the oscillation frequency, R is arbitrarily set to 10 k, and C is then determined.

$$f_o = \frac{1}{2RC}$$

$$C = \frac{1}{2Rf_o}$$

$$C = \frac{1}{2 \times 10\text{ k } 2\text{ kHz}}$$

$$C = 25\text{ nF}$$

9.2 Single-Chip Oscillators and Frequency Generators

Since the generation of signals is a basic requirement for a wide variety of applications, a number of manufacturers produce a selection of single IC oscillators and frequency generators. The majority tend to work in the range below 1 MHz and usually require some form of external resistor/capacitor network to set the operating frequency. Highly specialized circuits for targeted applications are also available. In this section we will examine three popular ICs which lend themselves to a wide range of applications: the NE566 voltage-controlled oscillator, the NE565 *phase-locked loop*, and the ICL8038 waveform generator.

Voltage-Controlled Oscillators

A voltage-controlled oscillator (usually abbreviated as VCO) does not produce a fixed output frequency. As its name suggests, the output frequency of a VCO is dependent on a control voltage (V_c). There is a fixed relationship between the control voltage and the output frequency. Theoretically, just about any oscillator can be turned into a VCO. For example, if a resistor is used as part of the tuning circuit, it can be replaced with some form of voltage-controlled resistor, such as an FET or a light-dependent resistor/lamp combination. By doing this, an external potential can be used to set the frequency of oscillation. This is very useful if the frequency needs to be changed quickly, or accurately swept through some range.

A classic example of the usefulness of a VCO is shown in Figure 9.25. This is a simplified schematic of a monophonic musical keyboard synthesizer. The keys on the synthesizer are little more than switches. These switches tap potentials off of a voltage divider. As the musician plays higher and higher on the keyboard, the switches engage higher and higher potentials. These levels are used to control a very accurate VCO. The higher the control voltage, the higher the output frequency or pitch will be. VCOs can be used for a number of other applications, including swept frequency spectrum analyzers, frequency modulation and demodulation, and control systems. The VCO is also an integral part of the phase-locked loop.

The NE566 is one popular IC VCO. Its block diagram is shown in Figure 9.26 (page 418). In general, it is not very different from the square/triangle generator examined in the last section. The major functional difference is the

Figure 9.25
Simplified music synthesizer using VCO

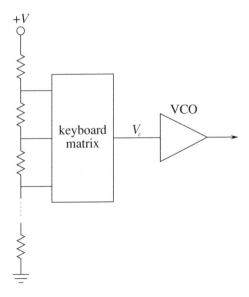

Figure 9.26   NE566 VCO block diagram

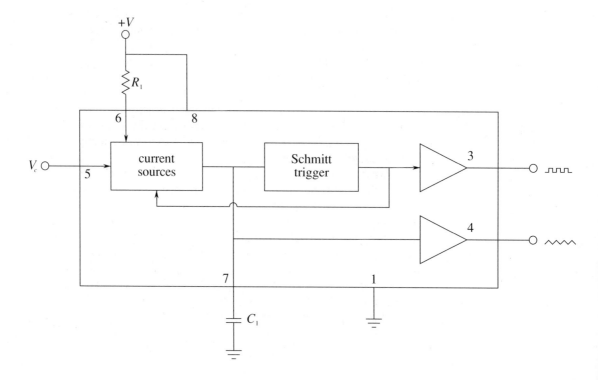

inclusion of the modulation, or control voltage, input. This serves to adjust the capacitor's charge/discharge current. By altering this rate, the frequency can be changed. As with most specialty ICs, the manufacturer gives specific design guidelines. For the Signetics NE566,

1. f_o should be less than 1 MHz

2. V_+ should be between 6 and 12 V

3. R_1 should be between 2 k and 20 k

4. V_c (the control voltage) should be between V_+ and three-fourths of V_+

The free-running, or center, frequency is found from

$$f_o = 2 \frac{V_+ - V_c}{V_+ R_1 C_1}$$

(9.37)

For a 12 V supply, the typical triangle wave output will be 1 V peak riding on a 5 V DC offset. The typical square wave output will be 3 V peak riding on an 8.5 V DC offset.

Example 9.7 The circuit of Figure 9.27 is connected as a fixed-frequency oscillator. Determine the output frequency.

The control voltage is set by the 2 k/20 k voltage divider.

$$V_c = 12 \text{ V} \frac{20 \text{ k}}{2 \text{ k} + 20 \text{ k}}$$

$$V_c = 10.91 \text{ V}$$

Note that this conforms to design guideline number 4.

$$f_o = 2 \frac{V_+ - V_c}{V_+ R_1 C_1}$$

$$f_o = 2 \frac{12 \text{ V} - 10.91 \text{ V}}{12 \text{ V} \times 4.7 \text{ k} \times 1.1 \text{ nF}}$$

$$f_o = 35.14 \text{ kHz}$$

Figure 9.27
NE566 VCO fixed
frequency oscillator for
Example 9.7

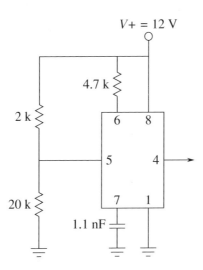

Example 9.8   If the circuit of Figure 9.28 is used to couple a control signal to the NE566, determine the maximum and minimum output frequencies if the control signal is 1 V peak to peak.

Since the DC bias on the control voltage input has not changed, the center frequency will remain at 35.14 kHz. Equation 9.37 indicates that the maximum output frequency will occur when V_c is at a minimum. This signal will be the addition of the 10.91 V DC bias and the negative .5 V peak from the control source.

$$V_{c(min)} = 10.91 \text{ V} + (-.5 \text{ V})$$

$$V_{c(min)} = 10.41 \text{ V}$$

Note that this conforms to design guideline number 4.

$$f_o = 2 \frac{V_+ - V_c}{V_+ R_1 C_1}$$

$$f_o = 2 \frac{12 \text{ V} - 10.41 \text{ V}}{12 \text{ V} \times 4.7 \text{ k} \times 1.1 \text{ nF}}$$

$$f_o = 51.26 \text{ kHz}$$

For the minimum frequency, use the maximum V_c.

$$V_{c(max)} = 10.91 \text{ V} + .5 \text{ V}$$

$$V_{c(max)} = 11.41 \text{ V}$$

Note that this conforms to design guideline number 4.

Figure 9.28
Circuit for Example 9.8:
applying a control voltage
to the NE566

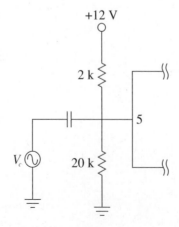

$$f_o = 2 \frac{V_+ - V_c}{V_+ R_1 C_1}$$

$$f_o = 2 \frac{12 \text{ V} - 11.41 \text{ V}}{12 \text{ V} \times 4.7 \text{ k} \times 1.1 \text{ nF}}$$

$$f_o = 19 \text{ kHz}$$

Thus, a range of 51.26:19, or 2.7:1, is achieved. Another way of looking at this is to note that a 1 V change in control voltage produced a 32.26 kHz change in frequency. The sensitivity of this circuit may be stated as 32.26 kHz per volt. If values are properly chosen, the NE566 can produce a 10:1 frequency sweep. Control signals close to the limits set in design guideline number 4 will produce some nonlinearity: the relationship between the control voltage and the resulting frequency will no longer be a straight line. In this example, the sensitivity will deviate from the 32.26 kHz per volt ideal.

In closing, note that the way in which the frequency sweeps depends on the wave shape of V_c. If a sinusoid is used, the output frequency will vary smoothly between the stated limits. On the other hand, if the wave shape for V_c is a ramp, the output frequency will start at one extreme, and then move smoothly to the other limit as the V_c ramp continues. When the ramp resets itself, the output frequency will jump back to its starting point. An example of this is shown in Figure 9.29. Finally, if the control wave shape is a square, the output

Figure 9.29
VCO frequency sweep using a ramp

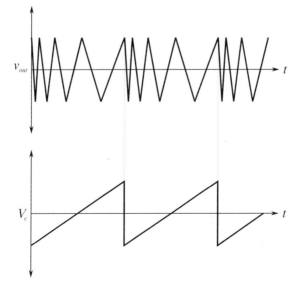

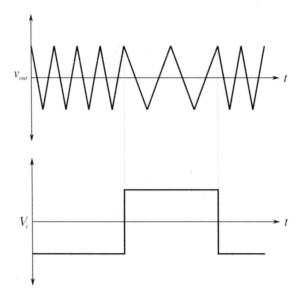

Figure 9.30
VCO 2-tone output using
a square wave

frequency will abruptly jump from the minimum to the maximum frequency and back. This effect is shown in Figure 9.30, and can be used to generate FSK (frequency shift key) signals. FSK is used in the communications industry to transmit binary information.

The Phase-Locked Loop

One step up from the VCO is the **p**hase-locked **l**oop, or PLL. The PLL is a self-correcting circuit; it can lock onto an input frequency and adjust to track changes in the input. PLLs are used in modems, for FSK systems, frequency synthesis, tone decoders, FM signal demodulation, and other applications. A block diagram of a basic PLL is shown in Figure 9.31.

Figure 9.31   Phase-locked loop

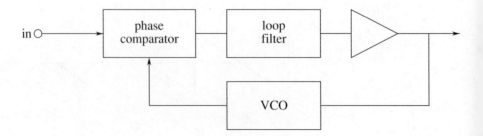

In essence, the PLL uses feedback in order to lock an oscillator to the phase and frequency of an incoming signal. It consists of three major parts: a *phase comparator*, a *loop filter* (typically, a lag network of some form), and a VCO. An amplifier may also exist within the loop. The phase comparator is driven by the input signal and the output of the VCO. It produces an error signal which is proportional to the phase difference between its inputs. This error signal is then filtered in order to remove spurious high-frequency signals and noise. The resulting error signal is used as the control voltage for the VCO, and as such, sets the VCO's output frequency. As long as the error signal is not too great, the loop will be self-stabilizing. In other words, the error signal will eventually drive the VCO to be in perfect frequency and phase synchronization with the input signal. When this happens, the PLL is said to be *in lock* with the input. The range of frequencies over which the PLL can stay in lock as the input signal changes is called the *lock range*. Normally, the lock range is symmetrical about the VCO's *free running*, or center, frequency. The deviation from the center frequency out to the edge of the lock range is called the *tracking range*, and is therefore one-half of the lock range. This is illustrated in Figure 9.32.

Although a PLL may be able to track changes throughout the lock range, it may not be able to initially acquire sync with frequencies at the range limits. A somewhat narrower band of frequencies, called the *capture range*, indicates frequencies which the PLL will always be able to lock onto. Again, the capture range is usually symmetrical about f_o. The deviation on either side of f_o is referred to as the *pull-in range*. For a PLL to function properly, the input frequency must first be within the capture range. Once the PLL has locked onto the signal, the input frequency can vary throughout the larger lock range. The VCO center frequency is usually set by an external resistor or capacitor. The loop filter may also require external components. Depending on the application, the desired output signal from the PLL can be either the VCO's output or the control voltage for the VCO.

Figure 9.32
Operating ranges for
phase-locked loop

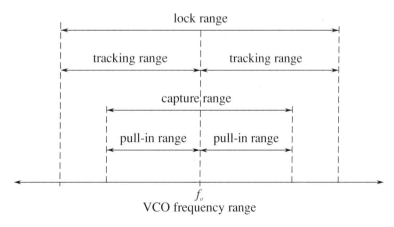

VCO frequency range

Figure 9.33   Block diagram of the NE565

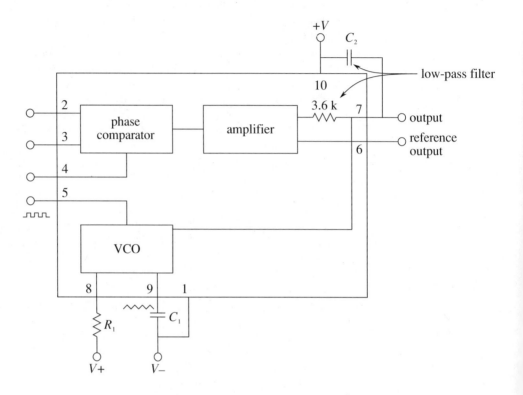

The block diagram for one popular PLL, the NE565, is shown in Figure 9.33. Not all of the items are internally connected as you might expect. By bringing specific signals out to IC pins (such as the VCO output), the circuit is more flexible and can be used in a wider variety of applications. The free-running frequency of the VCO is set by R_1 and C_1. The loop filter is controlled by C_2 and the internal 3.6 k resistor. The following design equations are given by the manufacturer.

VCO center frequency: $$f_o = \frac{.3}{R_1 C_1}$$ (9.38)

(Note, R_1 should be between 2 k and 20 k, with an optimal value of 4 k.)

Lock range: $$f_l = \pm \frac{8 f_o}{V_{cc}}$$ (in Hertz) (9.39)

Capture range: $$f_c = \pm \frac{\sqrt{\dfrac{2\pi f_l}{3600 C_2}}}{2\pi}$$ (9.40)

For the best lock and capture range, the input signal should be at least 100 mV. At this point, we will take a look at a couple of practical uses for the NE565.

One way to transmit binary signals is via FSK, which can be used to allow two computers to exchange data over telephone lines. Due to limited bandwidth, it is not practical to directly transmit the digital information in its normal pulse-type form. Instead, logic high and low can be represented by distinct frequencies. A square wave, for example, would be represented as an alternating set of two tones. FSK is very easy to generate. All you need to do is drive a VCO with the desired logic signal. To recover the data, the reception circuit needs to create a high or low level, depending on which tone is received. A PLL can be used for this purpose. The output signal will be the error signal which drives the VCO. The logic behind the circuit operation is deceptively simple. If the PLL is in lock, the output frequency of its VCO must be the same as the input signal. Remembering that the incoming FSK signal is itself derived from a VCO, for the VCOs to be in lock, they must be driven with identical control signals. Therefore, the control signal which drives the PLL's internal VCO must be the same as the control signal which originally generated the FSK signal. The PLL control signal can then be fed to a comparator in order to properly match the signal to the following logic circuitry. Figure 9.34

Figure 9.34 FSK circuit for 300 bits per second transmission

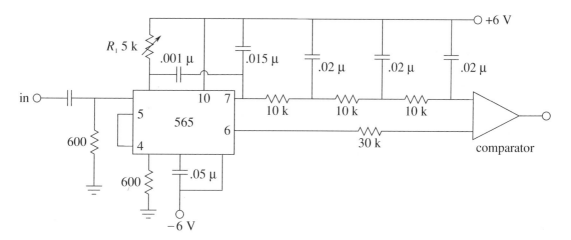

Figure 9.35   PLL frequency synthesizer

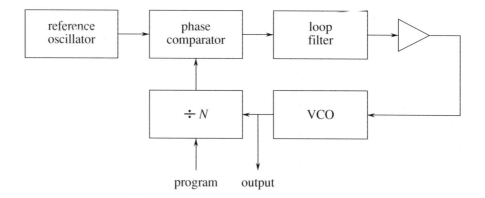

shows a circuit which works with the standard FSK tones of 1270 Hz and 1070 Hz. This circuit works with data rates up to 300 bits per second. The loop filter needs to be a bit more stringent for this application, and as such, a multistage filter is employed. Under free-running conditions, R_1 is adjusted so that the DC voltage at the output (pin 7) is the same as the reference (pin 6). By doing so, detection of the 1070 Hz tone will produce a positive differential, thus producing a logic high out of the comparator. A 1270 Hz tone will produce a lower potential at pin 7, thus causing the comparator to produce a logic low.

Along the same lines as the FSK demodulator is the standard FM signal demodulator. Again, the operational logic is the same. In order for the PLL to remain in lock, its VCO control signal must be the same as the original modulating signal. In the case of typical radio broadcasts, the modulating signal is either voice or music. The output signal will need to be AC-coupled and amplified further. The PLL serves as the intermediate frequency amplifier, limiter, and demodulator. The result is a very cost-effective system.

Another usage for the PLL is in frequency synthesis. From a single accurate signal reference, a PLL can be used to derive a number of new frequencies. A block diagram is shown in Figure 9.35. The major change is in the addition of a programmable divider between the VCO and the phase comparator. The PLL can remain in lock with the reference oscillator only by producing the same frequency out of the divider. This means that the VCO must generate a frequency N times higher than the reference oscillator. We can use the VCO output as desired. In order to change the output frequency, all that needs to be changed is the divider ratio. Normally, a highly accurate and stable reference, such as a quartz crystal oscillator, is used. In this way, the newly synthesized frequencies will also be very stable and accurate.

Example 9.9   For the circuit of Figure 9.36 (page 428), determine the free-running frequency, the lock range, and the capture range.

$$f_o = \frac{.3}{R_1 C_1}$$

$$f_o = \frac{.3}{4\,k \times 330\ pF}$$

$$f_o = 227.3\ kHz$$

$$f_l = \pm \frac{8 f_o}{V_+}$$

$$f_l = \pm \frac{8 \times 227.3\ kHz}{6\ V}$$

$$f_l = \pm 303\ kHz$$

$$f_c = \pm \frac{\sqrt{\dfrac{2\pi f_l}{3600 C_2}}}{2\pi}$$

$$f_c = \pm \frac{\sqrt{\dfrac{2\pi \times 303\ kHz}{3600 \times 270\ pF}}}{2\pi}$$

$$f_c = \pm 222.7\ kHz$$

Under free-running conditions, the DC level at pin 7 should be around 4.25 V, according to the manufacturer's data sheets. As the input frequency rises, the voltage at this pin drops. Conversely, as the frequency falls, the output voltage increases.

Waveform Generators

Given the usefulness of general-purpose waveform generators, it is no surprise that semiconductor manufacturers have created entire systems on a single monolithic IC. One good example is the ICL8038 from Harris. This device is capable of producing simultaneous sine, square, and triangle waves. Output frequencies can range from as low as .001 Hz to as high as 300 kHz. Sine wave distortion can be trimmed to under 1% THD. The unit also has provision for

Figure 9.36   Phase-locked loop for Example 9.9

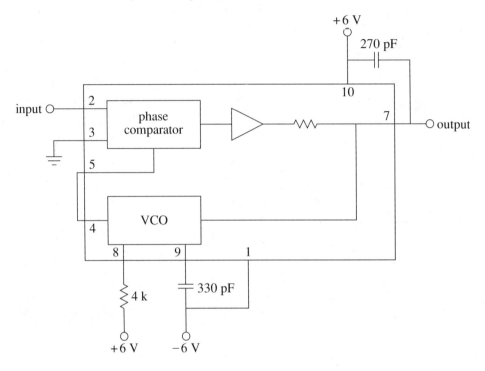

variable duty cycle and swept frequency operation. A block diagram of the ICL8038 is shown in Figure 9.37.

The operation of the generator revolves around two current sources and an associated comparator/flip-flop section. A single external capacitor is used to establish the base operating frequency. The technique used here is similar to that established earlier with the triangle/square wave generator. In this circuit, current source 1 charges the timing capacitor, producing a positive ramp. Eventually, this ramp will trigger the positive comparator, which in turn, toggles the flip-flop. The flip-flop now connects the capacitor to current source 2. Current source 2 actually *sinks* current, thus forcing the capacitor to discharge. Eventually, this negative-going ramp will trip the negative comparator, which will again toggle the flip-flop, causing current source 1 to charge the capacitor. This process repeats indefinitely, producing a triangle wave across the capacitor, and a square wave from the flip-flop. These signals are buffered, and then sent to the appropriate output pins. The triangle wave is also passed through a *sine shaper*. The sine shaper is a function synthesizer, as studied in Chapter 7. This produces a decreasing gain function which "rounds off" the positive and negative extremes of the triangle wave, via a piecewise

Figure 9.37   ICL8038 block diagram

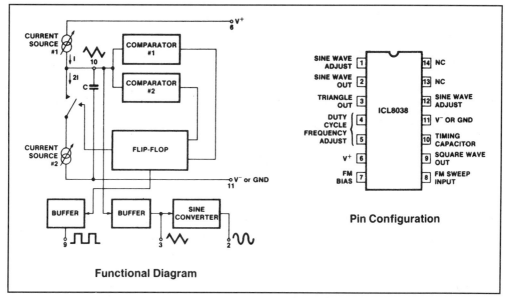

Functional Diagram

Pin Configuration

Reprinted with permission of Harris Semiconductor

linear sine approximation. The resulting sine wave is of medium quality. It is not suitable for critical applications such as a distortion test source, but is adequate for a number of general-purpose uses. While not perfect, the function synthesis approach has the advantage of reliable operation over a very wide frequency range. It is worth noting though that sine wave distortion begins to increase rapidly above about 20 kHz.

Oscillation frequency is set by the external capacitor and the two associated current sources. These current sources are independent and programmed with separate resistors. It is quite possible to set the charge and discharge currents to different values by using unequal resistors. This has the effect of altering the duty cycle of the square wave, and changing the symmetry of the triangle and sine waves. This effect can be seen in Figure 9.38.

Output connections are fairly straightforward with the ICL8038. If high impedance loads are utilized, they can be connected directly to the sine and triangle outputs. Lower impedance loads should be buffered, since the output impedance of the sine wave section is on the order of 1 kΩ. The square-wave section is an open collector type, and thus, requires a pull-up resistor for proper operation. This has the advantage of allowing you to set the upper limit of the voltage swing by tying the pull-up resistor to a desired potential, which makes logic circuit interfacing much easier. The upper limit of this potential

Figure 9.38   Effect of changing duty cycle

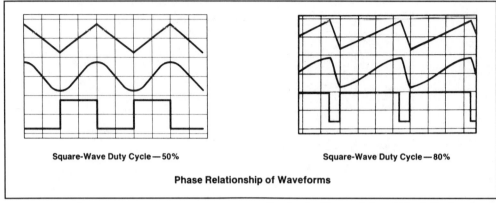

Square-Wave Duty Cycle — 50%   Square-Wave Duty Cycle — 80%

Phase Relationship of Waveforms

is set by the internal breakdown limit of the ICL8038, which is approximately 30 V. If a split-polarity supply is used, the output waveforms will be centered around ground. For a single-polarity supply (i.e., V_- is 0 V), the signals will be centered around one-half of the supply potential. The amplitudes of each section are not the same. The square wave will swing from just below the pull-up potential down to just above the negative supply potential. The triangle wave's peak-to-peak output will be approximately 30% of the total supplied potential, while the sine wave's peak-to-peak swing will be about 20% of the total supplied potential.

Typical circuit connections are shown in Figure 9.39. Figure 9.39a shows the most basic connection. Resistors R_A and R_B individually set the charge and discharge currents. Capacitor C is the external timing capacitor, R_L is the pull-up resistor for the square-wave section, and the 82 kΩ resistor is part of the sine wave converter. The disadvantage of this configuration is that frequency selection and duty cycle adjustment are interdependent. A better scheme is shown in Figure 9.39b. Here, fixed resistors are used for R_A and R_B, and a small potentiometer is tied between them. This potentiometer may be adjusted for the desired duty cycle. The effective value of R_A or R_B is the fixed value plus the portion of pot between the resistor and the pot's wiper. If a 1 k pot is used and is set at its midpoint, then 500 Ω is effectively added to both R_A and R_B. Another item worth noting is the replacement of the 82 k resistor with a 100 k potentiometer. This allows you to adjust the circuit for minimum sine wave distortion. Another variation (Figure 9.39c) shows R_A and R_B combined into a single potentiometer. This is useful if duty cycle adjustment is not required. Due to internal offsets, however, it is quite likely that this arrangement will not result in a perfect 50% duty cycle wave. The remaining pins 7 and 8

Figure 9.39   Typical connections

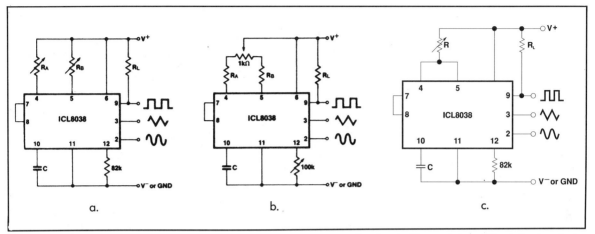

Reprinted with permission of Harris Semiconductor

are used for *frequency modulation.* This allows you to turn the ICL8038 into a VCO.

Due to this circuit's rather specialized nature, the manufacturer gives the following set of equations:

The general oscillation frequency is

$$f_o = \frac{1}{\frac{5}{3} R_A C \left(1 + \frac{R_B}{2R_A - R_B}\right)}$$

If $R_A = R_B$, this reduces to

$$f_o = \frac{.3}{RC}$$

If a single timing resistor is used, as in Figure 9.39c, this reduces to

$$f_o = \frac{.15}{RC}$$

When setting values for the timing resistors, the charge and discharge currents should be kept within 10 μA to 1 mA. The currents are found by

$$I_A = \frac{V_+}{5R_A}$$

and

$$I_B = \frac{V_+}{5R_B}$$

Finally, when using the FM capability, the voltage at pin 8 must always be kept within these limits:

$$V_+ > V_{pin8} > \frac{2}{3} V_{total} + (V_-) + 2\text{ V}$$

where V_{total} is the total supplied potential from the positive to the negative supply, and V_- is the negative supply potential *including sign, not just magnitude*. (Remember, it is possible to set V_- to 0 for single-supply applications.)

Example 9.10 A waveform generator with adjustable frequency, amplitude, and wave shape is shown in Figure 9.40. Determine the range of output frequencies.

This circuit has one of three ranges, which are set by the three timing capacitors. The largest capacitor will produce the lowest frequency. The single resistor approach is used here in place of separate values for R_A and R_B. The minimum value is 10 k (pot shorted) and the maximum value is 110 k (pot open). The maximum resistance will give the lowest frequency.

For the low range with lowest frequency:

$$f_{o(min)} = \frac{.15}{RC}$$

$$f_{o(min)} = \frac{.15}{110\text{ k} \times .1\ \mu\text{F}}$$

$$f_{o(min)} = 13.6\text{ Hz}$$

For the low range with highest frequency:

$$f_{o(max)} = \frac{.15}{RC}$$

$$f_{o(max)} = \frac{.15}{10\text{ k }.1\ \mu\text{F}}$$

$$f_{o(max)} = 150\text{ Hz}$$

In like manner, the middle range, using $C = .01\ \mu\text{F}$ produces

$$f_{o(min)} = 136\text{ Hz}$$

$$f_{o(max)} = 1500\text{ Hz}$$

Figure 9.40   Waveform generator for Example 9.10

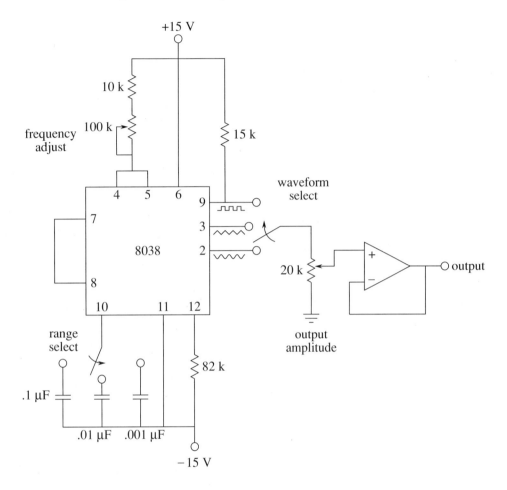

The high range, using $C = .001$ μF produces

$$f_{o(min)} = 1360 \text{ Hz}$$

$$f_{o(max)} = 15 \text{ kHz}$$

The output buffer serves to isolate the ICL8038 from low impedance loads. The 20 k pot acts as an output level control by tapping off a portion of the output signal. For reasonable square wave outputs, a high slew rate op amp is desired. By itself, the square wave section exhibits rise and fall times of 180 nS and 40 nS, respectively.

Summary

Oscillators and frequency generators find use in a wide variety of applications. They may be realized from simple single op amp topologies, or use more elaborate special-purpose integrated circuits. Basic op amp oscillators are usually constrained to the frequency range below 1 MHz. Two op amp–based sine wave oscillators are the Wien bridge and phase shift types. Both oscillators rely on positive feedback in order to create their outputs. To maintain oscillation, the amplifier/feedback loop must conform to the Barkhausen criterion. This states that in order to maintain oscillation, the loop phase must be 0°, or an integer multiple of 360°. Also, the product of the positive feedback loss and the forward gain must be greater than unity to start oscillations, and revert to unity to maintain oscillation. In order to make the gain fall back to unity, some form of gain-limiting device, such as a diode or lamp, is included in the amplifier's negative feedback loop. The oscillation frequency is usually set by a simple resistor/capacitor network. As such, the circuits are relatively easy to tune. Their ultimate accuracy will depend on the tolerance of the tuning components, and to a lesser degree, the characteristics of the op amp used.

Besides sine waves, shapes such as triangles and squares can be produced. A simultaneous square/triangle generator can be formed from a ramp generator/comparator combination. Only two op amps are needed for this design.

The voltage-controlled oscillator, or VCO, produces an output frequency which is dependent on an external control voltage. The free-running, or center, frequency is normally set via a resistor/capacitor combination. The control voltage can then be used to increase or decrease the frequency about the center point. The VCO is an integral part of the phase-locked loop, or PLL. The PLL has the ability to lock onto an incoming frequency. That is, its internal VCO frequency will match the incoming frequency. Should the incoming frequency change, the internal VCO frequency will change along with it. Two important parameters of the PLL are the capture range and the lock range. Capture range is the range of frequencies over which the PLL can acquire lock. Once lock is achieved, the PLL can maintain lock over a somewhat wider range of frequencies called the lock range. The PLL is in wide use in the electronics industry, and is found in such applications as FM demodulation, FSK-based communication systems, and frequency synthesis.

Self Test Questions

1. How does positive feedback differ from negative feedback?
2. Define the Barkhausen criterion.
3. Explain the operation of the Wien bridge op amp oscillator.

4. Detail the operation of the phase shift op amp oscillator.

5. How might a square wave be generated from a sinusoidal or triangular source?

6. Give two ways to make the output frequency of a Wien bridge oscillator user-adjustable.

7. What factors contribute to the accuracy of a Wien bridge oscillator's output frequency?

8. What is a VCO, and how does it differ from a fixed-frequency oscillator?

9. Draw a block diagram of a PLL and explain its basic operation.

10. What is the difference between capture range and lock range for a PLL?

11. Give at least two applications for a fixed frequency oscillator or VCO.

12. Give at least two applications for a PLL.

13. What is the purpose of a sine shaper in a waveform generator?

Problem Set

▷ Analysis Problems

Note: Unless otherwise specified, all circuits in the following problems use ± 15 V power supplies.

1. Given the circuit of Figure 9.41, determine the frequency of oscillation if $R_1 = 1.5$ k, $R_2 = R_3 = R_4 = 3$ k, and $C_1 = C_2 = .022\ \mu\text{F}$.

——— Figure 9.41

Figure 9.42

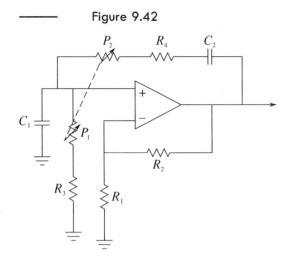

Figure 9.43

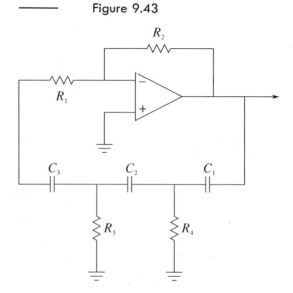

Figure 9.44

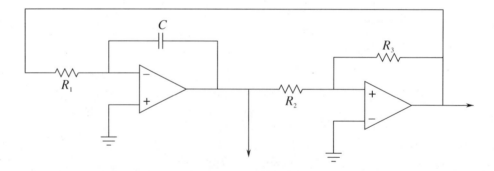

2. Given the circuit of Figure 9.41, determine the frequency of oscillation if $R_2 = 22$ k, $R_1 = R_3 = R_4 = 11$ k, and $C_1 = C_2 = 33$ nF.

3. Given the circuit of Figure 9.42, determine the maximum and minimum f_o if $R_1 = 5.6$ k, $R_2 = 12$ k, $R_3 = R_4 = 1$ k, $P_1 = P_2 = 10$ k, $C_1 = C_2 = 39$ nF.

4. Given the circuit of Figure 9.43, determine f_o if $R_4 = 2$ k, $R_3 = 20$ k, $R_1 = 200$ k, $R_2 = 1.6$ M, $C_1 = 30$ nF, $C_2 = 3$ nF, $C_3 = .3$ nF.

5. Given the circuit of Figure 9.43, determine f_o if $R_1 = R_3 = R_4 = 3.3$ k, $R_2 = 100$ k, $C_1 = C_2 = C_3 = 86$ nF.

6. Given the circuit of Figure 9.44, determine f_o if $R_1 = R_2 = 22$ k, $R_3 = 33$ k, $C = 3.3$ nF.

7. Using the circuit of Figure 9.45, find f_o if $R_3 = 3.3$ k, $R_2 = 30$ k, $R_1 = 10$ k, and $C = 2$ nF.

—————— Figure 9.45

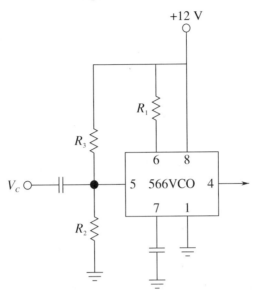

8. Give two different ways of moving the f_o (i.e., the free-running, or center, frequency) of the circuit of Figure 9.45 to 30 kHz.

9. If a control voltage of $.4 \sin 2\pi 60t$ is used for the circuit of Figure 9.45, find the resulting maximum and minimum output frequencies.

10. Sketch the output waveform for Problem 9.

11. If $R = 8$ k and $C = 800$ pF in Figure 9.46 (page 438), determine the output frequency if $V_c = -50$ mV DC.

12. For Problem 11, determine the output frequencies if V_c is a 1 kHz square wave, .1 V peak.

13. Sketch the output waveform for Problem 12.

14. Determine f_o, f_l, and f_c for the NE565 PLL of Figure 9.47 (page 438). Use $C_1 = 390$ pF, $C_2 = 680$ pF, and $R_1 = 10$ k.

15. Given the circuit of Figure 9.44, determine f_o if $R_1 = R_2 = 22$ k, $R_3 = 33$ k, and $C = 3.3$ nF.

16. Determine f_o, f_l, and f_c for the PLL of Figure 9.47. Use $C_1 = 410$ pF, $C_2 = 750$ pF, and $R_1 = 5.6$ k.

17. Determine the output frequency range of the circuit of Figure 9.40 if the potentiometer is lowered to 50 k and the series resistor is increased from 10 k to 60 k.

———   Figure 9.46

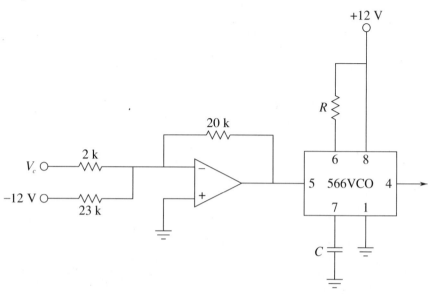

———   Figure 9.47

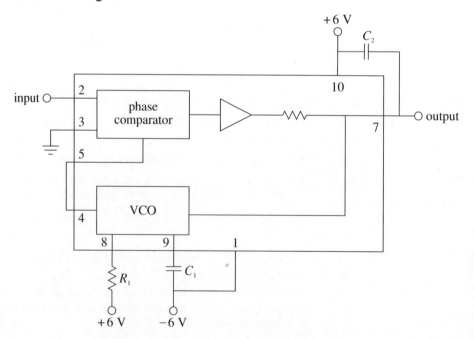

⊳ **Design Problems**

18. For the circuit of Figure 9.41, determine values for C_1 and C_2 if $R_1 = 6.8$ k, $R_2 = R_3 = R_4 = 15$ k, and $f_o = 30$ kHz.

19. For the circuit of Figure 9.41, determine values for R_3 and R_4 if $R_1 = 2.2$ k, $R_2 = 4.7$ k, $C_1 = C_2 = .047$ μF, and $f_o = 400$ Hz.

20. For the circuit of Figure 9.41, determine values for R_2, C_1, and C_2 if $R_1 = 7.2$ k, $R_3 = R_4 = 3.9$ k, and $f_o = 19$ kHz.

21. Determine the values required for R_3, R_4, P_1, and P_2 in Figure 9.42 if $C_1 = C_2 = 98$ nF, $R_1 = 5.6$ k, $R_2 = 12$ k, $f_{o(min)} = 2$ kHz, and $f_{o(max)} = 20$ kHz.

22. Repeat Problem 21 for $f_{o(min)} = 10$ kHz and $f_{o(max)} = 30$ kHz.

23. Redesign the circuit of Problem 1 so that exact gain resistors are not needed. Use Figure 9.6 as a model.

24. Redesign the circuit of Problem 3 so that clipping does not occur. Use Figure 9.6 as a model.

25. Determine new values for the capacitors of Problem 4 if f_o is changed to 10 kHz.

26. Given the circuit of Figure 9.43, determine values for the capacitors if $R_1 = R_3 = R_4 = 3.3$ k, $R_2 = 100$ k, and $f_o = 7.6$ kHz.

27. Given the circuit of Figure 9.43, determine values for the resistors if the capacitors all equal 1100 pF and $f_o = 15$ kHz.

28. Determine the capacitor and resistor values for the circuit of Figure 9.43 if $R_2 = 56$ k and $f_o = 1$ kHz.

29. Find C in Figure 9.44 if $f_o = 5$ kHz, $R_1 = R_3 = 39$ k, $R_2 = 18$ k.

30. Determine the resistor values in Figure 9.44 if $f_o = 20$ kHz and $C = 22$ nF. Set $R_1 = R_2$ and $R_2 = R_3/2$. Sketch the output waveforms as well.

31. Determine the required ratio for R_2/R_3 to set the triangle wave output to 5 V peak in Figure 9.44.

32. Using the circuit of Figure 9.45, find f_o if $R_3 = 3.3$ k, $R_2 = 30$ k, $R_1 = 10$ k, and $C = 2$ nF.

33. Determine values for R and C in Figure 9.46 to set the center frequency to 50 kHz with $V_c = 0$.

34. Design a square wave generator which is adjustable from 5 kHz to 20 kHz.

35. For the circuit of Figure 9.47, determine the capacitor values if R_1 is set to its optimal value (4 k), and the free-running frequency is 65 kHz. Try to make the capture range at least half as large as the lock range.

36. Redesign the circuit of Figure 9.40 for a frequency range spanning 50 Hz to 50 kHz.

▷ **Challenge Problems**

37. Using Figure 9.9 as a guide, design a sine wave oscillator that will operate from 2 Hz to 20 kHz, in decade ranges.

38. Design a 10 kHz TTL-compatible square wave oscillator using a Wien bridge oscillator and a 311 comparator.

39. Using a triangle or sine wave oscillator and a comparator, design a variable duty cycle pulse generator. *Hint*: consider varying the comparator reference.

40. Using a function synthesizer (Chapter 7) and the oscillator of Figure 9.19, outline a simple laboratory frequency generator with sine, triangle, and square wave outputs.

41. For the preceding problem, outline how amplitude and DC offset controls could be implemented as well.

42. Assuming an accurate 50 kHz oscillator as a source, generate a stable 400 kHz square wave using a PLL.

43. Using Figures 9.39 and 9.40 as guides, extend the ICL8038 circuit to include a span from 1 Hz to 100 kHz (in decade ranges), duty cycle adjustment from at least 55% to 45%, DC offset, and equal output amplitude when switching between the three waveforms.

▷ **SPICE Problems**

44. Perform a SPICE analysis for the circuit of Problem 1. Perform a frequency domain analysis of the positive feedback loop's gain and phase, and verify that the Barkhausen criterion is met.

45. Perform a SPICE analysis for the circuit of Problem 4. Perform a frequency domain analysis of the positive feedback loop's gain and phase, and verify that the Barkhausen criterion is met.

46. Perform a time domain SPICE analysis for the circuit of Problem 6. Make sure that you check both outputs (a simultaneous plot would be best).

10 Integrators and Differentiators

After completing this chapter, you should be able to:

❏ Describe the fundamental usefulness and operation of an integrator.

❏ Describe the fundamental usefulness and operation of a differentiator.

❏ Detail the modifications required in order to make a practical op amp integrator or differentiator.

❏ Plot the useful frequency range of a given integrator or differentiator.

❏ Analyze the operation of integrator circuits using both time-continuous and time-discrete methods.

❏ Analyze the operation of differentiator circuits using both time continuous and time discrete methods.

❏ Explain the operation of an *analog computer*.

10.1 Introduction

Up to this point, we have examined a number of different op amp circuits and applications. Viewed from a purely mathematical perspective, the circuits perform basic functions. Amplifiers multiply an input quantity by a constant. Voltage-controlled or transconductance amplifiers can be used to multiply a quantity by a variable. Absolute value and logarithmic functions also can be produced. There is no reason to stop with just these tools; higher mathematical functions can be enlisted. In this chapter we will examine circuits which perform integration and differentiation. Although these circuits may appear to be some-what esoteric at first glance, they can prove to be quite useful. *Integrators* perform the function of summation over time. They may be used whenever an integration function is required; for example, to solve differential equations. The *differentiator* is the mirror of the integrator, and can be used to find rates of change. One possible application is finding acceleration when the input voltage represents a velocity.

Integrators and differentiators can be combined with summing amplifiers and simple gain blocks to form *analog computers*, which can be used to model physical systems, and thus are valuable aids in the initial design and testing of such things as mechanical suspension systems or loudspeakers. Unlike their digital counterparts, analog computers do not require the use of a programming language, per se. Also, they can respond in real-time to an input stimulus.

Besides the obvious use as a direct mathematical tool, integrators and differentiators can be used for wave shaping. As an example, the square/triangle generator of Chapter 9 used an integrator. With the exception of simple sin-usoids, both the integrator and differentiator will change the fundamental shape of a waveform. This might be contrasted with a simple amplifier which only makes a wave larger, and does not alter the basic shape.

For practical work, certain alterations will be required to the ideal circuits. The effects of these changes, both good and bad, will also be studied.

10.2 Integrators

The basic operation of an integrator is shown in Figure 10.1. The output voltage is the result of the definite integral of v_{in} from time = 0 to some arbitrary time t. Added to this will be a constant which represents the output of the network at $t = 0$. (Remember, integration is basically the process of summation. You can also think of this as finding the "area under the curve.") The output of this circuit always represents the sum total of the input values up to that precise instant in time. Consequently, if a static value (nonzero) is used as an input, the output will continually grow over time. If this growth continues unchecked, output saturation will occur. If the input quantity changes polarity,

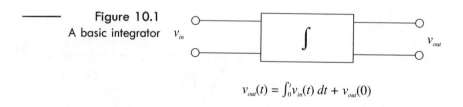

Figure 10.1
A basic integrator

$$v_{out}(t) = \int_0^t v_{in}(t)\, dt + v_{out}(0)$$

the output may also change in polarity. Having established the basic operation of the circuit, we are left with the design realization and limitations to be worked out.

As noted in earlier work, the response of an op amp circuit with feedback will reflect the characteristics of the feedback elements. If linear elements are used, the resulting response will be linear. If a logarithmic device is used in the feedback loop, the resulting response will have a log or antilog character. In order to achieve integration, the feedback network requires the use of an element which exhibits this characteristic. In other words, the current through the device must be proportional to either the integral or differential of the voltage across it. Inductors and capacitors answer these requirements, respectively.

It should be possible, then, to create an integrator with either an inductor or a capacitor. Capacitors tend to behave in a more ideal fashion than do inductors. Capacitors exhibit virtually no problems with stray magnetic field interference, do not have the saturation limitations of inductors, are relatively inexpensive to manufacture, and operate reliably over a wide frequency range. Because of these factors, most integrators are made using capacitors. The characteristic equation for a capacitor is

$$i(t) = C \frac{dv(t)}{dt} \tag{10.1}$$

This tells us that the current charging the capacitor is proportional to the differential of the input voltage. By integrating Equation 10.1, it can be seen that the integral of the capacitor current is proportional to the capacitor voltage:

$$v(t) = \frac{1}{C} \int_0^t i(t) \, dt \tag{10.2}$$

Assume for a moment that the capacitor voltage is the desired output voltage, as in Figure 10.2. If the capacitor current can be derived from the input

Figure 10.2
Capacitor as integration element

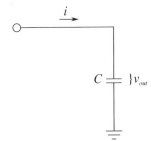

Figure 10.3
A simple op amp
integrator

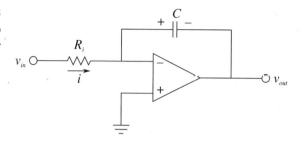

voltage, the output voltage will be proportional to the integral of the input. The circuit of Figure 10.3 will satisfy our requirements. Note that it is based on the parallel-parallel inverting amplifier studied earlier. The derivation of its characteristic equation follows.

First, note that the voltage across the capacitor is equal to the output voltage. This is due to the virtual ground at the inverting op amp terminal. Given the polarities marked, the capacitor voltage is of opposite polarity.

$$v_{out}(t) = -v_c(t) \tag{10.3}$$

Due to the virtual ground at the op amp's inverting input, all of v_{in} drops across R_i. This creates the input current, i.

$$i(t) = \frac{v_{in}(t)}{R_i} \tag{10.4}$$

Since the current drawn by the op amp's inputs is negligible, all of the input current flows into the capacitor, C. Combining Equations 10.2 and 10.4 yields

$$v_c(t) = \frac{1}{C} \int \frac{v_{in}(t)}{R_i} \, dt$$

$$v_c(t) = \frac{1}{R_i C} \int v_{in}(t) \, dt \tag{10.5}$$

Combining Equation 10.5 with Equation 10.3 produces the characteristic equation for the circuit:

$$v_{out}(t) = -\frac{1}{R_i C} \int v_{in}(t) \, dt \tag{10.6}$$

The only difference between this equation and the general equation as presented in Figure 10.1 is the preceding constant $-(1/R_iC)$. Also, note that Equation 10.6 assumes that the circuit is initially relaxed (i.e., there is no charge on the capacitor at time $= 0$). If the inversion or magnitude of this constant creates design problems, it can usually be corrected by gain/attenuation networks and/or inverting buffers. Finally, it is worth noting that due to the virtual ground at the inverting input, the input impedance of this circuit is approximately equal to R_i, as is the case with the ordinary parallel-parallel inverting amplifier.

Accuracy and Usefulness of Integration

As long as the circuit operation follows the model presented above, the accuracy of Equation 10.6 is very high. Small errors occur due to the approximations made. One possible source of error is the op amp's input bias current. If the input signal current is relatively low, the idealization that all of the input current will bypass the op amp and flow directly into the capacitor is no longer realistic. One possible solution is to use an FET input op amp. The limited frequency response and noise characteristics of the op amp will also play a role in the ultimate circuit accuracy. Generally, slew rate performance is not paramount as integrators tend to "smooth out" signal variations. As long as the input signal stays within the op amp's frequency limits, well above the noise floor, and the output does not saturate, accurate integration results.

Circuits of this type can be used to model any number of physical processes. For example, the heat stress on a given transducer might be found by integrating the signal across it. The resulting signal could then be compared to a predetermined maximum. If the temperature proved to be too high, the circuit could be powered down as a safety measure. While our initial approach might be to directly measure the temperature with some form of thermal sensor, this is not always practical. In such a case, simulation of the physical process is the only reasonable route.

Optimizing the Integrator

For best performance, high-quality parts are a must. Low offset and drift op amps are needed since the small DC values which they produce at the input will eventually force the circuit into saturation. Although they are not shown in Figure 10.3, bias and offset compensation additions are usually required. Very low offset devices such as chopper-stabilized (or commutating auto-zero) op amps can be used for lower frequency applications. For maximum accuracy,

Figure 10.4
Response of a simple
integrator

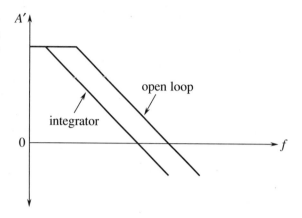

a small input bias current is desired, and thus, FET input op amps should be considered. Also, the capacitor should be a stable low-leakage type, such as polypropylene film. Electrolytic capacitors are generally a poor choice for these circuits.

Even with high-quality parts, the basic integrator can still prove susceptible to errors caused by small DC offsets. The input offset will cause the output to gradually move toward either positive or negative saturation. When this happens, the output signal is distorted, and therefore, useless. To prevent this it is possible to place a shorting switch across the capacitor, which can be used to re-initialize the circuit from time to time. While this approach does work, it is hardly convenient. Before we can come up with a more effective cure, we must first examine the cause of the problem.

We can consider the circuit of Figure 10.3 to be a simple inverting amplifier, where R_f is replaced by the reactance of the capacitor, X_C. In this case, we can see that the magnitude of the voltage gain is

$$A_v = -\frac{X_C}{R_i} = -\frac{1}{2\pi fCR_i}$$

Since X_C is frequency-dependent, it follows that the voltage gain must be frequency-dependent. Since X_C increases as the frequency drops, the voltage gain must increase as the frequency decreases. To be more specific, the gain curve will follow a -6 dB per octave slope. Eventually, the amplifier's response reaches the open-loop response, as shown in Figure 10.4. Since the DC gain of the system is at the open-loop level, it is obvious that even very small DC inputs can play havoc with the circuit response. If the low-frequency gain is limited to a more modest level, saturation problems can be reduced, if not

Figure 10.5
A practical integrator

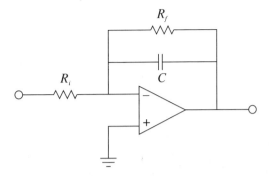

Figure 10.5

A practical integrator

completely eliminated. Gain limiting can be produced by shunting the integration capacitor with a resistor, as shown in Figure 10.5. This resistor sets the upper limit voltage gain to

$$A_{max} = -\frac{R_f}{R_i}$$

As always, there will be a trade-off with modification. By definition, integration occurs where the amplitude response rolls off at -6 dB per octave, as this is the response of our idealized capacitor model. Any alteration to the response curve will affect the ultimate accuracy of the integration. By limiting the low-frequency gain, the Bode amplitude response will no longer maintain a constant -6 dB per octave rolloff. Instead, the response will flatten out below the critical frequency set by C and R_f, as shown in Figure 10.6. This critical

Figure 10.6
Response of a practical integrator

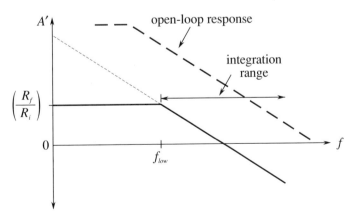

frequency, f_{low}, is found in the standard manner:

$$f_{low} = \frac{1}{2\pi R_f C}$$

(10.7)

This is a very important frequency. It tells us where the useful integration range starts. If the input frequency is less than f_{low}, the circuit acts like a simple inverting amplifier and no integration results. The input frequency must be higher than f_{low} for useful integration to occur. The question then, is how much higher? A few general rules of thumb are useful. First, if the input frequency equals f_{low}, the resulting calculation will only be about 50% accurate.[1] This means that significant signal amplitude and phase changes will exist relative to the ideal. Second, at about 10 times f_{low}, the accuracy is about 99%. At this point, the agreement between the calculated and actual measured values will be quite solid.

Example 10.1   Determine the equation for v_{out} and find the lower frequency limit of integration for the circuit of Figure 10.7.

Figure 10.7
Integrator for
Example 10.1

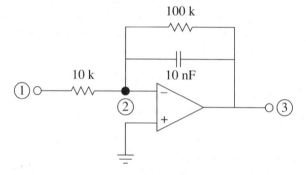

The general form of the output equation is given by Equation 10.6.

$$v_{out}(t) = -\frac{1}{R_i C} \int v_{in}(t)\, dt$$

$$v_{out}(t) = -\frac{1}{10\text{ k} \times 10\text{ nF}} \int v_{in}(t)\, dt$$

$$v_{out}(t) = -10^4 \int v_{in}(t)\, dt$$

---

[1] The term *50% accurate* is chosen for convenience. The actual values are 50% of the calculated value for power and phase angle, and 70.7% of the calculated value for voltage.

The lower limit of integration is set by f_{low}.

$$f_{low} = \frac{1}{2\pi R_f C}$$

$$f_{low} = \frac{1}{2\pi 100 \text{ k} \times 10 \text{ nF}}$$

$$f_{low} = 159 \text{ Hz}$$

Note that our solution to the example above represents our 50% accuracy point. For 99% accuracy, the input frequency should be **at least** one decade above f_{low}, or 1.59 kHz. Accurate integration will continue to higher and higher frequencies. The upper limit to useful integration is set by two factors: the frequency response of the op amp, and the signal amplitude versus noise. Obviously, at much higher frequencies, the basic assumptions of the circuit's operation will no longer be valid. The use of wide-bandwidth op amps reduces this limitation but cannot eliminate it. All amplifiers will eventually reach an upper frequency limit. Perhaps not so obvious is the limitation caused by signal strength. As noted earlier, the response of the integrator progresses at -6 dB per octave, which is equivalent to -20 dB per decade. In other words, a tenfold increase in input frequency results in a tenfold reduction in output amplitude. For higher frequencies, the net attenuation can be very great. At that point, the output signal runs the risk of being lost in the output noise. The only way around this is to ensure that the op amp is a low-noise type, particularly if a wide range of input frequencies will be used. For integration over a narrow band of frequencies, the RC integration constant can be optimized to prevent excessive loss.

Now that we have examined the basic structure of the practical integrator, it is time to analyze its response to different input waveforms. There are two basic ways of calculating the output: time-continuous or time-discrete. The time-continuous method involves the use of the indefinite integral, and is well suited for simple sinusoidal inputs. Waveforms which require a more complex time domain representation, such as a square wave, may be analyzed with the time-discrete method. This corresponds to the definite integral.

Analyzing Integrators with the Time-Continuous Method

The time-continuous analysis approach involves finding a time domain representation of the input waveform and then inserting it into Equation 10.6. The indefinite integral is taken, and the result is the time domain representation of the output waveform. The constant term produced by the indefinite integral is ignored. As long as the input waveform can be written in a time domain form, this method can be used.

While any waveform can be expressed in this way, it is not always practical, or the most expedient route. Indeed, relatively common waveforms such as square waves and triangle waves require an infinite series time domain representation. We shall not deal with these waveforms in this manner. The time-continuous integration of these functions is left as an exercise in the Challenge Problems at the end of this chapter.

Example 10.2 Using the circuit of Figure 10.7, determine the output if the input is a 1 V peak sine wave at 5 kHz.

First, write the input signal as a function time:

$$v_{in}(t) = 1 \, \sin(2\pi 5 \times 10^3 t) = 1 \, \sin(2\pi 5000 t)$$

Next, substitute this input into Equation 10.6:

$$v_{out}(t) = -\frac{1}{R_i C} \int v_{in}(t) \, dt$$

$$v_{out}(t) = -\frac{1}{10 \, k \times 10 \, nF} \int 1 \, \sin(2\pi 5000 t) \, dt$$

$$v_{out}(t) = -10^4 \int 1 \, \sin(2\pi 5000 t) \, dt$$

$$v_{out}(t) = -10^4 \frac{1}{2\pi 5000} \int (2\pi 5000) \, \sin(2\pi 5000 t) \, dt$$

$$v_{out}(t) = -.318 \left[-\cos(2\pi 5000 t) \right]$$

$$v_{out}(t) = .318 \, \cos(2\pi 5000 t)$$

Thus the output is a sinusoidal wave which is leading the input by 90° and is .318 V peak. Again, note that the constant produced by the integration is ignored. In Example 1 it was noted the f_{low} for this circuit is 159 Hz. Since the input frequency is over ten times f_{low}, the accuracy of the result should be better than 99%. Note that if the input frequency is changed, both the output frequency and amplitude change. For example, if the input were raised by a factor of 10 to 50 kHz, the output would also be at 50 kHz, and the output amplitude would be reduced by a factor of 10, to 31.8 mV. The output would still be a cosine, though.

The SPICE simulation of this circuit's steady-state response is shown in Figure 10.8. Note that in order to achieve steady-state results, the output is plotted only after several cycles of the input have passed through. Even after some 20 cycles, the output is not perfectly symmetrical. In spite of this, the

(Continued)

Figure 10.8
SPICE simulation of the practical integrator of Example 10.2

```
Practical Integrator, sine wave input.
*
*******Start of UA741 op amp*********
.SUBCKT UA741 1    2    3    4    5    6
*             IN+ IN- GND V+ V- OUT
Q1  11   1  13 UA741QA
Q2  12   2  14 UA741QB
RC1  4  11  5.305165E+03
RC2  4  12  5.305165E+03
C1  11  12  5.459553E-12
RE1 13  10  2.151297E+03
RE2 14  10  2.151297E+03
IEE 10   5  1.666000E-05
CE  10   3  3.000000E-12
RE  10   3  1.200480E+07
GCM  3  21  10 3 5.960753E-09
GA  21   3  12 11 1.884655E-04
R2  21   3  1.000000E+05
C2  21  22  3.000000E-11
GB  22   3  21 3 2.357851E+02
RO2 22   3  4.500000E+01
D1  22  31  UA741DA
D2  31  22  UA741DA
EC  31   3  6 3 1.0
RO1 22   6  3.000000E+01
D3   6  24  UA741DB
VC   4  24  2.803238E+00
D4  25   6  UA741DB
VE  25   5  2.803238E+00
.ENDS
.MODEL UA741DA D (IS=9.762287E-11)
.MODEL UA741DB D (IS=8.000000E-16)
.MODEL UA741QA NPN (IS=8.000000E-16 BF=9.166667E+01)
.MODEL UA741QB NPN (IS=8.309478E-16 BF=1.178571E+02)
*****End of UA741 op amp******
*
***Main circuit description***
Ri    1  2  10K
Rf    2  3  100K
Cf    2  3  10NF
VCC   4  0  DC 15
VEE   5  0  DC -15
X1    0  2  0 4 5 3  UA741
VIN   1  0  SIN (0 1 5KHZ)
*
***Delay the output plot until the circuit has settled.
*
.TRAN 4US 4400US 4000US
*
***Plot the input and output potentials.
*
.PLOT TRAN V(1) V(3)
*
*``To design with an op amp is to not design with an op amp.''
*   - Zen and The Art of Op Amp Design
*
.END
```

a. SPICE input

Figure 10.8 *(Continued)*

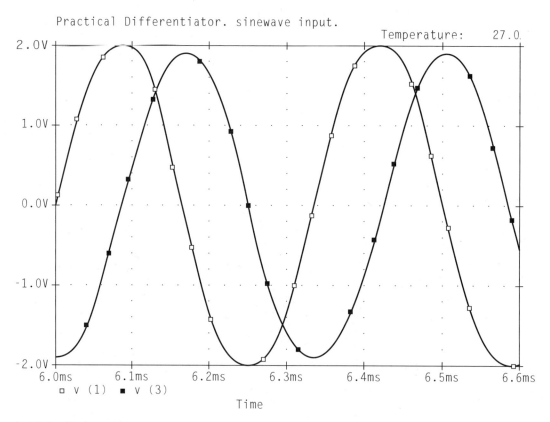

b. PSpice/Probe output

accuracy of both the amplitude and phase is quite good when compared to the calculated results. This further reinforces the fact that the circuit is operating within its useful range. It can be instructive to rerun the simulation to investigate the initial response, as well.

Analyzing Integrators with the Time-Discrete Method

Unlike the time-continuous approach, the time-discrete method uses the definite integral, and is used to find an output level at specific instances in time. This is useful if the continuous time domain representation is somewhat complex, and yet the wave shape is relatively simple, as with a square wave. Often, a little logic can be used to ascertain the shape of the resulting wave. The integral is then used to determine the exact amplitude. The time-discrete method also proves useful for more complex waves when modeled within a computer program. Here, several calculations can be performed per cycle, with the results joined together graphically to form the output signal.

The basic technique revolves around finding simple time domain representations of the input for specific periods of time. A given wave might be modeled as two or more sections. The definite integral is then applied over each section, and the results joined.

Example 10.3 Sketch the output of the circuit shown in Figure 10.7 if the input signal is a 10 kHz 2 V peak square wave with no DC component.

The first step is to break down the input waveform into simple-to-integrate components, and ascertain the basic shape of the result. The input is sketched in Figure 10.9. The input can be broken into two parts: a nonchanging 2 V potential from 0 to 50 μS, and a nonchanging -2 V potential from 50 μS to 100 μS. The sequence then repeats.

$$v_{in}(t) = 2 \text{ from } t = 0 \text{ to } t = 50 \ \mu S$$

$$v_{in}(t) = -2 \text{ from } t = 50 \ \mu S \text{ to } t = 100 \ \mu S$$

In essence, this means that the input may be treated as either ± 2 V DC for short periods of time. Since integration is a summing operation, as time progresses, the area swept out from underneath the input signal increases, and then decreases due to the polarity changes. It follows that the output will grow (or shrink) in a linear fashion since the input is constant. In other words, straight ramps will be produced during these time periods. The only difference between them will be the polarity of the slope. From this we find that the

Figure 10.9
Input waveform for
Example 10.3

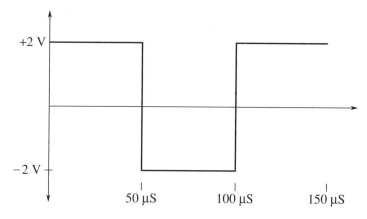

——— **Figure 10.10**
Output waveform for
Example 10.3

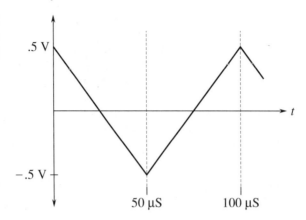

expected output waveform is a triangle wave. All we need to do now is determine the peak value of the output. To do this, we perform the definite integral using the first portion of the input wave:

$$v_{out}(t) = -\frac{1}{R_i C} \int v_{in}(t)\, dt$$

$$v_{out}(t) = -\frac{1}{10\text{ k} \times 10\text{ nF}} \int_0^{50\mu} 2\, dt$$

$$v_{out}(t) = -10^4 \times 2 \times t\Big|_{t=0}^{t=50\mu}$$

$$v_{out} = -2 \times 10^4 \times 50\mu$$

$$v_{out} = -1\text{ V}$$

This represents the total change over the 50 μS half-cycle interval. This is a peak-to-peak change: the output is 1 V negative relative to its value at the end of the preceding half-cycle. Integration for the second half-cycle is similar, and produces a positive change. The resulting waveform is shown in Figure 10.10.

Note that the wave is effectively inverted. For positive inputs, the output slope is negative. Since the input frequency is much higher than f_{low}, we can once again expect high accuracy. While the solution seems to imply that the output voltage should swing between 0 and -1 V, a real-world integrator does indeed produce the indicated $\pm.5$ V swing. This is due to the fact that the integration capacitor in this circuit will not be able to maintain the required DC offset indefinitely.

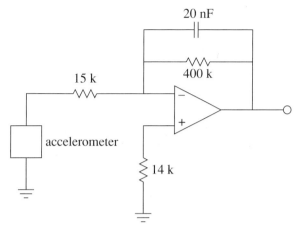

Figure 10.11
Accelerometer with
integrator

a. Circuit

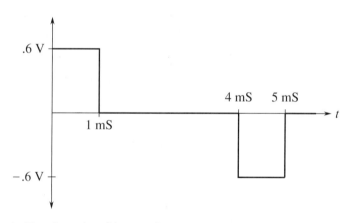

b. Signal produced by accelerometer

Example 10.4   Figure 10.11a shows an integrator connected to an *accelerometer*. This device produces a voltage which is proportional to the acceleration it experiences.[2] Accelerometers can be fastened to a variety of physical devices in order to determine how the devices respond to various mechanical inputs. This would be useful, for example, in experimentally determining the mechanical resonance characteristics of a surface. Another application is the determination of the

---

[2] A mechanical accelerometer consists of a small mass, associated restoring springs, and some form of transducer which is capable of reading the mass's motion.

lateral acceleration of an automobile as it travels through a corner. By integrating this signal, it is possible to determine velocity, and a further integration will produce position. If the accelerometer produces the voltage shown in Figure 10.11b, determine the shape of the velocity curve. Assume a nonrepetitive input wave.

First, check the frequency limit of the integrator in order to see if high accuracy can be maintained. If the input waveform was repetitive, it would be approximately 200 Hz (1/5 mS).

$$f_{low} = \frac{1}{2\pi R_f C}$$

$$f_{low} = \frac{1}{2\pi 400 \text{ k} \times 20 \text{ nF}}$$

$$f_{low} = 19.9 \text{ Hz}$$

The input signal is well above the lower limit.

Note that the input waveshape can be analyzed in piecewise fashion, as if it were a square wave. The positive and negative portions are both 1 mS in duration, and .6 V peak. The only difference is the polarity. Since these are square pulses, we expect ramp sections for the output. For the positive pulse ($t = 0$ to $t = 1$ mS),

$$v_{out}(t) = -\frac{1}{R_i C} \int v_{in}(t) \, dt$$

$$v_{out}(t) = -\frac{1}{15 \text{ k} \times 20 \text{ nF}} \int_0^{10^{-3}} .6 \, dt$$

$$v_{out}(t) = -3333 \times .6 \times t \Big|_{t=0}^{t=10^{-3}}$$

$$v_{out} = -2000 \times 10^{-3}$$

$$v_{out} = -2 \text{ V}$$

This tells us that we will see a negative-going ramp with a -2 V change. For the time period between 1 mS and 4 mS, the input signal is zero, and thus the change in output potential will be zero. This means that the output will remain at -2 V until $t = 4$ mS.

For the period between 4 mS and 5 mS, a positive-going ramp will be produced. Since only the polarity is changed relative to the first pulse, we can quickly find that the change is $+2$ V. Since the input waveform is nonrepetitive, the output waveform appears as shown in Figure 10.12. (A repetitive input would naturally cause the integrator to "settle" around ground over time,

Figure 10.12
Integrator output

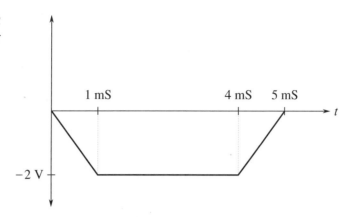

producing the same basic waveshape, but shifted positively.) This waveform tells us that the velocity of the device under test increases linearly up to a point. After this point, the velocity remains constant until a linear decrease in velocity occurs. (Remember, the integrator inverts the signal, so the output curve is effectively upside down. If the input wave indicates an initial positive acceleration, the actual output is an initial positive velocity.) On a different time scale, these are the sort of waveforms which might be produced by an accelerometer mounted on an automobile which starts at rest, smoothly climbs to a fixed speed, and then smoothly brakes to a stop.

10.3 Differentiators

Differentiators perform the complementary function to the integrator. The base form of the differentiator is shown in Figure 10.13. The output voltage is the differential of the input voltage, which is very useful for finding the rate at which a signal varies over time. For example, it is possible to find velocity given distance, and acceleration given velocity. This can be very useful in process control work.

Figure 10.13
A basic differentiator

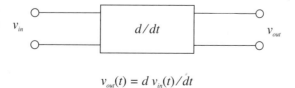

$$v_{out}(t) = d\, v_{in}(t)/dt$$

Essentially, the differentiator tends to reinforce fast signal transitions. If the input waveform is nonchanging (i.e., DC), the slope is zero, and thus the output of the differentiator is zero. On the other hand, an abrupt signal change such as the rising edge of a square wave produces a very large slope, and thus the output of the differentiator will be large. In order to create the differentiation, an appropriate device needs to be associated with the op amp circuit. This was the approach taken with the integrator, and it remains valid here. In fact, we are left with the same two options: using either an inductor or a capacitor. Again, capacitors tend to be somewhat easier to work with than inductors and are preferred. The only difference between the integrator and the differentiator is the position of the capacitor. Instead of placing it in the R_f position, the capacitor will be placed in the R_i position. The resulting circuit is shown in Figure 10.14. The analysis starts with the basic capacitor equation (Equation 10.1):

$$i(t) = C \frac{dv(t)}{dt}$$

We already know from previous work that the output voltage appears across R_f, though inverted.

$$v_{out} = -v_{R_f}$$

Also, by Ohm's Law,

$$v_{R_f} = iR_f$$

By using the approximation that all input current flows through R_f (since the op amp's input current is zero), and then substituting Equation 10.1 for the current, we find

$$v_{out}(t) = -R_f C \frac{dv(t)}{dt}$$

Figure 10.14
A simple op amp
differentiator

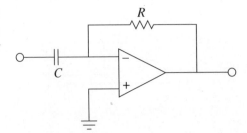

A quick inspection of the circuit shows that all of the input voltage drops across the capacitor, since the op amp's inverting input is a virtual ground. Bearing this in mind, we arrive at the final output voltage equation,

$$v_{out}(t) = -R_f C \frac{dv_{in}(t)}{dt} \qquad (10.8)$$

As with the integrator, a leading constant is added to the fundamental form. Again, it is possible to scale the output as required through the use of gain or attenuation networks.

Accuracy and Usefulness of Differentiation

Equation 10.8 is an accurate reflection of the circuit response as long as the base assumptions remain valid. As with the integrator, practical considerations tend to force limits on the circuit's operating range. If the circuit is analyzed at discrete points in the frequency domain, it can be modeled as an inverting amplifier with the following gain equation:

$$A_v = -\frac{R_f}{X_C} = -2\pi f C R_f$$

Note that as the frequency decreases, X_C grows, thus reducing the gain. Conversely, as the input frequency is raised, X_C falls in value, causing the gain to rise. This rise will continue until it intersects the open-loop response of the op amp. The resulting amplitude response is shown in Figure 10.15. This response is the mirror image of the basic integrator response and exhibits a slope of 6 dB per octave. Note that DC gain is zero, and therefore the problems

Figure 10.15
Response of a simple
differentiator

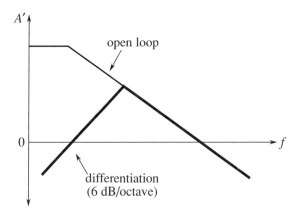

created by input bias and offset currents are not nearly as troublesome as in the integrator. Because of this, there is no limit as to how low the input frequency may be, excluding the effects of signal-to-noise ratio. Things are considerably different on the high end, though. Once the circuit response breaks away from the ideal 6 dB per octave slope, differentiation no longer takes place.

Optimizing the Differentiator

There are a couple of problems with the general differentiator of Figure 10.14. First of all, it is quite possible that the circuit may become unstable at higher frequencies. Also, the basic shape of the amplitude response suggests that high frequencies are accentuated, thus increasing the relative noise level. Both of these problems can be reduced by providing an artificial upper-limit frequency, f_{high}. This tailoring can be achieved by shunting R_f with a small capacitor. This reduces the high frequency gain, and thus reduces the noise. The resulting response is shown in Figure 10.16.

We find f_{high} in the standard manner:

$$f_{high(f\,dbk)} = \frac{1}{2\pi R_f C_f} \tag{10.9}$$

Note that f_{high} represents the highest frequency for differentiation. It is the 50% accuracy point. For higher accuracy, the input frequency must be kept *well below* f_{high}. At about .1 f_{high}, the accuracy of Equation 10.8 is about 99%. Generally, you have to be somewhat more conservative in the estimation of accuracy than with the integrator. This is because complex waves contain harmonics which are higher than the fundamental. Even though the fundamental may be well within the high accuracy range, the upper harmonics may not be.

Figure 10.16
Response of a partially optimized differentiator

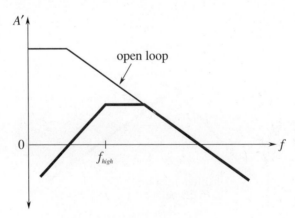

Figure 10.17
Response of a practical
differentiator

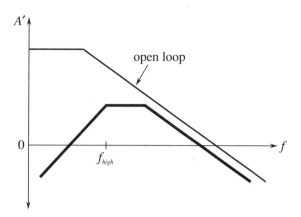

The other major problem of the basic differentiator circuit is that the input impedance is inversely proportional to the input frequency. This is because X_C is the sole input impedance factor. This may present a problem at higher frequencies because the impedance will approach zero. To circumvent this problem, a resistor can be placed in series with the input capacitor in order to establish a minimum impedance value. Unfortunately, this will also create an upper break frequency, f_{high}.

$$f_{high(in)} = \frac{1}{2\pi R_i C} \qquad (10.10)$$

The resulting response is shown in Figure 10.17. The effective f_{high} for the system will be the lower of Equations 10.9 and 10.10. The completed, practical differentiator is shown in Figure 10.18. Note that a bias compensation resistor may be required at the noninverting input, although it is not shown in the figure.

Figure 10.18
A practical differentiator

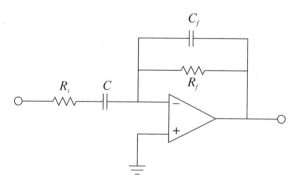

Analyzing Differentiators with the Time-Continuous Method

The time-continuous method will be used when the input signal can be easily written in the time domain (e.g., sine waves). For more complex waveforms, such as triangle waves, a discrete time method will be used. The continuous method will lead directly to a time domain representation of the output waveform. Specific voltage/time coordinates will not be evaluated.

Example 10.5 Determine the useful range for differentiation in the circuit of Figure 10.19. Also determine the output voltage if the input signal is a 2 V peak sine wave at 3 kHz.

The upper limit of the useful frequency range will be determined by the lower of the two *RC* networks.

$$f_{high(fdbk)} = \frac{1}{2\pi R_f C_f}$$

$$f_{high(fdbk)} = \frac{1}{2\pi \times 5\,k \times 100\,pF}$$

$$f_{high(fdbk)} = 318.3\;kHz$$

$$f_{high(in)} = \frac{1}{2\pi R_i C}$$

$$f_{high(in)} = \frac{1}{2\pi \times 100 \times .01\,\mu F}$$

$$f_{high(in)} = 159.2\;kHz$$

Therefore, the upper limit is 159.2 kHz. Remember, the accuracy at this upper limit is relatively low, and normal operation will typically be several octaves lower than this limit. Note that the input frequency is 3 kHz, so high

Figure 10.19
Differentiator for
Example 10.5

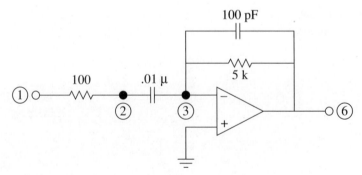

accuracy should result. First, write v_{in} as a time domain expression:

$$v_{in}(t) = 2 \sin 2\pi 3000t$$

$$v_{out}(t) = -R_f C \frac{dv_{in}(t)}{dt}$$

$$v_{out}(t) = -5 \text{ k} \times .01 \ \mu\text{F} \frac{d2 \sin 2\pi 3000t}{dt}$$

$$v_{out}(t) = -10^{-4} \frac{d \sin 2\pi 3000t}{dt}$$

$$v_{out}(t) = -1.885 \cos 2\pi 3000t$$

This tells us that the output waveform is also sinusoidal, but it lags the input by 90°. Note that the input frequency has not changed, but the amplitude has. Since the differentiator operates with a 6 dB per octave slope, it can be seen that the output amplitude is directly proportional to the input frequency. If this example is rerun with a frequency of 6 kHz, the output amplitude will be double the present value.

The SPICE simulation for this circuit is shown in Figure 10.20 (pages 464–465). Note the excellent correlation for both the phase and amplitude of the output. As was the case with the integrator simulation, the output plot is started after the initial conditions have settled.

Analyzing Differentiators with the Time-Discrete Method

For more complex waveforms, it is sometimes expedient to break the waveform into discrete chunks, differentiate each portion, and then combine the results. The idea is to break the waveform into equivalent straight line segments. Differentiation of a straight line segment will result in a constant (i.e., the slope, which does not change over that time). The process is repeated until one cycle of the input waveform is completed. The resultant levels are then joined together graphically to produce the output waveform.

Often, waveforms are symmetrical, and only part of the calculations need to be performed: a sign change is all that will be needed for the mirror image portions. As an example, a triangle wave can be broken into a positive-going line segment and a negative-going line segment. The slopes should be equal; only the direction (i.e., sign) has changed. A square wave can be broken into four parts: a positive-going edge, a static positive value, a negative-going edge, and a static negative value. Since the "flat" portions have a slope of zero, only one calculation must be performed, and that's the positive-going edge. We will take a look at both of these waveforms in the next two examples.

—— **Figure 10.20** SPICE simulation of the differentiator of Example 10.5

```
Practical Differentiator, sine wave input.

*******Start of UA741 op amp*********
.SUBCKT UA741 1    2    3    4    5    6
*              IN+ IN- GND  V+   V-  OUT
Q1   11   1   13 UA741QA
Q2   12   2   14 UA741QB
RC1   4  11   5.305165E+03
RC2   4  12   5.305165E+03
C1   11  12   5.459553E-12
RE1  13  10   2.151297E+03
RE2  14  10   2.151297E+03
IEE  10   5   1.666000E-05
CE   10   3   3.000000E-12
RE   10   3   1.200480E+07
GCM   3  21   10 3 5.960753E-09
GA   21   3   12 11 1.884655E-04
R2   21   3   1.000000E+05
C2   21  22   3.000000E-11
GB   22   3   21 3 2.357851E+02
RO2  22   3   4.500000E+01
D1   22  31   UA741DA
D2   31  22   UA741DA
EC   31   3   6 3 1.0
RO1  22   6   3.000000E+01
D3    6  24   UA741DB
VC    4  24   2.803238E+00
D4   25   6   UA741DB
VE   25   5   2.803238E+00
.ENDS
.MODEL UA741DA D (IS=9.762287E-11)
.MODEL UA741DB D (IS-8.000000E-16)
.MODEL UA741QA NPN (IS=8.000000E-16 BF=9.166667E+01)
.MODEL UA741QB NPN (IS=8.309478E-16 BF=1.178571E+02)
*****End of UA741 op amp******

***Main circuit description***

Ri    1   2   100
Rf    3   6   5K
Ci    2   3   .01UF
Cf    3   6   100PF
VCC   4   0   DC 15
VEE   5   0   DC -15
X1    0   3   0 4 5 6   UA741
VIN   1   0   SIN (0 2 3KHZ)

***Delay the output plot until the circuit has settled.
*
.TRAN 6US 6600US 6000US
*
***Plot the input and output potentials.
*
.PLOT TRAN V(1) V(6)
.END
```

a. SPICE input

Figure 10.20 (*Continued*)

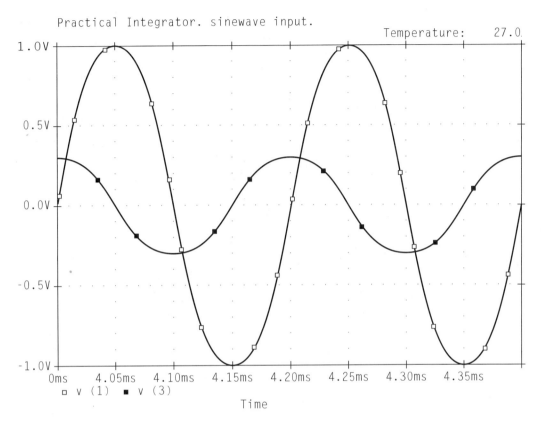

Practical Integrator. sinewave input.

b. PSpice/Probe output

Example 10.6 Sketch the ouput waveform for the circuit of Figure 10.19 if the input is a 3 V peak triangle wave at 4 kHz.

First, note that the input frequency is well within the useful range of this circuit, as calculated in Example 10.5. (Note that the highest harmonics will still be out of range, but the error introduced will be minor.)

The triangle wave can be broken into a positive-going portion and a negative-going portion. In either case, the total voltage change will be 6 V in one half-cycle. The period of the waveform is

$$T = \frac{1}{4 \text{ kHz}}$$

$$T = 250 \ \mu S$$

Figure 10.21   Input/output waveforms

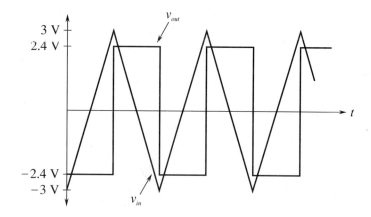

Therefore, for the positive-going portion, a 6 V change will be seen in 125 μS (-6 V in 125 μS for the negative-going portion). The slope is

$$\text{slope} = \frac{6 \text{ V}}{125 \text{ } \mu\text{S}}$$

$$\text{slope} = 48 \times 10^3 \text{ V/S}$$

which, as a time domain expression, is

$$v_{in}(t) = 48{,}000t$$

Substituting this equation into Equation 10.8 yields

$$v_{out}(t) = -R_fC \frac{dv_{in}(t)}{dt}$$

$$v_{out}(t) = -5 \text{ k} \times .01 \text{ } \mu\text{F} \frac{d(48{,}000t)}{dt}$$

$$v_{out}(t) = -2.4 \text{ V}$$

During $t = 0$ through $t = 125$ μS, the output is -2.4 V. Differentiation of the second half of the wave is similar, but produces a positive output ($+2.4$ V). The result is a 4 kHz square wave which is 2.4 V peak. The resulting waveform is shown in Figure 10.21.

Example 10.7   Repeat Example 10.6 with a 3 V peak 4 kHz square wave as the input. Assume that the rising and falling edges of the square wave have been slew limited to 5 V/μS.

During the time periods that the input is at ± 3 V, the output will be zero. This is because the input slope is, by definition, zero when the signal is "flat." An output is only noted during the transitions between the ± 3 V levels. Therefore, we need to find the slope of the transitions. It was stated that due to slew rate limiting (perhaps from some previous amplifier stage), the transitions run at 5 V/μS. As a time domain expression, this is

$$v_{in}(t) = 5 \times 10^6 t$$

Substituting this equation into Equation 10.8 yields

$$v_{out}(t) = -R_f C \frac{dv_{in}(t)}{dt}$$

$$v_{out}(t) = -5\,\text{k} \times .01\,\mu\text{F}\,\frac{d(5 \times 10^6 t)}{dt}$$

$$v_{out}(t) = -250\,\text{V}$$

Obviously, when using a standard op amp and ± 15 V power supply, clipping will occur in the vicinity of -13.5 V. For the negative-going edge, a similar result will be seen ($+250$ V calculated, with clipping at $+13.5$ V). The resulting waveform is shown in Figure 10.22. Note that the output waveform spikes will also be limited by the slew rate of the differentiator's op amp.

Figure 10.22
Differentiated square
wave (Note output
clipping.)

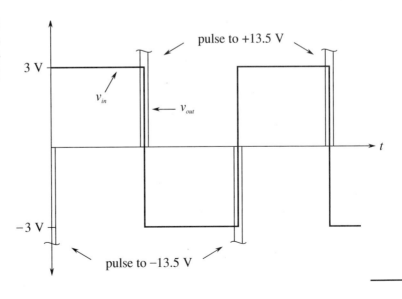

Example 10.8   Figure 10.23a shows a differentiator receiving a signal from an LVDT, or linear variable differential transformer.[3] An LVDT can be used to accurately measure the position of objects with displacements of less than one thousandth of an inch. This can be useful in a computer-aided manufacturing system. By differentiating this position signal, a velocity signal can be derived. A second differentiation produces an acceleration. If the LVDT produces the wave shown in Figure 10.23b, determine the velocity/time curve for the object being tracked.

Figure 10.23
Differentiator with LVDT

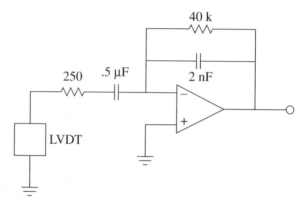

a. Circuit

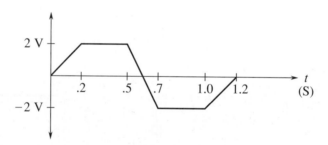

b. Signal produced by LVDT

---

[3] An LVDT is a transformer with dual secondary windings and a movable core. The core is connected to a shaft, which is in turn actuated by some external object. The movement of the core alters the mutual inductance between primary and secondary. A carrier signal is fed into the primary, and the changing mutual inductance alters the strength of the signal induced into the secondaries. This signal change is turned into a simple DC voltage by a demodulator. The resulting DC potential is proportional to the position of the core, and thus, proportional to the position of the object under measurement.

First, check the upper frequency limit for the circuit.

$$f_{high(f\,dbk)} = \frac{1}{2\pi R_f C_f}$$

$$f_{high(f\,dbk)} = \frac{1}{2\pi \times 40\text{ k} \times 2\text{ nF}}$$

$$f_{high(f\,dbk)} = 1.99\text{ kHz}$$

$$f_{high(in)} = \frac{1}{2\pi R_i C}$$

$$f_{high(in)} = \frac{1}{2\pi \times 250 \times .5\ \mu\text{F}}$$

$$f_{high(in)} = 1.273\text{ kHz}$$

The limit will be the lower of the two, or 1.273 kHz. This is well above the slowly changing input signal, and therefore, high accuracy should be possible.

This wave can be analyzed in piecewise fashion. The ramp portions will produce constant output levels and the flat portions will produce an output of 0 V (i.e., the rate of change is zero). For the first section,

$$\text{slope} = \frac{2\text{ V}}{.2\text{ S}}$$

$$\text{slope} = 10\text{ V/S}$$

which, as a time domain expression, is

$$v_{in}(t) = 10t$$

$$v_{out}(t) = -R_f C\,\frac{dv_{in}(t)}{dt}$$

$$v_{out}(t) = -40\text{ k} \times .5\ \mu\text{F}\,\frac{d(10t)}{dt}$$

$$v_{out}(t) = -.2\text{ V}$$

So, during $t = 0$ through $t = .2$ S, the output is $-.2$ V. The time period between .2 S and .5 S will produce an output of 0 V. For the negative-going portion,

$$\text{slope} = \frac{-4\text{ V}}{.2\text{ S}}$$

$$\text{slope} = -20\text{ V/S}$$

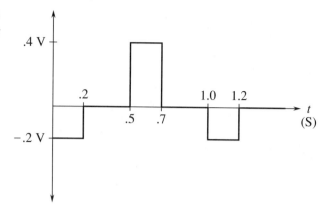

Figure 10.24
Differentiator output

Thus,

$$v_{in}(t) = -20t$$

$$v_{out}(t) = -R_f C \frac{dv_{in}(t)}{dt}$$

$$v_{out}(t) = -40 \text{ k} \times .5 \ \mu\text{F} \ \frac{d(20t)}{dt}$$

$$v_{out}(t) = .4 \text{ V}$$

The output is .4 V between .5 S and .7 S. The third section has the same slope as the first section, and will also produce a $-.2$ V level. The output waveform is drawn in Figure 10.24.

10.4 Analog Computers

Analog computers are used to simulate physical systems. These systems may be electrical, mechanical, acoustical, or what have you. An analog computer is basically a collection of integrators, differentiators, summers, and amplifiers. Due to their relative stability, integrators are favored over differentiators. It is not uncommon for analog computers to be made without any differentiators. Since physical systems can be described in terms of differential equations, analog computers can be used to solve these equations, thus producing as output some system parameter.

The basic advantage of simulation is that several variations of a given system can be examined in real time without actually constructing the system. For a large project this is particularly cost-efficient. The process starts by writing a differential equation (first-, second-, or third-order) which describes the system in question. The equation is then solved for its highest-order element and the result used to create a circuit.

Example 10.9 Let's investigate the system shown in Figure 10.25. This is a simple mechanical system which might represent (to a rough approximation) a variety of physical entities, including the suspension of an automobile. This system is comprised of a body with mass M, which is suspended from a spring. The spring has a spring constant, K. The mass is also connected to a shock absorber which produces damping, R. If an external force, F, excites the mass, it will move, producing some displacement, X. This displacement depends on the mass, force, spring constant, and damping. Essentially, the spring and shock absorber will create reactionary forces. From basic physics, $F = MA$, where A is the acceleration of the body. If X is the position of the body, then dX/dt is its velocity, and d^2X/dt^2 is its acceleration. Therefore, we can say

$$F = M \frac{d^2 X}{dt^2}$$

In this system, the total force is comprised of the excitation force F and the forces produced by the spring and shock absorber:

$$F - F_{\text{SPRING}} - F_{\text{SHOCK}} = M \frac{d^2 X}{dt^2}$$

Figure 10.25
Mechanical system

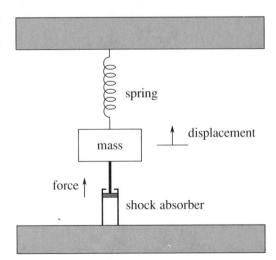

The spring's force is equal to the displacement times the spring constant:

$$F_{\text{SPRING}} = KX$$

The shock absorber's force is equal to the damping constant times the velocity of the body:

$$F_{\text{SHOCK}} = R\,\frac{dX}{dt}$$

By substituting and rearranging the above elements, we find

$$F = M\,\frac{d^2X}{dt^2} + R\,\frac{dX}{dt} + KX$$

Here F is seen as the input signal, and X as the output signal. A somewhat less busy notation form is the dot convention. A single dot represents the first derivative with respect to time, two dots represent the second derivative, and so on. The above equation may be rewritten as

$$F = M\ddot{X} + R\dot{X} + KX$$

This is the final differential equation. Note how it contains only derivatives and no integrals. The last step is to solve the equation for the highest-order differential. By setting it up in this form, the simulation circuit can be realized without using differentiators. The solution will indicate how many integrators will be required.

$$\ddot{X} = \frac{F}{M} - \frac{R}{M}\dot{X} - \frac{K}{M}X$$

This equation says that the second differential of X is the sum of three components. To realize the circuit, start with a summing amplifier with the three desired signals as inputs, as shown in block form in Figure 10.26. Note that two of the inputs use X and the first derivative of X. These elements may

Figure 10.26
Circuit realization
(block form)

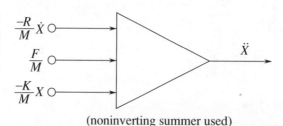

(noninverting summer used)

Figure 10.27   Circuit realization using function blocks

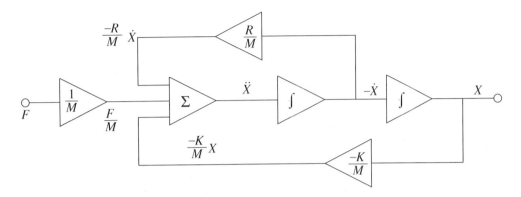

be produced by integrating the output of Figure 10.26. Appropriate constants can be used to achieve the desired signal levels. This is shown in Figure 10.27. Certain elements may be combined; for example, a weighted summing amplifier can be used to eliminate unneeded amplifiers.

In use, the constants R, K, and M are set by potentiometers (they are essentially nothing more than scaled gain factors). A voltage representing the excitation force is applied to the circuit, and the desired output quantity is recorded. Note that the output of interest could be the acceleration, velocity, or displacement of the body. In order to test the system with a new spring or damping constant, all that is needed is to adjust the appropriate potentiometer. In this manner, a large number of combinations can be tried quickly. The most successful combinations can then be built and tested for the final design. An analog computer such as this would be very useful in testing such items as the suspension of an automobile or a loudspeaker system. To ease the design of the simulation circuit, commercial analog computers are available. Construction (or programming) of the circuit involves wiring integrator, amplifier, and summer blocks together, with the appropriate potentiometers. In this way, the details of designing and optimizing individual integrators or amplifiers is bypassed.

10.5 Alternatives to Integrators and Differentiators

There are alternatives to using integrators and differentiators based on op amps. As long as systems can be described by a reasonable set of equations, simulation using digital computers is possible. The primary advantage of a simulation scheme based on digital computers is that it is very flexible. It can also be very accurate.

Digital simulation does have limits, though. The first limitation is the speed of response. Analog computers can be configured as real-time devices; they respond at the same speed as the system which is being simulated. The typically heavy computation load of the digital computer requires prodigious calculation speed to keep up with reasonably fast processes. For real-time work, digital simulation may be rather expensive. The other advantage of the analog computer is its immediacy. It is by nature, interactive, which means that an operator can change simulation parameters and immediately see the result. If a simulation does not require real-time performance and interactive adjustment, a digital-based simulation may, in fact, be preferred. The idea is to calculate the response for several closely spaced points in time. The results can be used to create graphs which can then be studied at leisure. Fundamentally, this process is the same as that used by the SPICE simulation program. Indeed, it is quite possible to design a simulator based on op amps, and then simulate its response by using SPICE. As you might guess, this double simulation is not particularly efficient, especially with large systems.

10.6 Extended Topic: Other Integrator and Differentiator Circuits

The basic integrator and differentiator circuits examined earlier can be extended into other forms. Perhaps the most obvious extension is to add multiple inputs, as in an ordinary summing amplifier. In complex systems, this approach may save the use of several op amps. A summing integrator is shown in Figure 10.28. Note its similarity to a normal summing amplifier. In this circuit, the input currents are summed at the inverting input of the op amp. If the resistors are all set to the same value, we can quickly derive the output equation by

Figure 10.28
Summing integrator

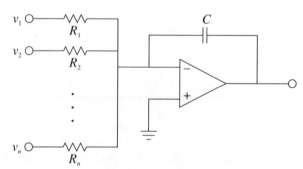

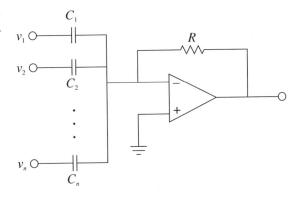

Figure 10.29
Summing differentiator

following the original derivation. The result is

$$v_{out} = -\frac{1}{RC} \int v_1 + v_2 + \cdots + v_n \, dt$$

The output is the negative integral of the sum of the inputs.

In a similar vein, a summing differentiator can be formed. This is shown in Figure 10.29. Again, the proof of its output equation follows the original differentiator derivation:

$$v_{out} = -RC_1 \frac{dv_1}{dt} - RC_2 \frac{dv_2}{dt} - \cdots - RC_n \frac{dv_n}{dt}$$

Another interesting adaptation of the integrator is the augmenting integrator. This circuit adds a constant gain portion to the output equation. An aumenting integrator is shown in Figure 10.30. The addition of the feedback resistor R_f provides the augmenting action. As you might surmise, the gain portion is directly related to R_f and R_i:

$$v_{out} = -v_{in} \frac{R_f}{R_i} - \frac{1}{RC} \int v_{in} \, dt$$

Figure 10.30
Augmenting integrator

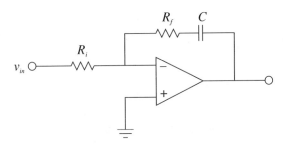

Figure 10.31
Double integrator

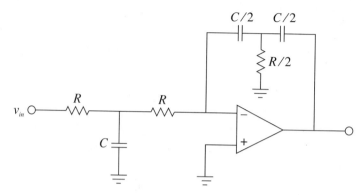

Figure 10.31
Double integrator

The augmenting integrator can also be turned into a summing/augmenting integrator by adding extra input resistors as in Figure 10.28. Note that the gain portion will be the same for all inputs if the input summing resistors are of equal value.

The final variant which we shall note is the double integrator. This design requires two reactive portions in order to achieve double integration. One possible configuration is shown in Figure 10.31. In this circuit, a pair of RC "Tee" networks are used. The output equation is

$$v_{out} = -\frac{4}{(RC)^2} \iint v_{in}\, dt$$

Properly used, the double integrator can cut down the parts requirement of larger circuit designs.

Summary

In this chapter we have examined the structure and use of integrator and differentiator circuits. Integrators produce a summing action while differentiators find the slope of the input. Both types are based on the general parallel-parallel inverting voltage feedback model. In order to achieve integration and differentiation, a capacitor is used in the feedback network in place of the standard resistor. Since the capacitor current is proportional to the rate of change of the capacitor's voltage, a differential or integral response is possible. For the integrator, the capacitor is placed in the normal R_f position, while for the differentiator, the capacitor is placed in the normal R_i position. As a result, the integrator exhibits a -6 dB per octave slope through the useful integration range. The differentiator exhibits the mirror image, or $+6$ dB per octave slope, throughout its useful range.

In both circuits, practical limitations require the use of additional components. In the case of the integrator, small DC offsets at the input can force the output into saturation. To avoid this, a resistor is placed in parallel with the integration capacitor in order to limit the low-frequency gain. This has the unfortunate side effect of limiting the useful integration range to higher frequencies. In the case of the differentiator, noise, stability, and input impedance limits can pose problems. In order to minimize noise and aid in stability, a small capacitor can be placed in parallel with R_f. This reduces the high-frequency gain. In order to place a lower limit on the input impedance, a resistor can be placed in series with the differentiation capacitor. The addition of either component will limit the upper range of differentiation.

We examined two general techniques for determining the output signal. The first form is referred to as the time-continuous method, and while it may be used with virtually any waveform, our use was with simple sine waves only. In the case of the integrator, it corresponds to the indefinite integral—a time domain equation for the output is the result. The second form is the time-discrete method and is useful for waves which can be easily broken into segments. Each segment is analyzed, and the results joined graphically. This has the advantage of an immediate graphical result, whereas the shape of the time-continuous result may not be immediately apparent. The time-discrete method is also useful for producing output tables and graphs with a digital computer. In the case of the integrator, the time-discrete method corresponds to the definite integral.

Integrators and differentiators can be used in combination with summers and amplifiers to form analog computers. Analog computers can be used to model a variety of physical systems in real-time. Unlike their digital counterparts, programming an analog computer requires only the proper interconnection of the various building blocks, and appropriate settings for the required physical constants.

Finally, it is noted that integrators and differentiators can be used as waveshaping circuits. Integrators can be used to turn square waves into triangle waves. Differentiators can be used to turn triangle waves into square waves. Indeed, since differentiators tend to "seek" rapid changes in the input signal, they can be quite useful as edge detectors.

Self Test Questions

1. What is the basic function of an integrator?

2. What is the basic function of a differentiator?

3. What is the function of the capacitor in the basic integrator and differentiator?

4. Why are capacitors used in favor of inductors?

5. What practical modifications need to be made to the basic integrator, and why?

6. What practical modifications need to be made to the basic differentiator, and why?

7. What are the negative side effects of the practical versus basic integrator and differentiator?

8. What is an analog computer and what is it used for?

9. What are some of the advantages and disadvantages of the analog computer versus the digital computer?

10. How might integrators and differentiators be used as wave-shaping circuits?

Problem Set

 Analysis Problems

1. Sketch the output waveform for the circuit of Figure 10.32 if the input is a 4 V peak square wave at 1 kHz.

Figure 10.32

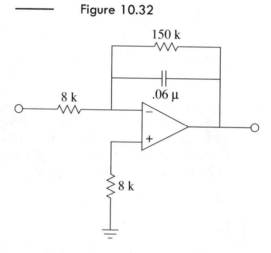

2. Repeat Problem 1 if $v_{in}(t) = 5 \sin 2\pi318t$.

3. Repeat Problem 1 if $v_{in}(t) = 20 \cos 2\pi10{,}000t$.

4. Using the circuit of Problem 1, if the input is a ramp with a slope of 10 volts per second, find the output after 1 mS, 10 mS, and 4 S. Sketch the resulting wave.

5. Determine the low-frequency gain for the circuit of Figure 10.33.

——— Figure 10.33

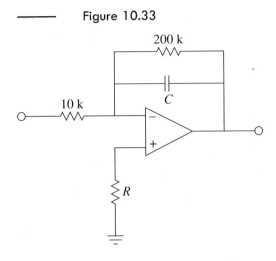

6. If $C = .033 \ \mu F$ in Figure 10.33, determine v_{out} if v_{in} is a 200 mV peak sine wave at 50 kHz.

7. Repeat Problem 6 using a 200 mV peak square wave at 50 kHz.

8. Sketch the output of the circuit of Figure 10.7 with the input signal given in Figure 10.34.

——— Figure 10.34

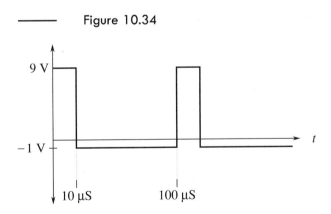

9. Assume that the input to the circuit of Figure 10.7 is 100 mV DC. Sketch the output waveform, and determine if it goes to saturation.

10. Sketch the output waveform for Figure 10.7, given an input of $v_{in}(t) = .5 \cos 2\pi 9000t$.

11. Given the circuit of Figure 10.35, sketch the output waveform if the input is a 100 Hz, 1 V peak triangle wave.

——————   Figure 10.35

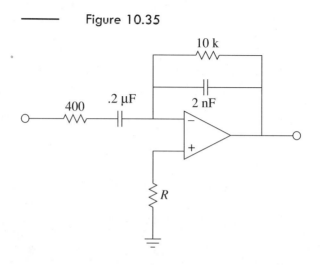

12. Repeat Problem 11 if the input is a 2 V peak square wave at 500 Hz. Assume that the rise and fall times are 1 μS, and are linear (versus exponential).

13. Repeat Problem 11 for the following input: $v_{in}(t) = 3 \cos 2\pi 60t$.

14. Repeat Problem 11 for the following input: $v_{in}(t) = .5 \sin 2\pi 1000t$.

15. Repeat Problem 11 for the following input: $v_{in}(t) = 10t^2$.

16. Given the input shown in Figure 10.36, sketch the output of the circuit of Figure 10.19.

——————   Figure 10.36

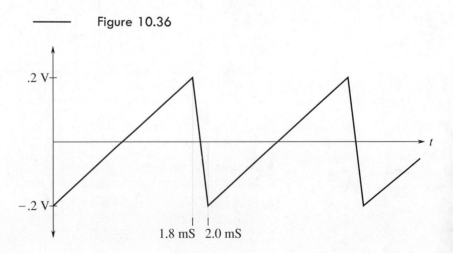

17. Given the input shown in Figure 10.37, sketch the output of the circuit of Figure 10.19.

——— Figure 10.37

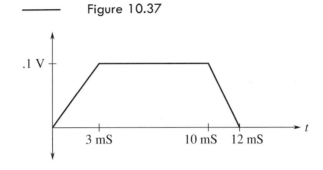

18. Sketch the output waveform for Figure 10.19, given an input of $v_{in}(t) = .5 \cos 2\pi 4000t$.

19. Sketch the output waveform for Figure 10.19, if v_{in} is a 300 mV peak triangle wave at 2500 Hz.

▷ Design Problems

20. Given the circuit of Figure 10.33,
   a. Determine the value of C required in order to yield an integration constant of -2000.
   b. Determine the required value of R for minimum integration offset error.
   c. Determine f_{low}.
   d. Determine the 99% accuracy point.

21. Given the differentiator of Figure 10.38, determine the value of R which will set the differentiation constant to $-.001$.

——— Figure 10.38

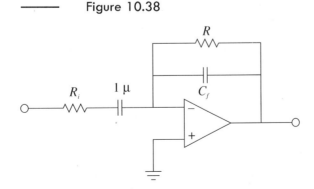

22. Determine the value of R_i such that the maximum gain is 20 for the circuit of Figure 10.38. (Use the values of Problem 21.)

23. Determine the value of C_f in Problem 22 such that noise above 5 kHz is attenuated.

24. Design an integrator to meet the following specifications: integration constant of -4500, f_{low} no greater than 300 Hz, Z_{in} at least 6 k, and DC gain no more than 32 dB.

25. Design a differentiator to meet the following specifications: differentiation constant of -1.2×10^{-4}, f_{high} at least 100 kHz, and a minimum Z_{in} of 50 Ω.

⇒ **Challenge Problems**

26. Sketch the output waveform for the circuit of Figure 10.7 if the input is the following damped sinusoid: $v_{in}(t) = 3\varepsilon^{-200t} \sin 2\pi 1000t$.

27. A 2 V peak 500 Hz square wave may be written as the following infinite series:

$$v_{in}(t) = 2 \sum_{n=1}^{\infty} \frac{1}{2n - 1} \sin 2\pi 500(2n - 1)t$$

Determine the infinite series output equation for the circuit of Figure 10.7.

28. Sketch the output for the preceding problem. Do this by graphically adding the first few terms of the output.

29. Remembering that the voltage across an inductor is equal to the inductance times the rate of change of current, determine the output equation for the circuit of Figure 10.39.

——— Figure 10.39

30. Repeat the preceding problem for the circuit of Figure 10.40.

Figure 10.40

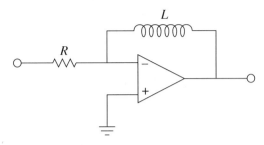

31. A 2 V peak, 500 Hz triangle wave may be written as the following infinite series:

$$v_{in}(t) = 2 \sum_{n=1}^{\infty} \frac{1}{(2n-1)^2} \cos 2\pi 500(2n-1)t$$

Determine the infinite series output equation for the circuit of Figure 10.19.

32. Sketch the output for the preceding problem. Do this by graphically adding the first few terms of the output.

SPICE Problems

33. Model Problem 1 using SPICE, and determine the steady-state response.

34. Model Problem 8 using SPICE. Determine both the steady-state and initial outputs.

35. Model Problem 14 using SPICE. Determine the steady-state output.

36. Model Problem 22 using SPICE and determine the output signal.

37. Compare the resulting waveform produced by the circuit of Problem 8 using both the 741 op amp and the medium-speed LF351. Does the choice of op amp make a discernable difference in this application?

11

Active Filters

After completing this chapter, you should be able to:

❑ Describe the four main types of filters.

❑ Detail the advantages and disadvantages of active versus passive filters.

❑ Describe the importance of filter *order* on the shape of a filter's response.

❑ Describe the importance of filter *alignment* on the shape of a filter's response.

❑ Compare the general gain and phase response characteristics of the popular filter alignments.

❑ Analyze high- and low-pass Sallen and Key filters.

❑ Analyze both low- and high-Q bandpass filters.

❑ Analyze state-variable filters.

❑ Analyze notch filters.

❑ Analyze cascaded filters of high order.

❑ Detail the operation of basic bass and treble audio equalizers.

❑ Explain the operation of switched-capacitor filters, and list their relative advantages and disadvantages compared to ordinary op amp filters.

11.1 Introduction

Generally speaking, a filter is a circuit which inhibits the transfer of a specific range of frequencies. Conversely, you can think of a filter as a circuit which allows only certain frequencies to pass through. Filters are used to remove undesirable frequency components from a complex input signal. The uses for this operation are many, including the suppression of power-line hum, reduction of very low- or high-frequency interference and noise, and specialized spectral shaping. One very common use for filters is bandwidth limiting, which, as you'll see in Chapter 12, is an integral part of any analog-to-digital conversion system.

There are numerous variations on the design and implementation of filters. Indeed, an in-depth discussion of filters could easily fill more than one text book. Our discussion must by necessity be of a limited and introductory nature. This chapter deals with the implementation of a number of popular op amp filter types. Due to the finite space available, every mathematical proof for the design sequences will not be detailed here, but can be found in the list of references for further reading at the end of the chapter.

Filter implementations can be classified into two very broad, yet distinct, camps: *digital filters* and *analog filters*. Digital filters work entirely in the digital domain, using numeric data as the input signal. The design of digital filters is an advanced topic and will not be examined here. The second category, analog filters, utilize standard linear circuit techniques for their construction.

Analog filter implementations can be broken into two subcategories: *passive* and *active*. Passive filters utilize only resistors, inductors, and capacitors, while active filters make use of active devices (i.e., discrete transistors or op amps) as well. While we will be examining only one subcategory in the world of filters, it is important to note that many of the circuits which we will design also can be realized through passive analog filters or digital filters.

11.2 Filter Types

No matter how a filter is realized, it will usually conform to one of four basic response types: *high-pass*, *low-pass*, *band-pass*, and *band-reject* (also known as a *notch* type). A high-pass filter allows only frequencies above a certain break point to pass through. In other words, it attenuates low frequency components. Its general amplitude frequency response is shown in Figure 11.1 (page 486). A low-pass filter is the logical mirror of the high-pass; that is, it allows only low frequency signals to pass through, while suppressing high-frequency

Figure 11.1
High-pass response

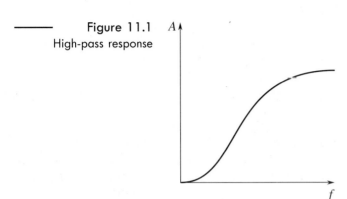

Figure 11.2
Low-pass response

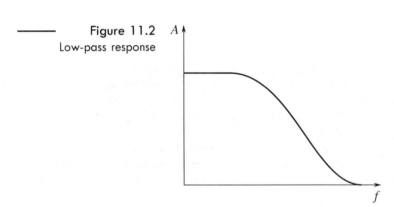

components. Its response is shown in Figure 11.2. The band-pass filter can be thought of as a combination of high- and low-pass filters. It allows only frequencies within a specified range to pass through. Its logical inverse 'is the band-reject filter, which allows everything to pass through with the exception of a specific range of frequencies. The amplitude frequency response plots for these last two types are shown in Figures 11.3 and 11.4, respectively. In complex

Figure 11.3
Band-pass response

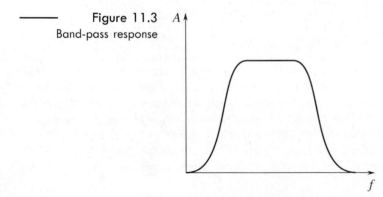

Figure 11.4
Notch response

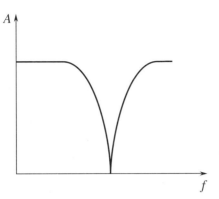

systems, it is possible to combine several different types together in order to achieve a desired overall response characteristic.

In each plot, there are three basic regions. The flat area where the input signal is allowed to pass through is known as the *pass band*. The edge of the pass band is denoted by the *break frequency*. The break frequency is usually defined as the point at which the response has fallen 3 dB from its pass-band value. Consequently, it is also referred to as the *3 dB down frequency*. It is important to note that the break frequency is **not** always equal to the natural critical frequency, f_c. The area where the input signal is fully suppressed is called the *stop band*. The section between the pass band and the stop band is referred to as the *transition band*. These areas are shown in Figure 11.5 for the low-pass filter. Note that plots in Figures 11.1 through 11.4 only define the general response of the filters. The actual shape of the transition band for a given design will be refined in following sections.

These response plots can be achieved through the use of simple *RLC* circuits. For example, the simple lag network shown in Figure 11.6 (page 488) can be classified as a low-pass filter. You know from earlier work that lag networks attenuate high frequencies. In a similar vein, a band-pass filter can be

Figure 11.5
Filter regions

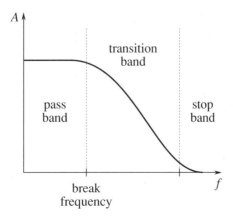

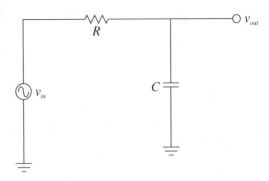

Figure 11.6
Passive low-pass filter

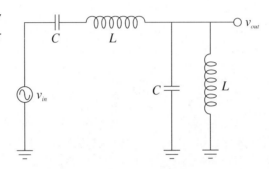

Figure 11.7
Passive band-pass filter

constructed as shown in Figure 11.7. Note that at resonance, the tank circuit exhibits an impedance peak, while the series combination exhibits its minimum impedance. Therefore, a large portion of v_{in} will appear at v_{out}. At very high or low frequencies the situation is reversed (the parallel combination shows a low impedance while the series combination shows a high impedance), and very little of v_{in} appears at the output. Note that no active components are used in this circuit. This is a passive filter.

11.3 The Use and Advantages of Active Filters

If filters can be made with only resistors, capacitors, and inductors, you might ask why anyone would want to design a variation which required the use of an op amp. This is a good question. Obviously, there must be certain short-comings or difficulties associated with passive filter designs, or active filters would not exist. Active filters offer many advantages over passive implemen-tations. First of all, active filters do not exhibit *insertion loss*. This means that the pass-band gain will equal 0 dB. Passive filters always show some signal loss in the pass band. Active filters can be made with pass-band gain, if desired. Active filters also allow for interstage isolation and control of input and output impedances. This alleviates problems with interstage loading, and simplifies complex designs. It also produces modest component sizes (i.e., capacitors tend to be smaller for a given response). Another advantage of the active approach is that complex filters can be realized without using inductors. This is desirable since practical inductors tend to be far less ideal than typical resistors and capacitors, and are generally more expensive. The bottom line is that the active approach allows for the rapid design of stable economical filters in a variety of applications.

Active filters are not without their drawbacks. First, by their very nature, active filters require a DC power supply while passives do not. This is usually not a problem, as the remainder of the circuit will probably require a DC supply anyway. Active filters are also limited in their frequency range. An op amp has a finite gain-bandwidth product, and the active filter produced can certainly not be expected to perform beyond it. For example, it would be impossible to design a filter which only passes frequencies above 10 MHz when using a standard μA741. Passive circuits do not have this limitation and can work well into the hundreds of megahertz. Finally, active filters are not designed to handle large amounts of power. They are low signal-level circuits. With appropriate component ratings, passive filters can handle hundreds of watts of input power. A classic example of this is the crossover network found in most home loudspeaker systems. The crossover network splits the music signal into two or more bands, and routes the results to individual transducers which are optimized to work within a given frequency range. Since the input to the loudspeaker may be as high as a few hundred watts, a passive design is in order.[1] Consequently, we can say that active filters are appropriate for designs at low-to-moderate frequencies (generally no more than 1 MHz with typical devices) which do not have to handle large amounts of power. As you might guess, that specification covers a great deal of territory, and therefore, active filters based on op amps have become rather popular.

11.4 Filter Order and Poles

The rate at which a filter's response falls in the transition band is determined by the filter's *order*. The higher the order of a filter, the faster its rolloff rate is. The order of a filter is given as an integer value and is derived from the filter's transfer function. As an example, all other factors being equal, a fourth-order filter will roll off twice as fast as a second-order filter, and four times faster than a first-order unit. The order of a filter also indicates the minimum number of reactive components which the filter will require. For example, a third-order filter requires at least three reactive components: one capacitor

---

[1] In more advanced playback systems, active filters can be used. Examples include recording studio monitors and public address systems. We'll take a look at just how this is done in one of the upcoming examples.

and two inductors, two capacitors and one inductor, or in the case of an active filter, three capacitors. Related to this is the number of *poles* which a filter utilizes. It is common to hear descriptions such as "a four-pole filter." For most general-purpose high- or low-pass filters, the terms pole and order can be used interchangeably, and completely describe the rolloff rate. For more complex filters this isn't quite the case, and you may also hear descriptions such as "a six-pole two-zero filter." Since this chapter is an introduction to filters, we shall not detail the operation of these more esoteric types. Suffice it to say that when a circuit is described as an Nth-order filter, you may assume that it is an N-pole filter, as well.

A general observation can be given that the rolloff rate of a filter will eventually approach 6 dB per octave per pole (20 dB per decade per pole). Therefore, a third-order filter (i.e., three-pole) eventually rolls off at a rate of 18 dB per octave (60 dB per decade). We say "eventually" because the response around the break frequency may be somewhat faster or slower than this value.

Figure 11.8   Effect of order on low-pass filters

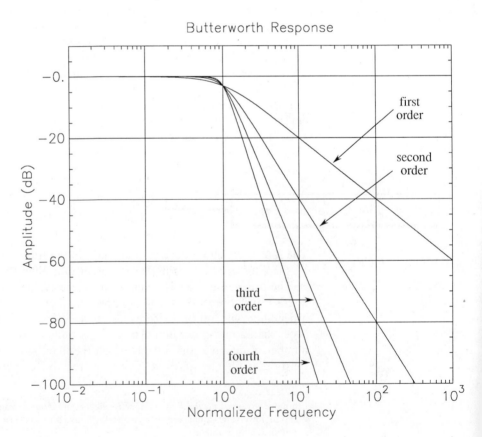

Figure 11.8 compares the effect of order on four otherwise identical low-pass filters. Note that the higher-order filters offer greater attenuation at any frequency beyond the break point. As with most response plots, Figure 11.8 utilizes decibel instead of ordinary gain. Also, these filters are shown with unity gain in the pass band, although this doesn't have to be the case. High-order filters are used when the transition band needs to be as narrow as possible. It is not uncommon to see twelfth-order and higher filters used in special applications. As you might guess, higher-order filters are more complex and costly to design and build. For many typical applications, orders in the range of two to six are common.

11.5 Filter Class or Alignment

Besides order, the shape of the transition band is determined by a filter's *class* or *alignment*. These terms are synonymous, and reflect the filter's *damping factor*. Damping factor is the reciprocal of Q, the quality factor. (You should be familiar with Q from earlier work with inductors and resonant circuits.) The symbol for damping factor is alpha, α. Alignment plays a key role in determining the shape of the transition region, and in some cases, the pass-band or stop-band shape, also. There are a great number of possible filter alignments. We shall look at a few of the more popular types. A graph comparing the relative responses of the major types is shown in Figure 11.9. For simplicity, only second-order types are shown. Figure 11.9a (page 492) shows filters with the same critical frequency and identical DC gains. In Figure 11.9b the responses have been adjusted for a peak gain of 0 dB and identical break frequencies ($f_{3 \text{ dB}}$).

The Butterworth Alignment

Perhaps the most popular alignment type is the Butterworth. The Butterworth is characterized by its moderate amplitude and phase response. It exhibits the fastest rolloff of any monotonic (i.e., single slope, or smooth) filter. In the time domain, moderate ringing on pulses may be observed. This is also the **only** filter whose 3 dB down frequency equals its critical frequency ($f_{3 \text{ dB}} = f_c$). The Butterworth makes an excellent general-purpose filter, and is widely used.

The Bessel Alignment

Like the Butterworth, the Bessel is also monotonic, so it shows a smooth pass-band response. The transition region is somewhat elongated though, and the initial rolloff is less than 6 dB per octave per pole. The Bessel does exhibit a

Figure 11.9 A comparison of the major filter alignments

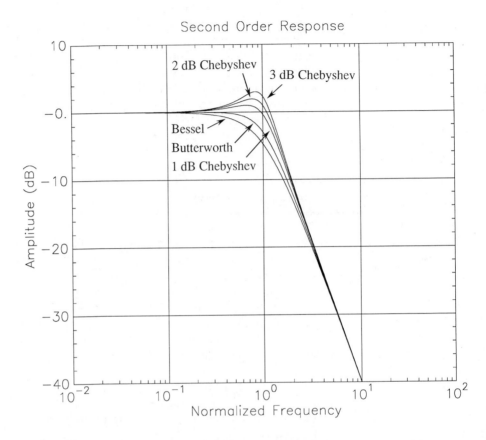

a. Response

linear phase response which produces little ringing in the time domain. It is therefore a good choice for filtering pulses when the overall shape of the pulse must remain coherent (i.e., smooth and undistorted in time).

The Chebyshev Alignment

The Chebyshev is actually a class of filters all its own. There are many possible variations on this theme. In general, the Chebyshev exhibits initial rolloff rates in excess of 6 dB per octave per pole. This extra-fast transition is paid for in two ways: first, the phase response tends to be rather poor, resulting in a great deal of ringing when filtering pulses or other fast transients. The second effect is that the Chebyshev is nonmonotonic. The pass-band response is not smooth; instead, ripples may be noticed. In fact, the height of the ripples defines a

Figure 11.9 *(Continued)*

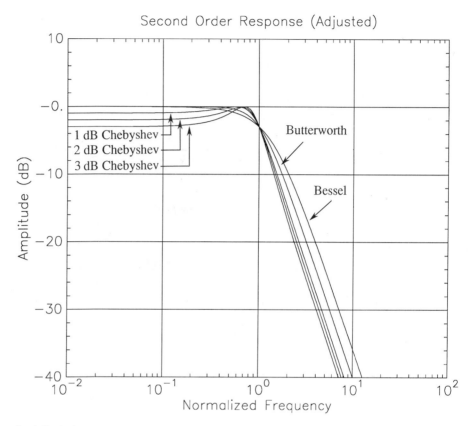

b. Adjusted response

particular Chebyshev response. It is possible to design an infinite number of variations from less than .1 dB ripple to more than 3 dB ripple. Generally, the more ripple that can be tolerated, the greater the rolloff will be, and the worse the phase response will be. The choice is obviously one of compromise. The basic differences between the various Chebyshev types are characterized in Figure 11.10 (pages 494–495). Figure 11.10a compares two different low-pass Chebyshev filters of the same order. Note that only the height of the ripples is different. Figure 11.10b compares equal ripple Chebyshevs, but of different orders. Note that the higher-order filter exhibits a greater number of ripples. Also, note that even-order Chebyshevs exhibit a dip at DC while odd order units show a crest at DC. The number of ripples in the pass band is equal to the order of the filter divided by 2. When compared to the Butterworth and Bessel alignments in Figure 11.9, it is apparent that heavy damping produces

Figure 11.10   Variation of Chebyshev parameter

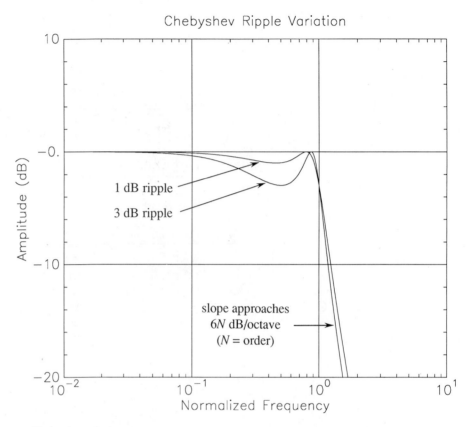

a. Chebyshev ripple variation

the smoothest curves. The final note on the Chebyshev concerns, of all things, its spelling. You will often see "Chebyshev" spelled in a variety of ways, including "Chebycheff" and "Tschebycheff." It merely depends on how the name of the Russian mathematician is transliterated from Russian Cyrillic into English. The spellings all refer to the same filter.

The Elliptic Alignment

The elliptic is also known as the *Cauer* alignment. It is a somewhat more advanced filter. It achieves very fast initial rolloff rates. Unlike the other alignments noted above, the elliptic does not "roll off forever." After its initial transition, the response rises back up, exhibiting ripples in the stop band. A typical response is shown in Figure 11.11. The design of elliptics is an advanced

Figure 11.10 (*Continued*)

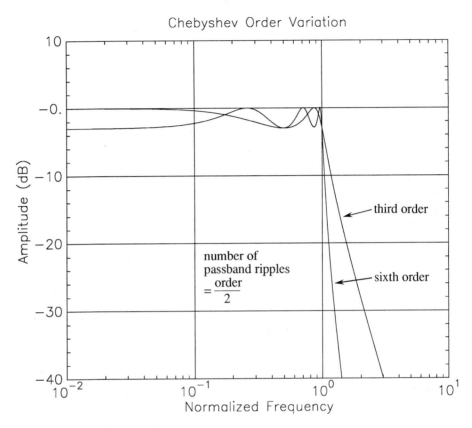

b. Chebyshev order variation

Figure 11.11
Elliptic response

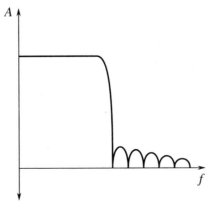

topic and will not be discussed further. It should be noted though that they are very popular in analog-to-digital conversion systems which require very narrow transition bands, such as those used to create audio CDs.

Other Possible Alignments

To optimize the time domain and frequency domain characteristics for specific applications, a number of other alignments can be used. These include such alignments as *Paynter* and *Linkwitz-Reilly*, and are often treated as being midway between Bessel and Butterworth or Butterworth and Chebyshev.

11.6 Realizing Practical Filters

Usually, filter design starts with a few basic desired parameters. This usually includes the break frequency, the amount of ripple which can be tolerated in the pass band (if any), and desired attenuation levels at specific points in the transition and stop bands. Phase response and associated time delays can also be specified. If phase response is paramount, the Bessel is normally chosen. Likewise, if pass-band ripple cannot be allowed, Chebyshevs are not considered. With the use of comparative curves such as those found in Figures 11.8 through 11.10, the filter order and alignment can be determined from the required attenuation values. At this point, the filter performance will be fully specified, for example, a 1 kHz low-pass third-order Butterworth. There are many ways in which this specification can be realized.

Sallen and Key VCVS Filters

There are many possible ways to create an active filter. Perhaps the most popular forms for realizing active high- and low-pass filters are the *Sallen and Key Voltage Controlled Voltage Source* models. As the name implies, the Sallen and Key forms are based on a VCVS. In other words, they use series-parallel negative feedback. A general circuit for these models is shown in Figure 11.12. This circuit is a two-pole (second-order) section and can be configured for either high- or low-pass filtering. For our purposes, the amplifier block will utilize an op amp, although a discrete amplifier is possible. Besides the amplifier, there are four general impedances in the circuit. Usually, each element is a single resistor or capacitor. As we shall see, the selection of the component type will determine the type of filter.

At this point, we need to derive the general transfer equation for the circuit. Once the general equation is established, we will be able to refine the circuit

Figure 11.12   General VCVS 2-pole filter section

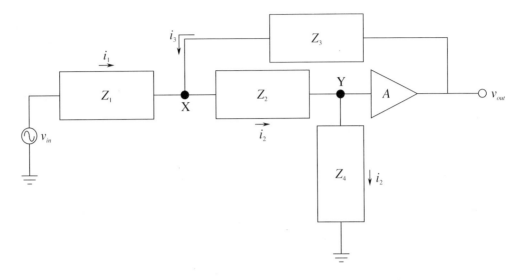

for special cases. Being a higher-order active circuit, this procedure will naturally be somewhat more involved than the derivations found in Chapter 1 for the simple first-order lead and lag networks. The concepts are consistent, though. This derivation utilizes a nodal analysis. In order to make the analysis a bit more convenient, it will help if we declare the voltage v_y as 1 V. This will save us from carrying an input voltage factor through our calculations, which would need to be factored out of the general equation anyway. By inspection we note the following.

$$v_{out} = Av_y = A \tag{11.1}$$

$$i_1 = \frac{v_{in} - v_x}{Z_1}$$

$$i_2 = \frac{1}{Z_4} \tag{11.2}$$

$$i_3 = \frac{v_{out} - v_x}{Z_3} = \frac{A - v_x}{Z_3}$$

$$v_x = i_2 Z_2 + v_y = i_2 Z_2 + 1 \tag{11.3}$$

By substituting Equation 11.2 into Equation 11.3, v_x can be expressed as

$$v_x = \frac{Z_2}{Z_4} + 1$$

We now sum the currents according to the figure,

$$i_2 = i_1 + i_3$$

substitute our current equivalences,

$$\frac{1}{Z_4} = \frac{v_{in} - v_x}{Z_1} + \frac{A - v_x}{Z_3}$$

and solve the equation in terms of v_{in}:

$$\frac{v_{in} - v_x}{Z_1} = \frac{1}{Z_4} - \frac{A - v_x}{Z_3}$$

$$v_{in} - v_x = \frac{Z_1}{Z_4} - \frac{Z_1(A - v_x)}{Z_3}$$

$$v_{in} = \frac{Z_1}{Z_4} - \frac{Z_1(A - v_x)}{Z_3} + v_x$$

$$v_{in} = \frac{Z_1}{Z_4} - \frac{Z_1(A - v_x)}{Z_3} + \frac{Z_2}{Z_4} + 1$$

$$v_{in} = \frac{Z_1}{Z_4} - \frac{Z_1}{Z_3}\left(A - \frac{Z_2}{Z_4} - 1\right) + \frac{Z_2}{Z_4} + 1$$

$$v_{in} = \frac{Z_1}{Z_4} - \frac{Z_1}{Z_3}A + \frac{Z_1 Z_2}{Z_3 Z_4} + \frac{Z_1}{Z_3} + \frac{Z_2}{Z_4} + 1$$

$$Z_{in} = \frac{Z_1}{Z_4} + \frac{Z_1}{Z_3}(1 - A) + \frac{Z_1 Z_2}{Z_3 Z_4} + \frac{Z_2}{Z_4} + 1 \qquad (11.4)$$

We can now write our general transfer equation using Equations 11.1 and 11.4.

$$\frac{v_{out}}{v_{in}} = \frac{A}{\dfrac{Z_1}{Z_4} + \dfrac{Z_1}{Z_3}(1 - A) + \dfrac{Z_1 Z_2}{Z_3 Z_4} + \dfrac{Z_2}{Z_4} + 1} \qquad (11.5)$$

Equation 11.5 is applicable to any variation on Figure 11.12 we wish to make. All we need to do is substitute the appropriate circuit elements for Z_1 through Z_4. As you might guess, direct substitution of resistance and reactance values would make for considerable work. In order to alleviate this difficulty, designers rely on the *Laplace transform* technique. This is also known as the *s domain* technique. A detailed analysis of the Laplace transform is beyond the scope of this text, but some familiarity will prove helpful. The basic idea is to remove complex terms with simpler ones. This is performed by setting the variable s equal to $j\omega$. As you will see, this substitution leads to far simpler and more generalized circuit derivations and equations. A capacitive reactance can be reduced as follows.

$$\text{capacitive reactance} = -jX_C$$

$$\text{capacitive reactance} = \frac{-j}{\omega C}$$

$$\text{capacitive reactance} = \frac{1}{j\omega C}$$

$$\text{capacitive reactance} = \frac{1}{sC}$$

Therefore, whenever a capacitive reactance is needed, the expression $1/sC$ is used. Equations can then be manipulated using basic algebra.

Sallen and Key Low-Pass Filters

Let's derive a general expression based on Equation 11.5 for low-pass filters. A low-pass filter is a lag network, so to echo this, we will use resistors for the first two elements and capacitors for the third and fouth. Using the s operator, we find $Z_1 = R_1$, $Z_2 = R_2$, $Z_3 = 1/sC_1$, and $Z_4 = 1/sC_2$.

$$\frac{v_{out}}{v_{in}} = \frac{A}{\dfrac{Z_1}{Z_4} + \dfrac{Z_1}{Z_3}(1 - A) + \dfrac{Z_1 Z_2}{Z_3 Z_4} + \dfrac{Z_2}{Z_4} + 1}$$

$$\frac{v_{out}}{v_{in}} = \frac{A}{sR_1 C_2 + sR_1 C_1(1 - A) + s^2 R_1 R_2 C_1 C_2 + sR_2 C_2 + 1}$$

$$\frac{v_{out}}{v_{in}} = \frac{A}{s^2 R_1 R_2 C_1 C_2 + s(R_1 C_2 + R_2 C_2 + R_1 C_1(1 - A)) + 1}$$

The highest power of s in the denominator determines the number of pole. in the filter. Since the highest power here is 2, the filter must be a two-pol type (second order), as expected. Usually, it is most convenient if the denominator coefficient for s^2 is unity. This makes the equation easier to factor.

$$\frac{v_{out}}{v_{in}} = \frac{A/R_1 R_2 C_1 C_2}{s^2 + s\left(\dfrac{1}{R_2 C_1} + \dfrac{1}{R_1 C_1} + \dfrac{1}{R_2 C_2}(1 - A)\right) + \dfrac{1}{R_1 R_2 C_1 C_2}} \tag{11.6}$$

Second-order systems appear in a variety of areas including mechanical, acoustical, hydraulic, and electrical applications. They have been widely studied, and a generalized form of one group of second-order responses is given by

$$G = \frac{A\omega^2}{s^2 + \alpha\omega s + \omega^2} \tag{11.7}$$

where A is the gain of the system, ω is the resonant frequency in radians, α is the damping factor, and G is the complex system response.

By comparing the general form of Equation 11.7 to our low-pass filter Equation 11.6, we find that

$$\omega^2 = \frac{1}{R_1 R_2 C_1 C_2} \tag{11.8}$$

$$\alpha\omega = \frac{1}{R_2 C_1} + \frac{1}{R_1 C_1} + \frac{1}{R_2 C_2}(1 - A) \tag{11.9}$$

Based on the general expression of Equation 11.7, we can derive equations for the gain magnitude versus frequency, and phase versus frequency, as we did in Chapter 1 for the first-order systems. For most filter work, it is convenient to work with *normalized frequency* instead of a true frequency. This means that the critical frequency will be set to 1 radian per second, and a generalized equation developed. For circuits using other critical frequencies, the general equation is used and its results are simply scaled by a factor equal to this new frequency. The normalized version of Equation 11.7 is

$$G = \frac{A}{s^2 + \alpha s + 1} \tag{11.10}$$

In order to determine the gain and phase expressions, Equation 11.10 must be split into its real and imaginary components. The first step is to replace s with its equivalent, $j\omega$, and then group the real and imaginary components.

$$G = \frac{A}{(j\omega)^2 + j\alpha\omega + 1}$$

$$G = \frac{A}{-\omega^2 + j\alpha\omega + 1}$$

$$G = \frac{A}{(1 - \omega^2) + j\alpha\omega}$$

To split this into separate real and imaginary components, we must multiply the numerator and denominator by the complex conjugate of the denominator, $(1 - \omega^2) - j\alpha\omega$.

$$G = \frac{A}{(1 - \omega^2) + j\alpha\omega} \frac{(1 - \omega^2) - j\alpha\omega}{(1 - \omega^2) - j\alpha\omega}$$

$$G = \frac{A((1 - \omega^2) - j\alpha\omega)}{(1 - \omega^2)^2 + \alpha^2\omega^2}$$

$$G = \frac{A(1 - \omega^2)}{(1 - \omega^2)^2 + \alpha^2\omega^2} - j\frac{A\alpha\omega}{(1 - \omega^2)^2 + \alpha^2\omega^2} \qquad (11.11)$$

For the gain magnitude, recall that $\text{mag} = \sqrt{\text{real}^2 + \text{imaginary}^2}$. Applying Equation 11.11 to this relation yields

$$\text{mag} = \sqrt{\left(\frac{A(1 - \omega^2)}{(1 - \omega^2)^2 + \alpha^2\omega^2}\right)^2 + \left(-\frac{A\alpha\omega}{(1 - \omega^2)^2 + \alpha^2\omega^2}\right)^2}$$

$$\text{mag} = \sqrt{\frac{A^2(1 - \omega^2)^2}{((1 - \omega^2)^2 + \alpha^2\omega^2)^2} + \frac{A^2\alpha^2\omega^2}{((1 - \omega^2)^2 + \alpha^2\omega^2)^2}}$$

$$\text{mag} = \sqrt{\frac{A^2((1 - \omega^2)^2 + \alpha^2\omega^2)}{((1 - \omega^2)^2 + \alpha^2\omega^2)^2}}$$

$$\text{mag} = \frac{A\sqrt{(1 - \omega^2)^2 + \alpha^2\omega^2}}{(1 - \omega^2)^2 + \alpha^2\omega^2}$$

$$\text{mag} = \frac{A}{\sqrt{(1 - \omega^2)^2 + \alpha^2\omega^2}}$$

$$\text{mag} = \frac{A}{\sqrt{1 + \omega^2(\alpha^2 - 2) + \omega^4}} \qquad (11.12)$$

For the phase response, recall that $\theta = \arctan(\text{imaginary}/\text{real})$. Applying Equation 11.11 to this relation yields

$$\theta = \arctan \frac{-\dfrac{A\alpha\omega}{(1 - \omega^2)^2 + \alpha^2\omega^2}}{\dfrac{A(1 - \omega^2)}{(1 - \omega^2)^2 + \alpha^2\omega^2}}$$

$$\theta = \arctan \frac{-A\alpha\omega}{A(1 - \omega^2)}$$

$$\theta = -\arctan \frac{\alpha\omega}{1 - \omega^2} \qquad (11.13)$$

To use Equations 11.12 and 11.13 with a particular alignment, substitute the appropriate damping value and simplify. Derivation of specific damping factors is beyond the scope of this chapter, but can be found in texts specializing in filter design. For our purposes, tables of damping factors for specific alignments will be presented. As one example, the damping factor for a second-order Butterworth alignment is $\sqrt{2}$. Substituting this into Equation 11.12 produces

$$\text{mag} = \frac{A}{\sqrt{1 + \omega^2(\alpha^2 - 2) + \omega^4}}$$

$$\text{mag} = \frac{A}{\sqrt{1 + \omega^2((\sqrt{2})^2 - 2) + \omega^4}}$$

$$\text{mag} = \frac{A}{\sqrt{1 + \omega^4}}$$

There are a large number of ways of configuring a low-pass filter given the above equations. So that we might put some order into what appears to be a chaotic mess, we'll look at two distinct and useful variations. They are the *equal component* realization and the *unity gain* realization.

The equal-component version. Here, we will set $R_1 = R_2$ and $C_1 = C_2$. To keep the resulting equation generic, we will use a *normalized frequency* of 1 radian per second. For other tuning frequencies, we will just scale our results to the desired value. From Equation 11.8, if $\omega = 1$, then $R_1 R_2 C_1 C_2 = 1$, and the most straightforward solution would be to set $R_1 = R_2 = C_1 = C_2 = 1$ (units of ohms and farads). Equation 11.6 simplifies to

$$\frac{v_{out}}{v_{in}} = \frac{A}{s^2 + s(1 + 1 + 1(1 - A)) + 1}$$

Note the damping factor is now given by

$$\alpha = 1 + 1 + 1(1 - A)$$

$$\alpha = 3 - A$$

We see that the gain and damping of the filter are linked together. Indeed, for a certain damping factor, only one specific gain will work properly:

$$A = 3 - \alpha$$

Since the gain of a noninverting amplifier is

$$A = 1 + \frac{R_f}{R_i}$$

Figure 11.13
Low-pass equal-
component VCVS

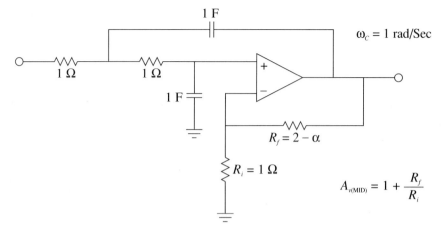

we may find the required value for R_f by combining these two equations:

$$R_f = 2 - \alpha$$

The finished prototype is shown in Figure 11.13.

The unity-gain version. Here we will set $A = 1$ and $R_1 = R_2$. From Equation 11.8, if $\omega = 1$, then $R_1 R_2 C_1 C_2 = 1$, and therefore $C_1 = 1/C_2$. In effect, the ratio of the capacitors will set the damping factor for the system. Equation 11.6 may be simplified to

$$\frac{v_{out}}{v_{in}} = \frac{A}{s^2 + s\left(\dfrac{1}{C_1} + \dfrac{1}{C_1}\right) + 1}$$

The damping factor is now given by

$$\alpha = \frac{1}{C_1} + \frac{1}{C_1}$$

$$\alpha = \frac{2}{C_1}$$

or

$$C_1 = \frac{2}{\alpha}$$

Figure 11.14
Low-pass unity-gain VCVS

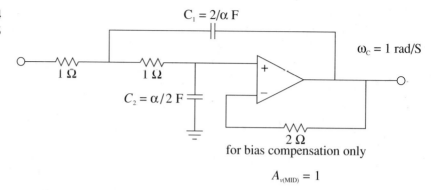

Since $C_1 = 1/C_2$, we find

$$C_2 = \frac{\alpha}{2}$$

The finished prototype is shown in Figure 11.14.

As you can see, there is quite a bit of similarity between the two versions. It is important to note that the inputs to these circuits must return to ground via a low-impedance DC path. If the signal source is capacitively coupled, the op amp's input bias current cannot be set up properly, and thus, some form of DC return resistor must be used at the source. Also, you can see that the damping factor (alignment) of the filter plays a role in setting component values. If the values shown are taken as having units of ohms and farads, the critical frequency will be 1 radian per second. It is an accepted practice to normalize the basic forms of circuits such as these, so that the critical frequency works out to this convenient value. This makes it very easy to scale the component values to fit your desired critical frequency. Since the critical frequency is inversely proportional to the tuning resistor and capacitor values, you only need to shrink R or C in order to increase f_c. [Remember, $f_c = 1/(2\pi RC)$.] A second scaling step normally follows this, in order to create practical values for R and C. This procedure is best shown with an example.

Example 11.1   Design a 1 kHz low-pass second-order Butterworth filter. Examine both the equal component and the unity gain forms as drawn in Figures 11.13 and 11.14, respectively. The required damping factor is 1.414.

Let's start with the equal component version. First, find the required value for R_f from the damping factor, as given on the diagram.

$$R_f = 2 - \alpha$$
$$R_f = 2 - 1.414$$
$$R_f = .586 \ \Omega$$

Figure 11.15
Initial damping calculation
for Example 11.1 (equal-
component version)

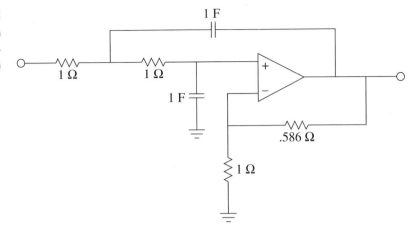

Note that this will produce a pass-band gain of

$$A_v = 1 + \frac{R_f}{R_i}$$

$$A_v = 1 + \frac{.586}{1}$$

$$A_v = 1.586$$

Figure 11.15 shows the second-order Butterworth low-pass filter. Its critical frequency is 1 radian per second. We need to scale this to 1 kHz.

$$\omega_c = 2\pi f_c$$

$$\omega_c = 2\pi 1 \text{ kHz}$$

$$\omega_c = 6283 \text{ rad/S}$$

Our desired critical frequency is 6283 times higher than the normalized base. Since $\omega_c = 1/(RC)$, to translate the frequency up, all we need to do is divide R or C by 6283. It doesn't really matter which one you choose, although it is generally easier to find "odd" sizes for resistors than capacitors, so we'll use R.

$$R = \frac{1}{6283}$$

$$R = 1.59 \times 10^{-4} \, \Omega$$

Figure 11.16 (page 506) shows our 1 kHz low-pass second-order Butterworth filter. As you can see though, the component values are not very practical. It is therefore necessary to perform the final scaling operations. First, consider multiplying R_f and R_i by 10 k. Note that this will have no effect on the damping

————— **Figure 11.16**
The filter for Example 11.1
after frequency scaling
(equal-component version)

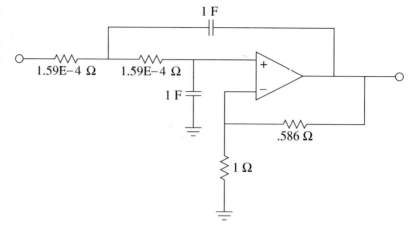

since it is the ratio of these two elements which determines damping. This scaling will not affect the critical frequency either, since f_c is set by the tuning resistors and capacitors. Second, we need to increase R to a reasonable value. A factor of 10^7 will place it at 1.59 k. In order to compensate, the tuning capacitors must be dropped by an equal amount, which brings them to 100 nF. The completed design is shown in Figure 11.17. Other scaling factors could also be used. Also, if bias compensation is important, the R_i and R_f values will need to be scaled further, in order to balance the resistance seen at the noninverting input.

The approach for the unity-gain version is similar. First, adjust the capacitor values in order to achieve the desired damping, as specified in Figure 11.14.

$$C_1 = \frac{2}{\alpha}$$

$$C_1 = \frac{2}{1.414}$$

$$C_1 = 1.414 \text{ F}$$

$$C_2 = \frac{\alpha}{2}$$

$$C_2 = \frac{1.414}{2}$$

$$C_2 = .707 \text{ F}$$

The resulting circuit is shown in Figure 11.18. Once again, the circuit must be scaled to the desired f_c. The factor is 6823 once again, and the result is shown in Figure 11.19. The final component scaling is seen in Figure 11.20.

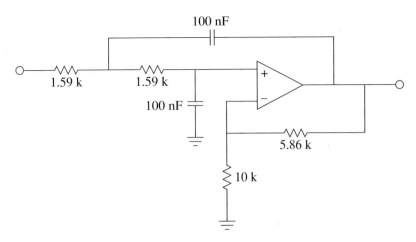

Figure 11.17
Final impedance scaling for Example 11.1 (equal-component version)

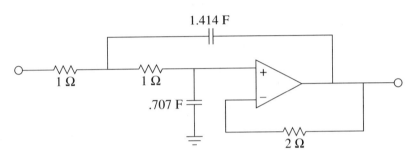

Figure 11.18
Initial damping calculation for the filter of Example 11.1 (unity-gain version)

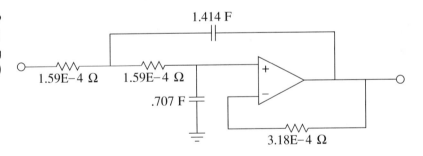

Figure 11.19
The filter for Example 11.1 after frequency scaling (unity-gain version)

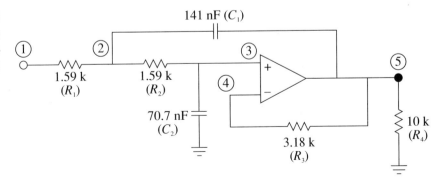

Figure 11.20
Final impedance scaling for the filter of Example 11.1 (unity-gain version)

Figure 11.21
SPICE simulation of the
circuit for Example 11.1

```
Low pass 1 KHz unity gain Sallen and Key filter.
*******Start of UA741 op amp*******
.SUBCKT UA741 1    2    3    4    5    6
*                 IN+ IN- GND V+  V-  OUT
Q1   11   1   13 UA741QA
Q2   12   2   14 UA741QB
RC1   4  11   5.305165E+03
RC2   4  12   5.305165E+03
C1   11  12   5.459553E-12
RE1  13  10   2.151297E+03
RE2  14  10   2.151297E+03
IEE  10   5   1.666000E-05
CE   10   3   3.000000E-12
RE   10   3   1.200480E+07
GCM   3  21   10 3  5.960753E-09
GA   21   3   12 11 1.884655E-04
R2   21   3   1.000000E+05
C2   21  22   3.000000E-11
GB   22   3   21 3 2.357851E+02
RO2  22   3   4.500000E+01
D1   22  31   UA741DA
D2   31  22   UA741DA
EC   31   3   6 3 1.0
RO1  22   6   3.000000E+01
D3    6  24   UA741DB
VC    4  24   2.803238E+00
D4   25   6   UA741DB
VE   25   5   2.803238E+00
.ENDS
.MODEL UA741DA D (IS=9.762287E-11)
.MODEL UA741DB D (IS=8.000000E-16)
.MODEL UA741QA NPN (IS=8.000000E-16 BF=9.166667E+01)
.MODEL UA741QB NPN (IS=8.309478E-16 BF=1.178571E+02)

*****End of UA741 op amp*****

*****Start of circuit description*****

R1    1 2   1.59K
R2    2 3   1.59K
R3    4 5   3.18K
R4    0 5   10K
C1    2 5   141NF
C2    3 0   70.7NF
VCC   6 0   DC 15
VEE   7 0   DC -15
X1    3 4 0 6 7 5  UA741
VIN   1 0   AC 1

*****Output control directives*****

.AC DEC 10 50 20K
.PLOT AC VDB(5)
.PROBE
*Read Me Doctor Memory . . .

.END
```

a. Input file

Figure 11.21   (Continued)

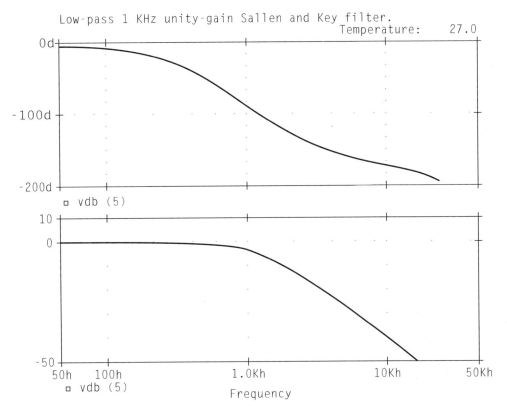

b. PSpice/Probe output

A SPICE simulation of the filter design is shown in Figure 11.21. The analysis shows the Bode plot ranging from 50 Hz to better than 20 kHz. This yields over one decade on either side of the 1 kHz critical frequency. The graph clearly shows the -3 dB point at approximately 1 kHz, with an attenuation slope of -12 dB per octave. Since this is the unity-gain version, the low frequency gain is set at 0 dB. Also, note that no peaking is evident in the response curve, as is expected for a Butterworth alignment. It is worth noting, however, that a greater density of points per decade is desirable in order to verify that the response really is monotonic. (Increase the number of points per decade for .AC from 10 to perhaps 50.) The phase response is also shown. Note that the phase shift starts to increase at the highest frequencies instead of leveling off. This is due to the extra phase shift produced by the op amp as the operating frequency approaches f_{unity}. A simpler op amp model would not create this real-world effect.

As you can see, the realization process is little more than a scaling sequence. This makes filter design very rapid. The operation for high-pass filters is essentially the same.

Sallen and Key High-Pass Filters

We can derive a general expression for high-pass filters based on Equation 11.5. A high-pass filter is a lead network, so to echo this, we will use capacitors for the first two elements and resistors for the third and fourth. Using the s operator, we find $Z_1 = 1/sC_1$, $Z_2 = 1/sC_2$, $Z_3 = R_1$, and $Z_4 = R_2$.

$$\frac{v_{out}}{v_{in}} = \frac{A}{\dfrac{Z_1}{Z_4} + \dfrac{Z_1}{Z_3}(1 - A) + \dfrac{Z_1 Z_2}{Z_3 Z_4} + \dfrac{Z_2}{Z_4} + 1}$$

$$\frac{v_{out}}{v_{in}} = \frac{A}{sR_2 C_1 + sR_1 C_1(1 - A) + s^2 R_1 R_2 C_1 C_2 + sR_2 C_2 + 1}$$

$$\frac{v_{out}}{v_{in}} = \frac{A}{s^2 + s\left(\dfrac{1}{R_2 C_1} + \dfrac{1}{R_2 C_2} + \dfrac{1}{R_1 C_1}(1 - A)\right) + 1} \tag{11.14}$$

As with the low-pass filters, we have two basic realizations: equal component and unity gain. In both cases, we start with Equation 11.14 and use normalized frequency (1 radian per second). The derivations are very similar to the low-pass case, and the results are summarized below.

The equal-component version: $\alpha = 3 - A$

The unity-gain version: $R_2 = 2/\alpha$

$$R_1 = \alpha/2$$

These forms are shown in Figures 11.22 and 11.23. You may at this point ask two questions: one, how do you find the damping factor, and two, what about higher-order filters?

Figure 11.22
High-pass unity-gain
VCVS

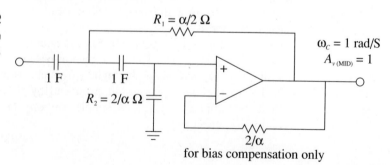

$R_1 = \alpha/2\ \Omega$

$\omega_c = 1$ rad/S
$A_{v\,(MID)} = 1$

1 F   1 F

$R_2 = 2/\alpha\ \Omega$

$2/\alpha$
for bias compensation only

Figure 11.23
High-pass
equal-component VCVS

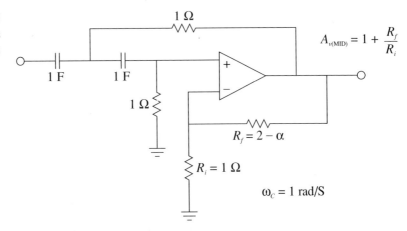

$$A_{v(MID)} = 1 + \frac{R_f}{R_i}$$

$$R_f = 2 - \alpha$$

$$R_i = 1 \; \Omega$$

$$\omega_C = 1 \; rad/S$$

In order to find the damping factor needed, a chart such as Figure 11.24 can be consulted. This chart also introduces a new item, and that is the *frequency factor*, k_f. Normally, the critical frequency and 3 dB down frequency (break frequency) of a filter are not the same value. They are identical only for the Butterworth alignment. For any other alignment, the desired break frequency must first be translated to the appropriate critical frequency before scaling is performed. This is illustrated in the following example.

Figure 11.24
Second-order filter
parameters

Type	Damping	f_c Factor (k_f)
Bessel	1.732	1.274
Butterworth	1.414	1.0
1 dB Chebyshev	1.045	.863
2 dB Chebyshev	.895	.852
3 dB Chebyshev	.767	.841

$f_c = f_{3\;dB}/f_c$ Factor for high-pass

$f_c = f_{3\;dB} \times f_c$ Factor for low-pass

From Lancaster, Don. *Active Filter Cookbook.* Howard W. Sams, 1982. Reprinted with permission.

Example 11.2 Design a second-order high-pass Bessel filter with a break frequency $(f_{3\;dB})$ of 5 kHz.

For this example, let's use the unity gain form shown in Figure 11.22. First, obtain the damping and frequency factors from Figure 11.24.

$$k_f = 1.274, \; damping = 1.732.$$

Using the damping factor, the two tuning resistors may be found:

$$R_1 = \frac{\alpha}{2}$$

$$R_1 = \frac{1.732}{2}$$

$$R_1 = .866$$

$$R_2 = \frac{2}{\alpha}$$

$$R_2 = \frac{2}{1.732}$$

$$R_2 = 1.155$$

The intermediate result is shown in Figure 11.25. In order to do the frequency scaling, the desired break frequency of 5 kHz must first be translated into the required critical frequency. Since this is a high-pass filter,

$$f_c = \frac{f_{3\,dB}}{k_f}$$

$$f_c = \frac{5\,kHz}{1.274}$$

$$f_c = 3925\,Hz$$

$$\omega_c = 2\pi f_c$$

$$\omega_c = 2\pi 3925$$

$$\omega_c = 24.66\,k\,rad/S$$

Figure 11.25
Initial damping calculation
for Example 11.2

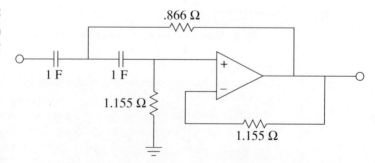

Figure 11.26
Frequency scaling for
Example 11.2

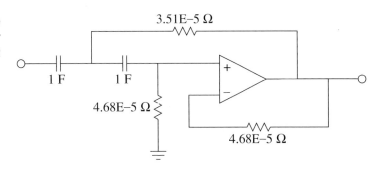

Figure 11.26
Frequency scaling for
Example 11.2

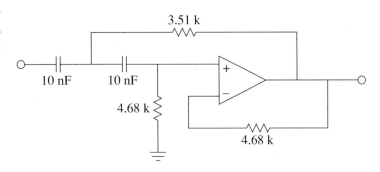

Figure 11.27
Impedance scaling for
Example 11.2

Either the tuning resistors or capacitors can now be scaled.

$$R_1 = \frac{.866}{24.66 \text{ k}}$$

$$R_1 = 3.51 \times 10^{-5}$$

$$R_2 = \frac{1.155}{24.66 \text{ k}}$$

$$R_2 = 4.68 \times 10^{-5}$$

We now have a second-order high-pass 5 kHz Bessel filter. This is shown in Figure 11.26. A final scaling of 10^8 will give us reasonable values, and is shown in Figure 11.27.

Filters of Higher Order

There is a common misconception among novice filter designers that higher-order filters can be produced by cascading a number of lower-order filters of the same type. This is not true. For example, cascading three second-order 10 kHz Butterworth filters will **not** produce a sixth-order 10 kHz Butterworth filter. A quick inspection reveals why this is not the case: A single filter of any

———— **Figure 11.28**   Single-pole sections

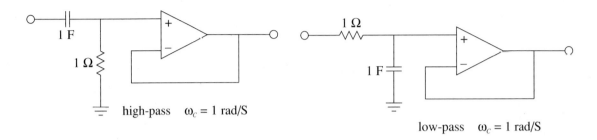

high-pass $\omega_c = 1$ rad/S

low-pass $\omega_c = 1$ rad/S

order will show a 3 dB loss at its break frequency, by definition (in this case, 10 kHz). If three filters of the same type are cascaded, each filter will produce a 3 dB loss at the break frequency, which means an overall loss of 9 dB occurs.

However, this much is true: a higher-order filter will require a number of individual sections, each with specific damping and frequency factors. Each section will be based on the second-order forms already examined.[2] In order to make odd-ordered filters, we will introduce a simple single-pole filter. The high- and low-pass versions of this unit are shown in Figure 11.28. The damping factor for this circuit is always unity. When working with the circuit, you need only worry about the frequency factor.

Designing higher-order filters is conceptually no different from designing second-order filters. The reality is that new charts are needed for the required damping and frequency factors. A set of compatible charts is shown in Figure 11.29 (pages 515–519) for orders 3 through 6. To show the design sequence flows, let's look at an example.

———— **Example 11.3**   We wish to design a filter suitable for removing subsonic tones from a stereo system. This could be used to reduce turntable rumble in a home system, or to reduce stage vibration in a public address system. The filter should attenuate frequencies below the lower limit of human hearing (about 20 Hz) while allowing all higher frequencies to pass. Transient response may be important here, so we'll choose a Bessel alignment. We will also specify a fifth-order system. This will create an attenuation of about 15 dB one octave below the break frequency.

First, note that the specification requires the use of a high-pass filter. This filter can be realized with either the equal component or the unity gain forms.

---

[2] The transfer function of a higher-order filter contains a high-order polynomial in the denominator of the form $s^n + b_{n-1}s^{n-1} + \cdots + b_1s + b_0$, where n indicates the order of the filter and the b coefficients determine the alignment. This polynomial is factored into a product of second-order expressions (with a possible first-order unit for odd-ordered systems). Each of these expressions corresponds to a single section in the larger filter.

Figure 11.29   Filter design tools

THIRD-ORDER FILTER PARAMETERS

Type	Second-order section		First-order section
	Damping	f_c Factor	f_c Factor
Bessel	1.447	1.454	1.328
Butterworth	1.0	1.0	1.0
1 dB Chebyshev	.496	.911	.452
2 dB Chebyshev	.402	.913	.322
3 dB Chebyshev	.326	.916	.299

FOURTH-ORDER FILTER PARAMETERS

Type	Second-order section		Second-order section	
	Damping	f_c Factor	Damping	f_c Factor
Bessel	1.916	1.436	1.241	1.610
Butterworth	1.848	1.0	.765	1.0
1 dB Chebyshev	1.275	.502	.281	.943
2 dB Chebyshev	1.088	.466	.224	.946
3 dB Chebyshev	.929	.443	.179	.950

FIFTH-ORDER FILTER PARAMETERS

Type	Second-order section		Second-order section		First-order section
	Damping	f_c Factor	Damping	f_c Factor	f_c Factor
Bessel	1.775	1.613	1.091	1.819	1.557
Butterworth	1.618	1.0	.618	1.0	1.0
1 dB Chebyshev	.714	.634	.180	.961	.280
2 dB Chebyshev	.578	.624	.142	.964	.223
3 dB Chebyshev	.468	.614	.113	.967	.178

SIXTH-ORDER FILTER PARAMETERS

Type	Second-order section		Second-order section		Second-order section	
	Damping	f_c Factor	Damping	f_c Factor	Damping	f_c Factor
Bessel	1.959	1.609	1.636	1.694	.977	1.910
Butterworth	1.932	1.0	1.414	1.0	.518	1.0
1 dB Chebyshev	1.314	.347	.455	.733	.125	.977
2 dB Chebyshev	1.121	.321	.363	.727	.0989	.976
3 dB Chebyshev	.958	.298	.289	.722	.0782	.975

$f_c = f_{3\,dB}/f_c$ Factor for high-pass

$f_c = f_{3\,dB} \times f_c$ Factor for low-pass

From Lancaster, Don. *Active Filter Cookbook.* Howard W. Sams, 1982. Reprinted with permission.

(Continued)

a. Parameters for third-through sixth-order filters

———— **Figure 11.29** *(Continued)*

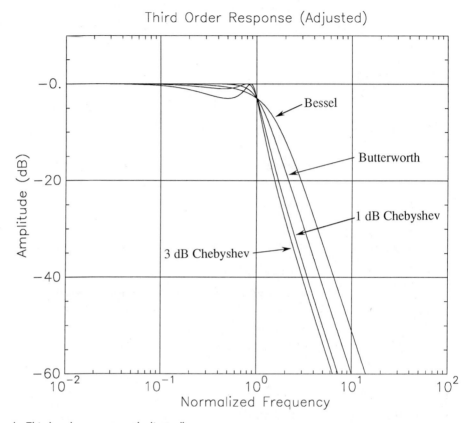

b. Third order response (adjusted)

Since this design requires multiple sections, excessive gain may result from the equal component version. Our fifth-order system will be comprised of two second-order sections and a first-order section. An overview of the design is shown in Figure 11.30 (page 519), with the appropriate damping and frequency factors taken from Figure 11.29. We'll break the analysis down stage by stage.

First, find the desired break frequency in radians.

$$\omega_{3\ dB} = 2\pi f_{3\ dB}$$

$$\omega_{3\ dB} = 2\pi 20\ Hz$$

$$\omega_{3\ dB} = 125.7\ rad/S$$

Figure 11.29   (Continued)

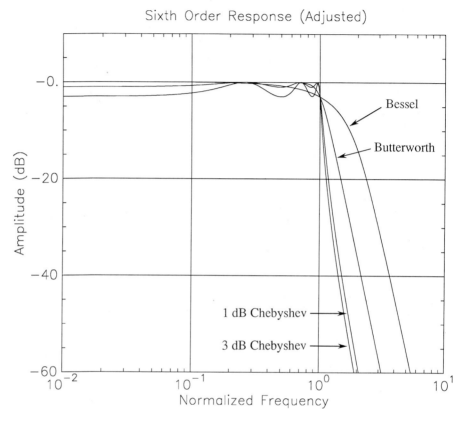

c. Fourth order response (adjusted)

Stage 1. The break frequency must be translated to the required critical frequency. Since this is a high-pass filter, we need to divide by the frequency factor.

$$\omega_c = \frac{\omega_{3\ dB}}{k_f}$$

$$\omega_c = \frac{125.7}{1.557}$$

$$\omega_c = 80.7 \ \text{rad/S}$$

——————   **Figure 11.29**   *(Continued)*

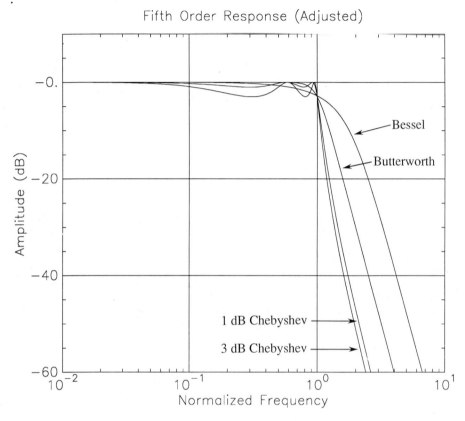

d. Fifth order response (adjusted)

We will scale the resistor by 80.7 to achieve this tuning frequency.

$$R = \frac{1}{80.7}$$

$$R = .0124$$

The R and C values must now be scaled for practical values. A factor of 10^6 would be reasonable. The final result is

$$R = 12.4 \text{ k}$$

$$C = 1 \ \mu F$$

Figure 11.29 (*Continued*)

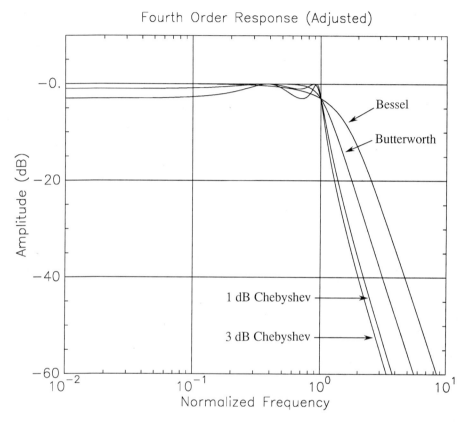

Fourth Order Response (Adjusted)

e. Sixth order response (adjusted)

Figure 11.30   Circuit outline (Note: all values in ohms and farads)

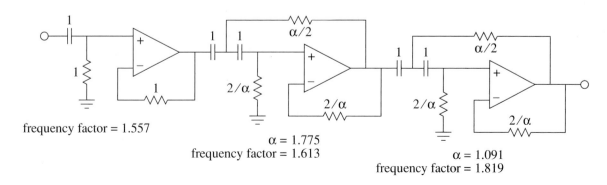

frequency factor = 1.557

$\alpha = 1.775$
frequency factor = 1.613

$\alpha = 1.091$
frequency factor = 1.819

Stage 2.   First, determine the values for the two resistors from the given damping factor.

$$R_1 = \frac{\alpha}{2}$$

$$R_1 = \frac{1.775}{2}$$

$$R_1 = .8875$$

$$R_2 = \frac{2}{\alpha}$$

$$R_2 = \frac{2}{1.775}$$

$$R_2 = 1.127$$

Now, the break frequency must be translated to the required critical frequency.

$$\omega_c = \frac{\omega_{3\,dB}}{k_f}$$

$$\omega_c = \frac{125.7}{1.613}$$

$$\omega_c = 77.9 \text{ rad/S}$$

We will scale the resistors by 77.9 to achieve this tuning frequency.

$$R_1 = \frac{.8875}{77.9}$$

$$R_1 = .0114$$

$$R_2 = \frac{1.127}{77.9}$$

$$R_2 = .0145$$

Again, R and C must be scaled for practical values. A factor of 10^6 would be reasonable. The final result is

$$R_1 = 11.4 \text{ k}$$

$$R_2 = 14.5 \text{ k}$$

$$C = 1 \ \mu F$$

Stage 3. First, determine the values for the two resistors from the given damping factor.

$$R_1 = \frac{\alpha}{2}$$

$$R_1 = \frac{1.091}{2}$$

$$R_1 = .5455$$

$$R_2 = \frac{2}{\alpha}$$

$$R_2 = \frac{2}{1.091}$$

$$R_2 = 1.833$$

Now, the break frequency must be translated to the required critical frequency.

$$\omega_c = \frac{\omega_{3\,dB}}{k_f}$$

$$\omega_c = \frac{125.7}{1.819}$$

$$\omega_c = 69.1 \text{ rad/S}$$

We will scale the resistors by 69.1 to achieve this tuning frequency.

$$R_1 = \frac{.5455}{69.1}$$

$$R_1 = 7.89 \times 10^{-3}$$

$$R_2 = \frac{1.833}{69.1}$$

$$R_2 = .0265$$

Again, R and C must be scaled for practical values. A factor of 10^6 would be reasonable. The final result is

$$R_1 = 7.89 \text{ k}$$

$$R_2 = 26.5 \text{ k}$$

$$C = 1 \ \mu F$$

Figure 11.31   Completed filter for Example 11.3

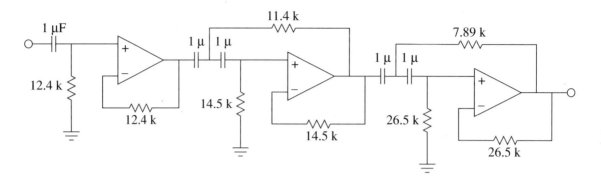

The completed design is shown in Figure 11.31. Note that all of the capacitors are set at 1 μF. This certainly helps to cut inventory and parts placement costs.

A commercial example of low-pass active filter use is shown in Figure 11.32. This is the stereo audio output section of the Commodore Amiga 2000® personal computer. It is used along with the operating system's text-to-speech facilities, which allow the computer to recite text files. It can also be used for musical and instrument control applications. The computer uses a device which turns digital words into analog voltages, in essentially the same fashion that a CD player reproduces music. During this process, a certain amount of high frequency noise may be generated. The filters are used to remove this distracting signal. The noise components lie above the human voice spectrum, so filtering them does not affect voice quality. For applications which require an extended frequency response, the filters can be bypassed under software control. To do this, a logic signal is sent to the gates of the MPF102 FETs. This will cause the drain-to-source resistance of the FET to drop to a very low value, effectively shorting out the tuning resistors. This results in a simple buffer, and no filtering action takes place.

We shall complete our discussion of high- and low-pass VCVS filters with the following example.

Example 11.4   As mentioned earlier, it is common for loudspeaker systems to rely on passive filters to create their crossover networks. More demanding applications such as recording studio monitoring or large public address systems (i.e., concert systems) cannot afford the losses associated with passive crossovers. Instead,

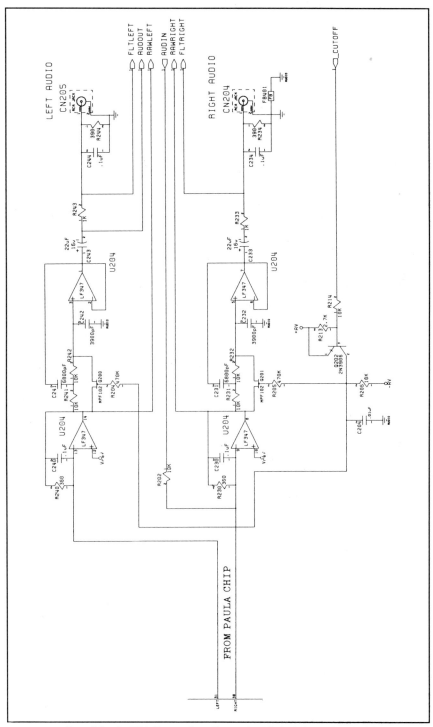

Figure 11.32

────── ‣Figure 11.33

Electronic crossover system

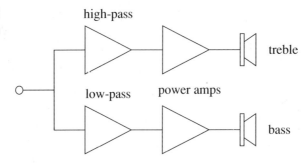

these applications utilize an active crossover, which is composed of active filters. Before the audio signal is fed to a power amplifier, it is split into two or more frequency bands. The resulting signals each feed their own power amplifier/loudspeaker section. A block diagram of this approach is shown in Figure 11.33. This one is a two-way system. Large concert sound reinforcement systems may break the audio spectrum into four or five segments. The resulting system will be undoubtedly expensive, but will show lower distortion and higher output levels than a passively-crossed system. A typical two-way system might be crossed at 800 Hz. In other words, frequencies above 800 Hz will be sent to a specialized high-frequency transducer, while frequencies below 800 Hz will be sent to a specialized low-frequency transducer. In essence, the crossover network is a combination of an 800 Hz low-pass filter, and an 800 Hz high-pass filter. The filter order and alignment vary considerably depending on the application. Let's design an 800 Hz crossover with second-order Butterworth filters.

The basic circuit layout is shown in Figure 11.34. We're using the equal component value version here. Exact gain is normally not a problem in this case since some form of volume control needs to be added anyway, in order to compensate between the sensitivity of the low and high frequency transducers. (This is most easily produced by adding a simple voltage-divider/potentiometer at the output of the filters).

For second-order Butterworth filters, the damping factor is found to be 1.414, and the frequency factor is unity (indicating that f_c and $f_{3\text{ dB}}$ are equal). Note that the design for both halves is almost the same. Both sections show an f_c of 800 Hz and a damping factor of 1.414. With identical characteristics, it follows that the component values will be the same in both circuits.

The required value for R_f is

$$R_f = 2 - \alpha$$
$$R_f = 2 - 1.414$$
$$R_f = .586$$

—— **Figure 11.34**
Basic filter sections for
crossover of Example 11.4

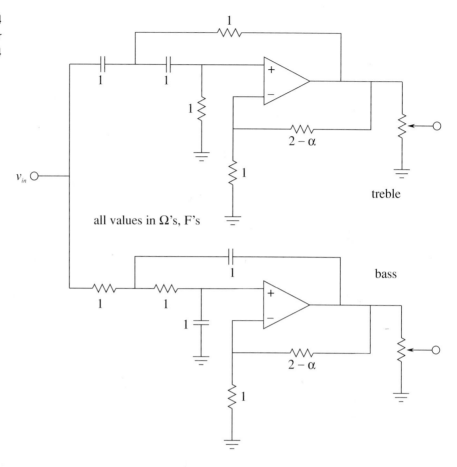

all values in Ω's, F's

treble

bass

The critical frequency in radians is

$$\omega_c = 2\pi f_c$$

$$\omega_c = 2\pi 800 \text{ Hz}$$

$$\omega_c = 5027 \text{ rad/S}$$

Again, we shall scale the tuning resistors to yield

$$R = \frac{1}{5027}$$

$$R = 1.989 \times 10^{-4}$$

A final RC scaling by 10^8 produces

$$R = 19.9 \text{ k}$$

$$C = 10 \text{ nF}$$

Figure 11.35
Completed crossover
design for Example 11.4

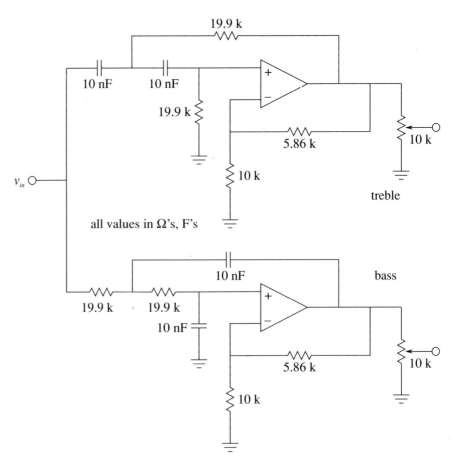

R_i and R_f are scaled by 10 k, and 10 k log taper potentiometers can be used for the output volume trimmers. The resulting circuit is shown in Figure 11.35.

11.7 Band-Pass Filter Realizations

There are many ways to form a band-pass filter. Before we introduce a few of the possibilities, we must define a number of important parameters. As in the case of the high- and low-pass filters, the concept of damping is important. For historical reasons, band-pass filters are normally specified with the parameter Q, the quality factor, which is the reciprocal of the damping factor.

Comparable to the break frequency is the center, or peak, frequency of the filter. This is the point of maximum gain. In *RLC* circuits, it is usually referred to as the resonance frequency. The symbol for center frequency is f_o. Since a band-pass filter produces attenuation on either side of the center frequency, there are two "3 dB down" frequencies. The lower frequency is normally given the name f_1, while the upper is given f_2. The difference between f_2 and f_1 is called the bandwidth of the filter. The ratio of center frequency to bandwidth (BW) is equal to the filter's Q:

$$BW = f_2 - f_1 \tag{11.15}$$

$$Q = \frac{f_o}{BW} \tag{11.16}$$

It is important to note that the center frequency is not equal to the arithmetic average of f_1 and f_2. Instead, it is equal to the geometric average of f_1 and f_2.

$$f_o = \sqrt{f_1 f_2} \tag{11.17}$$

These parameters are shown graphically in Figure 11.36. If a filter requires a fairly low Q, say unity or less, the filter is best realized as a cascade of separate low- and high-pass filters. For higher Q's, we shall examine two possible realizations. Multiple-feedback filters will be used for Q's up to about 10. For Q's above 10, the state-variable filter is presented.

Figure 11.36
Band-pass response

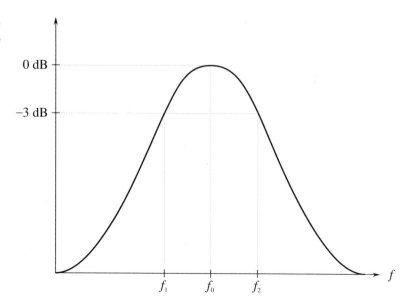

Multiple-Feedback Filters

The basic multiple-feedback filter is a second-order type. It contains two re-active elements as shown in Figure 11.37. One pair of elements creates the low-pass response $(R_1 C_1)$, and the other pair creates the high-pass response $(R_2 C_2)$. Because of this, the ultimate attenuation slopes are ± 6 dB. As with the VCVS high- and low-pass designs, the circuit of Figure 11.37 is normalized to a center frequency of 1 radian per second. Extrapolation to new center frequencies is performed in the same manner as shown earlier. The peak gain for this circuit is

$$A_v = -2Q^2 \qquad (11.18)$$

You can see from Equation 11.18 that higher Q's will produce higher gains. For a Q of 10, the voltage gain will be 200. For this circuit to function properly, the open-loop gain of the op amp used must be greater than 200 at the chosen center frequency. Usually, a safety factor of 10 is included in order to keep stability high and distortion low. By combining these factors, we may determine the minimum acceptable f_{unity} for the op amp:

$$f_{unity} \ge 10 f_o A_v \qquad (11.19a)$$

or more directly,

$$f_{unity} \ge 20 f_o Q^2 \qquad (11.19b)$$

For a Q of 10 and a center frequency of 2 kHz, the op amp will need an f_{unity} of at least 4 MHz. It is not possible to use this type of filter for high-

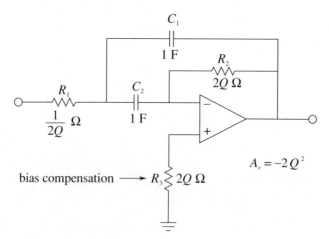

Figure 11.37
Multiple feedback
band-pass filter

frequency high-Q work, as standard op amps soon "run out of steam." This difficulty aside, the high gains produced by even moderate values for Q may well be impractical. For many applications, a unity-gain version would be preferred. This is not particularly difficult to achieve. All that we need to do is attenuate the input signal by a factor equal to the voltage gain of the filter. Since the gain magnitude of the filter is $2Q^2$, the attenuation should be

$$\text{attenuation} = \frac{1}{2Q^2} \qquad (11.20)$$

While it is possible to place a pair of resistors in front of the filter to create a voltage divider, there is a more efficient way: We can split R_1 into two components, as shown in Figure 11.38. As long as the Thevenin equivalent of R_{1a} and R_{1b} as seen from the op amp equals the value of R_1, the tuning frequency of the filter will not be changed. Also required is that the voltage divider ratio produced by R_{1a} and R_{1b} satisfies Equation 11.20. First, let's determine the ratio of the two resistors. We can start by setting R_{1b} to the arbitrary value K. Using the voltage divider rule and Equation 11.20, R_{1a} is found:

$$\text{attenuation} = \frac{R_{1b}}{R_{1a} + R_{1b}}$$

$$\frac{1}{2Q^2} = \frac{K}{R_{1a} + K}$$

$$R_{1a} + K = K2Q^2$$

$$R_{1a} = K(2Q^2 - 1) \qquad (11.21)$$

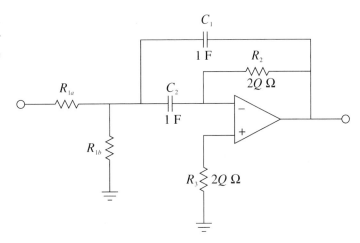

Figure 11.38
Multiple feedback filter
with unity-gain variation

So, we see that R_{1a} must be $2Q^2 - 1$ times larger than R_{1b}. Now we must determine the value of K which will set the parallel combination of R_{1a} and R_{1b} to the required value of $\frac{1}{2}Q$, as based on Figure 11.37.

$$R_{Thevenin} = R_{1a} \parallel R_{1b}$$

$$R_{Thevenin} = \frac{R_{1a}R_{1b}}{R_{1a} + R_{1b}}$$

$$R_{Thevenin} = \frac{K^2(2Q^2 - 1)}{K(2Q^2 - 1) + K}$$

$$R_{Thevenin} = \frac{K^2(2Q^2 - 1)}{K2Q^2}$$

$$R_{Thevenin} = \frac{K(2Q^2 - 1)}{2Q^2}$$

Since $R_{Thevenin} = \frac{1}{2}Q$,

$$\frac{1}{2}Q = \frac{K(2Q^2 - 1)}{2Q^2}$$

$$1 = \frac{K(2Q^2 - 1)}{Q}$$

$$K = \frac{Q}{2Q^2 - 1} \qquad (11.22)$$

Since R_{1b} was set to K,

$$R_{1b} = \frac{Q}{2Q^2 - 1}\ \Omega \qquad (11.23)$$

Substituting Equation 11.22 into Equation 11.21 yields

$$R_{1a} = Q\ \Omega \qquad (11.24)$$

By using these values for R_{1a} and R_{1b} the filter will have a peak gain of unity. Note that since this scheme only attenuates the signal prior to gain, the f_{unity} requirement set in Equation 11.19 still holds true.

─────── **Example 11.5**   Design a filter which will only pass frequencies from 800 Hz to 1200 Hz. Make sure that this is a unity-gain realization.

First, we must determine the center frequency, bandwidth, and Q.

$$BW = f_2 - f_1$$

$$BW = 1200\ Hz - 800\ Hz$$

$$BW = 400\ Hz$$

$$f_o = \sqrt{f_1 f_2}$$

$$f_o = \sqrt{800 \text{ Hz} \times 1200 \text{ Hz}}$$

$$f_o = 980 \text{ Hz}$$

$$Q = \frac{f_o}{\text{BW}}$$

$$Q = \frac{980 \text{ Hz}}{400 \text{ Hz}}$$

$$Q = 2.45$$

The Q is too high to use separate high- and low-pass filters, but sufficiently low so that a multiple-feedback type may be used. Before proceeding, we should check to make sure that the required f_{unity} for the op amp is reasonable.

$$A_v = -2Q^2$$

$$A_v = -2 \times 2.45^2$$

$$A_v = -12$$

$$f_{unity} \geq 10A_v f_o$$

$$f_{unity} \geq 10 \times 12 \times 980 \text{ Hz}$$

$$f_{unity} \geq 117.6 \text{ kHz}$$

Just about any modern op amp will exceed the f_{unity} specification. Since this circuit shows a gain of 12, the unity-gain variation shown in Figure 11.38 will be used. The calculations for the normalized components follow.

$$R_2 = 2Q$$

$$R_2 = 2 \times 2.45$$

$$R_2 = 4.9 \; \Omega$$

$$R_{1b} = \frac{Q}{2Q^2 - 1}$$

$$R_{1b} = \frac{2.45}{2 \times 2.45^2 - 1}$$

$$R_{1b} = .2226 \; \Omega$$

The resulting normalized circuit is shown in Figure 11.39 (page 532). We must now find the frequency scaling factor.

$$\omega_o = 2\pi f_o$$

$$\omega_o = 2\pi 980 \text{ Hz}$$

$$\omega_o = 6158 \text{ rad/S}$$

——————   **Figure 11.39**
Initial damping calculation
for Example 11.5

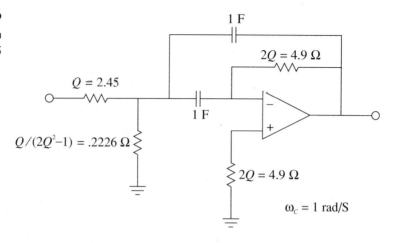

In order to translate our circuit to this frequency, we must divide either the resistors or the capacitors by 6158. In this example, let's use the capacitors.

$$C = \frac{1}{6158}$$

$$C = 162.4 \ \mu F$$

A further impedance scaling is needed for practical component values. A factor of a few thousand or so would be appropriate here. To keep the calculations simple, we'll choose 10 k. Each resistor will be increased by 10 k, and each capacitor will be reduced by 10 k. The final scaled filter is shown in Figure 11.40.

The SPICE simulation of the circuit is shown in Figure 11.41. Note that the gain is 0 dB at the approximate center frequency (about 1 kHz). Also,

——————   **Figure 11.40**
Final impedance and
frequency scaling for
Example 11.5

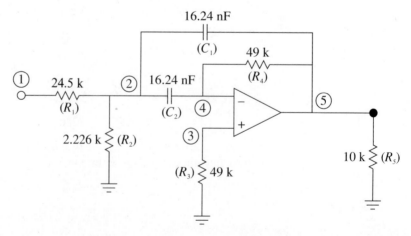

Figure 11.41
Spice simulation of the
circuit of Example 11.5

```
Band pass 800-1.2 KHz unity gain filter.
*******Start of UA741 op amp*******
.SUBCKT UA741 1     2     3   4   5    6
*                  IN+ IN- GND V+ V- OUT
Q1    11   1    13 UA741QA
Q2    12   2    14 UA741QB
RC1    4  11    5.305165E+03
RC2    4  12    5.305165E+03
C1    11  12    5.459553E-12
RE1   13  10    2.151297E+03
RE2   14  10    2.151297E+03
IEE   10   5    1.666000E-05
CE    10   3    3.000000E-12
RE    10   3    1.200480E+07
GCM    3  21    10 3  5.960753E-09
GA    21   3    12 11 1.884655E-04
R2    21   3    1.000000E+05
C2    21  22    3.000000E-11
GB    22   3    21 3 2.357851E+02
RO2   22   3    4.500000E+01
D1    22  31    UA741DA
D2    31  22    UA741DA
EC    31   3    6 3 1.0
RO1   22   6    3.000000E+01
D3     6  24    UA741DB
VC     4  24    2.803238E+00
D4    25   6    UA741DB
VE    25   5    2.803238E+00
.ENDS
.MODEL UA741DA D (IS=9.762287E-11)
.MODEL UA741DB D (IS=8.000000E-16)
.MODEL UA741QA NPN (IS=8.000000E-16 BF=9.166667E+01)
.MODEL UA741QB NPN (IS=8.309478E-16 BF=1.178571E+02)
*****End of UA741 op amp*****

*****Start of circuit description*****

R1     1   2    24.5K
R2     2   0    2.226K
R3     3   0    49K
R4     4   5    49K
R5     0   5    10K
C1     2   5    16.24NF
C2     2   4    16.24NF
VCC    6   0    DC 15
VEE    7   0    DC -15
X1     3   4    0  6   7   5   UA741
VIN    1   0    AC 1

*****Output control directives*****

.AC DEC 20 100 10K
.PLOT AC VDB(5)
.PROBE
.END
```

a. Input file

(Continued)

—— **Figure 11.41** *(Continued)*

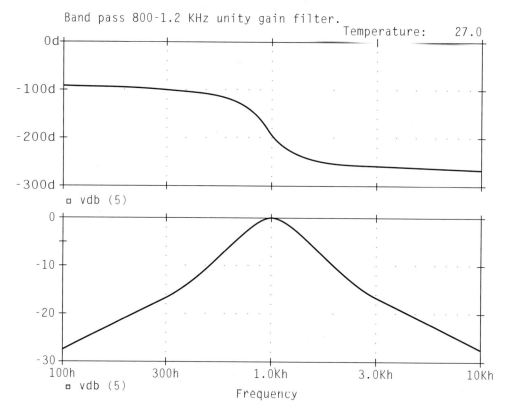

Band pass 800-1.2 KHz unity gain filter.

Temperature: 27.0

b. PSpice/Probe output

the −3 dB breakpoints of 800 Hz and 1200 Hz are clearly seen. The phase response of this filter is also plotted. Note the very fast phase transition in the area around f_o. If the Q of this circuit was increased, this transition would be faster still.

In circuits such as this, it is very important that realistic op amp models be employed. If an over-idealized version is used, non-ideal behavior due to a reduction of loop gain will go unnoticed. This error is most likely to occur in circuits with high center frequencies and/or high Q's. You can verify this by translating the filter to a higher frequency and rerunning the simulation. For example, if C_1 and C_2 are decreased by a factor of 1000, the center frequency should move to about 1 MHz. If the simulation is run with an appropriate range of test frequencies, you will see that the limited bandwidth of the μA741 op amp prematurely cuts off the filter response. The result is a peaking frequency

Figure 11.41   (*Continued*)

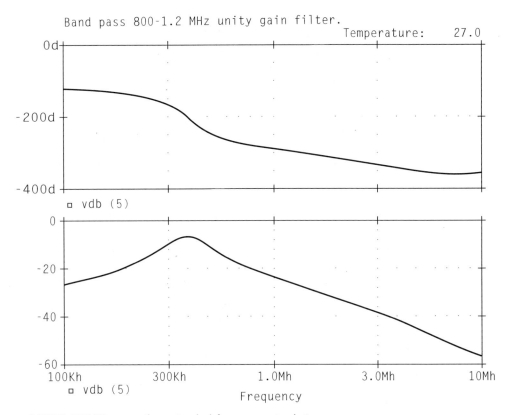

c. PSPICE PROBE output for extended frequency simulation

more than one octave below target, a maximum amplitude several dB below 0, and an asymmetrical response curve. This response graph is shown in Figure 11.41c. The accompanying phase plot also shows a great deviation from the ideal filter. An excessive phase shift at the middle and higher frequencies is clearly evident.

State-Variable Filters

As noted earlier, the multiple-feedback filter is not suited to high-frequency or high-Q work. For applications requiring Q's of about 10 or more, the state-variable filter is the form of choice. The state-variable is often referred to as the universal filter since band-pass, high-pass, and low-pass outputs are all available. With additional components, a band-reject output may be formed as well. Unlike the earlier filter forms examined, the basic state-variable filter requires three op amps. Also, it is a second-order type, although higher-order types are possible. This form gets its name from state-variable analysis. One of the earliest uses for op amps was in the construction of analog computers (see Chapter 10). Interconnections of differentiators, amplifiers, summers, and integrators were used to electronically solve differential equations which described physical systems. State-variable analysis provides a technique for solving involved differential equations. The equations can in fact describe a required filter's characteristics. While we will not examine state-variable analysis, this does not preclude a study of the state-variable filter. Designing with state-variable filters is really no more complex than our previous work.

Besides its ability to provide stable filters with relatively high Qs, the state-variable filter has other unique characteristics:

1. It is relatively easy to tune electronically over a broad frequency range.

2. It is possible to independently adjust the Q and tuning frequency.

3. It offers the ability to create other, more complex, filters since it has multiple outputs.

The state-variable filter is based on integrators. The general form utilizes a summing amplifier and two integrators, as shown in Figure 11.42. To un-

Figure 11.42   Block diagram of state-variable filter

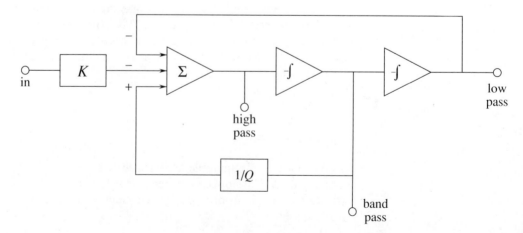

Figure 11.43   Fixed-gain version of state-variable filter

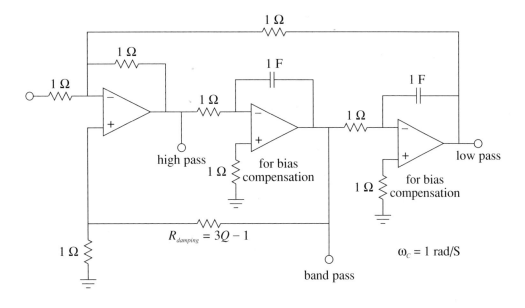

derstand on an intuitive level how this circuit works, recall that integrators are basically first-order low-pass filters. As you can see, the extreme right side output has passed through the integrators and produces a low-pass response. If the low-pass output is summed out of phase with the input signal, the low frequency information will cancel, leaving just the high frequency components. Therefore, the output of the summer is the high-pass output of the filter. If the high-pass signal is integrated (using the same critical frequency), the result will be a band-pass response. This is seen at the output of the first integrator. The band-pass signal is also routed back to the input summing amplifier. By changing the amount of the signal which is fed back, the response near the critical frequency can be altered, effectively setting the filter Q. Finally, the loop is completed by integrating the band-pass response which yields the low-pass output. In effect, the second integrator's -6 dB per octave rolloff perfectly compensates for the rising band-pass response below f_o. This produces flat response below f_o. Above f_o, the combination of the two falling response curves produces the expected second-order low-pass response.

　　Two popular ways of configuring the state-variable filter are the fixed-gain and adjustable-gain forms. The fixed-gain form is shown in Figure 11.43. This circuit uses a total of three op amps. The Q of the circuit is set by a single resistor, $R_{damping}$. Q's up to 100 are possible with state-variable filters. For the high- and low-pass outputs, the gain of this circuit is unity. For the band-pass

Figure 11.44   Variable-gain version of state-variable filter

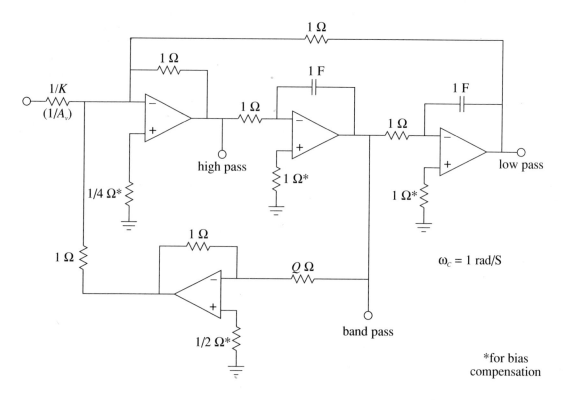

output, the gain is equal to Q. Figure 11.44 shows an adjustable gain version. For high- or low-pass use, the gain is equal to the arbitrary value K, while for band-pass use, the gain is equal to KQ. This variation requires a fourth op amp in order to isolate the Q and gain settings. While four op amps may sound like a large number of devices, remember that a variety of quad op amp packages exist, indicating that the actual physical layout may be quite small. Also, even though three different outputs are available, it is not possible to individually optimize each one for simultaneous use. Consequently, the state-variable is most often used as a stable and switchable high/low pass filter, or as a high-Q band-pass filter. Finally, in keeping with our previous work, the circuits are shown normalized to a critical frequency of one radian per second. While we shall concentrate on band-pass design in this section, it is possible to use these circuits to realize various high- and low-pass filters, such as those generated with the Sallen and Key forms. The procedure is nearly identical, and uses the same frequency and damping factors (see Figures 11.24 and 11.29).

Example 11.6   Design a band-pass filter with a center frequency of 4.3 kHz and a Q of 25. Use the fixed-gain form.

First, determine the damping resistor value. Then, scale the components for the desired center frequency. Note that a Q of 25 produces a bandwidth of only 172 Hz for this filter (4.3 kHz/25).

$$R_{damping} = 3Q - 1$$

$$R_{damping} = 3 \times 25 - 1$$

$$R_{damping} = 74 \ \Omega$$

$$\omega_o = 2\pi \times 4.3 \ \text{kHz}$$

$$\omega_o = 27.02 \ \text{k rad/S}$$

In order to translate the filter to our desired center frequency, we need to divide either the resistors or the capacitors by 27,020. For this example, we'll use the capacitors.

$$C = \frac{1}{27.02 \ \text{k}}$$

$$C = 37 \ \mu F$$

A final impedance scaling is required to achieve reasonable component values. A reasonable value might be a factor of 5000.

$$C = \frac{37 \ \mu F}{5000}$$

$$C = 7.4 \ \text{nF}$$

$$R_{damping} = 74 \times 5000$$

$$R_{damping} = 370 \ \text{k}$$

All remaining resistors will equal 5 kΩ.

Since this is a band-pass filter,

$$A_v = Q$$

$$A_v = 25$$

The completed filter is shown in Figure 11.45 (page 540). The value for $R_{damping}$ is considerably larger than the other resistors. This effect gets worse as the required Q is increased. If this value becomes too large for practical

Figure 11.45 Completed design of band-pass filter for Example 11.6

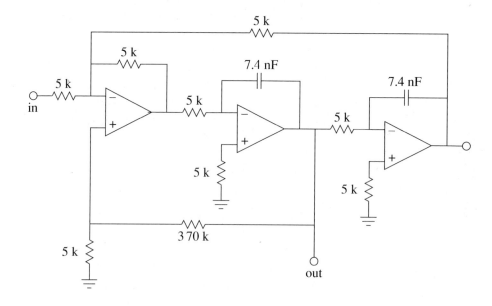

components, it can be reduced to a more reasonable value as long as the associated divider resistor (from the noninverting input to ground) is reduced by the same amount. The ratio of these two resistors is what sets the filter Q, not their absolute values. Lowering these values will upset the ideal input bias current compensation, but this effect can be ignored in many cases, or reduced through the use of FET input op amps.

Altering this circuit for a variable-gain configuration requires the addition of a fourth amplifier as shown in Figure 11.44. The calculation for the damping resistor is altered, and a value for the input gain determining resistor is needed. The remaining component calculations are unchanged from the example above. Note that by setting the gain constant K to $1/Q$, the final filter gain can be set to unity.

11.8 Notch Filter (Band-Reject) Realizations

By summing the high-pass and low-pass outputs of the state-variable filter, a notch, or band-reject, filter can be formed. Filters of this type are commonly used to remove interference signals. The summation is easily performed with

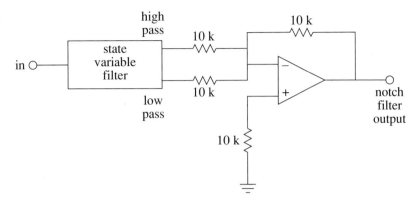

Figure 11.46
Notch filter

a simple parallel-parallel–based summing amplifier, as shown in Figure 11.46. For reasonable Q values, there will be tight correlation between the calculated band-pass center and -3 dB frequencies, and the notch center and -3 dB frequencies. The component calculations proceed as in the band-pass filter.

Example 11.7   A filter is needed to remove induced 60 Hz hum from a transducer's signal. The rejection bandwidth of the filter should be no more than 2 Hz.

From the specifications, we know that the center frequency is 60 Hz and the Q is $\frac{60}{2}$, or 30. For simplicity, we shall use the fixed-gain form. (Note that the gain of the filter on either side of the notch will be unity.)

$$R_{damping} = 3Q - 1$$

$$R_{damping} = 3 \times 30 - 1$$

$$R_{damping} = 89 \ \Omega$$

$$\omega_o = 2\pi 60$$

$$\omega_o = 377 \ \text{rad/S}$$

Scaling C produces

$$C = \frac{1}{377}$$

$$C = 2.65 \ \text{mF}$$

A practical value scaling of 10^4 produces the circuit shown in Figure 11.47 (page 542). Note that the damping resistors have only been scaled by 10^3 since

Figure 11.47   Completed notch filter for Example 11.7

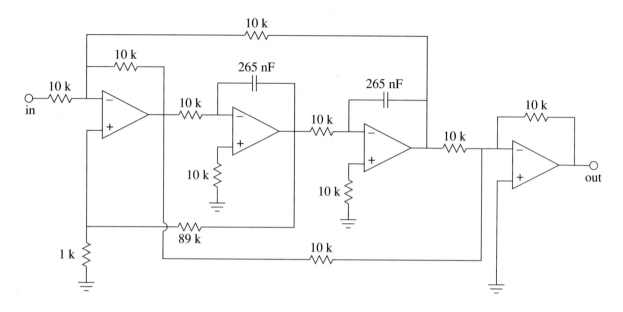

an $R_{damping}$ value of 890 k might be excessive. Remember, it is the ratio of these two resistors which is important, not their absolute values.

A Note on Component Selection

Ideally, the circuit of Example 11.7 will produce -3 dB points at approximately 59 Hz and 61 Hz, and will infinitely attenuate 60 Hz tones. In reality, component tolerances may alter the response and, therefore, high-quality parts are required for accurate high-Q circuits such as this. Even simpler, less-demanding circuits such as a second-order Sallen and Key filter may not perform as expected if lower-quality parts are used. As a general rule, component accuracy and stability becomes more important as filter Q and order increase. One percent tolerance metal film resistors are commonly used, with five percent carbon film types being satisfactory for the simpler circuits. For capacitors, film types such as polyethylene (mylar) are common for general-purpose work, with polycarbonate, polystyrene, polypropylene and teflon being used for the more stringent requirements. For small capacitance values (<100 pF), NPO ceramics can be used. Generally, large ceramic disc and aluminum electrolytic capacitors are avoided due to their wide tolerance and instability with temperature, applied voltage, and other factors.

Another range of circuits which fall under the heading of filters are equalizers. Actually, equalizers are a class of adjustable filters which can produce gain as well as attenuation. Perhaps the most common uses for equalizers are in the audio, music, and communications areas. As the name suggests, equalizers are used to adjust or balance the input frequency spectrum. Equalizers range from complex *1/3 octave* and *parametric* types for use in recording studios and large public address systems, to the simpler bass and treble controls found on virtually all home and car stereos. Many of the more complex equalizers are based on extensions of the state-variable filter. On the other hand, some equalizers are little more than modified amplifiers. We'll take a look at the very common bass and treble controls, which adjust low and high audio frequencies, respectively.

The purpose of bass and treble controls is to allow the listener some control of the balance of high and low tones. They may be used to help compensate for the acoustical shortcomings of a loudspeaker, or perhaps solely to compensate for personal taste. Unlike the filters previously examined, these controls will be manipulated by the user and must provide for signal boost as well as cut. A typical response curve is shown in Figure 11.48. Typically, bass and treble circuits are realized with parallel-parallel inverting amplifiers. In essence, the feedback network will change with frequency.

A simple bass control is shown in Figure 11.49 (page 544). To understand how this circuit works, let's look at what happens at very low and at very high frequencies. First of all, at very high frequencies, capacitor C is ideally

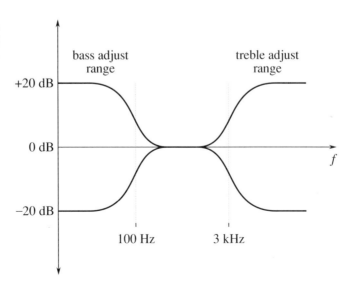

Figure 11.48
Response of general
bass/treble equalizer

——— **Figure 11.49**
Simple bass section

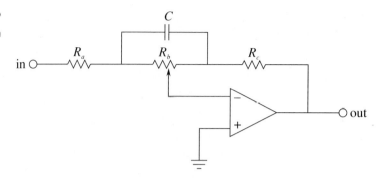

shorted. Thus, the setting of potentiometer R_b is inconsequential. In this case, the gain magnitude of the amplifier will be set at R_c/R_a. Normally, R_c is equal to R_a, so the gain is unity. At very low frequencies the exact opposite happens; the capacitor is seen as an open. Under this condition the gain of the amplifier depends on the setting of potentiometer R_b. If the wiper is set to the extreme right, the gain becomes $R_c/(R_a + R_b)$. If the wiper is moved to the extreme left, the gain becomes $(R_c + R_b)/R_a$. If R_b is set to nine times R_a, the total gain range will vary from .1 to 10 (-20 dB to $+20$ dB). A similar arrangement may be used to adjust the treble range.

When the bass and treble controls are combined, component loading makes the circuit somewhat more difficult to design. Basic models have already been derived by a number of sources, though, including the one shown in Figure 11.50. The following equations are used to design the desired equalizer. (Refer to Figure 11.51.)

——— **Figure 11.50**
Bass/treble equalizer

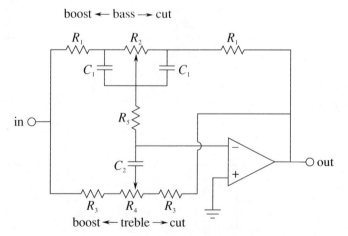

Figure 11.51
Response of bass/treble
equalizer

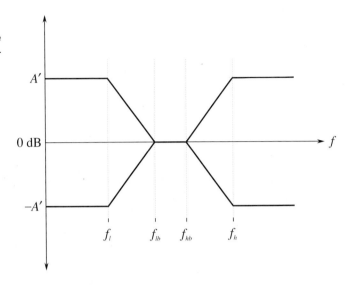

Figure 11.51
Response of bass/treble
equalizer

Bass section (assumes $R_2 \gg R_1$):

$$f_h = \frac{1}{2\pi R_3 C_3}$$

$$f_{hb} = \frac{1}{2\pi(R_1 + R_3 + 2R_5)C_3}$$

$$A_{vt} = 1 + \frac{R_1 + 2R_5}{R_3}$$

Treble section (assumes $R_4 \gg R_1 + R_3 + 2R_5$):

$$f_l = \frac{1}{2\pi R_2 C_1}$$

$$f_{lb} = \frac{1}{2\pi R_1 C_1}$$

$$A_{vb} = 1 + \frac{R_2}{R_1}$$

In actuality, it is very common to design these sorts of circuits empirically. In other words, the given equations are used as a starting point, and then component values are adjusted in the laboratory until the desired response range is obtained. Circuits like this may be altered further to include a midrange control. Generally, three adjustments is considered to be the maximum for this type of circuit. An example of a bass-midrange-treble equalizer is shown in the schematic for the Pocket Rockit amplifier back in Chapter 6.

11.10 Switched-Capacitor Filters

Our final topic is the class of ICs known as *switched-capacitor filters*. These are just specific realizations of the types of filters which we have already examined. Generally, switched-capacitor filters come in two types: fixed order and alignment, and universal (state-variable based). A typical fixed IC might offer a sixth-order low-pass Butterworth filter. The number of external components required is minimal. The universal types offer most of the flexibility of the state-variable designs discussed above. Both types are tunable and are relatively easy to use. Tuning is accomplished by adjusting the frequency of an external clock signal. The higher the clock frequency, the higher the resulting critical frequency. These ICs offer a convenient "black box" approach to general-purpose filter design. As is the case with most special-purpose devices, individual manufacturer's data sheets will give the specific application and design procedures for their parts.

The concept behind the switched-capacitor filter is quite interesting. The basic idea is to mimic a resistor through the use of a capacitor and a pair of alternating switches. As an example, a simple integrator is shown in Figure 11.52. As you have already seen, integrators are little more than first-order low-pass filters. In this circuit, the input resistor has been replaced with a capacitor, C_{in}, and a pair of switches. These switches are controlled by a non-overlapping biphase clock. This means that when one switch is closed, the other will be open, and that during switching, one switch will break contact before the other switch makes contact. (This is sometimes referred to as a "break before make" switch.)

During the first half of the clock cycle, C_{in} charges to the value of v_{in}. During the second half of the clock cycle, this charge is transferred to the integration capacitor. Therefore, the total charge transferred during one clock cycle is

$$Q = C_{in}v_{in} \qquad (11.25)$$

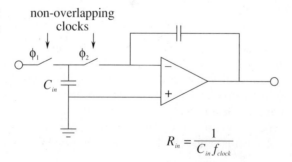

Figure 11.52
Switched capacitor circuit

non-overlapping clocks

$$R_{in} = \frac{1}{C_{in}f_{clock}}$$

The flow of charge versus time defines current, so the average input current is

$$i_{in} = \frac{Q}{T_{clock}} \qquad (11.26)$$

Substituting Equation 11.25 into Equation 11.26 yields

$$i_{in} = C_{in} \frac{v_{in}}{T_{clock}} \qquad (11.27)$$

Since the input resistance is defined as the ratio of v_{in} to i_{in}, and recognizing that f_{clock} is the reciprocal of T_{clock}, Equation 11.27 is used to find R_{in}:

$$R_{in} = \frac{1}{C_{in}f_{clock}} \qquad (11.28)$$

Equation 11.28 says two important things: first, since R_{in} sets the input impedance, it follows that input impedance is inversely proportional to the clock frequency. Second, since R_{in} is used to determine the corner frequency (in conjunction with the integration/feedback capacitor), it follows that the critical frequency of this circuit is directly proportional to the clock frequency. In other words, a doubling of clock frequency will halve the input impedance, and double the critical frequency. Depending on the actual design of the IC, there will be a constant ratio between the clock frequency and the critical frequency. This is a very useful attribute. It means that you can make a tunable/sweepable filter by using one of these ICs and an adjustable square wave generator. For that matter, anything which can produce a square wave (such as a personal computer), can be used to control the filter response.

Typically, the ratio of clock frequency to critical frequency will be in the range of 50 to 100. The lower limit of clock frequency is controlled by internal leakage paths which create offset errors. The upper limit is controlled by switch settling time, propagation delays, and the like. A range of 100 Hz to 1 MHz is reasonable. This means that the entire audio frequency range is covered by these devices. For use at the highest frequencies, or with high impedance sources, an input buffer amplifier should be used.

Two examples of switched-capacitor filter ICs are the MF4 and MF5, shown in Figures 11.53 and 11.54, respectively, on page 548. The MF4 is a fourth-order Butterworth low-pass filter. It comes in two variations: the MF4-50, which has a 50:1 clock to critical frequency ratio, and the MF4-100, which has a 100:1 ratio. Both variations produce unity gain in the pass band. The input capacitor for the MF4-50 is 2 pF, which, from Equation 11.28, produces an input impedance of 500 k for a critical frequency of 20 kHz. The external component count is minimal (excluding the clock source). Input clock and level shift pins are provided for use with TTL or CMOS clocks. For less

Figure 11.53 Block diagram of the MF4

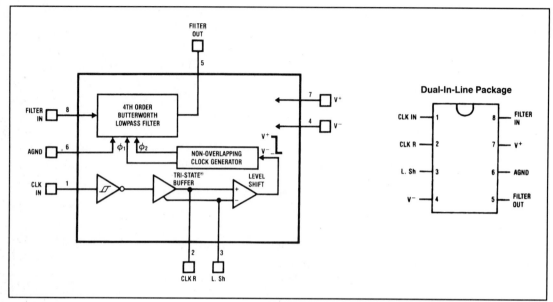

Reprinted with permission of National Semiconductor Corporation

Figure 11.54 Block diagram of the MF5

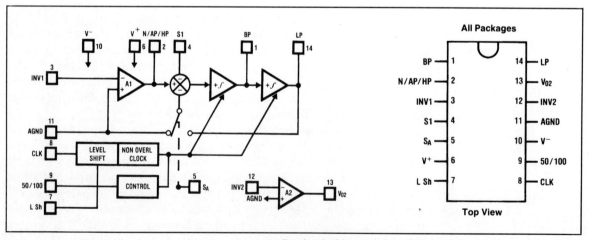

Reprinted with permission of National Semiconductor Corporation

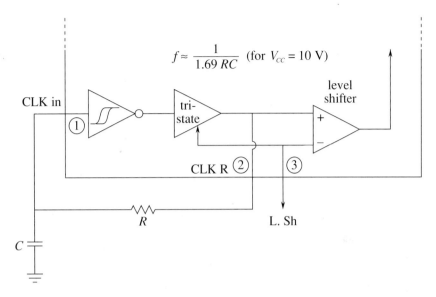

Figure 11.55
MF4 equivalent internal
circuit for oscillator

$$f \approx \frac{1}{1.69\,RC} \quad \text{(for } V_{cc} = 10 \text{ V)}$$

demanding fixed-frequency work, it is possible to configure a simple RC oscillator with the internal Schmitt trigger as shown in Figure 11.55. An example of a TTL clocked filter is shown in Figure 11.56. As you can see, the design is sparse, at best. Besides fourth-order Butterworth, filters of higher order and other alignments are available. In short, these devices are good choices for general-purpose filter work, particularly when space and tuning considerations are important.

For designs requiring a bit more flexibility, universal filters such as the MF5 are available. Since this filter is basically little more than a switched-capacitor version of the state-variable, a wide range of response types are possible. These include the high-, low-, and band-pass outputs, as well as a variety of alignments including Bessel, Butterworth, and Chebyshev. The MF5 also includes a separate op amp which can be used as a gain block, or for creating a notch output

Figure 11.56
TTL clock input

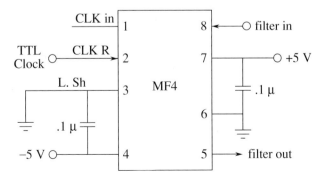

Figure 11.57
Second-order filter using
the MF5

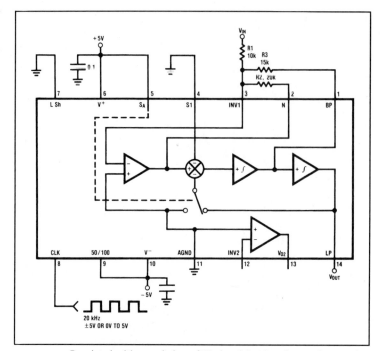

Figure 11.57
Second-order filter using the MF5

Reprinted with permission of National Semiconductor Corporation

(as seen in Example 11.7). Normally, no more than four or five external components (resistors and capacitors) are needed to realize a given filter function. Component calculation procedures are specified by the manufacturer. As an example of the low parts count required, a second-order Butterworth low-pass design is shown in Figure 11.57.

While switched capacitor filters offer relatively quick and physically small realizations, they are not without problems. First of all, clock feedthrough is typically in the range of 10 mV, meaning that 10 mV of clock signal "leaks" into the output. Fortunately, this signal is much higher than the critical frequency, but may cause some problems for low-noise applications. Another problem arises from the fact that switched-capacitor filters are actually sampled data devices. As you will see in the next chapter, sampled data devices may suffer from a distortion producing phenomenon called *aliasing*. In order to avoid aliasing, the input signal must not contain any components which are greater than one-half of the clock frequency. For example, if the MF4-50 is used to create a 1 kHz filter, a 50 kHz clock is required. No component of the input signal may exceed 25 kHz (one half of the clock) if aliasing is to be avoided. If this requirement cannot be guaranteed, some form of prefiltering is needed. Within these limits though, switched-capacitor filters make light work out of many general-purpose applications.

11.11 Extended Topic: Voltage-Controlled Filters

A voltage-controlled filter, or VCF, is nothing more than a standard filter whose tuning frequency is controlled by an external voltage. You might think of this concept as an extension of the clock control aspects of the switched-capacitor filter. VCFs are used in a wide range of applications including instrumentation devices such as swept frequency analyzers, and music synthesizers. Any application which requires precise or rapid control of tuning frequency calls for a VCF. Virtually any of the filters presented in this chapter may be turned into VCFs. All you need to do is substitute the tuning elements of the filter with a voltage-controlled version. Typically, this means replacing the tuning resistors with voltage-controlled resistances.

Two popular ways of creating a voltage-controlled resistance include the photoresistor/lamp combination (see Figure 11.58), and the use of an FET in its ohmic region (see Figure 11.59 on page 552). To use these items, simply remove the tuning resistor(s), and replace them with a voltage-controlled resistance. As an example, a simple single-pole high-pass VCF is shown in Figure 11.60. As the control voltage (V_c) increases, the lamp brightness increases, causing the photoresistor's value to drop. Since the photoresistor sets the tuning frequency, the net result is an increase in f_c. The FET version produces a resistance which is proportional to the magnitude of the gate voltage (V_c).

These two solutions are not without their problems. In the case of the lamp/photoresistor, response time is not very fast and the lamp portion requires a fairly large drive current. The FET circuit eliminates these problems, but requires that the voltage across it remain fairly low (usually less than 100 mV). Larger signal swings will drive the FET out of the ohmic region, and distortion will increase dramatically. Also, the popular N-channel variety requires a negative gate potential, which is generally not preferred. In both cases, one more problem remains: it is difficult to create a wide linear control range.

Figure 11.58
A photoresistor/lamp used as a variable resistance

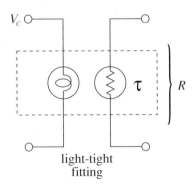

light-tight fitting

Figure 11.59
Using a JFET in the ohmic
region

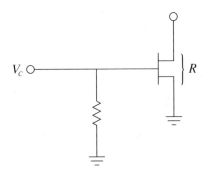

Figure 11.59
Using a JFET in the ohmic
region

Another way to create the effect of a voltage-variable tuning element is through the use of an operational transconductance amplifier, or OTA (see Chapter 6). Remember, this device is essentially a voltage-to-current converter. Its output current is a function of its control current. (The control current is easily derived from a control voltage and resistor.) This device is ideally suited to "inverting" inputs, where an input resistor is used as a voltage-to-current converter. One possible example is shown in Figure 11.61, a state-variable VCF. The boxed sections show where an OTA has replaced a standard single resistor. In this circuit, a large control voltage creates a large control current, thus increasing transconductance. This simulates a smaller tuning resistor value, and thus creates a higher tuning frequency. The OTA approach proves to be reliable, repeatable, and generally low in cost. It also offers a fairly wide linear tuning range.

For more exacting applications, or those applications requiring logarithmic control over the tuning frequency (music synthesis, for example), specialized VCF ICs may be purchased. One example of the breed is the SSM2044 from Precision Monolithics. The SSM2044 is a four-pole (24 dB per octave) low-pass filter. It offers a 90 dB dynamic range with a minimum tuning sweep range of 10,000:1.

Figure 11.60
Variable resistance
connection for VCF

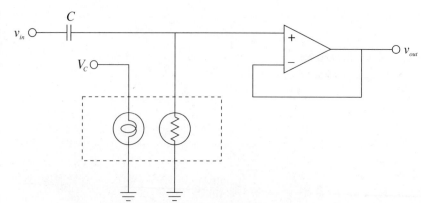

——— Figure 11.61   Using OTAs as a controlled element in a VCF

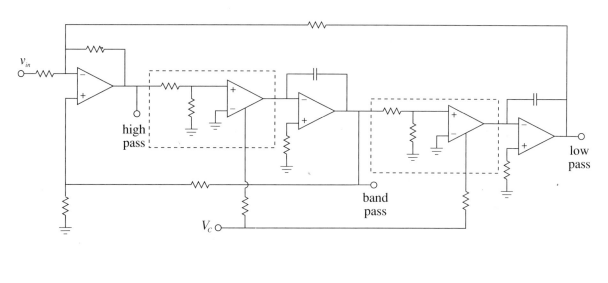

——— Summary

Filters are frequency-selective circuits. The basic forms are high-pass, low-pass, band-pass, and band-reject. While filters can be constructed solely from resistors, capacitors, and inductors, active filters using op amps offer many advantages. These advantages include: modest component size, control over impedances and loading effects, elimination of inductors, and gain (if desired). The negative aspects include: frequency range limited by op amps used, power supply required, and the inability to handle large input/outout powers. For many applications the advantages far outweigh the disadvantages, and therefore, active filters are used in a wide variety of modern products.

Filters are further defined by order and alignment. Order indicates the steepness of the attenuation slope. As a general rule, the eventual rolloff rate will equal 6 dB times the order, per octave. Order also indicates the minimum number of reactive elements needed to realize the filter. Alignment indicates the shape of the filter response in the frequency domain. Popular alignments include Bessel (constant time delay), Butterworth (maximally flat response in the pass band), and Chebyshev (ripples in the pass band, but with faster rolloff rates). There is generally a trade-off between fast attenuation rates and smooth phase response. Alignment is indicated by the damping or Q of the filter. Q is the reciprocal of damping. Filters with low damping factors (i.e., high Q) tend to be "peaky" in the frequency domain, and produce ringing on pulse-type inputs. (Chebyshevs would be in this category.) The filter's critical frequency and 3 dB down frequency are not the same for alignments other than the

Butterworth. The actual amount of "skew" depends on the alignment and order of the filter.

Once filter performance is specified, there are a number of ways in which the circuit can be physically realized. Common high- and low-pass realizations use the Sallen and Key VCVS approach. There are two variations on this theme: the unity-gain form and the equal-component form. Both forms use a second-order building block section. For higher orders, several second-order sections (and optionally, a first-order section) are combined in order to produce the final filter. It is important to remember that higher-order filters are not simple combinations of identical lower-order filters. For example, a fourth-order 1 kHz Butterworth filter is not made by cascading a pair of identical second-order 1 kHz Butterworth filters; rather, each section requires specific damping and frequency factors. A common design procedure utilizes look-up tables for these factors. The filters are designed by first scaling the general filter to the desired cutoff frequency, and then scaling the components for practical values.

For relatively low Qs (<1), band-pass filters are best realized as a cascade of high- and low-pass filters. For higher Qs, this technique is not satisfactory. Moderate Qs (up to 10) can be realized with the multiple-feedback filter. Very high-Q applications (up to 100) may be realized with the state-variable filter. The state-variable filter is often known as the universal filter, as it produces high-, low-, and band-pass outputs. With the addition of a fourth amplifier, a notch (band-reject) filter can be formed. Fixed and adjustable gain versions of the state-variable can be utilized by the designer.

A somewhat more specialized group of filters are the equalizers commonly employed in audio recording and playback equipment. Unlike traditional filters, equalizers offer both boost and attenuation of frequencies. Generally, these circuits are based on parallel-parallel inverting amplifiers, utilizing an adjustable frequency-selective feedback network.

Switched-capacitor filter ICs offer the designer expedient solutions to general-purpose filter design. They are generally suited to the audio frequency range and require very few external components. The critical frequency is set by a clock input. The order and alignment may be either factory-set or user-adjustable (as in the universal state-variable types).

For Further Reading

Active Filter Cookbook, Don Lancaster. Howard W. Sams, 1982.

Audio Handbook, Dennis Bohn, Editor. National Semiconductor Corporation, 1976.

Audio Handbook, Precision Monolithics Inc., 1990.

Audio IC Op Amp Applications, 2nd Edition, Walter G. Jung. Howard W. Sams, 1978.

The Active Filter Handbook, Frank P. Tedeschi. TAB Books, 1979.

— Self Test Questions ┃━━━━━━

1. What are the four main types of filters?
2. What are the advantages of active filters versus passive filters?
3. What are the disadvantages of active filters versus passive filters?
4. What do the characteristics of order and poles indicate?
5. Name several popular filter alignments.
6. How do the popular filter alignments differ from one another in terms of phase and magnitude response?
7. Which alignment should be used if linear phase response is of particular importance?
8. Which alignment should be used if fastest rolloff rate is of particular importance?
9. When are the "3 dB down" and critical frequencies of a filter identical?
10. Outline the process for creating high-order filters and explain why cascades of similar lower-order filters do not give the appropriate results.
11. Explain where each of the following band-pass filters would be appropriate: high-pass/low-pass cascade, multiple-feedback, and state-variable.
12. Why is the state-variable often referred to as the universal filter?
13. Briefly explain the operation of an adjustable bass equalizer.
14. Briefly explain the concept behind the switched-capacitor filter.
15. What are the advantages and disadvantages of switched-capacitor ICs versus the more traditional op amp approach?

— Problem Set ┃━━━━━━

▷ Analysis Problems

1. Using Figure 11.9b, determine the loss for a 1 kHz Butterworth second-order low-pass filter at 500 Hz, 1 kHz, 2 kHz, and 4 kHz.
2. Using Figure 11.29, determine the loss for a 2 kHz 3 dB–ripple Chebyshev third-order low-pass filter at 500 Hz, 1 kHz, 2 kHz, 4 kHz, and 6 kHz.
3. Using Figure 11.9b, determine the loss for a 500 Hz Bessel second-order high-pass filter at 200 Hz, 500 Hz, and 2 khz.

4. Using Figure 11.29, determine the loss one octave above the cutoff frequency for a fourth-order low-pass filter of the following alignments: Butterworth, Bessel, 1 dB–ripple Chebyshev.

5. Repeat Problem 4 for high-pass filters.

6. Using Figures 11.9b and 11.29, determine the loss at 8 kHz for 3 kHz low-pass Butterworth filters of orders 2 through 6.

7. Using Figures 11.9b and 11.29, determine the loss at 500 Hz for 1.5 kHz high-pass Bessel filters of orders 2 through 6.

8. Repeat Problem 7 using 3 dB–ripple Chebyshevs.

9. Using Figures 11.9b and 11.29, determine the loss at 500 Hz for 200 Hz high-pass 3 dB–ripple Chebyshev filters of orders 2 through 6.

10. Using Figure 11.29, determine the loss one octave below the cutoff frequency for a third-order high-pass filter of the following alignments: Butterworth, Bessel, 1 dB–ripple Chebyshev.

11. Repeat Problem 10 for low-pass filters.

12. An application requires that the stop-band attenuation of a low-pass filter be at least −15 dB at 1.5 times the critical frequency. Determine the minimum order required for the Butterworth and 1 dB–ripple and 3 dB–ripple Chebyshev alignments.

13. An application requires that the stop-band attenuation of a high-pass filter be at least −20 dB one octave below the critical frequency. Determine the minimum order required for the Butterworth and 1 dB–ripple and 3 dB–ripple Chebyshev alignments.

14. A band-pass filter has a center frequency of 1020 Hz and a bandwidth of 50 Hz. Determine the filter Q.

15. A band-pass filter has upper and lower break frequencies of 9.5 kHz and 8 kHz. Determine the center frequency and Q of the filter.

⊳ **Design Problems**

16. Design a second-order Butterworth low-pass filter with a −3 dB frequency of 125 Hz. The pass-band gain should be unity.

17. Repeat Problem 16 for a high-pass filter.

18. Repeat Problem 16 using a Bessel alignment.

19. A particular application requires that all frequencies below 400 Hz should be attenuated. The attenuation should be at least −22 dB at 100 Hz. Design a filter to meet this requirement.

20. Repeat Problem 19 for an attenuation of at least −35 dB at 100 Hz.

21. Audiophile-quality stereo systems often use subwoofers to reproduce the lowest possible musical tones. These systems typically use an electronic crossover approach as explained in Example 11.4. Design an electronic crossover for this application using third-order Butterworth filters. The crossover frequency should be set at 65 Hz.

22. Explain how the design sequence of Problem 21 is altered if either a new crossover frequency is chosen, or a different alignment is specified.

23. Design a band-pass filter which will only allow frequencies between 150 Hz and 3 kHz. The attenuation slopes should be at least 40 dB per decade. (A filter such as this is useful for "cleaning up" recordings of human speech.)

24. Design a band-pass filter with a center frequency of 2040 Hz and a bandwidth of 400 Hz. The circuit should have unity gain. Also, determine the f_{unity} requirement of the op amp(s) used.

25. Repeat Problem 24 for a center frequency of 440 Hz and a bandwidth of 80 Hz.

26. Design a band-pass filter with upper and lower break frequencies of 700 Hz and 680 Hz.

27. Design a notch filter to remove 19 kHz tones. The Q of the filter should be 25. (This filter is useful in removing the stereo "pilot" signal from FM radio broadcasts.)

28. Design a second-order low-pass filter with a critical frequency of 30 kHz. Use a state-variable filter. The circuit should have a gain of +6 dB in the pass band.

29. Design a bass/treble equalizer to meet the following specification: maximum cut and boost = 25 dB below 50 Hz and above 10 kHz.

30. Using the MF4, design a fourth-order low-pass Butterworth filter with a critical frequency of 3.5 kHz. Do not use an external oscillator.

31. Using the MF4, design a low-pass filter which is adjustable from 200 Hz to 10 kHz. Do not ignore the oscillator design.

⊳ Challenge Problems

32. Design a low-pass second-order filter which can be adjusted by the user from 200 Hz to 2 kHz. Also, make the circuit switchable between Butterworth and Bessel alignments.

33. Design a subsonic filter which will be 3 dB down from the pass-band response at 16 Hz. The attenuation at 10 Hz must be at least 40 dB. While pass-band ripple is permissible, the gain should be unity.

34. Design an adjustable band-pass filter with a Q range from 10 to 25, and a center frequency range from 1 kHz to 5 kHz.

35. Modify the design of the previous problem so that as the Q is varied, the pass-band gain remains constant at unity.

⊳ SPICE Problems

36. Verify the magnitude response of the circuit designed in Problem 16 by using SPICE. Check both the critical frequency and the rolloff rate.

37. Verify the magnitude response of the electronic crossover designed in Problem 21 by using SPICE. Plot both outputs simultaneously on one graph.

38. Verify the magnitude and phase of the filter designed in Problem 24 by using SPICE.

39. Compare the simulations of the circuit designed in Problem 28 using the relatively slow LM741 versus the medium-speed LF351. Is there any noticeable change? What can you conclude from this? Would the results be similar if the break frequency was increased by a factor of 50?

40. It is very common to plot the adjustment range of equalizers on a single graph, as shown in Figure 12.46. Use SPICE to create a plot of the adjustment range of the equalizer designed in Problem 29.

41. Verify the design of Problem 32 using SPICE.

42. Verify the design of Problem 33 using SPICE.

43. Verify the design of Problem 34 using SPICE. Include four separate plots, showing maximum and minimum Q with maximum and minimum center frequency.

44. Verify the design of Problem 35 using SPICE. Include two simultaneous plots, one showing minimum Q with maximum and minimum center frequency, and the other showing maximum Q with maximum and minimum center frequency.

12 Analog-to-Digital-to-Analog Conversion

After completing this chapter, you should be able to:

❏ Outline the concept of *pulse code modulation*.

❏ Detail the advantages and disadvantages of signal processing in the digital domain.

❏ Define the terms *resolution*, *quantization*, and *Nyquist frequency*.

❏ Define an *alias*, and detail how it is produced, and subsequently avoided.

❏ Explain the operation of an *R/2R* digital-to-analog converter.

❏ Explain the need for anti-alias and reconstruction filters.

❏ Explain the operation of a successive-approximation analog-to-digital converter.

❏ Detail the need for, and operation of, a track-and-hold amplifier.

❏ Explain the operation of a flash analog-to-digital converter.

❏ Compare the different analog-to-digital converters in terms of speed, size, and complexity, and detail typical applications for each.

12.1 Introduction

Up to now, all of the circuits which you have studied in this book were analog circuits. That is, the input waveforms were time-continuous and had infinite *resolution* along the time and amplitude axes. That is, you could discern increasingly smaller and finer changes as you examined a particular section. No matter whether the circuit was a simple amplifier, function synthesizer, integrator, filter, or what have you, the analog nature of the signal was always true. Fundamentally, the universe is analog in nature (at least as far as we can tell—until someone discovers a quantum time particle). Our only real deviation from the pure analog system was the use of the comparator. While the input to the comparator was analog, the output was decidedly digital; its output was either a logic high ($+V_{sat}$) or a logic low ($-V_{sat}$). You can think of the comparator's output as having very low resolution since only two states are possible. The comparator's output is still time-continuous in that a logic transition can occur at any time. This is in contrast to a pure digital system where transitions are time-discrete, which means that logic levels can only change at specific times, usually controlled by some form of master clock. A purely digital system then, is the antithesis of a pure analog system. An analog system is time-continuous and has infinite amplitude resolution. A digital system is time-discrete and has finite amplitude resolution (two states in our example).

As you have no doubt noticed in your parallel work, digital systems have certain advantages and benefits relative to analog systems. These advantages include noise immunity, storage capability, and available numeric processing power. It makes sense, then, that a combination of analog and digital systems could offer the best of both worlds. This chapter examines the processes of converting analog signals into a digital format, and turning digital words into an analog signal. A few representative examples of processing the signal in the digital domain are presented as well. Some examples which you might already be familiar with include the stereo compact disk (CD) and the digital-storage oscilloscope. We will break down this topic into two broad sections: analog-to-digital conversion (AD) and digital-to-analog conversion (DA). Since many AD systems require digital-to-analog convertors, we shall examine DA systems first.

The Advantages and Disadvantages of Working in the Digital Domain

Given enough time, an analog circuit can be designed and manufactured for virtually any application. Why then, would anyone desire to work in the digital domain? Perhaps the major reason is flexibility. Once signals are represented in a digital form, they can be manipulated by various means, including software programs. You have probably discovered that replicating a computer program is far easier than replicating an analog circuit. What's more, a program is much easier to update and customize than a hardware circuit. Because of this, it is possible to manipulate a signal in many different ways with the same digital/computer hardware; all that needs to be altered is the manipulation instructions (i.e., the program). The analog circuit, in contrast, needs to be rewired, extra components need to be added, or old portions removed. This can be far more costly and time-intensive than just updating software. By working in the digital domain, processing circuits do not exist per se; rather, a generic IC such as a CPU is used to create a "virtual circuit." With a certain amount of intelligence in the system design, the virtual circuit may be able to alter its own performance in order to precisely adapt to various signals. This all boils down to the fact that a digital scheme may offer much greater flexibility for involved tasks, and allows a streamlined, generic hardware solution for complex applications. Because of this attribute, the digital solution may be significantly less expensive than its analog counterpart.

When an analog signal is transferred to the digital domain, it is represented as a series of numbers (usually high/low binary logic levels). One nice property of this representation is that it is exactly repeatable. In other words, an infinite number of copies of the data can be generated, and no distortions or deviations from the original will appear. The last copy will be identical to the first. Compare this to a simple analog copy. For example, if you were to tape-record a song and then make a copy of the tape, the second-generation copy would suffer from increased noise and distortion. A copy of the second copy would produce even worse results. Every time the signal is copied, some corruption occurs. It is for this reason that early long-distance telephone calls were of such low quality. Modern communications systems employ digital techniques which allow much higher quality, even when one person is in New York and the other is in Australia, halfway around the planet.

Besides being a desirable mathematical attribute, repeatability also lends itself to the problem of long-term storage. A storage medium for a binary signal only needs to resolve two levels, whereas the analog medium needs to resolve very fine changes in signal strength. As you might guess, deterioration of the analog medium is a serious problem, and results in information loss. The digital medium can theoretically survive a much higher level of deterioration without information loss. In a computerized system, data can be stored in a variety of formats including RAM (random access memory) and magnetic tape or disk. For playback only (i.e., read only), data can be stored in ROM (read only memory) or laser-disk formats (such as video disk or audio CD).

As always, the benefits of the digital scheme arrive with specific disadvantages. First, for simpler applications, the cost of the digital approach is very high and cannot be justified. Second, the process of converting a signal between the analog and digital domains is an inexact one. Some information about the signal will be lost during the conversion. This is because the digital representation has finite resolution. This means that only signal changes larger than a certain minimum size (the resolution step size) are discernable, and therefore, some form of round-off error is inevitable. This characteristic helps determine the range of allowable signals, from the smallest detectable signal to the maximum signal before overload occurs. Third, analog systems are inherently faster than digital systems. Analog solutions can process input signals at much higher frequencies than digital schemes. Also, analog systems work in real-time, while digital systems generally do not. Digital systems can only perform in real-time if the input signal is a fairly low frequency, if the processing task is relatively simple, or if very specialized and expensive processing circuits are added. Not all applications require real-time performance, so this limitation is not always a problem.

12.2 The Sampling Theorem

There are many different ways in which an analog signal can be turned into a digital form. This process is referred to as *AD conversion*, or more simply, as *digitization*. We will examine only the most popular method, called *pulse code modulation*, or PCM for short. The reverse of this process, or turning the digital information back into analog form, is called *DA conversion*. The two most important characteristics in the conversion process are *sampling frequency* and *amplitude resolution*. Let's look at the basic idea behind PCM.

In essence, PCM measures and encodes the value of the input signal at specific points in time. Normally, the time spacing is constant, and several points are used over the length of one input cycle. In other words, the process involves taking representative samples of the input signal over time. This is shown graphically in Figure 12.1, and is referred to as *sampling*. The result of the sampling procedure is a list of times and corresponding amplitude values. A sequence may look something like this: at $T = 1$ mS, $v_{in} = 23$ mV; at $T = 2$ mS, $v_{in} = 45$ mV; at $T = 3$ mS, $v_{in} = -15$ mV; etc. If the time interval between samples is held constant (i.e., constant rate of sampling), then all we need to know is a starting time and the sampling rate in order to reconstruct the actual sample times.[1] This is much more efficient than recording each sample time. The resulting amplitude values can be manipulated in a variety of ways since they are now in a numeric form. The ultimate accuracy of this conversion will depend on two primary factors: how often we sample the signal, and the *accuracy* and *resolution* of the sample measurement. Theoretically, the conversion will never be 100% accurate, that is, once converted, a finite amount

---

[1] The terms *sample rate* and *sample frequency* are often used interchangeably, and are usually denoted by f_s.

Figure 12.1
Sampling input signals
over time

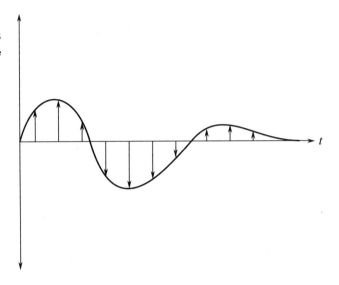

of information will be lost forever. Another way of stating this is that when the digital representation is converted back to analog, the result will not be identical to the original waveform. In practical terms, it is possible to reduce the error to such small values that it may be ignored in many applications. It is important then, that we investigate the implications of sample rate and accuracy/resolution on the quality of conversion.

12.3 Resolution and Sampling Rate

Perhaps the most obvious source of error is the finite measurement accuracy of the individual signal levels. The major problem here is one of resolution. Resolution represents the finest discernible change in the signal and is often specified in terms of a *number of bits*, although a voltage specification is also possible. Since the signal level is represented with a binary number, it follows that a large number of bits are needed in order to achieve fine resolution. For example, if an 8-bit word is used, there will be 256 distinct values available. If the maximum peak-to-peak value of the input signal is 1 V, it works out that each *step* in the word represents about 3.9 mV (1 V/256). Under these conditions, it would be impossible to perfectly encode a value of 14 mV. The nearest values available would be binary 11, which yields 11.7 mV (3 × 3.9 mV), and binary 100, which yields 15.6 mV (4 × 3.9 mV). Obviously, the resulting round-off creates some error in the digital representation. This error can be reduced by increasing resolution so that finer steps may be detected. If 16 bits are used for the same 1 V range, a total of 65,536 values are available, with each step working out to 15.26 μV. Now, while we still may not be able to exactly represent the 14 mV level, we are guaranteed to be within ± 7.63 μV, instead of ± 1.95 mV as in the 8-bit case. Since the round-off errors tend to be random in magnitude and polarity, this effect may be viewed as a noise source. In other words, lower resolutions (i.e., fewer bits) produce noisier signals. The number of bits required for a particular application can vary from fewer than six for high-speed video applications to more than 16 for high-quality audio or measurement purposes.

It is important to note that once the size of the digital word is chosen, and the peak amplitude fixed, the input signal must stay within specific bounds or gross distortion will occur. For example, if a peak input of 1 V produces the maximum numeric value, there is no way that the digital word can represent a level greater than 1 V. Likewise, if the step size is set at 1 mV, any signals less than 1 mV are lost. Also, low-level signals will suffer from reduced resolution. For best results, the signal peak should produce the maximum numeric value. If the peak is significantly less, the result is akin to using fewer bits in the representation. Finally, even if the resolution is adequate, the absolute accuracy of the conversion must be considered, as it is in any measurement device.

Example 12.1

A certain system uses a 12-bit word to represent the input signal. If the maximum peak-to-peak signal is set for 2 V, determine the resolution of the system and its dynamic range.

A *12-bit word* means that 2^{12}, or 4096, levels are possible. Since these levels are equally spaced across the 2 V range, each step is

$$\text{step size} = \frac{2\ \text{V}}{4096}$$

$$\text{step size} = 488\ \mu\text{V}$$

Therefore, the system can resolve changes as small as 488 μV.

Dynamic range represents the ratio of the largest value possible to the smallest.

$$\text{dynamic range} = \frac{2\ \text{V}}{488\ \mu\text{V}}$$

$$\text{dynamic range} = 4096$$

$$\text{dynamic range} = 20 \log_{10} 4096$$

$$\text{dynamic range} = 72\ \text{dB}$$

Note that the voltage range affects the step size, but does not affect the bit resolution. The actual number of discrete steps which can be resolved is set by the number of bits available. You may notice that each additional bit adds approximately 6 dB of range (a doubling of voltage). Consequently, the dynamic range calculation may be streamlined to

$$\text{dynamic range} \approx 6 \text{ dB} \times \text{number of bits} \qquad (12.1)$$

Example 12.2   Audio compact disks use a 16-bit representation of the music signal. Determine the dynamic range. Also, if the maximum output level is .775 V peak, determine the step size.

$$\text{dynamic range} \approx 6 \text{ dB} \times \text{number of bits}$$

$$\text{dynamic range} \approx 6 \text{ dB} \times 16$$

$$\text{dynamic range} \approx 96 \text{ dB}$$

For 16 bits, the total number of steps is 2^{16} or 65,536. Assuming that the signal is bipolar, the total signal range will be from $-.775$ V to $+.775$ V, or 1.55 V.

$$\text{step size} = \frac{1.55 \text{ V}}{65,536}$$

$$\text{step size} = 23.65 \ \mu\text{V}$$

At this point, we must consider the effect of sampling rate on the quality of the signal. It should be intuitively obvious that higher sampling rates afford greater overall conversion accuracy. Of course, there is a trade-off associated with high sampling rates, and that is the accompanying high data rate. In other words, greater resources will be required to store and process the larger volume of digital information. The real question is, just how fast does the sampling rate need to be for optimum efficiency? The Nyquist sampling theorem states that at least two samples are needed per cycle for proper signal conversion. If the signal is not a sinusoid, then at least two samples are required per cycle of the highest-frequency component. For example, if a range of signals up to 10 kHz needs to be digitized, then a sample rate of at least 20 kHz is required. Normally, a certain amount of "breathing room" is added to this figure. Another way of looking at this relationship is to state that the highest input-frequency component can be no more than one-half the sampling rate.

Since this is such an important parameter, the value of one-half the sampling rate is given the name *Nyquist frequency.*

$$\text{Nyquist frequency} = \frac{f_s}{2} \tag{12.2}$$

If an input frequency component is greater than the Nyquist frequency, a unique form of distortion called *alias distortion* is produced. The resulting distortion product, called an *alias*, is a new signal at a frequency which is equal to the difference between the input and Nyquist frequencies. Normally, this new signal is not harmonically related to the input signal, and thus is easily detected. The aliasing effect is shown graphically in Figure 12.2. Here we see a sampling rate that is only about 1.5 times the input frequency, rather than the required factor of 2 times minimum. In Figure 12.3 the sample points are redrawn and connected as simply as possible. Note that the resulting outline is that of a lower-frequency wave. What we notice here is that the data points produced in Figure 12.2 are identical to the points produced by a lower-frequency input wave. When these data points are converted back to analog form, the DA converter will produce this lower frequency wave. Oddly enough, the original waveform has completely disappeared; hence the term *alias*. Any signal component which is greater than the Nyquist frequency will produce aliases.

When considering the possibility of alias distortion, it is worth repeating that the components of the signal must be investigated, not just the base frequency. For example, a 1 kHz square wave has a 1 kHz fundamental, and an infinite series of odd-numbered harmonics (3 kHz, 5 kHz, 7 kHz, 9 kHz,

Figure 12.2
The aliasing effect
(sampling rate too low)

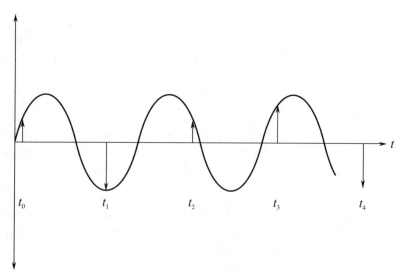

Figure 12.3
Alias production

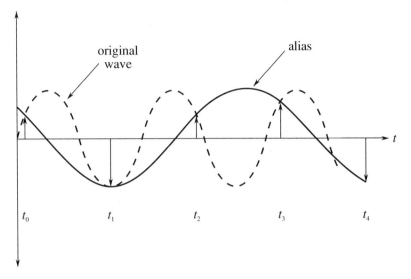

Figure 12.3
Alias production

etc.). If a 12 kHz sampling rate is used to digitize this signal, a fair amount of alias distortion will be seen. In this example, the Nyquist rate is 6 kHz. All of the harmonics above 6 kHz will produce an alias. In order to prevent this, the input signal must be frequency-band limited. That is, a low-pass filter must be used to attenuate all components above the Nyquist frequency before AD conversion takes place. The amount of filtering required depends on the resolution of the conversion and the relative strength of the above-band signals. Since filters cannot roll off infinitely fast, as noted in Chapter 11, sampling rates are normally set more than twice as high as the highest needed input component. In this way, the Nyquist frequency will be somewhat greater than the maximum desired input frequency. The low-pass filter (often referred to as an *anti-alias filter*) will use this frequency range as its transition band. Even though the attenuation in the transition band is less than optimum, alias distortion will not be a problem. This is shown graphically in Figure 12.4 (page 568). As you can see, high-rolloff-rate filters are desirable in order to attenuate the out-of-band signals as quickly as possible. Very fast filter rolloff rates mean that the sampling frequency need be as little as only 10% greater than the theoretical minimum in order to maintain sufficient alias rejection.

Example 12.3   Suppose that you need to digitize telephone signals. Assuming that you would like to maintain a dynamic range of at least 50 dB, with an upper frequency limit of 3 kHz, determine the minimum acceptable sampling rate and number of bits required.

The minimum Nyquist rate is equal to the highest desired input frequency. In this case, that's 3 kHz. Since the sampling rate is twice the Nyquist frequency, the sampling rate must be at least 6 kHz. In reality, if input components above

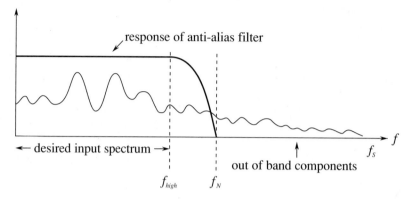

a. Spectrum before filtering

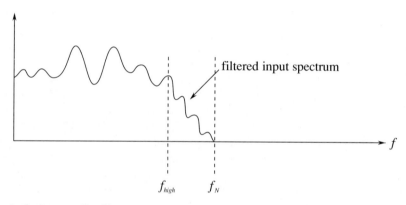

b. Spectrum after filtering

3 kHz exist, an anti-alias filter will be needed, and the sampling rate will have to be increased somewhat.

Since dynamic range is set by the number of bits used, we find that

$$\text{bits required} = \frac{\text{dynamic range}}{6 \text{ dB}}$$

$$\text{bits required} = \frac{50 \text{ dB}}{6 \text{ dB}}$$

$$\text{bits required} = 9$$

Since we cannot have a fractional bit, the value is rounded up. The final system specification is a minimum rate of 6 kHz with a 9-bit resolution. Note that this represents a data rate of 9 bits per sample × 6000 samples per second, or 54,000 bits per second (6750 bytes per second).

12.4 Digital-to-Analog Conversion Techniques

The basic digital-to-analog converter is little more than a weighted summing amplifier. Each successive bit in the digital word represents a level that is twice as large as the preceding bit. If each bit is taken as a given current or voltage, the increasing levels can be produced by using different gains in the summing inputs. A simple 4-bit converter is shown in Figure 12.5. This system can represent 2^4, or 16, different levels. Each input is driven by a simple high/low logic level which represents a 1 or 0 for that particular bit. Note that the input resistors vary by factors of 2. The gain for the uppermost path is R_f/R_f, or unity. This input is used for the **m**ost significant **b**it (MSB) of the input word. The next input shows a gain of $R_f/2R_f$, or .5. The third input shows a gain of .25, while the final input shows a gain of .125. The final input has the lowest gain and is used for the **l**east significant **b**it (LSB) of the input word. If the input word had a higher resolution (i.e., more bits), extra channels would be added, each having half the gain of the preceding input. To better understand the conversion process, let's take a look at a few representative inputs and outputs.

The circuit of Figure 12.5 can be driven by simple 5 V TTL-type logic circuits. 5 V represents a logic high while 0 V represents a logic low. What is the output level if the input word is 0100? Since a logic high represents 5 V, 5 V is being applied to the second input. All other inputs receive a logic low, or 0 V. The output is the summation of the input signals. (Remember, this is an inverting summer, so the final output should have its sign reversed.)

$$v_{out} = -(v_{in1}A_1 + v_{in2}A_2 + v_{in3}A_3 + v_{in4}A_4)$$

$$v_{out} = -(0 \times 1 + 5 \text{ V} \times .5 + 0 \times .25 + 0 \times .125)$$

$$v_{out} = -2.5 \text{ V}$$

Figure 12.5
A simple 4-bit converter

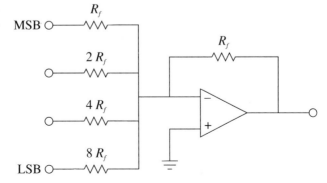

Figure 12.6
Output with four digital words

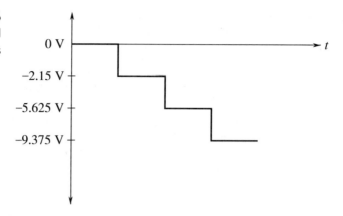

So, a value of 4 (binary 100) is equivalent to a potential of 2.5 V. If we increase the word value to 9 (binary 1001), we see

$$v_{out} = -(v_{in1}A_1 + v_{in2}A_2 + v_{in3}A_3 + v_{in4}A_4)$$

$$v_{out} = -(5 \text{ V} \times 1 + 0 \times .5 + 0 \times .25 + 5 \text{ V} \times .125)$$

$$v_{out} = -5.625 \text{ V}$$

The minimum output occurs at binary 0000 (0 V) and the maximum at binary 1111 (-9.375 V). The step size is equal to the logic level times the minimum gain: in this case that's .625 V. Notice that the output value can be found by simply multiplying the value of the input word by the minimum step size. Also, it is important to note that the output signal is unipolar (in this example, always negative).

A digital representation, of course, is made up of a sequence of words, not just one word. In reality, the logic circuits are constantly feeding the summing amplifier new words at a predetermined rate. Because of the changing inputs, the output of the converter is constantly changing as well. Using our previously calculated values, if the converter is fed the sequence 0000, 0100, 1001, 1111, the output will move from 0 V to -2.5 V, to -5.625 V, to a final value of -9.375 V. This output is graphed in Figure 12.6. If this sequence is repeated over and over, the waveform of Figure 12.7 is the result. Note that a "stair-step" type wave is created. You might also think of this as a very rough form of a ramp function. A better ramp would be produced if we used all of the available values for the input sequence, as in 0000, 0001, 0010, 0011, ..., 1111. In order to remove the negative DC offset and make the signal bipolar, all we need to do is pass the signal through a coupling capacitor. The frequency of this waveform is controlled by the rate at which the words are fed to the converter. Note that by increasing the resolution and the number of words fed to the converter per cycle, a very close approximation to the ideal ramp

Figure 12.7
Cycled output

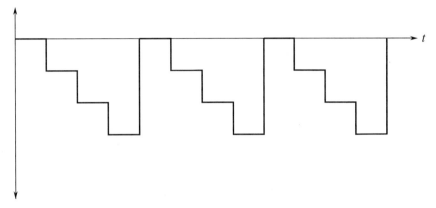

function can be achieved. For that matter, by changing the input words to other sequences, we can create a wide variety of output waveshapes. This is the concept behind the digital arbitrary function generator. An arbitrary function generator allows you to create wave shapes beyond the simple sine/square/triangle found on the typical laboratory function generator. We'll take a closer look at this particular piece of test equipment a little later.

In order to increase resolution, it appears that all you need to do to the summing amplifier is add extra channels with larger and larger resistors. Unfortunately, the resistor sizes soon become impractical, and another approach is required. For example, a 16-bit system would require that the LSB resistor be equal to $65,536R_f$. One problem is that the resulting small input current may be dwarfed by input bias and offset currents. Also, high component accuracy is needed for the more significant inputs in terms of the input resistors and the drive signals. The excessively large resistors may also contribute added noise. The standard solution to this problem involves the use of an $R/2R$ resistive divider network.

An $R/2R$ network is shown in Figure 12.8. This circuit exhibits the unique attribute of constant division by 2 for each stage. You may think of this as either a division of voltage at each successive node, or a division of current in each successive leg. An example of a 4-stage (i.e., 4-bit) network is shown

Figure 12.8
$R/2R$ ladder network

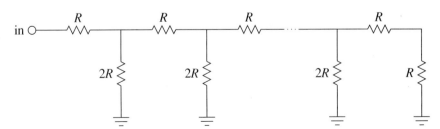

Figure 12.9
A 4-stage ladder

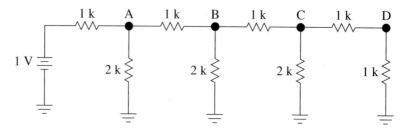

in Figure 12.9. In order to find the voltage at any given node, the loading effects of the following stages must be taken into account. This is much easier to do than it first appears. If we need to find the voltage at point A, we must first find the resistance in parallel with the initial 2 k resistor. Since a quick inspection shows that each stage is loaded by the following stages, it is easiest if we start at the last stage and work toward the input. The effective resistance to the right of node C is 1 k in series with 1 k, or 2 k. This resistance is placed in parallel with the 2 k resistor seen from node C to ground. The result is 1 k. In other words, from C to ground, we see 1 k. This creates a 2:1 voltage divider with the 1 k resistor placed from B to C, so the voltage at C must be half the voltage at B. This also points out the fact that the current entering node C splits into two equal portions: one which travels towards point D, and the other which travels through the 2 k resistor to ground. This is shown graphically in Figure 12.10. If you look at the equivalent circuit section of Figure 12.10, you will notice that this portion now looks exactly like the final portion

Figure 12.10
Ladder analysis

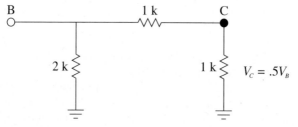

a. Equivalent circuit

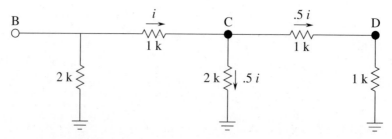

b. Current division

of the original network. That is, every time a section is simplified and analyzed, the result will be a halving of voltage and current. It is already apparent
that the voltage at D must be half the voltage at C, which in turn, must be
half the voltage at B. As you can now prove, it follows that the voltage at B
must be half the voltage at A. In a similar fashion, the current passing through
each $2R$ leg is half the preceding current. (For current division, the final section
is not used to derive a current since it will be equal to the value in the preceding
stage.) The halving of current is just what is needed for the binary representation
of the digital input word.

Adapting the $R/2R$ network to the DA converter is relatively easy. The
network is fed from a stable current source, with each $2R$ element feeding into
a summing amplifier. In series with each $2R$ element is a solid-state switch,
which sets the appropriate logic level. This is shown in Figure 12.11, with the
network effectively on its side. When a logic high is presented to a given bit,
the switch is closed and current flows through the $2R$ element and into the
op amp. Note that the right end of the resistor is effectively at ground since
the summing node of the op amp is a virtual ground. If a logic low is presented,
the switch shunts the current to ground, bypassing the op amp. In this
way, the appropriately weighted currents are summed, and used to produce
the output voltage.

This technique offers several advantages over the simpler weighted gain
version. First, all branches are fed by one common current source. Because of

Figure 12.11
Converter with $R/2R$
ladder

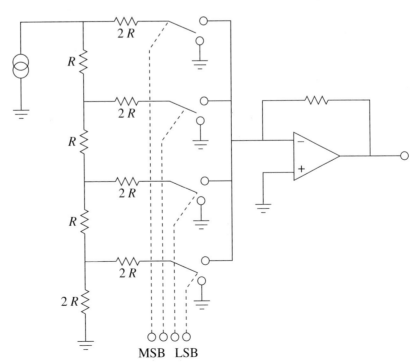

MSB LSB

this, there is no need for output level matching. Second, only two different values of resistors are required for any number of bits used, rather than the impractically wide range seen earlier. It is more economical to control the tolerance of just two different parts than 12 or 16. Note that small input currents are still generated for the least significant bits, so attention to input bias and offset currents remains important.

Practical Digital-to-Analog Converter Limits

Perhaps the most obvious limit associated with the DA converter is its speed. The op amp used in the DAC must be much faster than the final signals it is meant to produce. A given output waveform may contain several dozen individual sample points per cycle. The op amp must respond to each sample point. Consequently, wide bandwidth and high slew rates are required.

Integrated DAC spec sheets offer a few important parameters of which you should be aware. First of all, there is *conversion speed*. This figure tells how long it takes the DAC to turn the digital input word into a stable analog output voltage. This sets the maximum data rate. Next come the *accuracy* and *resolution*. Resolution indicates the number of discrete steps which can be produced at the output, and is set by the number of bits available. This is not the same as accuracy.

Accuracy is actually comprised of several different factors including *offset error*, *gain error*, and *nonlinearity*. Offset error is normally measured by applying the all-zero input word, and then measuring the output signal. Ideally, this signal will be zero volts. The deviation from zero is taken as the offset error. This has the effect of making all output levels inaccurate by a constant voltage. Offset error is relatively easy to compensate for in many applications by applying an equal offset of opposite polarity. Gain error is a deviation which affects each output level by a constant percentage. It is as if the signal were passed through a small amplifier or attenuator. This error can be compensated for by using an amplifier with a gain equal to the reciprocal of the error. The two gains will effectively cancel. The effect of offset and gain error are shown in Figures 12.12 and 12.13.

Nonlinearity errors can be broken into two forms, *integral nonlinearity* and *differential nonlinearity*. Integral nonlinearity details the maximum offset between the ideal outputs and the actual outputs for all possible inputs. Differential nonlinearity details the maximum output deviation relative to one LSB caused by two adjacent input words. If differential nonlinearity is beyond ± 1 LSB, the system may be nonmonotonic. In other words, a higher digital input word may actually produce a lower analog output voltage. These two forms of error are shown in Figure 12.14 (page 576). Note that it is possible to have high integral nonlinearity and yet still have modest differential nonlinearity, as is the case in Figure 12.14b.

As you can see, accuracy is dependent on rather complex factors. In an effort to boil this down to a single number, some manufacturers give an *effective*

Figure 12.12
Offset error

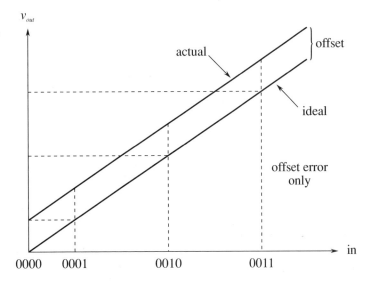

Figure 12.13
Gain error only

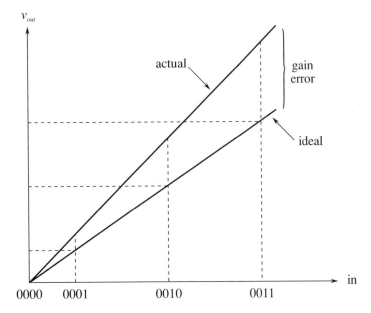

number of bits specification. For example, a 16-bit DAC may be specified as having 14-bit accuracy. This means that the 14 most significant bits behave in the idealized fashion, while the lowest 2 bits may be swamped out by linearity errors. Another spec which you will sometimes see is *no missing codes*. This means that for every increase in the input word, there will be an appropriate positive output level change.

In practice, the standard DA converter is used with an output filter. As you can see from the previous figures, the waveforms produced by the DAC contain

Figure 12.14
Linearity error

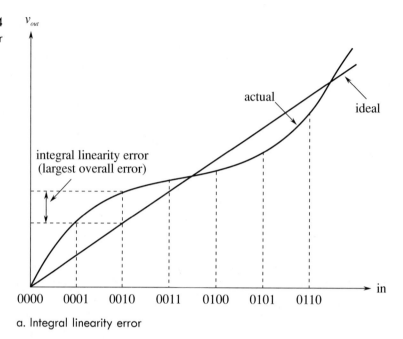

a. Integral linearity error

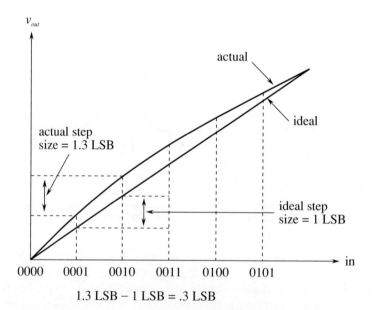

$$1.3 \text{ LSB} - 1 \text{ LSB} = .3 \text{ LSB}$$

b. Differential linearity error (relative-adjacent error)

a stair-step side effect. Generally, this is not desirable. The abrupt changes in output level indicate that higher frequency components are present. All components above the Nyquist rate should be filtered out with an appropriate low-pass filter. This filter is sometimes referred to as a *reconstruction* or *smoothing* filter. In an improperly designed system, the reconstruction filter will remove some of the highest in-band frequency components (i.e., components immediately below the Nyquist frequency). To compensate for this, logic levels are often latched to the DAC for shortened periods, thus creating a more spiked appearance, rather than the stair-step form. This effect is shown in Figure 12.15. While this spiked waveform appears to be less desirable than the stair-step form, it creates higher levels for the uppermost components, and after filtering, the result is a smoother overall frequency response.

Figure 12.15
Output reconstruction

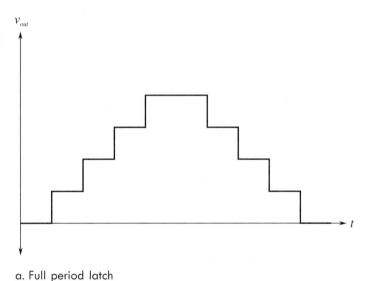

a. Full period latch

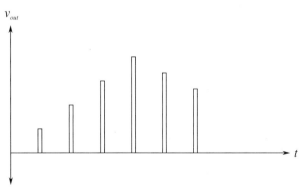

b. Partial period latch

Figure 12.16
Oversampled output

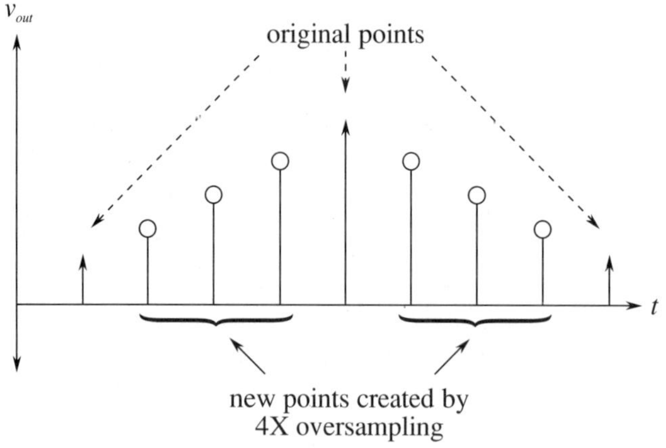

new points created by
4X oversampling

To further increase the quality of the output waveform a technique known as *oversampling* is sometimes employed. The basic idea is to create new sample points in between the existing ones. The result is a much denser data rate, which we hope will yield more exacting results after filtering. Also, the higher data rate may loosen the requirements of the reconstruction filter. A typical system might use 4 times oversampling, meaning that the output data rate is 4 times the original. Therefore, for each input word, three new words have to be added. This effect is shown in Figure 12.16. There are a number of ways to create the new sample points. The most obvious way is via simple interpolation, but this does not achieve the best results. Another technique involves initializing the new values to zero and then passing the data stream through a digital low-pass filter, which effectively calculates the proper values. This approach is fairly advanced and will not be pursued further here. Suffice to say that oversampling can increase the quality of the output signal and is widely used in applications such as audio compact disc players.

Digital-to-Analog Converter Integrated Circuits

There are many possible applications for digital-to-analog converters, and a number of different chips have evolved to meet specific needs. Generally, you can group these into specific classes, such as high-speed, high-resolution, or low-cost. We shall examine three representative types. The devices we will look at are the DAC88 (an 8-bit *compressed* unit), the DAC7545 (a microprocessor-compatible 12-bit unit), and the PCM56 (a 16-bit high-quality converter used in the audio industry).

The DAC88. This IC is used primarily in telecommunications, and is shown in Figure 12.17. While it has only 8-bit resolution, a unique *compressed* coding

Figure 12.17   The DAC88

DAC-88
COMDAC® COMPANDING
D/A CONVERTER (μ-255 LAW)

Precision Monolithic Inc.

FEATURES

- **IMPROVED ACCURACY over DAC-86**
- **IMPROVED SPEED over DAC-86**
- **Conforms With Bell System μ-255 Companding Law**
- **Meets D3 Compandor Tracking Specifications**
- **Both Encode and Decode Capability**
- **Tight Full-Scale Tolerance Eliminates Calibration**
- **Low Full-Scale Drift Over Temperature**
- **Extremely Low Noise Contribution**
- **Multiplying Reference Inputs**
- **Simplifies PCM System Design**
- **High Reliability**
- **Low Power Consumption and Low Cost**
- **Fully Specified Dice Available**

GENERAL DESCRIPTION

The DAC-88 monolithic COMDAC® D/A Converter provides a 15 segment linear approximation to the Bell System μ-255 companding law. The law is implemented by using three bits to select one of eight binarily-related chords (or segments) and four bits to select one of sixteen linearly-related steps within each chord. A sign bit determines signal polarity, and an encode/decode input determines the mode of operation.

Accuracy is assured by specifying chord end point values, step nonlinearity, and monotonicity over the full operating temperature range. Typical applications include PCM carrier systems, digital PBX's, intercom systems, and PCM recording. For CCITT "A" Law models, refer to the DAC-89 data sheet.

PIN CONNECTIONS & ORDERING INFORMATION

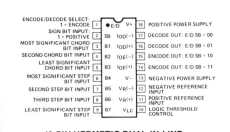

18-PIN HERMETIC DUAL-IN-LINE
(X-Suffix)

Grade	Temp. Range	Accuracy
DAC-88EX†	−25°C/+85°C	±1/4 Step

† Burn-in is available on commercial and industrial temperature range parts in CerDIP, plastic DIP, and TO-can packages. For ordering information, see 1990/91 Data Book, Section 2.

GAIN TRACKING

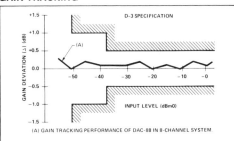

(A) GAIN TRACKING PERFORMANCE OF DAC-88 IN 8-CHANNEL SYSTEM

BELL μ-255 LAW TRANSFER CHARACTERISTIC

The DAC-88 transfer characteristic is a piecewise linear approximation to the Bell System μ255 law expressed by:

$$Y(\chi) = \text{sgn}(\chi) \frac{\ln(1 + \mu \,|\chi|)}{\ln(1 + \mu)} \quad -1 \le \chi \le 1$$

for a normalized coding range of ±1
where: χ = input signal level
   Y = output compressed signal level
   μ = 255

This law is implemented with an eight chord (or segment) piecewise linear approximation with 16 linear steps in each chord. Dynamic range of 72dB in both polarities is achieved with eight-bit coding.

EQUIVALENT CIRCUIT

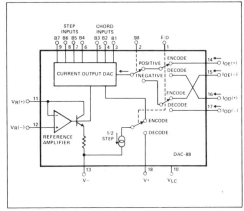

Courtesy of Analog Devices, Inc., Precision Monolithics Division

Figure 12.18   COMDAC transfer characteristic

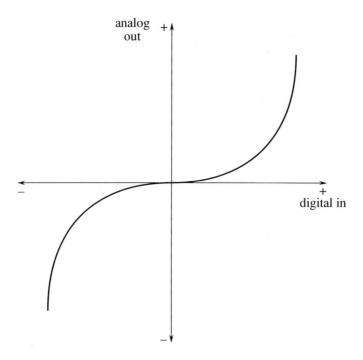

format allows it to have a much wider dynamic range than might be expected. By adopting this technique, higher apparent sound quality is available than with the standard 8-bit linear coding. Indeed, a theoretical dynamic range of 72 dB is possible, which is equivalent to a standard 12-bit linear converter. The downside is that unlike a linear converter, the noise floor varies with the signal level. Instead of having a straight line input/output characteristic, the response of the DAC88 is noticeably altered, as seen in Figure 12.18. As you can see, the output step size increases as the digital input word increases. A closer inspection of the transfer characteristic reveals that the curve is actually made up of a number of straight line segments, called chords (usually 13 or 15). Within each chord the step size is the same. Adjacent chords show a step size increase of 2 times. This is shown in Figure 12.19. Note that each chord contains 16 steps, so this requires 4 bits. The unit will have 8 positive and 8 negative chords, for 16 total chords, also requiring 4 bits. Therefore, the resulting word needs 8 bits.

By keeping the step size small near the origin, low-amplitude signals can be represented with increased accuracy. Since high-amplitude signals utilize

Figure 12.19   COMDAC zoom-in

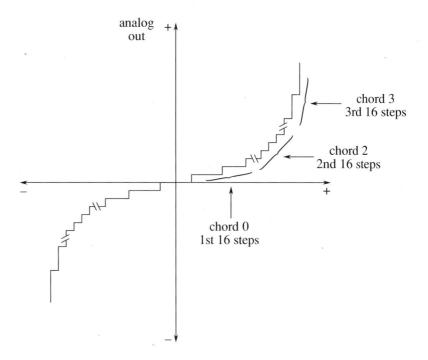

the larger step sizes, signals which would be beyond the range of a linear 8-bit system can be handled. This does mean though that a somewhat coarser representation results at the higher levels. This translates as an increase in noise. If you were to listen to a conversation coded using this scheme and you could remove the words, the result would be a noise signal which increased and decreased in step with the original conversation. As odd as this may seem, the effect is quite useful since the human ear tends ignore the variation, and zeroes in on the lack of noise between the words. This psycho-acoustic phenomenon is called *noise masking*, and is responsible for the subjective improvement in sound quality.

The DAC7545.   The DAC7545 is a fairly standard 12-bit linear converter, and is shown in Figure 12.20 (page 582). Its interesting aspects are that it is a multiplying converter and that it is microprocessor-compatible. The multiplying effect comes from the fact that a reference is used to drive the $R/2R$ ladder network. If the reference is changed, the output is effectively rescaled. Consequently, you can think of the output signal as equal to the reference value times the digital input word. You may also think of this as a form of "digital volume control."

Figure 12.20   The DAC7545

DAC7545

CMOS 12-Bit Multiplying
DIGITAL-TO-ANALOG CONVERTER
Microprocessor Compatible

FEATURES

- FOUR-QUADRANT MULTIPLICATION
- LOW GAIN TC: 2PPM/°C typ
- MONOTONICITY GUARANTEED OVER TEMPERATURE
- SINGLE 5V TO 15V SUPPLY

- TTL/CMOS LOGIC COMPATIBLE
- LOW OUTPUT LEAKAGE: 10nA max
- LOW OUTPUT CAPACITANCE: 70pF max
- DIRECT REPLACEMENT FOR AD7545, PM-7545

DESCRIPTION

The DAC7545 is a low-cost CMOS, 12-bit four-quadrant multiplying, digital-to-analog converter with input data latches. The input data is loaded into the DAC as a 12-bit data word. The data flows through to the DAC when both the chip select ($\overline{CS}$) and the write ($\overline{WR}$) pins are at a logic low.

Laser-trimmed thin-film resistors and excellent CMOS voltage switches provide true 12-bit integral and differential linearity. The device operates on a

single +5V to +15V supply and is available in 20-pin side-brazed DIP, 20-pin plastic DIP or a 20-lead plastic SOIC package. Devices are specified over the commercial, industrial, and military temperature ranges and are available with additional reliability screening.

The DAC7545 is well suited for battery or other low power applications because the power dissipation is less than 0.5mW when used with CMOS logic inputs and $V_{DD} = +5V$.

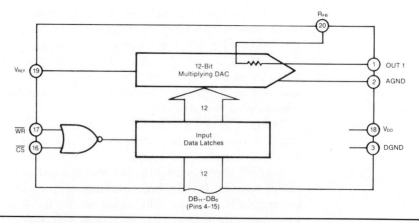

Figure 12.21
Microprocessor to
DAC7545 interface

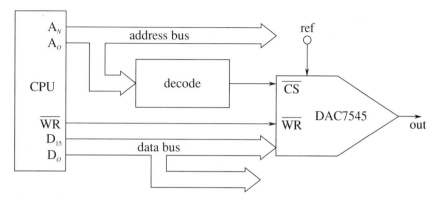

With the inclusion of a few extra logic lines, the IC has become micropro-cessor-compatible. This means that the DAC7545 has chip select and read/write lines along with the 12 data input lines. This allows the converter to be connected directly to the microprocessor data bus. By using memory mapped I/O, the microprocessor can write data to the converter just as it writes data to memory. While a 16-bit microprocessor system can present the converter with all of the data it needs during one write cycle, an 8-bit microprocessor will need two write cycles and some form of latch. One address can be used for the lower 8 bits, and another address for the remaining 4 bits. A simplified system is shown in Figure 12.21 using a 16-bit microprocessor.

The PCM56. The PCM56 is a linear 16-bit converter designed specifically for digital audio applications. It comes in a standard 16-pin DIP package and is relatively low in cost considering its performance. It is shown in Figure 12.22 (pages 584–585). Unlike the other converters, the PCM56 features serial input of data, not parallel. It includes its own on-board serial-to-parallel shift register/conversion circuitry and latch. This technique helps to reduce system cost. It is also surprisingly convenient since many specialized digital-signal-processing ICs which might be used with the PCM56 utilize a serial-type output. This may be fed directly into the PCM56. In order to synchronize the data stream, a bit clock and a latch-enable signal are used. The bit clock runs at 16 times the data rate (one pulse for each data bit). The latch enable is used to signify the end of one data word. The timing diagram is shown in Figure 12.23 on page 586.

The PCM56 offers settling times of only 1.5 μS and a total harmonic dis-tortion of .0025% maximum for a full-scale input, and .02% for an input 20 dB below full scale. One PCM56 is fast enough to handle two channels of 4-times-oversampled data for an audio compact disk (CD) playback system. Due to its high resolution and 96 dB dynamic range, extra care must be taken during circuit layout to avoid hum pickup and RF interference.

Figure 12.22   The PCM56

PCM56P

DESIGNED FOR AUDIO

Serial Input 16-Bit Monolithic
DIGITAL-TO-ANALOG CONVERTER

FEATURES

- SERIAL INPUT
- LOW COST
- NO EXTERNAL COMPONENTS REQUIRED
- 16-BIT RESOLUTION
- 15-BIT MONOTONICITY, TYP
- 0.001% OF FSR TYP DIFFERENTIAL LINEARITY ERROR
- 0.0025% MAX THD (FS Input, K Grade, 16 Bits)
- 0.02% MAX THD (−20dB Input, K Grade, 16 Bits)
- 1.5μs SETTLING TIME, TYP (Voltage Out)
- 96dB DYNAMIC RANGE
- ±3V or ±1mA AUDIO OUTPUT
- EIAJ STC-007-COMPATIBLE
- OPERATES ON ±5V to ±12V SUPPLIES
- PINOUT ALLOWS I$_{OUT}$ OPTION
- PLASTIC DIP PACKAGE

DESCRIPTION

The PCM56P is a state-of-the-art, fully monotonic, digital-to-analog converter that is designed and specified for digital audio applications. This device employs ultra-stable nichrome (NiCr) thin-film resistors to provide monotonicity, low distortion, and low differential linearity error (especially around bipolar zero) over long periods of time and over the full operating temperature.

This converter is completely self-contained with a stable, low noise, internal zener voltage reference; high speed current switches; a resistor ladder network; and a fast settling, low noise output operational amplifier all on a single monolithic chip. The converters are operated using two power supplies that can range from ±5V to ±12V. Power dissipation with ±5V supplies is typically less than 200mW. Also included is a provision for external adjustment of the MSB error (differential linearity error at bipolar zero) to further improve total harmonic distortion (THD) specifications if desired. Few external components are necessary for operation, and all critical specifications are 100% tested. This helps assure the user of high system reliability and outstanding overall system performance.

The PCM56P is packaged in a high-quality 16-pin molded plastic DIP package and has passed operating life tests under simultaneous high-pressure, high-temperature, and high-humidity conditions.

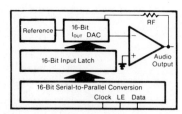

a. Block diagram and features

Figure 12.22   (Continued)

ORDERING INFORMATION

Model	THD at FS (%)
PCM56P	0.008 Max
PCM56P-J	0.004
PCM56P-K	0.0025

PIN ASSIGNMENTS

1	$-V_S$	Analog Negative Supply
2	LOG COM	Logic Common
3	$+V_L$	Logic Positive Supply
4	NC	No Connection
5	CLK	Clock Input
6	LE	Latch Enable Input
7	DATA	Serial Data Input
8	$-V_L$	Logic Negative Supply
9	V_{OUT}	Voltage Output
10	RF	Feedback Resistor
11	SJ	Summing Junction
12	ANA COM	Analog Common
13	I_{OUT}	Current Output
14	MSB ADJ	MSB Adjustment Terminal
15	TRIM	MSB Trim-pot Terminal
16	$+V_S$	Analog Positive Supply

ABSOLUTE MAXIMUM RATINGS

DC Supply Voltages	±16VDC
Input Logic Voltage	$-1V$ to $+V_S/+V_L$
Power Dissipation	850mW
Operating Temperature	$-25°C$ to $+70°C$
Storage Temperature	$-60°C$ to $+100°C$
Lead Temperature During Soldering	10s at 300°C

DISCUSSION OF SPECIFICATIONS

The PCM56P is specified to provide critical performance criteria for a wide variety of applications. The most critical specifications for a D/A converter in audio applications are Total Harmonic Distortion, Differential Linearity Error, Bipolar Zero Error, parameter shifts with time and temperature, and settling time effects on accuracy.

The PCM56P is factory-trimmed and tested for all critical key specifications.

The accuracy of a D/A converter is described by the transfer function shown in Figure 1. Digital input to analog output relationship is shown in Table I. The errors in the D/A converter are combinations of analog errors due to the linear circuitry, matching and tracking properties of the ladder and scaling networks, power supply rejection, and reference errors. In summary, these errors consist of initial errors including Gain, Offset, Linearity, Differential Linearity, and Power Supply Sensitivity. Gain drift over temperature rotates the line (Figure 1) about the bipolar zero point and Offset drift shifts the line left or right over the operating temperature range. Most of the Offset and Gain drift with temperature

CONNECTION DIAGRAM

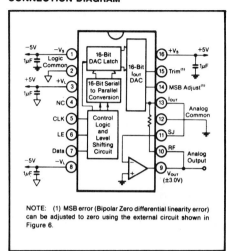

NOTE: (1) MSB error (Bipolar Zero differential linearity error) can be adjusted to zero using the external circuit shown in Figure 6.

or time is due to the drift of the internal reference zener diode. The converter is designed so that these drifts are in opposite directions. This way the Bipolar Zero voltage is virtually unaffected by variations in the reference voltage.

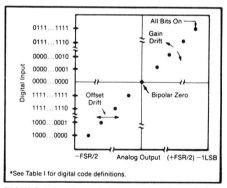

*See Table I for digital code definitions.

FIGURE 1. Input vs Output for an Ideal Bipolar D/A Converter.

TABLE I. Digital Input to Analog Output Relationship.

Digital Input	Analog Output		
Binary Twos Complement (BTC)	DAC Output	Voltage (V), V_{OUT} Mode	Current (mA), I_{OUT} Mode
7FFF Hex	+ Full Scale	+2.999908	−0.999970
8000 Hex	− Full Scale	−3.000000	+1.000000
0000 Hex	Bipolar Zero	0.000000	0.000000
FFFF Hex	Zero − 1LSB	−0.000092	+0.030500µA

b. Connection diagram and discussion

Figure 12.23
PCM56 timing

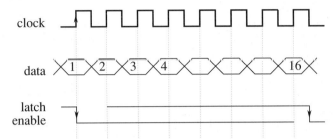

Applications of Digital-to-Analog Converter Integrated Circuits

Example 12.4   Perhaps the first thing many people think of when they hear the terms "digital" or "digitized" is the audio CD. Home CD players are excellent examples of the use of precision DA circuitry in our everyday lives. Music data is stored on the CD with 16-bit resolution and a sampling rate of 44.1 kHz. This produces a Nyquist frequency of 22.05 kHz, which is high enough to encompass the hearing range of most humans. Error correction and auxiliary data is stored with the music data on the disk in the form of very tiny pits, which are read by a laser. The signal is then converted into the common electronic logic form where it is checked for error and adjusted as need be. The data stream is then fed to the DA converter for audio reconstruction. A single converter may be multiplexed between the two stereo channels, or two dedicated converters may be used. Oversampling in the range of 2X to 8X is often used for improved signal quality. A block diagram of the system is shown in Figure 12.24. The actual DAC portion seems almost trivial when compared to some of the more sophisticated elements.

The storage density of the optical CD is quite remarkable. This small disk (less than 5 inches in diameter) can hold 70 minutes of music. Ignoring the auxiliary data, we can quickly calculate the total storage. We have two channels of 16-bit data, or 32 bits (4 bytes) per sample point. There are 44,100 samples per second for 70 minutes, yielding 185.22 megasamples. The total data storage is 5.927 gigabits, or 741 megabytes.

Figure 12.24
Audio compact disk
playback system

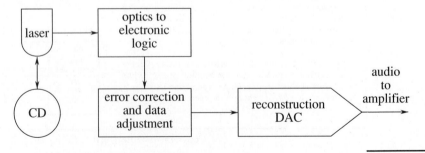

—————— **Example 12.5** As we have already mentioned, it is possible to connect DACs directly to microprocessor systems. Furthermore, the microprocessor can write to the DAC with no more effort than writing to a memory location. The microprocessor can write any series of data words we desire out to the DAC, and can repeat a sequence virtually forever. Given this ability, we can make an arbitrary waveform generator. Instead of being locked into a set of predefined wave shapes as on ordinary function generators, this system allows for all manner of wave shapes. The accuracy and flexibility of the system will depend on its speed and the available DAC resolution.

The basic idea is one of table look-up. For example, let's say that we have a 16-bit system. We will create a table of data values for one cycle of the output waveform. For the sake of convenience, we might make the table size a handy power of 2, such as 256. In other words, a single output cycle will be chopped into 256 discrete time chunks. It is obvious then that the converter must be a few hundred times faster than the highest fundamental which we wish to produce. By increasing or decreasing the output data rate, we can change the frequency of the output fundamental. This is known as a *variable sample rate* technique. It is also possible to change the fundamental frequency with a *fixed rate* technique. (This is somewhat more complex, but does offer certain advantages.) An output flow chart is shown in Figure 12.25a (page 588). Upon initialization, an address pointer is set to the starting address of the data table. The CPU reads the data from the table via the pointer. The pointer is incremented so that it now points to the next element in the table. (Some CPUs such as the Motorola 68000 series offer a post-increment addressing mode so that both steps can be performed in a single instruction.) Next the CPU writes the data to the special DAC address. At this point, some form of software/hardware delay is invoked which sets the output data rate. After the delay, the CPU reads the next data element via the pointer and continues as in the first run. Once the 256th element is produced, the pointer is reset to the start of the table and the process continues on. In this way, the table can be thought of as circular or never-ending. If the system software is written in a higher-level language, the pointer/data table can be implemented as a simple array where the array index is set by a counter. This will not be as efficient as a direct assembly level approach, though.

The real beauty of this system is that the data table can contain virtually any sequence of data. The data could represent a sine, pulse, triangle, or other standard function. More important, the data could just as easily represent a sine wave with an embedded noise transient or a signal containing a hum component. This data can come from three basic sources. First, the data table can be filled through direct computation if the time domain equation of the desired function is known. Second, the data can be manufactured by the user through some form of interaction with a computer, perhaps with a mouse or drawing pad. Finally, the data can be derived from a real-world signal. That is, an analog-to-digital converter can be used to record the signal in digital

Figure 12.25 Arbitrary waveform generator

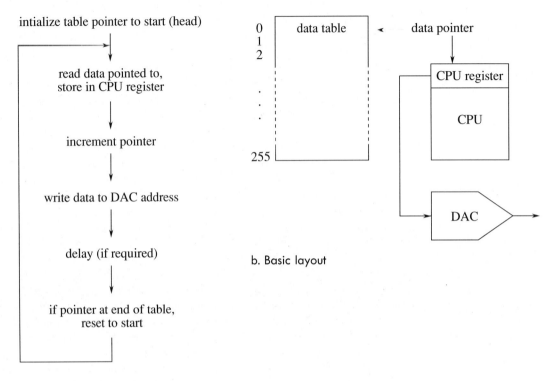

a. Output flowchart

form. The data then can be loaded into the table and played back repeatedly. The arbitrary waveform generator allows its user to test circuits and systems with a range of wave shapes which would be impossible or impractical to generate otherwise.

Example 12.6 Under computer control, DA converters can be used as part of an automated test equipment system. In order to fully characterize an electronic product, a number of individual tests need to be run. Setting up each individual test can be somewhat time-consuming, and is subject to operator error. Automating this procedure can improve repeatability and decrease testing time. There are many ways in which this process can be automated. We'll look at one approach.

Let's assume that we would like to make frequency response measurements for an amplifier at 20 different frequencies. A circuit test system appropriate

Figure 12.26
Simple test set-up

signal source → amp → voltmeter

for this job is shown in Figure 12.26. To perform this test manually requires 20 distinct settings of the source signal, and 20 corresponding output readings. This can prove to be rather tedious if many units are to be tested. It would be very handy if there were some way in which the source frequency could be automatically changed to preset values. This is not particularly difficult. Most modern sources have control voltage inputs which may be used to set the frequency. The required control voltage can be created and accurately set through the use of a computer and DA converter. The computer can be programmed to send specific digital words to the DAC, which in turn feeds the signal source control input. In other words, the data word directly sets the frequency of the signal source. The computer can be programmed to produce virtually any sequence of data words at almost any rate, and to do it all without operator intervention. All the operator needs to do is start the process. Test repeatability is very high with a system like this. A block diagram of this system is shown in Figure 12.27.

In order to record the data, the voltmeter can be connected to a strip chart recorder, or better yet, back to the computer. The data can be sent to the computer in digital form if the voltmeter is of fairly advanced design, or, with the inclusion of an analog-to-digital converter, the output signal can be directly sampled and manipulated by the computer. In either case, data files can be created for each unit tested and stored for later use. Also, statistical analysis can be conveniently and quickly performed at the end of a test batch. Note that since the DA converter generates only a control signal, very high resolution and low distortion is normally not required. If a high-resolution converter is used, it is possible to create the test signals in the computer (as in the arbitrary function generator), and dispense with the signal source.

Figure 12.27
Automated test set-up

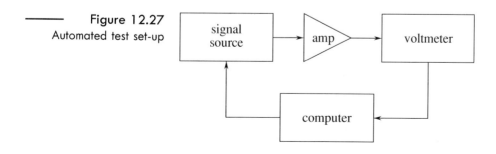

signal source → amp → voltmeter → computer → signal source

Figure 12.28 Computer-controlled lighting

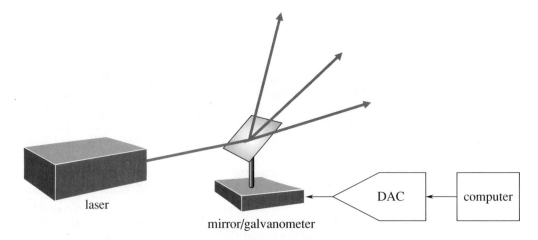

laser

mirror/galvanometer

DAC

computer

The automated test system is only one possible application of instrument control. Another interesting example is in the generation of laser "light shows." A block diagram of a simplified system is shown in Figure 12.28. In order to create the complex patterns seen by the audience, a laser beam is bounced off tiny moving mirrors. The mirrors can be mounted on something as simple as a galvanometer. The galvanometer is fed by a DAC. The pattern which the laser beam makes is dependent on how the galvanometer moves the mirror, which is, in turn, controlled by the data words fed to the DAC. In practice, several mirrors can be used to deflect the beam along 3 axes.

DA converters can be used to adjust any device with a control-voltage type input. Also, they can be used to control electromechanical devices which respond to an applied voltage. Their real advantage is the repeatability and flexibility they offer.

12.5 Analog-to-Digital Conversion

Now that you know the basics behind digital-to-analog conversion, we can examine the converse system, analog-to-digital conversion. The analog-to-digital conversion process is sometimes referred to as *quantization*, implying the individual discrete steps which the output assumes. There are several techniques to produce the conversion. Some techniques are optimized for fastest possible conversion speed, and some for highest accuracy. We shall investigate the more popular types.

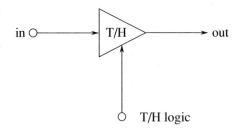

Figure 12.29
Symbol for the track-and-hold amplifier

The concept of AD conversion is simple enough: you wish to measure the voltage of the incoming waveform at specific instances in time. This measurement will be translated into a digital word. One practical limitation is the fact that the conversion circuitry may require a small amount of computation or translation time. For ultimate accuracy then, it is important that the measured waveform not change during the conversion interval. To ensure this, specialized subcircuits called *track-and-hold* (or *sample-and-hold*) amplifiers are used. They are usually abbreviated as T/H or S/H. Their job is to capture an input potential and produce a steady output to feed to the AD converter. The T/H schematic symbol is shown in Figure 12.29. When the T/H logic is in track mode, the circuit acts as a simple buffer, so its output voltage equals its input. When the logic goes to the hold state, the output voltage locks at its present potential and stays there until the circuit is switched back to track mode. A representation of this operation is shown in Figure 12.30.

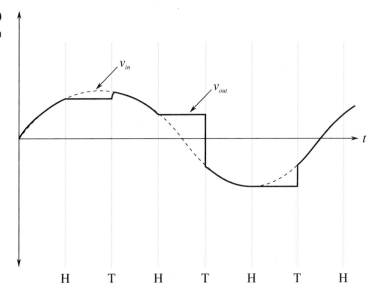

Figure 12.30
Track-and-hold operation

Figure 12.31
Track-and-hold errors

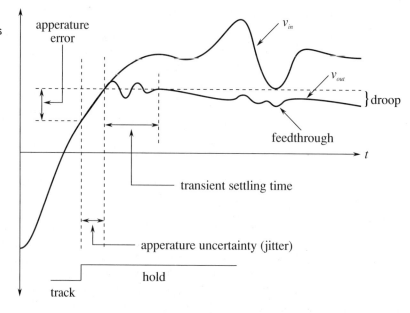

In reality, the track-and-hold process is not perfect, and errors can arise. These errors are shown magnified in Figure 12.31. First of all, there is a small delay between the time the logic signal changes and the time the T/H starts to react. The exact amount of time is variable and is responsible for *apperature error*. This is equal to the voltage difference between the signals at the desired and actual times. Also, due to the dynamic nature of the process of switching from track to hold, some initial ringing may occur in the hold waveform. You can see that as time progresses, the hold waveform tends to decay toward zero. This is because the held voltage is usually formed across a capacitor. While the discharge time constant can be very long, it cannot be infinite, and thus, the charge eventually bleeds off. This parameter is measured by the *droop rate* (fundamental units of volts per second). Finally, the possibility of *feedthrough error* exists. If a large change occurs on the input waveform during the hold period, it is possible that a portion of the signal may "leak through" to the T/H output. These errors are particularly troublesome when working with high-resolution converters. Lower-resolution systems may not be adversely affected by these relatively small abberations.

In order to create a T/H, a pair of high-impedance buffers are normally used, along with some form of switch element and a capacitor to hold the charge. Examples are shown in Figure 12.32. Figure 12.32a shows the general voltage type, open-loop, T/H. The buffers exhibit high input impedance. When the switch is closed, op amp 1 directly feeds op amp 2, and therefore the output voltage equals the input voltage. Normally, the hold capacitor is relatively

Figure 12.32
Track-and-hold circuits

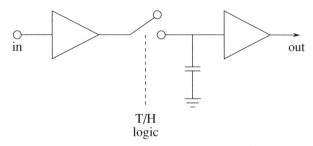

a. General open-loop voltage mode track-and-hold

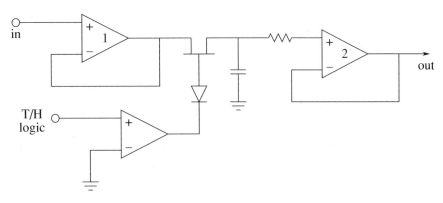

b. Op amp–based track-and-hold

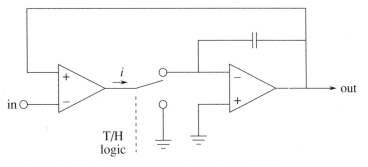

c. General closed-loop current-mode track-and-hold

small and does not adversely affect the drive capability of op amp 1. A more detailed version of this circuit is shown in Figure 12.32b. Each op amp is an FET input type for minimum input current draw. The switch is a simple JFET which is controlled by a comparator. When the T/H logic is high, the gate of the JFET is high, thus producing a low on-resistance (i.e., a closed switch). When the comparator output goes low, the JFET is turned off, creating a high impedance (i.e., open switch). In this state, op amp 2 is fed by the hold capacitor and buffers this potential to its output. For minimum droop, it is essential that the capacitor be a low leakage type and that op amp 2 have very low input bias current (e.g., FET input). The input resistor is used only to limit possible destructive discharge currents when the circuit is switched off. The diode positioned between the comparator and FET is used to prevent an excessively large positive comparator output potential from reaching the gate of the FET and possibly damaging it. (A comparator high will reverse-bias the series diode.) Figure 12.32c shows an alternate circuit using a closed-loop current-mode approach. Note that the hold capacitor is now forming part of an integrator. While the open-loop form offers faster acquisition and settling times, the closed-loop system offers improved signal tracking. A closed-loop voltage mode is also possible, but the current form generally offers fewer problems with leakage and switching transients. For general-purpose work, a variety of track-and-hold amplifiers are available in IC form from several manufacturers.

Analog-to-Digital Conversion Techniques

Several different ADC techniques have evolved for dealing with differing system requirements. We shall examine the *flash, staircase,* and *successive approximation* techniques.

Flash conversion is generally used for high-speed work, such as video applications. The circuits are usually low-resolution. A flash converter is made up of a string of comparators as shown in Figure 12.33. The input signal is applied to all of the comparators simultaneously. Each comparator is also tied into a reference ladder. Effectively, there is one comparator for each quantization step. When a given signal is applied, a number of comparators toward the bottom of the string will produce a high level since v_{in} will be greater than their references. Conversely, the comparators toward the top will indicate a low. The comparator at which the outputs shift from high to low indicates the step value closest to the input signal. The set of comparator outputs can be fed into a priority encoder which will turn this simple unweighted sequence into a normal binary word.

The only time delay involved in the conversion is that of logic propagation delay. Therefore, this conversion technique is quite useful for rapidly changing signals. Its downfall lies in the fact that one comparator is needed for each possible output-step change. An 8-bit flash converter requires 256 comparators, while a 16-bit version requires 65,536. Obviously, this is rather excessive, and units in the 4- to 6-bit range are common. While 6-bit resolution may appear at first to be too coarse for any application, it is actually quite useful for video displays.

Figure 12.33
Flash converter

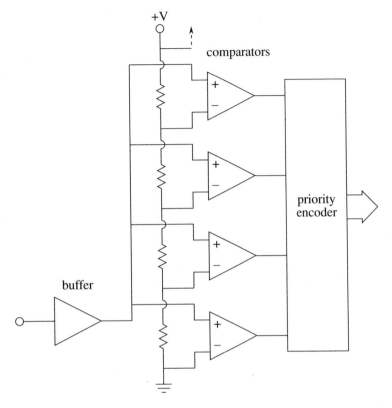

For high-resolution work, some other technique must be used. One possibility is shown in Figure 12.34 (page 596). This is called a staircase converter. Its operation is fairly simple. When a conversion is first started, the output of the counter will be all zeroes. This produces a DAC output of zero, and thus, the comparator output will be high. The next clock cycle will increment the up-counter, causing the DAC output to increase by one step-level. This signal is compared against the input, and if the input is greater, the output will remain high. The clock will continue to increment the counter in this fashion until the DAC output just exceeds the input level. At this point the comparator output drops low, indicating that the conversion is complete. This signal can then be used to latch the output of the counter.

This circuit gets its name from the fact that the waveform produced by the DAC looks like a staircase. The staircase technique can be used for very high-resolution conversion, as long as an appropriate high-resolution DAC is used. The major problem with this form is its very low conversion speed, due to the fact that there must be time to test every possible signal level. Consequently, a 16-bit system requires 65,536 comparisons. Even if a fast one-microsecond DAC is used, this limits the sampling interval to nearly 66 milliseconds. This

Figure 12.34
Staircase converter

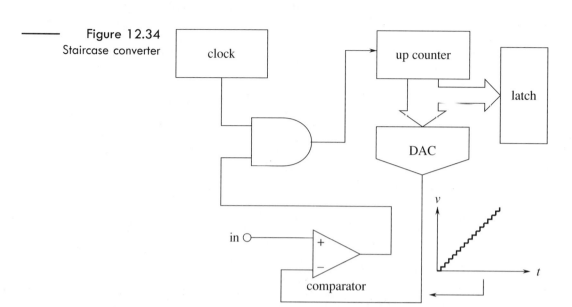

translates to a maximum input frequency before aliasing of only 7 Hz. Noting that the lowest frequency which most humans can hear is about 20 Hz, this technique is hardly suitable for something like digital audio recording. In fact, for general-purpose work, the staircase system is avoided in favor of the successive approximation technique.

Successive approximation is a good general-purpose solution suitable for systems requiring resolutions in the 16-bit area. Instead of trying to convert the input signal at one instant like the flash converter, this technique creates the output bit by bit. Unlike the staircase converter, each individual level does not need to be tested. You might think of it as making a series of guesses, each time getting a little closer to the result. Each guess results in a simple comparison. For an n-bit output, n comparisons need to be made.

A block diagram of a successive approximation converter is shown in Figure 12.35. Instead of a simple up-counter, a circuit is used to implement the successive approximation algorithm. Here is how the circuit works: The first bit to be tested is the most significant bit. The DAC is fed a 1 with all of the remaining bits set to 0 (i.e., 100000 . . .). This word represents half the maximum value that the system is capable of. The comparator output indicates whether v_{in} is greater or less than the resulting DAC output. If the comparator output is high, it means that the required digital output must be greater than the present word. If the comparator output is low, then the present digital word is too large, so the MSB is set to 0. The next most significant bit is then tested by setting it to 1. The new word is fed to the DAC, and again, the comparator

Figure 12.35
Successive-approximation
converter

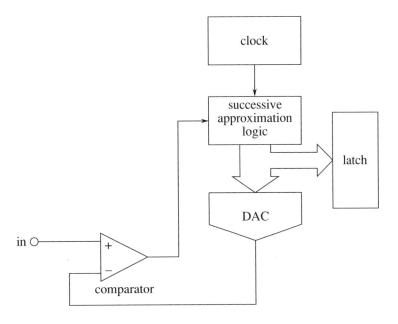

Figure 12.35
Successive-approximation
converter

is used to determine whether or not the bit under test should remain at 1 or be reset. At this point the two most significant bits have been determined. The remaining bits are individually set to 1 and tested in a similar manner until the least significant bit is determined. In this way, a 16-bit system only requires 16 comparisons. With a one-microsecond DAC, conversion takes only 16 microseconds. This translates to a sampling rate of 62.5 kHz; thus a maximum input frequency of 31.25 kHz is allowed. As you can see, the successive approximation converter is far more efficient than the staircase technique.

Analog-to-Digital Converter Integrated Circuits

In this section we will examine a few specific ADC ICs, along with selected applications. The ICs include the ADC0841 (an 8-bit microprocessor-compatible unit), the CS5412 (a very fast 12-bit converter utilizing flash techniques), and the PCM75 (a 16-bit converter designed primarily for audio applications).

The ADC0841.   A block diagram of the ADC0841 is shown in Figure 12.36 (page 598). Along with the built-in clock and 8-bit successive approximation register (SAR), the IC also includes tri-state latches and read, write, and chip-select logic pins. This means that the ADC0841 is easily interfaced to a microprocessor data bus and can be used as a memory-mapped I/O device. This IC is a moderate-speed device, showing a typical conversion speed of 40

Figure 12.36   The ADC0841

National Semiconductor Corporation

ADC0841   8-Bit μP Compatible A/D Converter

General Description

The ADC0841 is a CMOS 8-bit successive approximation A/D converter. Differential inputs provide low frequency input common mode rejection and allow offsetting the analog range of the converter. In addition, the reference input can be adjusted enabling the conversion of reduced analog ranges with 8-bit resolution.

The A/D is designed to operate with the control bus of a variety of microprocessors. TRI-STATE® output latches that directly drive the data bus permit the A/D to be configured as a memory location or I/O device to the microprocessor with no interface logic necessary.

Features

- Easy interface to all microprocessors
- Operates ratiometrically or with 5 V_{DC} voltage reference
- No zero or full-scale adjust required
- Internal clock
- 0V to 5V input range with single 5V power supply
- 0.3" standard width 20-pin package
- 20 Pin Molded Chip Carrier Package

Key Specifications

- Resolution   8 Bits
- Total Unadjusted Error   ± ½ LSB and ± 1 LSB
- Single Supply   5 V_{DC}
- Low Power   15 mW
- Conversion Time   40 μs

Block and Connection Diagrams

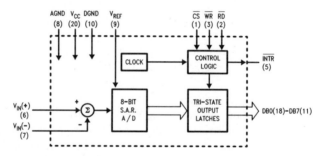

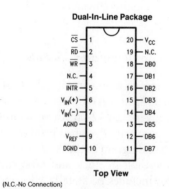

Dual-In-Line Package

Top View

(N.C.-No Connection)

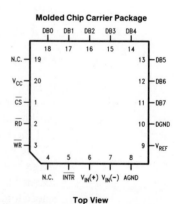

Molded Chip Carrier Package

Top View

Figure 12.37
Timing diagram for the
ADC0841

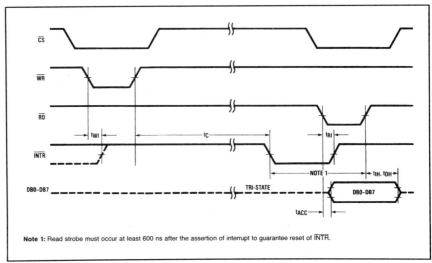

Note 1: Read strobe must occur at least 600 ns after the assertion of interrupt to guarantee reset of INTR.

Reprinted with permission of National Semiconductor Corporation

microseconds. The timing diagram for the ADC0841 is shown in Figure 12.37. A conversion is initiated by bringing both the chip-select and write-logic lines low. The falling edge of the write line resets the converter, and its rising edge starts the actual conversion. After the conversion period, which is set internally, the digital data can be transferred to the output latches with the read-logic line. Note that the pulse repetition rate of the write-logic line sets the sampling rate. Therefore, a small program running on the host microprocessor which reads and writes to the ADC can be used to control the sampling rate and store the data for later use.

The CS5412. The CS5412 is unique in that it offers fairly high resolution (12 bits) and fast conversion times (sampling rates to 1 MHz are possible). To achieve this level of performance, an interesting variation on the flash conversion technique is employed. For true flash conversion to 12 bits, a total of 2^{12}, or 4096, comparators must be used. This is impractical with present technology. To get around this problem, the CS5412 utilizes a two-step *subranging flash scheme*. A block diagram of this technique is shown in Figure 12.38 (pages 600–601). A single 6-bit flash converter is used to create the 12-bit output in two passes. In the first pass, the 6-bit converter is used to obtain an approximation of the input signal. In other words, the 6 most significant bits are found. This value is stored in a latch and will be combined with the results of the second pass. During the second pass, the 6 bits are fed to a digital-to-analog converter and the resulting signal is then subtracted from the

Figure 12.38   The CS5412

CS5412

Semiconductor Corporation

12-Bit, 1MHz Self-Calibrating A/D Converter

Features

- Monolithic CMOS Sampling ADC
  On-Chip Track and Hold Amplifier
  Microprocessor Interface

- Throughput Rates up to 1MHz

- True 12-Bit Accuracy over Temperature
  Typical Nonlinearity: 3/4 LSB
  No Missing Codes to 12 Bits

- Total Harmonic Distortion: 0.02%

- Dynamic Range: 72dB

- Self-Calibration Maintains Accuracy
  over Time and Temperature

- Low Power Dissipation: 750mW

General Description

The CS5412 CMOS analog to digital converter provides a true 12-bit representation of an analog input signal at sampling rates up to 1MHz. To achieve high throughput, the CS5412 uses pipelined acqusition and settling times as well as overlapped conversion cycles.

Unique self-calibration circuitry insures 12-bit accuracy over time and temperature. Also, a background calibration process constantly adjusts the converter's linearity, thereby insuring superior harmonic distortion and signal-to-noise performance throughout operating life.

The CS5412's advanced CMOS construction provides low power consumption of 750mW and the inherent reliability of monolithic devices.

An evaluation board is available which allows fast confirmation of performance, as well as example ground and layout arrangements.

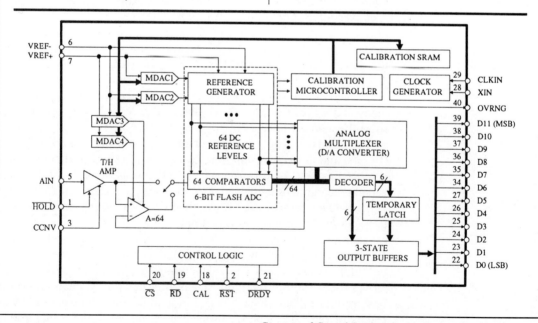

Courtesy of Crystal Semiconductor Corporation, Austin, Texas

a. Block diagram and features

——— Figure 12.38 *(Continued)*

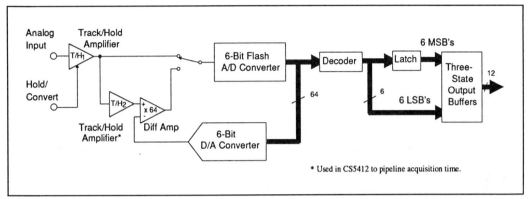

b. Block diagram of operation

Courtesy of Crystal Semiconductor Corporation, Austin, Texas

input signal. This difference represents the remaining error. The error is multiplied by 64 in order to scale the signal up to the flash converter's sensitivity ($2^6 = 64$). The scaled value is then fed to the flash converter, which produces a 6-bit output. These bits represent the 6 least significant bits of the final 12-bit word.

The CS5412 is used for applications requiring both high speed and high resolution. Other features of this IC include tri-state parallel data output, on-chip track-and-hold amplifier, and appropriate logic for microprocessor interfacing.

The PCM75. The PCM75 offers 16-bit resolution with a maximum conversion time of 17 microseconds. This is equivalent to a sampling rate of 58.8 kHz. This is ideal for both home and professional digital audio systems. A block diagram of the PCM75 is shown in Figure 12.39 (pages 602–603). This IC uses the successive-approximation conversion technique and offers an on-chip reference and clock. The PCM75 can be configured to produce either serial or parallel output data. The IC can also be *short cycled*. This means that if full 16-bit accuracy is not required, a shorter conversion time can be used. For 14-bit output, the maximum conversion time drops to 15 microseconds.

Due to the high resolution of the PCM75, you must be particularly careful with the circuit layout or excessive noise can result. The IC has separate digital and analog common pins which must be connected together as close as possible to the package, preferably via a large ground plane passing under the IC. Also, attention must be paid to proper power-supply bypassing. Note that ± 15 V supplies are used for the analog portions of the ADC while a separate $+5$ V supply is used for its digital segments.

Figure 12.39   The PCM75

PCM75
DESIGNED FOR AUDIO

16-Bit Hybrid
ANALOG-TO-DIGITAL CONVERTER

FEATURES

- 16-BIT RESOLUTION
- 90dB DYNAMIC RANGE
- 0.004% THD (FS Input, 16 Bits)
- 0.02% MAX THD (-15dB, 16 Bits)
- 17μs MAX CONVERSION TIME (16 Bits)
- 15μs MAX CONVERSION TIME (14 Bits)
- 10μs CONVERSION TIME (Reduced Specs)
- EIAJ STC-007-COMPATIBLE

DESCRIPTION

The PCM75 is a low cost, high quality, 16-bit successive approximation analog-to-digital converter. The PCM75 uses state-of-the-art IC and laser-trimmed thin-film components and is packaged in a bottom-brazed ceramic 32-pin dual-in-line package. The converter is complete with internal reference and clock.

The PCM75 is designed for PCM audio applications and is compatible with EIAJ STC-007 specifications.

The conversion time can be reduced from 15μs to 10μs with some increase in distortion. Distortion is specified on the data sheet to assure performance in critical audio applications.

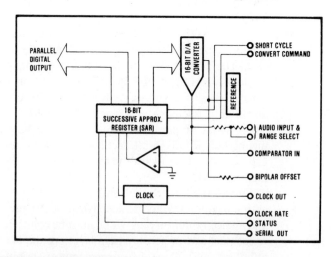

Figure 12.39
(*Continued*)

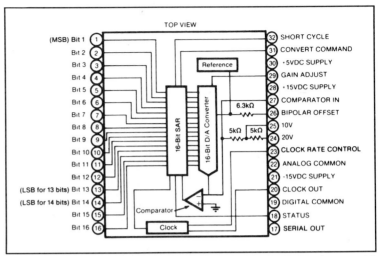

b. Connection diagram

Applications of Analog-to-Digital Converter Integrated Circuits

The following examples detail two specific—and very different—applications of ADC ICs. There are countless others which make use of the same principles.

Example 12.7   Perhaps the most straightforward application of analog-to-digital conversion is the acquisition of a signal. Once in the digital domain, the signal can be processed in a variety of ways. A good example of this is the digital storage oscilloscope (DSO). A simplified block diagram of a DSO is shown in Figure 12.40 (page 604). Since the final output of the system is a simple graph, it generally does not make sense to resolve the input beyond 8 bits (256 steps), and often, even fewer bits can be used. Typically, high sampling rates are more important than fine resolution in this application. Consequently, 12- and 16-bit converters are not found here. Instead, lower-resolution converters capable of sampling at tens or hundreds of MHz are used. A commercial DSO is shown in Figure 12.41 (page 605).

DSOs can usually be run in one of two modes: continuous, or single-shot. In single-shot mode, the ADC acquires the signal and stores it in memory. This data can then be routed to the display circuits continuously in order to create a trace on the CRT. Since the signal is captured and stored in computer memory, it can be replayed virtually forever without a loss of clarity. This is very useful for catching quick, nonrepetitive transients. Once the signal is

Figure 12.40   Basic digital storage oscilloscope

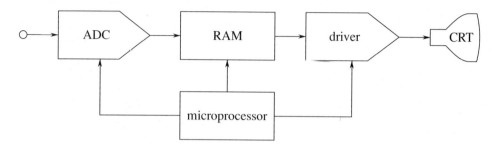

captured, it can be examined at leisure. For that matter, if some form of level-sensing logic is included, sampling can be initiated by specific transient events. (This is known as *baby-sitting*.) For example, you might suspect that a circuit you have designed occasionally emits an undesirable voltage spike. It is not practical for you to hook up the circuit and stare at the face of an oscilloscope, perhaps for hours, waiting for the system to misbehave. Instead, the DSO can be programmed to wait for the transient event before recording. In this way, you can leave the system, and when you return some time later, the spike will have been recorded and will be awaiting your inspection. An extension of this concept is pre-trigger recording. In this variation, the DSO is constantly re-cording and "throwing away" data. When the transient spike finally occurs, the DSO has a snapshot of the events leading up to the spike, as well as the spike itself. This can provide very useful information in some applications. In either case, since the data is in digital form, it may be off-loaded to a computer for further analysis (if the DSO has facility for this). Indeed, many DSOs offer some interesting on-board analysis functions, including signal-smoothing (noise reduction) and trace cursors which allow for easy delta-time/delta-voltage computation.

In continuous mode, a DSO appears to operate in much the same fashion as an ordinary analog oscilloscope. In this mode, the DSO samples the input signal and passes the data out to memory. From here the data is relayed to the CRT circuitry where a trace is produced. Since the trace is being constantly updated, any change of the input will be quickly displayed. Trace updates are not actually in real-time, and thus are not as well suited to rapidly changing waveforms as are analog oscilloscopes.

Example 12.8   Another interesting application of the analog-to-digital converter is in the digital sampling music keyboard or drum computer (generally referred to as a *sampler*). The usage of a sampler is quite straightforward, actually. The goal is to mimic the sound of a given instrument (such as a trumpet or flute) from the keyboard. Before the advent of the sampler, this was done by properly setting the filters, amplifiers, and oscillators of a keyboard synthesizer. Although

Figure 12.41   A commercial DSO

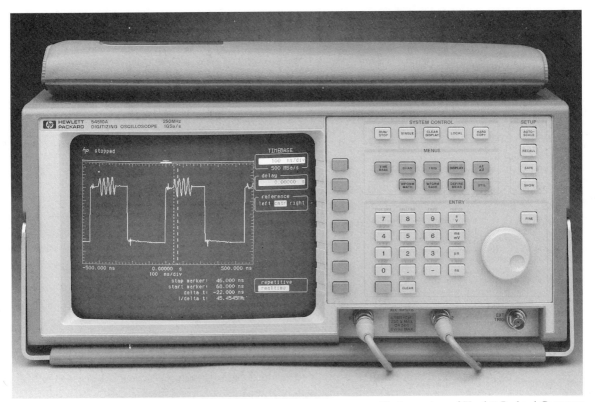

the resulting sounds were reasonably close to the desired instrument, they usually weren't close enough to fool the average listener. The sampler bypasses the problems of synthesizers by directly recording an instrument with an analog-to-digital converter. For example, a trumpet player might play the note A into a microphone which is connected to the sampler. This note is digitized and stored in RAM. Now, when the keyboard player hits the A key, the data is retrieved from RAM and fed to a DAC where it is reconstructed. The result is the exact same note that the trumpet player originally produced. By recording several different pitches from many different instruments, keyboard players literally have the versatility of an entire orchestra at their fingertips. Of course, there is no limit to the sounds which might be sampled, and a keyboardist could just as easily use the sampler to "play" a collection of dog barks, door slams, and bird calls.

A sampling drum computer is similar to a sampling keyboard, but replaces the standard musical keyboard with a series of buttons which allow the musician to create a programmed sequence of notes. This sequence can be played back

Figure 12.42
Block diagram of musical
instrument sampling and
playback device

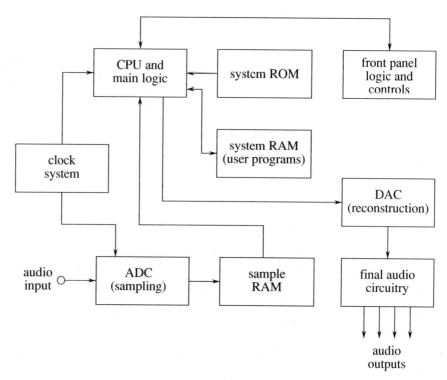

at any time, and at virtually any tempo. In this manner, an entire percussion
section can be simulated using digital recordings of real drums.

An example of a digital sampling drum computer is E-Mu System's SP-12.
A block diagram is shown in Figure 12.42. This is a relatively sophisticated
musical instrument and its design and construction owes more to personal
computers than to drums and cymbals. The system is, in fact, a self-contained
specialized computer. It is an excellent example of a complete analog-to-digital-
to-analog recording and processing system. Unlike a simple data-acquisition
system, the sampler requires several simultaneous signal outputs. This is man-
datory since several instruments can sound at the same instant. On the block
diagram, each output channel is referred to as a *voice*.

The SP-12 can record the input audio with 12-bit resolution at a sampling
rate of approximately 27.5 kHz. The input section is shown in Figure 12.43.
First, the audio signal is passed through a high-order anti-alias filter. The
filtered signal is fed to an LF398 sample-and-hold IC. At this point, an LM311
comparator drives a 74504 12-bit successive-approximation register, which in
turn drives a 12-bit DAC. The output of the DAC serves as the other input
to the LM311. This scheme is essentially the same as the one diagrammed in
Figure 12.35. Once a conversion is complete, the 12-bit data is loaded onto
the data bus. From there, the microprocessor system directs the data to RAM.

Figure 12.43 Input section of the E-Mu SP-12

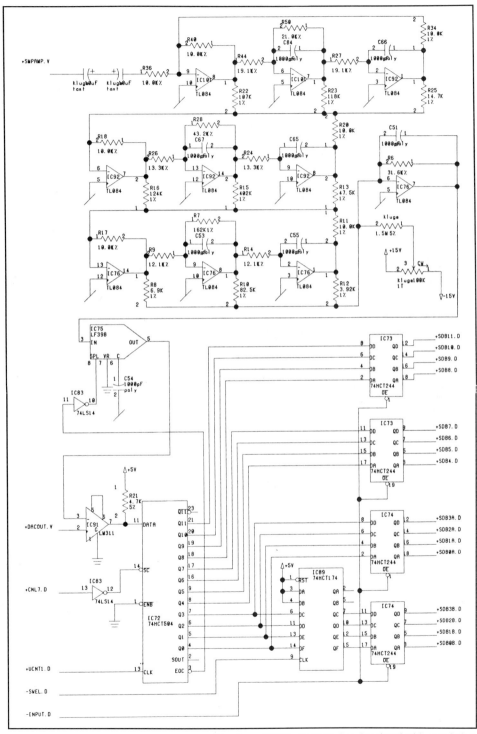

Courtesy of E-Mu Systems, Inc. Reprinted with permission.

Further enhancements to the digital sampling musical instrument include direct connections to personal computers. In this way, the computer can be used to analyze and augment the sound samples in new ways. For example, the personal computer can be used to create a graphic display of the waveform, or compute and display the spectral components of the sound using a popular technique called the *Fast Fourier Transform*. An example is shown in Figure

——————   Figure 12.44   Commercial waveform display and editing software

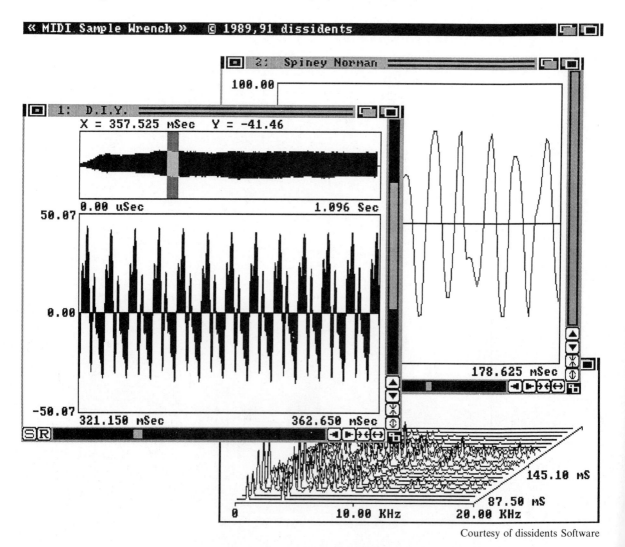

Courtesy of dissidents Software

12.44. Finally, the computer can be programmed to "play" the sampler. In this way, even non-keyboardists can take advantage of the inherent musical flexibility this system offers.

12.6 Extended Topic: Digital Signal Processing

Much has been said in this chapter concerning the ability of personal computers to store and manipulate data in the digital domain. This process is called *digital signal processing*, or DSP. The sampled data is generally stored in RAM as a large array of values. Normally, the data is in integer form, because microprocessors are generally faster at performing integer calculations than floating-point calculations. Also, AD converters produce integer values, so this is doubly convenient. As an example, if 8-bit resolution is used, each sample point will require one byte (8 bits). The byte is the fundamental unit of computer memory storage, and thus storage as a simple byte multiple (8 bits, 16 bits, 32 bits) is straightforward. The data can be stored in either unipolar (unsigned) or bipolar (signed) formats. For unsigned systems, the data value "0" means "most negative peak." In signed systems, a 2's complement form is normally used. Here, a data value of "0" means "0 volts." All negative values have their most significant bit set (for an 8-bit representation, $-1 = 11111111$, $-2 = 11111110$, etc.). Since input signals are often AC, signed representations are quite common.

In BASIC, the data array might be called data(), and individual elements are accessed by properly setting the array index. The first element is data(1), the second is data(2), and the Nth element is data(N). While BASIC is useful, it is not as powerful as Assembly, or languages such as C, which offer direct manipulation of memory addresses via pointers. In C, the starting, or base, address of the data might be given the name data_ptr. To access different elements in the array, an offset is added to the base. The address of the second element is data_ptr + 1, the third is data_ptr + 2, and so on. Direct manipulation of pointers is generally faster for the microprocessor than array indexing. [To find the value stored at a given address, C uses the indirection operator, *. To assign the second data value to the variable x, you would say x = *(data_ptr + 1). The BASIC equivalent is x = data(2).]

No matter how the data is accessed, any mathematical function can be applied to the data elements. In the following examples, BASIC is used to illustrate the concepts. Only the processing portion of the code is shown. The array is called data(), and is assumed to have TOTAL number of elements. Also, questions concerning integer versus floating-point representation are bypassed in order to keep the examples as straightforward as possible. In the real world, problems such as execution speed, round-off error, and overflow cannot be ignored.

Let's start with something simple. Suppose we would like to simulate the operation of an inverting buffer. An inverting buffer simply multiplies the input signal by -1. Here is how we do this in the digital domain: each point is individually multiplied by -1.

```
10 For element = 1 to TOTAL
20 data(element) = -1*data(element)
30 Next element
```

We can extend this concept a bit further by replacing "-1" with a variable called "gain." In this way, we can alter the signal amplitude digitally.

Another useful calculation is average value. For signed data, the average value is equal to the DC offset component.

```
10 sum = 0
20 For element = 1 to TOTAL
30 sum = sum + data(element)
40 Next element
50 average = sum/TOTAL
```

In order to add a DC offset, a constant is added to each data element in turn.

```
10 For element = 1 to TOTAL
20 data(element) = offset + data(element)
30 Next element
```

We might wish to combine two different waveforms into a third waveform. Simple addition is all that is required. (The example assumes that both arrays are the same size.)

```
10 For element = 1 to TOTAL
20 data3(element) = data1(element) + data2(element)
30 Next element
```

If the two source arrays are multiplied instead of added, signal modulation can be produced.

These examples are relatively simple. With proper programming, a wide variety of functions can be simulated, including spectral analysis, translation to new sampling rates, and the realization of complex filter functions which are impractical to implement in the analog domain. The major drawback of DSP techniques is the inherent computing power required. Thus, real-time application of DSP principles is not possible without using very powerful computers. Recently, specialized ICs have been designed to implement DSP subfunctions very quickly. An example of the breed is the Motorola 56000 DSP IC. These chips make real-time DSP practical. For example, an all-DSP graphic equalizer for home or professional audio use can be designed using the 56000. The spectral shaping is performed directly in the digital domain instead of using dedicated analog circuitry. Since the DSP ICs are programmable, they offer the distinct advantage of being able to produce different functions at different times by loading a new program. In this way, they are far more flexible than dedicated analog circuits.

Summary

In this chapter we have covered the basic concepts of analog-to-digital and digital-to-analog conversion. By placing a signal in the digital domain, a variety of new analysis, storage, and transmission techniques become available.

The process of transforming an analog signal into a digital representation is referred to as quantization or more simply, as digitizing or sampling. One of the most popular methods used is PCM, pulse code modulation. In this scheme, the input signal is measured, or sampled, at a constant rate. Each sample point is represented as a digital word. The sequence of words describes the input signal, and is used to recreate it, if necessary. The ultimate accuracy of the conversion is dependent on the resolution and sampling rate of the system. Resolution refers to the number of bits present in the digital word. An 8-bit word can represent 256 steps while a 16-bit word is much finer, offering 65,536 discrete steps. With so many steps, round-off error is much less of a problem in 16-bit systems than in say, 8-bit or 12-bit systems.

The minimum allowable sampling rate is twice the highest input frequency. In other words, at least two samples per cycle are required for the highest harmonic in the input signal. Another way of stating this is that no input frequency component can exceed the Nyquist frequency, which is defined as one-half of the sampling frequency. If the input exceeds this limit, alias distortion may occur. Aliases are nonharmonically-related frequencies that are effectively created by improper sampling. In order to remove all possibility of alias distortion, special high-order low-pass filters, called anti-alias filters, are normally placed before the sampling circuitry.

Two popular methods of analog-to-digital conversion are the flash and successive-approximation techniques. Flash converters are very fast, but require one comparator per output step, so they are not normally used where high resolution is required. Successive approximation takes longer than flash conversion, but can produce resolutions in excess of 16 bits. In either case, the conversion process is not instantaneous and any fluctuation of the input signal during the conversion can produce errors. In order to alleviate this difficulty, special track-and-hold amplifiers are used between the anti-alias filters and the AD converter to create a nonvarying signal.

Once the signal has been digitized, it may be stored in RAM or some other media for future use, or directly analyzed. Often, a personal computer can prove to be very useful for waveform analysis.

To reconstruct the waveform, a digital-to-analog converter is used. In essence, this is usually little more than a weighted summing amplifier. Due to accuracy and construction constraints, an $R/2R$ ladder technique is often employed. In order to smooth out the resulting waveform and remove any remaining digital "glitches," the signal is passed through a reconstruction filter. This is a low-pass filter, and is often called a smoothing filter. Generally, the process of digital-to-analog conversion is much faster than analog-to-digital conversion. Indeed, the successive-approximation analog-to-digital scheme requires an internal digital-to-analog converter.

Applications for AD and DA converters range from laboratory instruments such as digital sampling oscilloscopes and hand-held digital multimeters to industrial instrument and device control. Recently, with the continued cost decreases in computing power, AD and DA systems have found a large application market in the commercial sector. Uses include the popular compact disk player and musical keyboard equipment.

Self Test Questions

1. What is *PCM*?
2. Define the term *resolution*.
3. What is *quantization*?
4. What is an *alias*, and how is it produced? How is an alias avoided?
5. Define *Nyquist frequency*, and discuss the importance of this parameter.
6. Explain how a summing amplifier can be used to create a digital-to-analog converter.
7. Explain the difference between integral nonlinearity and differential nonlinearity.
8. What is a smoothing (reconstruction) filter?
9. What is the purpose of a track-and-hold amplifier?
10. Detail the differences between flash conversion and the successive-approximation technique. Where would each type be used? What are their limitations?
11. Give several examples of possible DSP functions.

Problem Set

▷ Analysis Problems

1. Determine the number of quantization steps for a 10-bit system.

2. A 14-bit converter produces a maximum peak-to-peak output of 2.5 V. What is the step size?

3. Determine the dynamic range of the converters in Problems 1 and 2.

4. We wish to resolve a 1 V peak-to-peak signal to at least 1 mV. What is the minimum allowable number of bits in the converted data?

5. We wish to generate analog signals using an arbitrary waveform generator. If we send out digital words at the rate of 50 kHz, what is the maximum allowable conversion speed for the DAC?

6. Assume that we are trying to digitize ultrasonic signals lying between 25 kHz and 45 kHz.
   a. What is the Nyquist frequency?
   b. What is the minimum acceptable sampling rate?

7. Determine the maximum allowable conversion time for the ADC of Problem 6.

8. Given a 14-bit ADC,
   a. Determine the number of comparators required for flash technique.
   b. Determine the number of comparisons required if the successive-approximation technique is used.
   c. Determine the number of comparators required if the two-step sub-ranging flash conversion technique is used.

9. Assume that a single 16-bit ADC is connected to a personal computer. The sampling rate is 10 kHz. Determine the data rate in bytes per second.

10. Referring to Problem 9, if the personal computer has 350 kilobytes of RAM available for data storage, how much data transfer time does this represent?

11. DAT (**d**igital **a**udio **t**ape) recorders normally use a 16-bit representation with a sampling rate of 48 kHz. If the unit is used to record a performance of Stravinsky's "Rite of Spring" (35 minutes total), what is the required storage capacity in bytes?

12. If the data is transferred serially from the DAT of Problem 11 to a digital signal processing IC in real time, what is the width of each individual pulse?

13. A 12-bit 2-microsecond DAC is used as part of a discrete successive-approximation analog-to-digital converter. Assuming that logic delays and signal settling times are negligible, determine

   a. The minimum time allowable between sample points

   b. The maximum input signal frequency without aliasing.

14. A 6-bit video DA converter produces a maximum output of 1.25 V. Determine the output voltage for the following digital input words.

   a. 000001   b. 100000   c. 111111   d. 011101

15. A 10-bit instrumentation DAC produces an output of 16 mV with an input of 0000000100. Determine

   a. The step size.   b The maximum output signal.

16. An 8 bit ADC produces a full-scale output of 11111111 with a 2 V input signal. Determine the output word given the following inputs. (Assume that this converter rounds to the nearest output value.)

   a. 100 mV   b. 10 μV   c. 0 V   d. 1.259 V

17. Assume that comparator/logic delays, amplifier settling times, and other factors require .4 μS total in a particular IC fabrication technique. If this technology is used to create AD converters, determine the maximum conversion time for the following 8-bit converters.

   a. Flash.   b. Successive approximation.   c. Ramp type.

▷ **Design Problems**

18. We wish to digitize human voice signals. The maximum input frequency is to be limited to 3 kHz, and resolution is required to be better than .5% of the maximum input value.

   a. Draw a block diagram of the complete system.

   b. Determine the minimum bit requirement.

   c. Determine the minimum sampling rate if the Nyquist rate is set to 25% greater than the theoretical minimum.

   d. Determine the anti-alias filter tuning frequency.

   e. Determine the preferred conversion technique.

   f. Determine which of the ICs presented in this chapter is best suited for this system.

19. We wish to design a system capable of digitizing complex signals with a spectrum ranging from DC to 400 kHz. Accuracy must be at least .2% of full scale.

   a. Draw a block diagram of the complete system.

   b. Determine the minimum bit requirement.

c. Determine the minimum sampling rate if the Nyquist rate is set to 20% greater than the theoretical minimum.

d. Determine the anti-alias filter tuning frequency.

e. Determine the preferred conversion technique.

f. Determine which of the ICs presented in this chapter is best suited for this system.

20. Write a computer algorithm which can be used to "flip" digital data back-to-front (i.e., play it backwards).

21. Write a computer algorithm which will determine the maximum peak value of the digital data.

22. Write a computer algorithm which will determine the RMS value of the digital data.

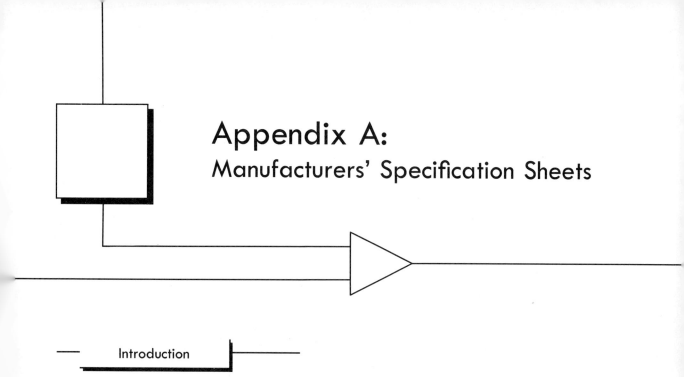

Appendix A:
Manufacturers' Specification Sheets

Introduction

For safety and reliability, be sure to obtain the most current release of product specifications before you use these (or any) products in an actual application. New products, and new versions of existing ones, are released often.

Manufacturers' manuals and technical guides are generally inexpensive. Some are even free. Since many companies have local offices throughout the United States and abroad, there may be a source of reference materials near you. If not, contact the main offices at the addresses given in this appendix.

Apex Microtechnology Corporation

WA01

The specification sheets for the WA01 Wideband Amplifier are reprinted, with permission, from the *Apex High Performance Amplifier Handbook, Volume IV*. For current product information, contact

Apex Microtechnology Corporation
5980 N. Shannon Road
Tucson AZ 85741

WA01 • WA01A
WIDEBAND AMPLIFIER

FEATURES

- SUPER SLEW – 5000V/µsec
- HIGH BANDWIDTH – 100MHz
- OUTPUT CURRENT TO 400mA
- PIN COMPATIBLE WITH 3554
- HIGH DC ACCURACY – ±5mV Vos
- LOW DISTORTION – 70db at 100KHz

APPLICATIONS

- LINE DRIVERS
- DATA ACQUISITION SYSTEMS
- SAMPLE AND HOLD CIRCUITS
- VIDEO PROCESSING
- FUNCTION GENERATORS
- ATE PIN DRIVER

DESCRIPTION

The WA01 uses low impedance push-pull circuits to achieve high speed amplification. The output node of each amplifier stage consists of two transistors, with the first transistor driving the signal in one direction and the second in the other. As a result, speed is enhanced and a lower, much more linear output impedance is obtained. This technique also exhibits low input impedance which is more compatible with high speed signal processing.

Unlike conventional op amps, the feedback resistor is included in the package. At 1.5K ohms, it provides a transimpedance function of 1.5V/1mA. Standard inverting and non-inverting op amp configurations may be implemented using fewer external components than would otherwise be required. The resultant feedback path is much shorter than when using a conventional external feedback element. As a result, summing node capacitance to ground is lower and thus high frequency characteristics are very stable. To enhance the input characteristics of this wideband amplifier, sophisticated bias current cancellation and voltage offset trim networks have been added to the input stage.

This hybrid circuit utilizes thick film (cermet) resistors, ceramic capacitors and silicon semiconductor chips to maximize reliability, minimize size and give top performance. Ultrasonically bonded aluminum wires provide reliable interconnections at all operating temperatures. The 8 pin TO-3 package is hermetically sealed and electrically isolated. The use of compressible isolation washers may void the warranty.

PATENT PENDING

NONINVERTING GAIN

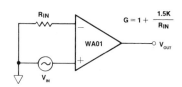

$$G = 1 + \frac{1.5K}{R_{IN}}$$

INVERTING GAIN

$$G = -\frac{1.5K}{R_{IN}}$$

EQUIVALENT SCHEMATIC

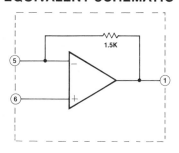

EXTERNAL CONNECTIONS

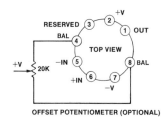

OFFSET POTENTIOMETER (OPTIONAL)

WA01 ABSOLUTE MAXIMUM RATINGS

SUPPLY VOLTAGE, +Vs to −Vs	32V
OUTPUT CURRENT, within SOA	400mA
POWER DISSIPATION, internal	10.5W
INPUT VOLTAGE, differential	±6V
INPUT VOLTAGE, common-mode	±Vs
TEMPERATURE, pin solder-10sec	300°C
TEMPERATURE, junction[1]	175°C
TEMPERATURE RANGE, storage	−65 to +150°C
OPERATING TEMPERATURE RANGE, case	−25 to +85°C

SPECIFICATIONS

PARAMETER	TEST CONDITIONS[2]	WA01 MIN	WA01 TYP	WA01 MAX	WA01A MIN	WA01A TYP	WA01A MAX	UNITS
INPUT								
OFFSET VOLTAGE, initial	$T_C = 25°C$		±4	±10		±.2	5	mV
OFFSET VOLTAGE, vs. temperature	Full temperature range		15	50		10	25	µV/°C
OFFSET VOLTAGE, vs. supply	$T_C = 25°C$		5	10		*	*	mV/V
OFFSET VOLTAGE, vs. power	Full temperature range		20			10		µV/W
BIAS CURRENT, initial, +In	$T_C = 25°C$		5	20		3	10	µA
BIAS CURRENT, vs. supply	$T_C = 25°C$		.01			*		pA/V
INPUT IMPEDANCE, dc	$T_C = 25°C$		200			*		kΩ
INPUT CAPACITANCE	$T_C = 25°C$		6			*		pF
COMMON-MODE VOLTAGE RANGE[3]	Full temperature range		Vs −7.5			*		V
COMMON-MODE REJECTION, dc[3]	Full temp. range, $V_{CM} = ±5V$	48	54		*	*		db
POWER SUPPLY REJECTION, dc[3]	Full temp. range, V_S = 24 to 30	60	75		*	*		db
GAIN								
ACCURACY	$T_C = 25°C$, F = DC		2	5		1	2	%
POWER BANDWIDTH	$T_C = 25°C$, I_O = .4A, V_O = 20V_{PP}		40			*		MHz
	$T_C = 25°C$, I_O = .05A, V_O = 4V_{PP}		80			*		MHz
GAIN FLATNESS	DC to 75MHz, G = −10		1			*		db
OUTPUT								
VOLTAGE SWING[3]	$T_C = 25°C$, I_O = .4A	±Vs-4.5			*			V
VOLTAGE SWING[3]	Full temp. range, I_O = .2A	±Vs-4			*			V
VOLTAGE SWING[3]	Full temp. range, I_O = .1A	±Vs-3.5			*			V
CURRENT, limit	$T_C = 25°C$		.6			*		A
SETTLING TIME to .1%	$T_C = 25°C$, 10V step		20			*		nS
SLEW RATE	$T_C = 25°C$		5000			*		V/µs
CAPACITIVE LOAD	Full temperature range, G = 1	22			*			pF
CAPACITIVE LOAD	Full temperature range, G = 30	47			*			pF
PROPAGATION DELAY	$T_C = 25°C$, G = 1		2.9			*		nS
POWER SUPPLY								
VOLTAGE	Full temperature range	±12	±15	±16	*		*	V
CURRENT, quiescent	$T_C = 25°C$		28	30		*	*	mA
THERMAL								
RESISTANCE, AC[4] junction to case	Full temp. range, F>60Hz		9	10		*	*	°C/W
RESISTANCE, DC junction to case	Full temp. range, F<60Hz		12	14		*	*	°C/W
RESISTANCE, junction to ambient	Full temp. range		30			*		°C/W
TEMPERATURE RANGE, case	Meets full range specification	−25		+85	*		*	°C

NOTES:
* The specification of WA01A is identical to the specification for WA01 in applicable column to the left.
1. Long term operation at the maximum junction temperature will result in reduced product life. Derate internal power dissipation to achieve high MTTF.
2. The power supply voltage for all specifications is the TYP rating unless noted as a test condition.
3. +Vs and −Vs denote the positive and negative supply rail respectively. Total Vs is measured from +Vs to −Vs.
4. Rating applies if the output current alternates between both output transistors at a rate faster than 60Hz.

CAUTION: The internal substrate contains berylia (BeO). Do not break the seal. If broken, do not crush, machine, or subject to temperatures in excess of 850°C to avoid generating toxic fumes.

WA01 TYPICAL PERFORMANCE GRAPHS

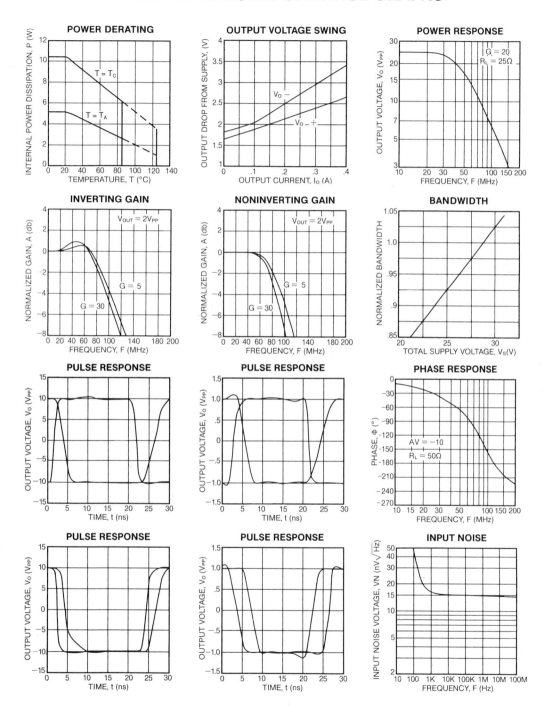

WA01 OPERATING CONSIDERATIONS

GENERAL

Please read the "General Operating Considerations" section, which covers stability, supplies, heatsinking, mounting, current limit, SOA interpretation, and specification interpretation. Additional information can be found in the applications notes. For information on the package outline, heatsinks, and mounting hardware, consult the "Accessory and Package Mechanical Data" section of the handbook.

BYPASSING OF SUPPLIES

Each supply rail must be bypassed to common with a tantalum capacitor in parallel with a ceramic capacitor directly connected from the power supply pins to the ground plane. The ceramic bypass capacitor should have leads as short as possible, be mounted as close to the supply pin as possible, and should have a series resonant frequency above 200MHz including lead inductance. A typical value range would be 2.2 to 10μF for the tantalum and 500pF to 3000pF for the ceramic.

LEADS

Keep the output, supply, and bypass leads as short as possible. In the video frequency range, even a few inches of wire has significant inductance, raising the interconnection impedance and limiting the output current slew rate. Furthermore the skin effect increases the resistance of heavy wires at high frequencies. Multistrand Litz wire is recommended for high frequency use with low losses.

GROUNDING

Single point grounding of the input resistors and input signal to a common ground plane will prevent undesired current feedback, which can cause errors and/or instabilities. Also the case is electrically isolated (floating) with respect to the internal circuit. It is recommended that the case be connected to the same common ground plane as the inputs.

SAFE OPERATING AREA (SOA)

The bipolar output stage of this wideband amplifier has two distinct limitations:
1. The internal current limit, which limits maximum available output current.
2. The junction temperature of the output devices.

Compliance within the power derating curve guarantees maximum junction temperature is met.

CURRENT LIMIT

Internal current limit is set using a 1.2Ω ±20% resistor and one transistor V_{BE} drop. Nominal current limit at T_c = 25°C is:
$$I_{LIM} = V_{BE}/R_{CL}; .54A = .65V/1.2\Omega.$$

Temperature variance of the current limit is dominated by the V_{BE} temperature coefficient of −2mV/°C. For a case temperature of +85°C the nominal current limit is:

$$.442A = [.65 - (60°C)(2mV/°C)]/[1.2\Omega]$$

GAIN SETTING

Unlike other APEX amplifiers the WA01's feedback resistor is inside the package. This reduces external part count as well as increases performance. To determine the value of the resistor required to achieve the desired gain the following formulas are necessary.

$$\begin{array}{ll} \text{NONINVERTING} & R_{IN} = 1.5k/(G-1) \\ \text{INVERTING} & R_{IN} = 1.5k/G \end{array}$$

BALANCE CONTROL

The voltage offset of the WA01 is laser trimmed at the factory. To externally zero residual errors in applications where offset is critical, a 20K ohm potentiometer may be installed between pins 4 and 8 and the wiper arm connected to the positive supply. If the optional adjust provision is not used, leave the pins floating (no connection). If adjust is not used and settling time is important, tie pin 8 to AC ground with 100-150pF.

STABILITY

The use of an internal feedback resistor of low impedance insures ease of use and stability of the WA01. Additionally, the architecture provides for a constant bandwidth at different gain settings without the need for external compensation. Although the WA01 is stable and well behaved at high frequency a good PC layout is essential for optimum performance. A layout that keeps inductances, capacitances, and trace lengths to a minimum will prevent oscillations. To avoid peaking at high frequency when driving a capacitive load, a small resistor (1 to 22 ohms) may be placed between the output and the load to lower the Q of any parasitic resonant circuit that might occur.

Harris Semiconductor

ICL8038

The specification sheets for the ICL8038 Precision Waveform Generator/Voltage Controlled Oscillator are reprinted, with permission, from *Linear ICs for Commercial Applications*. For current product information, contact

Harris Semiconductor Literature Department
P. O. Box 883, MS CB1-28
Melbourne FL 32901

Harris
SEMICONDUCTOR

HARRIS · RCA · GE · INTERSIL

ICL8038
Precision Waveform Generator/Voltage Controlled Oscillator

GENERAL DESCRIPTION

The ICL8038 Waveform Generator is a monolithic integrated circuit capable of producing high accuracy sine, square, triangular, sawtooth and pulse waveforms with a minimum of external components. The frequency (or repetition rate) can be selected externally from .001Hz to more than 300kHz using either resistors or capacitors, and frequency modulation and sweeping can be accomplished with an external voltage. The ICL8038 is fabricated with advanced monolithic technology, using Schottky-barrier diodes and thin film resistors, and the output is stable over a wide range of temperature and supply variations. These devices may be interfaced with phase locked loop circuitry to reduce temperature drift to less than 250ppm/°C.

FEATURES

- **Low Frequency Drift With Temperature — 250ppm/°C**
- **Simultaneous Sine, Square, and Triangle Wave Outputs**
- **Low Distortion — 1% (Sine Wave Output)**
- **High Linearity — 0.1% (Triangle Wave Output)**
- **Wide Operating Frequency Range — 0.001Hz to 300kHz**
- **Variable Duty Cycle — 2% to 98%**
- **High Level Outputs — TTL to 28V**
- **Easy to Use — Just A Handful of External Components Required**

ORDERING INFORMATION

Part Number	Stability	Temp. Range	Package
ICL8038CCPD	250ppm/°C typ	0°C to +70°C	14 pin DIP
ICL8038CCJD	250ppm/°C typ	0°C to +70°C	14 pin CERDIP
ICL8038BCJD	180ppm/°C typ	0°C to +70°C	14 pin CERDIP
ICL8038ACJD	120ppm/°C typ	0°C to +70°C	14 pin CERDIP
ICL8038BMJD*	350ppm/°C max	−55°C to +125°C	14 pin CERDIP
ICL8038AMJD*	250ppm/°C max	−55°C to +125°C	14 pin CERDIP

*Add /883B to part number if 883 processing is required.

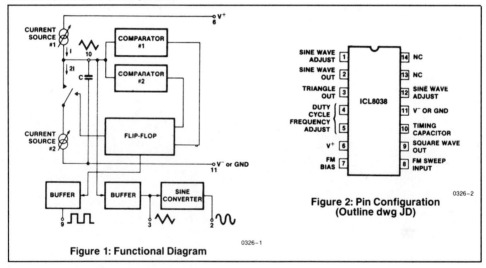

Figure 1: Functional Diagram

0326-1

Figure 2: Pin Configuration
(Outline dwg JD)

0326-2

NOTE: All typical values have been characterized but are not tested.

ABSOLUTE MAXIMUM RATINGS

Supply Voltage (V⁻ to V⁺) 36V
Power Dissipation[1] 750mW
Input Voltage (any pin) .... V⁻ to V⁺
Input Current (Pins 4 and 5) 25mA
Output Sink Current (Pins 3 and 9) 25mA

Storage Temperature Range .... −65°C to +150°C
Operating Temperature Range:
  8038AM, 8038BM .... −55°C to +125°C
  8038AC, 8038BC, 8038CC .... 0°C to +70°C
Lead Temperature (Soldering, 10sec) .... 300°C

NOTE: Stresses above those listed under "Absolute Maximum Ratings" may cause permanent damage to the device. These are stress ratings only and functional operation of the device at these or any other conditions above those indicated in the operational sections of the specifications is not implied. Exposure to absolute maximum rating conditions for extended periods may affect device reliability.

NOTE 1: Derate ceramic package at 12.5mW/°C for ambient temperatures above 100°C.

ELECTRICAL CHARACTERISTICS

(V$_{SUPPLY}$ = ±10V or +20V, T$_A$ = 25°C, R$_L$ = 10kΩ, Test Circuit Unless Otherwise Specified)

Symbol	General Characteristics	8038CC Min	Typ	Max	8038BC(BM) Min	Typ	Max	8038AC(AM) Min	Typ	Max	Units
V$_{SUPPLY}$	Supply Voltage Operating Range										
V⁺	Single Supply	+10		+30	+10		30	+10		30	V
V⁺, V⁻	Dual Supplies	±5		±15	±5		±15	±5		±15	V
I$_{SUPPLY}$	Supply Current (V$_{SUPPLY}$ = ±10V)[2]										
	8038AM, 8038BM					12	15		12	15	mA
	8038AC, 8038BC, 8038CC		12	20		12	20		12	20	mA
Frequency Characteristics (all waveforms)											
f$_{max}$	Maximum Frequency of Oscillation	100			100			100			kHz
f$_{sweep}$	Sweep Frequency of FM Input		10			10			10		kHz
	Sweep FM Range[3]		35:1			35:1			35:1		
	FM Linearity 10:1 Ratio		0.5			0.2			0.2		%
Δf/ΔT	Frequency Drift With Temperature[5] 8038 AC, BC, CC 0°C to 70°C		250			180			120		ppm/°C
	8038 AM, BM, −55°C to 125°C						350			250	
Δf/ΔV	Frequency Drift With Supply Voltage (Over Supply Voltage Range)		0.05			0.05			0.05		%/V
Output Characteristics											
I$_{OLK}$	**Square-Wave** Leakage Current (V$_9$ = 30V)			1			1			1	µA
V$_{SAT}$	Saturation Voltage (I$_{SINK}$ = 2mA)		0.2	0.5		0.2	0.4		0.2	0.4	V
t$_r$	Rise Time (R$_L$ = 4.7kΩ)		180			180			180		ns
t$_f$	Fall Time (R$_L$ = 4.7kΩ)		40			40			40		ns
ΔD	Typical Duty Cycle Adjust (Note 6)	2		98	2		98	2		98	%
V$_{TRIANGLE}$	**Triangle/Sawtooth/Ramp** Amplitude (R$_{TRI}$ = 100kΩ)	0.30	0.33		0.30	0.33		0.30	0.33		xV$_{SUPPLY}$
	Linearity		0.1			0.05			0.05		%
Z$_{OUT}$	Output Impedance (I$_{OUT}$ = 5mA)		200			200			200		Ω
V$_{SINE}$	**Sine-Wave** Amplitude (R$_{SINE}$ = 100kΩ)	0.2	0.22		0.2	0.22		0.2	0.22		xV$_{SUPPLY}$
THD	THD (R$_S$ = 1MΩ)[4]		2.0	5		1.5	3		1.0	1.5	%
THD	THD Adjusted (Use Figure 6)		1.5			1.0			0.8		%

NOTES: 2. R$_A$ and R$_B$ currents not included.
    3. V$_{SUPPLY}$ = 20V; R$_A$ and R$_B$ = 10kΩ, f ≈ 10kHz nominal; can be extended 1000 to 1. See Figures 7a and 7b.
    4. 82kΩ connected between pins 11 and 12, Triangle Duty Cycle set at 50%. (Use R$_A$ and R$_B$.)
    5. Figure 3, pins 7 and 8 connected, V$_{SUPPLY}$ = ±10V. See Typical Curves for T.C. vs V$_{SUPPLY}$.
    6. Not tested, typical value for design purposes only.

NOTE: All typical values have been characterized but are not tested.

TEST CONDITIONS

Parameter		R_A	R_B	R_L	C_1	SW_1	Measure
Supply Current		10kΩ	10kΩ	10kΩ	3.3nF	Closed	Current into Pin 6
Sweep FM Range[1]		10kΩ	10kΩ	10kΩ	3.3nF	Open	Frequency at Pin 9
Frequency Drift with Temperature		10kΩ	10kΩ	10kΩ	3.3nF	Closed	Frequency at Pin 9
Frequency Drift with Supply Voltage[2]		10kΩ	10kΩ	10kΩ	3.3nF	Closed	Frequency at Pin 9
Output Amplitude: (Note 4)	Sine	10kΩ	10kΩ	10kΩ	3.3nF	Closed	Pk-Pk output at Pin 2
	Triangle	10kΩ	10kΩ	10kΩ	3.3nF	Closed	Pk-Pk output at Pin 3
Leakage Current (off)[3]		10kΩ	10kΩ		3.3nF	Closed	Current into Pin 9
Saturation Voltage (on)[3]		10kΩ	10kΩ		3.3nF	Closed	Output (low) at Pin 9
Rise and Fall Times (Note 5)		10kΩ	10kΩ	4.7kΩ	3.3nF	Closed	Waveform at Pin 9
Duty Cycle Adjust: (Note 5)	MAX	50kΩ	~1.6kΩ	10kΩ	3.3nF	Closed	Waveform at Pin 9
	MIN	~25kΩ	50kΩ	10kΩ	3.3nF	Closed	Waveform at Pin 9
Triangle Waveform Linearity		10kΩ	10kΩ	10kΩ	3.3nF	Closed	Waveform at Pin 3
Total Harmonic Distortion		10kΩ	10kΩ	10kΩ	3.3nF	Closed	Waveform at Pin 2

NOTES: 1. The hi and lo frequencies can be obtained by connecting pin 8 to pin 7 (f_{hi}) and then connecting pin 8 to pin 6 (f_{lo}). Otherwise apply Sweep Voltage at pin 8 ($2/3\ V_{SUPPLY} + 2V) \le V_{SWEEP} \le V_{SUPPLY}$ where V_{SUPPLY} is the total supply voltage. In Figure 7b, pin 8 should vary between 5.3V and 10V with respect to ground.
    2. $10V \le V^+ \le 30V$, or $\pm 5V \le V_{SUPPLY} \le \pm 15V$.
    3. Oscillation can be halted by forcing pin 10 to +5 volts or −5 volts.
    4. Output Amplitude is tested under static conditions by forcing pin 10 to 5.0V then to −5.0V.
    5. Not tested; for design purposes only.

DEFINITION OF TERMS:

Supply Voltage (V_{SUPPLY}). The total supply voltage from V^+ to V^-

Supply Current. The supply current required from the power supply to operate the device, excluding load currents and the currents through R_A and R_B.

Frequency Range. The frequency range at the square wave output through which circuit operation is guaranteed.

Sweep FM Range. The ratio of maximum frequency to minimum frequency which can be obtained by applying a sweep voltage to pin 8. For correct operation, the sweep voltage should be within the range

$$(2/3\ V_{SUPPLY} + 2V) < V_{SWEEP} < V_{SUPPLY}$$

FM Linearity. The percentage deviation from the best-fit straight line on the control voltage versus output frequency curve.

Output Amplitude. The peak-to-peak signal amplitude appearing at the outputs.

Saturation Voltage. The output voltage at the collector of Q_{23} when this transistor is turned on. It is measured for a sink current of 2mA.

Rise and Fall Times. The time required for the square wave output to change from 10% to 90%, or 90% to 10%, of its final value.

Triangle Waveform Linearity. The percentage deviation from the best-fit straight line on the rising and falling triangle waveform.

Total Harmonic Distortion. The total harmonic distortion at the sine-wave output.

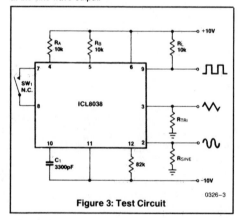

Figure 3: Test Circuit

0326–3

TYPICAL PERFORMANCE CHARACTERISTICS

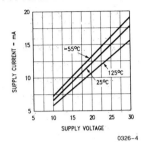

0326-4

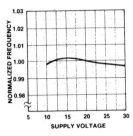

0326-5

Performance of the Square-Wave Output

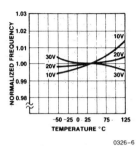

0326-6

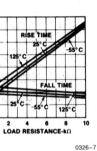

0326-7

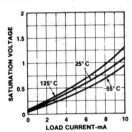

0326-8

Performance of Triangle-Wave Output

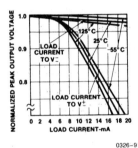

0326-9

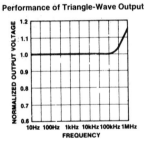

0326-10

0326-11

Performance of Sine-Wave Output

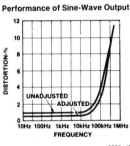

0326-12

0326-13

A11

National Semiconductor Corporation

LF351

LH0036

LM3080

LM311

LM317

LM318

LM325

LM386

LM3900

LM723

LM78XX

MF4

Life Support Policy

The specification sheets for National Semiconductor Corporation products are reprinted, with permission, from *Linear Databook 1*, *Linear Databook 2*, and *Linear Databook 3*. For current product information, contact

National Semiconductor Corporation
2900 Semiconductor Drive
P. O. Box 58090
Santa Clara CA 95052-8090

LF351 Wide Bandwidth JFET Input Operational Amplifier

General Description

The LF351 is a low cost high speed JFET input operational amplifier with an internally trimmed input offset voltage (BI-FET II™ technology). The device requires a low supply current and yet maintains a large gain bandwidth product and a fast slew rate. In addition, well matched high voltage JFET input devices provide very low input bias and offset currents. The LF351 is pin compatible with the standard LM741 and uses the same offset voltage adjustment circuitry. This feature allows designers to immediately upgrade the overall performance of existing LM741 designs.

The LF351 may be used in applications such as high speed integrators, fast D/A converters, sample-and-hold circuits and many other circuits requiring low input offset voltage, low input bias current, high input impedance, high slew rate and wide bandwidth. The device has low noise and offset voltage drift, but for applications where these requirements are critical, the LF356 is recommended. If maximum supply current is important, however, the LF351 is the better choice.

Features

- Internally trimmed offset voltage   10 mV
- Low input bias current   50 pA
- Low input noise voltage   25 nV/√Hz
- Low input noise current   0.01 pA/√Hz
- Wide gain bandwidth   4 MHz
- High slew rate   13 V/μs
- Low supply current   1.8 mA
- High input impedance $10^{12}\Omega$
- Low total harmonic distortion $A_V = 10$,   <0.02%
  $R_L = 10k$, $V_O = 20$ Vp-p, BW = 20 Hz–20 kHz
- Low 1/f noise corner   50 Hz
- Fast settling time to 0.01%   2 μs

Typical Connection

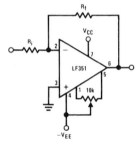

Simplified Schematic

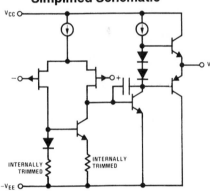

Connection Diagrams (Top Views)

Metal Can Package

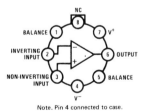

Note. Pin 4 connected to case.

Order Number LF351H
See NS Package Number H08C

Dual-In-Line Package

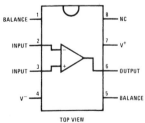

TOP VIEW

Order Number LF351J,
LF351M or LF351N
See NS Package Number J08A, M08A or N08E

TL/H/5648–1

Absolute Maximum Ratings

If Military/Aerospace specified devices are required, contact the National Semiconductor Sales Office/Distributors for availability and specifications.

Supply Voltage	± 18V
Power Dissipation (Notes 1 and 6)	670 mW
Operating Temperature Range	0°C to +70°C
$T_{j(MAX)}$	115°C
Differential Input Voltage	± 30V
Input Voltage Range (Note 2)	± 15V
Output Short Circuit Duration	Continuous
Storage Temperature Range	−65°C to +150°C
Lead Temp. (Soldering, 10 sec.)	
Metal Can	300°C
DIP	260°C

	H Package	N Package
θ_{jA}	225°C/W (Still Air)	120°C/W
	160°C/W	
	(400 LF/min Air Flow)	
θ_{jC}	25°C/W	

Soldering Information
Dual-In-Line Package
Soldering (10 sec.) 260°C
Small Outline Package
Vapor Phase (60 sec.) 215°C
Infrared (15 sec.) 220°C

See AN-450 "Surface Mounting Methods and Their Effect on Product Reliability" for other methods of soldering surface mount devices.

ESD rating to be determined.

DC Electrical Characteristics (Note 3)

Symbol	Parameter	Conditions	LF351 Min	LF351 Typ	LF351 Max	Units
V_{OS}	Input Offset Voltage	$R_S = 10$ kΩ, $T_A = 25$°C		5	10	mV
		Over Temperature			13	mV
$\Delta V_{OS}/\Delta T$	Average TC of Input Offset Voltage	$R_S = 10$ kΩ		10		μV/°C
I_{OS}	Input Offset Current	$T_j = 25$°C, (Notes 3, 4)		25	100	pA
		$T_j \le 70$°C			4	nA
I_B	Input Bias Current	$T_j = 25$°C, (Notes 3, 4)		50	200	pA
		$T_j \le \pm 70$°C			8	nA
R_{IN}	Input Resistance	$T_j = 25$°C		10^{12}		Ω
A_{VOL}	Large Signal Voltage Gain	$V_S = \pm 15$V, $T_A = 25$°C $V_O = \pm 10$V, $R_L = 2$ kΩ	25	100		V/mV
		Over Temperature	15			V/mV
V_O	Output Voltage Swing	$V_S = \pm 15$V, $R_L = 10$ kΩ	± 12	± 13.5		V
V_{CM}	Input Common-Mode Voltage Range	$V_S = \pm 15$V	± 11	+15		V
				−12		V
CMRR	Common-Mode Rejection Ratio	$R_S \le 10$ kΩ	70	100		dB
PSRR	Supply Voltage Rejection Ratio	(Note 5)	70	100		dB
I_S	Supply Current			1.8	3.4	mA

AC Electrical Characteristics (Note 3)

Symbol	Parameter	Conditions	LF351			Units
			Min	Typ	Max	
SR	Slew Rate	$V_S = \pm 15V$, $T_A = 25°C$		13		$V/\mu s$
GBW	Gain Bandwidth Product	$V_S = \pm 15V$, $T_A = 25°C$		4		MHz
e_n	Equivalent Input Noise Voltage	$T_A = 25°C$, $R_S = 100\Omega$, f = 1000 Hz		25		$nV/\sqrt{Hz}$
i_n	Equivalent Input Noise Current	$T_j = 25°C$, f = 1000 Hz		0.01		$pA/\sqrt{Hz}$

Note 1: For operating at elevated temperature, the device must be derated based on the thermal resistance, θ_{JA}.

Note 2: Unless otherwise specified the absolute maximum negative input voltage is equal to the negative power supply voltage.

Note 3: These specifications apply for $V_S = \pm 15V$ and $0°C \leq T_A \leq +70°C$. V_{OS}, I_B and I_{OS} are measured at $V_{CM} = 0$.

Note 4: The input bias currents are junction leakage currents which approximately double for every 10°C increase in the junction temperature, T_j. Due to the limited production test time, the input bias currents measured are correlated to junction temperature. In normal operation the junction temperature rises above the ambient temperature as a result of internal power dissipation, P_D. $T_j = T_A + \theta_{JA} P_D$ where θ_{JA} is the thermal resistance from junction to ambient. Use of a heat sink is recommended if input bias current is to be kept to a minimum.

Note 5: Supply voltage rejection ratio is measured for both supply magnitudes increasing or decreasing simultaneously in accordance with common practice. From $\pm 15V$ to $\pm 5V$.

Note 6: Max. Power Dissipation is defined by the package characteristics. Operating the part near the Max. Power Dissipation may cause the part to operate outside guaranteed limits.

Typical Performance Characteristics

Input Bias Current

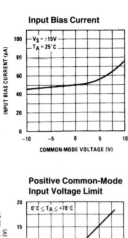

Input Bias Current

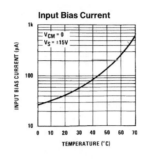

Supply Current

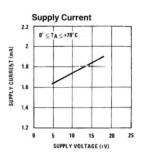

Positive Common-Mode Input Voltage Limit

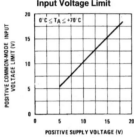

Negative Common-Mode Input Voltage Limit

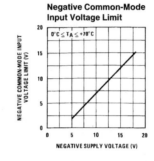

Positive Current Limit

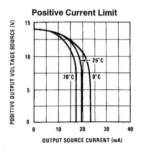

Negative Current Limit

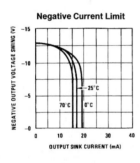

Voltage Swing

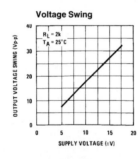

Output Voltage Swing

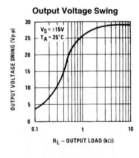

Gain Bandwidth

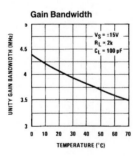

Bode Plot

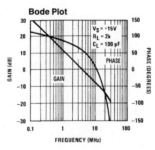

Slew Rate

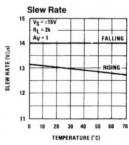

TL/H/5648–2

Typical Performance Characteristics (Continued)

Distortion vs Frequency

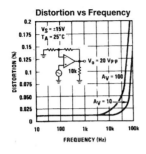

Undistorted Output Voltage Swing

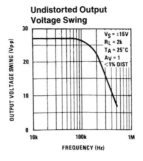

Open Loop Frequency Response

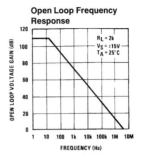

Common-Mode Rejection Ratio

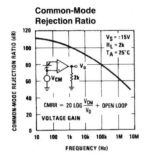

Power Supply Rejection Ratio

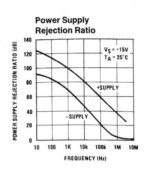

Equivalent Input Noise Voltage

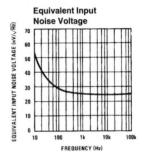

Open Loop Voltage Gain (V/V)

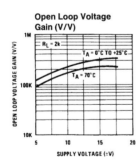

Output Impedance

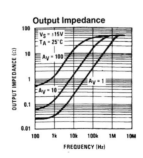

Inverter Settling Time

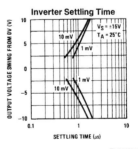

TL/H/5648–3

A17

LH0036/LH0036C Instrumentation Amplifier

General Description

The LH0036/LH0036C is a true micro power instrumentation amplifier designed for precision differential signal processing. Extremely high accuracy can be obtained due to the 300 MΩ input impedance and excellent 100 dB common mode rejection ratio. It is packaged in a hermetic TO-8 package. Gain is programmable from 1 to 1000 with a single external resistor. Power supply operating range is between ±1V and ±18V. Input bias current and output bandwidth are both externally adjustable or can be set by internally set values. The LH0036 is specified for operation over the −55°C to +125°C temperature range and the LH0036C is specified for operation over the −25°C to +85°C temperature range.

Features

- High input impedance   300 MΩ
- High CMRR   100 dB
- Single resistor gain adjust   1 to 1000
- Low power   90 μW
- Wide supply range   ±1V to ±18V
- Adjustable input bias current
- Adjustable output bandwidth
- Guard drive output

Equivalent Circuit and Connection Diagrams

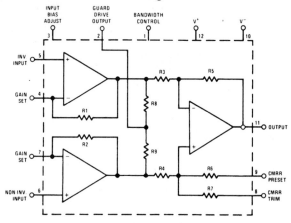

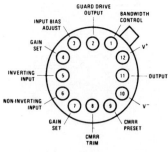

TOP VIEW

Order Number LH0036G or LH0036CG
See NS Package Number G12B

TL/H/5545–1

Absolute Maximum Ratings

If Military/Aerospace specified devices are required, contact the National Semiconductor Sales Office/ Distributors for availability and specifications. (Note 5)

Supply Voltage	±18V
Differential Input Voltage	±30V
Input Voltage Range	±V_S
Shield Drive Voltage	±V_S
CMRR Preset Voltage	±V_S

CMMR Trim Voltage	±V_S
Power Dissipation (Note 3)	1.5W
Short Circuit Duration	Continuous
Operating Temperature Range	
LH0036	−55°C to +125°C
LH0036C	−25°C to +85°C
Storage Temperature Range	−65°C to +150°C
Lead Temperature (Soldering, 10 sec.)	260°C
ESD rating to be determined.	

Electrical Characteristics (Notes 1 and 2)

Parameter	Conditions	Limits						Units
		LH00336			LH0036C			
		Min	Typ	Max	Min	Typ	Max	
Input Offset Voltage	$R_S = 1.0\ k\Omega$, $T_A = 25°C$		0.5	1.0		1.0	2.0	mV
(V_{IOS})	$R_S = 1.0\ k\Omega$			2.0			3.0	mV
Output Offset Voltage	$R_S = 1.0\ k\Omega$, $T_A = 25°C$		2.0	5.0		5.0	10	mV
(V_{OOS})	$R_S = 1.0\ k\Omega$			6.0			12	mV
Input Offset Voltage	$R_S \leq 1.0\ k\Omega$		10			10		$\mu V/°C$
Tempco ($\Delta V_{IOS}/\Delta T$)								
Output Offset Voltage			15			15		$\mu V/°C$
Tempco ($\Delta V_{OOS}/\Delta T$)								
Overall Offset Referred	$A_V = 1.0$		2.5			6.0		mV
to Input (V_{OS})	$A_V = 10$		0.7			1.5		mV
	$A_V = 100$		0.52			1.05		mV
	$A_V = 1000$		0.502			1.005		mV
Input Bias Current	$T_A = 25°C$		40	100		50	125	nA
(I_B)				150			200	nA
Input Offset Current	$T_A = 25°C$		10	40		20	50	nA
(I_{OS})				80			100	nA
Input Voltage Range	Differential	±10	±12		±10	±12		V
	Common Mode	±10	±12		±10	±12		V
Gain Nonlinearity			0.03			0.03		%
Deviation From Gain Equation Formula	$A_V = 1$ to 1000 (Note 4)		±0.3	±1.0		±1.0	±3.0	%

Electrical Characteristics (Notes 1 and 2) (Continued)

Parameter	Conditions	Limits						Units
		LH00336			LH0036C			
		Min	Typ	Max	Min	Typ	Max	
PSRR	$\pm 5.0V \leq V_S \leq \pm 15V$, $A_V = 1.0$		1.0	2.5		1.0	5.0	mV/V
	$\pm 5.0V \leq V_S \leq \pm 15V$, $A_V = 100$		0.05	0.25		0.10	0.50	mV/V
CMRR	$A_V = 1.0$ DC to		1.0	2.5		2.5	5.0	mV/V
	$A_V = 10$ 100 Hz		0.1	0.25		0.25	0.50	mV/V
	$A_V = 100$ $\Delta R_S = 1.0k$		50	100		50	100	μV/V
Output Voltage	$V_S = \pm 15V$, $R_L = 10 \text{ k}\Omega$	± 10	± 13.5		± 10	± 13.5		V
	$V_S = \pm 1.5V$, $R_L = 100 \text{ k}\Omega$	± 0.6	± 0.8		± 0.6	± 0.8		V
Output Resistance			0.5			0.5		Ω
Supply Current			300	400		400	600	μA
Small Signal Bandwidth	$A_V = 1.0$, $R_L = 10 \text{ k}\Omega$		350			350		kHz
	$A_V = 10$, $R_L = 10 \text{ k}\Omega$		35			35		kHz
	$A_V = 100$, $R_L = 10 \text{ k}\Omega$		3.5			3.5		kHz
	$A_V = 1000$, $R_L = 10 \text{ k}\Omega$		350			350		Hz
Full Power Bandwidth	$V_{IN} = \pm 10V$, $R_L = 10k$, $A_V = 1$		5.0			5.0		kHz
Equivalent Input Noise Voltage	$0.1 \text{ Hz} < f < 10 \text{ kHz}$, $R_S < 50\Omega$		20			20		μV/p-p
Slew Rate	$\Delta V_{IN} = \pm 10V$, $R_L = 10 \text{ k}\Omega$, $A_V = 1.0$		0.3			0.3		V/μS
Settling Time	To ± 10 mV, $R_L = 10 \text{ k}\Omega$, $\Delta V_{OUT} = 1.0V$							
	$A_V = 1.0$		3.8			3.8		μS
	$A_V = 100$		180			180		μS

Note 1: Unless otherwise specified, all specifications apply for $V_S = \pm 15V$, Pins 1, 3, and 9 grounded, $-25°C$ to $+85°C$ for the LH0036C and $-55°C$ to $+125°C$ for the LH0036.

Note 2: All typical values are for $T_A = 25°C$.

Note 3: The maximum junction temperature is 150°C. For operation at elevated temperature derate the G package on a thermal resistance of 90°C/W, above 25°C.

Note 4: $A_V = 1000$ guaranteed by design and testing at $A_V = 100$.

Note 5: Refer to RETS0036G for LH0036G military specifications.

Typical Performance Characteristics

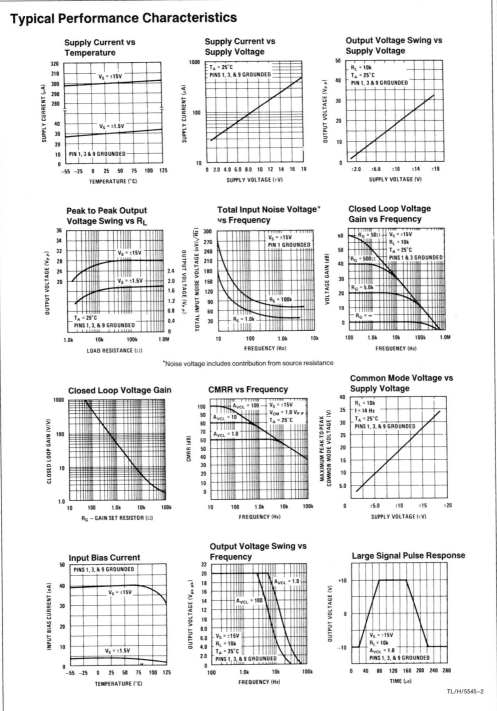

*Noise voltage includes contribution from source resistance

TL/H/5545-2

A21

LM3080/LM3080A
Operational Transconductance Amplifier

General Description

The LM3080 is a programmable transconductance block intended to fulfill a wide variety of variable gain applications. The LM3080 has differential inputs and high impedance push-pull outputs. The device has high input impedance and its transconductance (g_m) is directly proportional to the amplifier bias current (I_{ABC}).

High slew rate together with programmable gain make the LM3080 an ideal choice for variable gain applications such as sample and hold, multiplexing, filtering, and multiplying.

The LM3080N and LM3080AN are guaranteed from 0°C to +70°C.

Features

■ Slew rate (unity gain compensated): 50 V/μs
■ Fully adjustable gain: 0 to g_m • R_L limit
■ Extended g_m linearity: 3 decades
■ Flexible supply voltage range: ±2V to ±18V
■ Adjustable power consumption

Schematic and Connection Diagrams

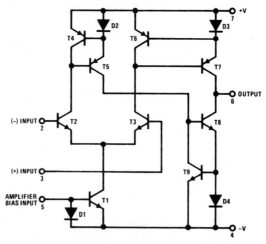

TL/H/7148–1

Dual-In-Line Package

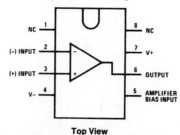

TL/H/7148–2

Top View
Order Number LM3080AN or LM3080N
See NS Package Number N08E

Absolute Maximum Ratings

If Military/Aerospace specified devices are required, contact the National Semiconductor Sales Office/Distributors for availability and specifications.

Supply Voltage (Note 2)	
LM3080	± 18V
LM3080A	± 22V
Power Dissipation	250 mW
Differential Input Voltage	± 5V
Amplifier Bias Current (I_{ABC})	2 mA
DC Input Voltage	$+V_S$ to $-V_S$
Output Short Circuit Duration	Indefinite
Operating Temperature Range	
LM3080N or LM3080AN	0°C to +70°C
Storage Temperature Range	-65°C to +150°C
Lead Temperature (Soldering, 10 sec.)	260°C

Electrical Characteristics (Note 1)

Parameter	Conditions	LM3080 Min	LM3080 Typ	LM3080 Max	LM3080A Min	LM3080A Typ	LM3080A Max	Units
Input Offset Voltage			0.4	5		0.4	2	mV
	Over Specified Temperature Range			6			5	mV
	$I_{ABC} = 5\ \mu A$		0.3			0.3	2	mV
Input Offset Voltage Change	$5\ \mu A \leq I_{ABC} \leq 500\ \mu A$		0.1			0.1	3	mV
Input Offset Current			0.1	0.6		0.1	0.6	μA
Input Bias Current			0.4	5		0.4	5	μA
	Over Specified Temperature Range		1	7		1	8	μA
Forward Transconductance (g_m)		6700	9600	13000	7700	9600	12000	μmho
	Over Specified Temperature Range	5400			4000			μmho
Peak Output Current	$R_L = 0$, $I_{ABC} = 5\ \mu A$		5		3	5	7	μA
	$R_L = 0$	350	500	650	350	500	650	μA
	$R_L = 0$ Over Specified Temperature Range	300			300			μA
Peak Output Voltage								
Positive	$R_L = \infty$, $5\ \mu A \leq I_{ABC} \leq 500\ \mu A$	+12	+14.2		+12	+14.2		V
Negative	$R_L = \infty$, $5\ \mu A \leq I_{ABC} \leq 500\ \mu A$	-12	-14.4		-12	-14.4		V
Amplifier Supply Current			1.1			1.1		mA
Input Offset Voltage Sensitivity								
Positive	$\Delta V_{OFFSET}/\Delta V+$		20	150		20	150	$\mu V/V$
Negative	$\Delta V_{OFFSET}/\Delta V-$		20	150		20	150	$\mu V/V$
Common Mode Rejection Ratio		80	110		80	110		dB
Common Mode Range		± 12	± 14		± 12	± 14		V
Input Resistance		10	26		10	26		$k\Omega$
Magnitude of Leakage Current	$I_{ABC} = 0$		0.2	100		0.2	5	nA
Differential Input Current	$I_{ABC} = 0$, Input $= \pm 4$V		0.02	100		0.02	5	nA
Open Loop Bandwidth			2			2		MHz
Slew Rate	Unity Gain Compensated		50			50		$V/\mu s$

Note 1: These specifications apply for $V_S = \pm 15$V and $T_A = 25$°C, amplifier bias current (I_{ABC}) = 500 μA, unless otherwise specified.

Note 2: Selection to supply voltage above ± 22V, contact the factory.

Typical Performance Characteristics

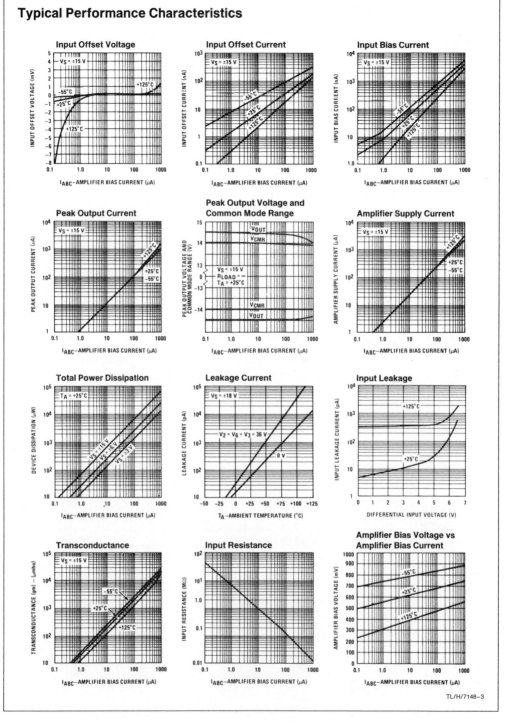

TL/H/7148–3

LM111/LM211/LM311 Voltage Comparator

General Description

The LM111, LM211 and LM311 are voltage comparators that have input currents nearly a thousand times lower than devices like the LM106 or LM710. They are also designed to operate over a wider range of supply voltages: from standard ±15V op amp supplies down to the single 5V supply used for IC logic. Their output is compatible with RTL, DTL and TTL as well as MOS circuits. Further, they can drive lamps or relays, switching voltages up to 50V at currents as high as 50 mA.

Both the inputs and the outputs of the LM111, LM211 or the LM311 can be isolated from system ground, and the output can drive loads referred to ground, the positive supply or the negative supply. Offset balancing and strobe capability are provided and outputs can be wire OR'ed. Although slower than the LM106 and LM710 (200 ns response time vs

40 ns) the devices are also much less prone to spurious oscillations. The LM111 has the same pin configuration as the LM106 and LM710.

The LM211 is identical to the LM111, except that its performance is specified over a −25°C to +85°C temperature range instead of −55°C to +125°C. The LM311 has a temperature range of 0°C to +70°C.

Features

- Operates from single 5V supply
- Input current: 150 nA max. over temperature
- Offset current: 20 nA max. over temperature
- Differential input voltage range: ±30V
- Power consumption: 135 mW at ±15V

Typical Applications**

**Note: Pin connections shown on schematic diagram and typical applications are for TO-5 package.

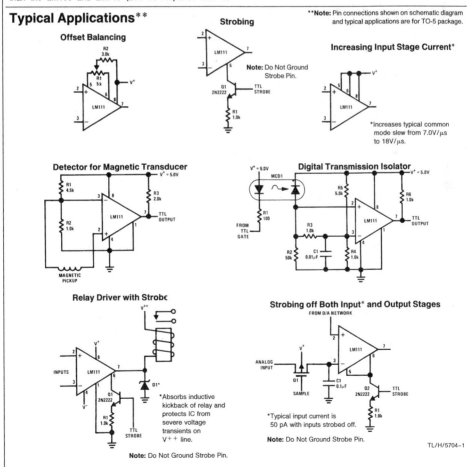

TL/H/5704–1

Absolute Maximum Ratings for the LM111/LM211

If Military/Aerospace specified devices are required, contact the National Semiconductor Sales Office/Distributors for availability and specifications. (Note 7)

Total Supply Voltage (V_{84})	36V
Output to Negative Supply Voltage (V_{74})	50V
Ground to Negative Supply Voltage (V_{14})	30V
Differential Input Voltage	±30V
Input Voltage (Note 1)	±15V
Power Dissipation (Note 2)	500 mW
Output Short Circuit Duration	10 sec

Operating Temperature Range LM111	$-55°C$ to $125°C$
LM211	$-25°C$ to $85°C$
Storage Temperature Range	$-65°C$ to $150°C$
Lead Temperature (Soldering, 10 sec)	260°C
Voltage at Strobe Pin	$V^+ - 5V$

Soldering Information
Dual-In-Line Package
Soldering (10 seconds) . . . . 260°C
Small Outline Package
Vapor Phase (60 seconds) . . . . 215°C
Infrared (15 seconds) . . . . 220°C

See AN-450 "Surface Mounting Methods and Their Effect on Product Reliability" for other methods of soldering surface mount devices.

ESD rating to be determined.

Electrical Characteristics for the LM111 and LM211 (Note 3)

Parameter	Conditions	Min	Typ	Max	Units
Input Offset Voltage (Note 4)	$T_A = 25°C$, $R_S \leq 50k$		0.7	3.0	mV
Input Offset Current (Note 4)	$T_A = 25°C$		4.0	10	nA
Input Bias Current	$T_A = 25°C$		60	100	nA
Voltage Gain	$T_A = 25°C$	40	200		V/mV
Response Time (Note 5)	$T_A = 25°C$		200		ns
Saturation Voltage	$V_{IN} \leq -5$ mV, $I_{OUT} = 50$ mA $T_A = 25°C$		0.75	1.5	V
Strobe ON Current (Note 6)	$T_A = 25°C$	2.0	3.0	5.0	mA
Output Leakage Current	$V_{IN} \geq 5$ mV, $V_{OUT} = 35V$ $T_A = 25°C$, $I_{STROBE} = 3$ mA		0.2	10	nA
Input Offset Voltage (Note 4)	$R_S \leq 50$ k			4.0	mV
Input Offset Current (Note 4)				20	nA
Input Bias Current				150	nA
Input Voltage Range	$V^+ = 15V$, $V^- = -15V$, Pin 7 Pull-Up May Go To 5V	-14.5	13.8,-14.7	13.0	V
Saturation Voltage	$V^+ \geq 4.5V$, $V^- = 0$ $V_{IN} \leq -6$ mV, $i_{SINK} \leq 8$ mA		0.23	0.4	V
Output Leakage Current	$V_{IN} \geq 5$ mV, $V_{OUT} = 35V$		0.1	0.5	µA
Positive Supply Current	$T_A = 25°C$		5.1	6.0	mA
Negative Supply Current	$T_A = 25°C$		4.1	5.0	mA

Note 1: This rating applies for ±15 supplies. The positive input voltage limit is 30V above the negative supply. The negative input voltage limit is equal to the negative supply voltage or 30V below the positive supply, whichever is less.

Note 2: The maximum junction temperature of the LM111 is 150°C, while that of the LM211 is 110°C. For operating at elevated temperatures, devices in the TO-5 package must be derated based on a thermal resistance of 150°C/W, junction to ambient, or 45°C/W, junction to case. The thermal resistance of the dual-in-line package is 110°C/W, junction to ambient.

Note 3: These specifications apply for $V_S = \pm15V$ and Ground pin at ground, and $-55°C \leq T_A \leq +125°C$, unless otherwise stated. With the LM211, however, all temperature specifications are limited to $-25°C \leq T_A \leq +85°C$. The offset voltage, offset current and bias current specifications apply for any supply voltage from a single 5V supply up to ±15V supplies.

Note 4: The offset voltages and offset currents given are the maximum values required to drive the output within a volt of either supply with a 1 mA load. Thus, these parameters define an error band and take into account the worst-case effects of voltage gain and input impedance.

Note 5: The response time specified (see definitions) is for a 100 mV input step with 5 mV overdrive.

Note 6: Do not short the strobe pin to ground; it should be current driven at 3 to 5 mA.

Note 7: Refer to RETS111X for the LM111H, LM111J and LM111J-8 military specifications.

Absolute Maximum Ratings for the LM311

If Military/Aerospace specified devices are required, contact the National Semiconductor Sales Office/Distributors for availability and specifications.

Total Supply Voltage (V_{84})	36V
Output to Negative Supply Voltage V_{74})	40V
Ground to Negative Supply Voltage V_{14})	30V
Differential Input Voltage	±30V
Input Voltage (Note 1)	±15V
Power Dissipation (Note 2)	500 mW

Output Short Circuit Duration	10 sec
Operating Temperature Range	0° to 70°C
Storage Temperature Range	−65°C to 150°C
Lead Temperature (soldering, 10 sec)	260°C
Voltage at Strobe Pin	$V^+ - 5V$

Soldering Information
Dual-In-Line Package
Soldering (10 seconds) . . . . 260°C
Small Outline Package
Vapor Phase (60 seconds) . . . . 215°C
Infrared (15 seconds) . . . . 220°C

See AN-450 "Surface Mounting Methods and Their Effect on Product Reliability" for other methods of soldering surface mount devices.

Electrical Characteristics for the LM311 (Note 3)

Parameter	Conditions	Min	Typ	Max	Units
Input Offset Voltage (Note 4)	$T_A = 25°C$, $R_S \leq 50k$		2.0	7.5	mV
Input Offset Current (Note 4)	$T_A = 25°C$		6.0	50	nA
Input Bias Current	$T_A = 25°C$		100	250	nA
Voltage Gain	$T_A = 25°C$	40	200		V/mV
Response Time (Note 5)	$T_A = 25°C$		200		ns
Saturation Voltage	$V_{IN} \leq -10$ mV, $I_{OUT} = 50$ mA $T_A = 25°C$		0.75	1.5	V
Strobe ON Current	$T_A = 25°C$	1.5	3.0		mA
Output Leakage Current	$V_{IN} \geq 10$ mV, $V_{OUT} = 35V$ $T_A = 25°C$, $I_{STROBE} = 3$ mA $V^- = V_{GRND} = -5V$		0.2	50	nA
Input Offset Voltage (Note 4)	$R_S \leq 50K$			10	mV
Input Offset Current (Note 4)				70	nA
Input Bias Current				300	nA
Input Voltage Range		−14.5	13.8, −14.7	13.0	V
Saturation Voltage	$V^+ \geq 4.5V$, $V^- = 0$ $V_{IN} \leq -10$ mV, $I_{SINK} \leq 8$ mA		0.23	0.4	V
Positive Supply Current	$T_A = 25°C$		5.1	7.5	mA
Negative Supply Current	$T_A = 25°C$		4.1	5.0	mA

Note 1: This rating applies for ±15V supplies. The positive input voltage limit is 30V above the negative supply. The negative input voltage limit is equal to the negative supply voltage or 30V below the positive supply, whichever is less.

Note 2: The maximum junction temperature of the LM311 is 110°C. For operating at elevated temperature, devices in the TO-5 package must be derated based on a thermal resistance of 150°C/W, junction to ambient, or 45°C/W, junction to case. The thermal resistance of the dual-in-line package is 100°C/W, junction to ambient.

Note 3: These specifications apply for $V_S = \pm 15V$ and the Ground pin at ground, and $0°C < T_A < +70°C$, unless otherwise specified. The offset voltage, offset current and bias current specifications apply for any supply voltage from a single 5V supply up to ±15V supplies.

Note 4: The offset voltages and offset currents given are the maximum values required to drive the output within a volt of either supply with 1 mA load. Thus, these parameters define an error band and take into account the worst-case effects of voltage gain and input impedance.

Note 5: The response time specified (see definitions) is for a 100 mV input step with 5 mV overdrive.

Note 6: Do not short the strobe pin to ground; it should be current driven at 3 to 5 mA.

LM111/LM211 Typical Performance Characteristics

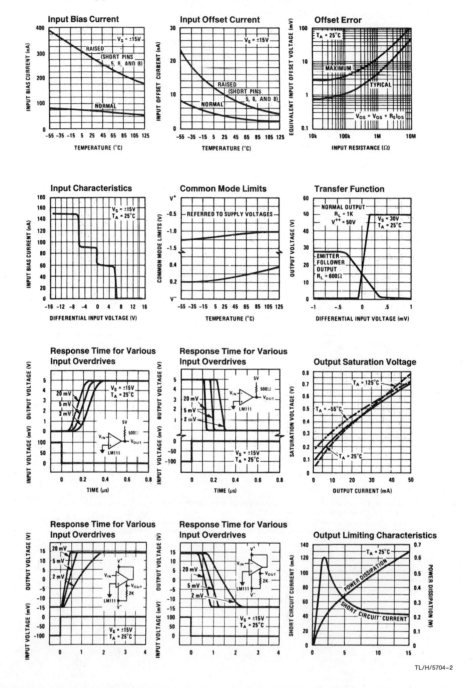

TL/H/5704–2

LM111/LM211 Typical Performance Characteristics (Continued)

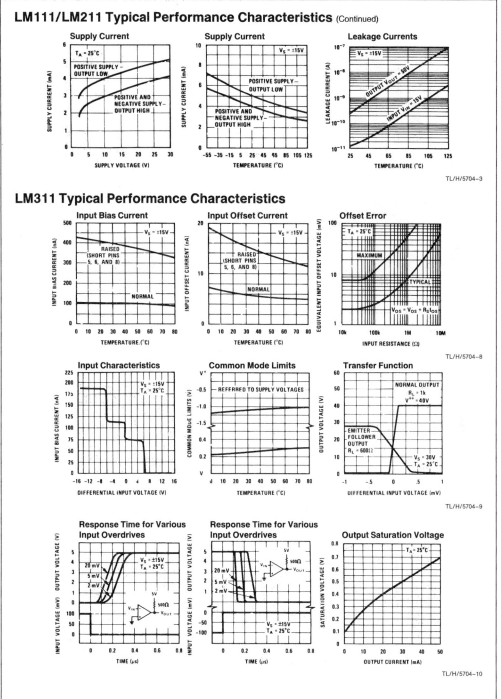

TL/H/5704–3

LM311 Typical Performance Characteristics

TL/H/5704–8

TL/H/5704–9

TL/H/5704–10

LM311 Typical Performance Characteristics (Continued)

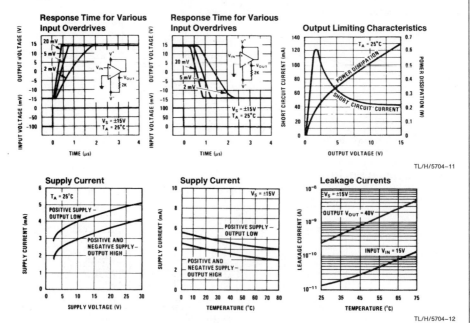

TL/H/5704–11

TL/H/5704–12

National Semiconductor Corporation

LM117/LM317 3-Terminal Adjustable Regulator

General Description

The LM117/LM317 are adjustable 3-terminal positive voltage regulators capable of supplying in excess of 1.5A over a 1.2V to 37V output range. They are exceptionally easy to use and require only two external resistors to set the output voltage. Further, both line and load regulation are better than standard fixed regulators. Also, the LM117 is packaged in standard transistor packages which are easily mounted and handled.

In addition to higher performance than fixed regulators, the LM117 series offers full overload protection available only in IC's. Included on the chip are current limit, thermal overload protection and safe area protection. All overload protection circuitry remains fully functional even if the adjustment terminal is disconnected.

Normally, no capacitors are needed unless the device is situated more than 6 inches from the input filter capacitors in which case an input bypass is needed. An optional output capacitor can be added to improve transient response. The adjustment terminal can be bypassed to achieve very high ripple rejections ratios which are difficult to achieve with standard 3-terminal regulators.

Besides replacing fixed regulators, the LM117 is useful in a wide variety of other applications. Since the regulator is "floating" and sees only the input-to-output differential voltage, supplies of several hundred volts can be regulated as long as the maximum input to output differential is not exceeded, i.e., avoid short-circuiting the output.

Also, it makes an especially simple adjustable switching regulator, a programmable output regulator, or by connecting a fixed resistor between the adjustment pin and output, the LM117 can be used as a precision current regulator. Supplies with electronic shutdown can be achieved by clamping the adjustment terminal to ground which programs the output to 1.2V where most loads draw little current.

The LM117K and LM317K are packaged in standard TO-3 transistor packages while the LM117H and LM317H are packaged in a solid Kovar base TO-39 transistor package. The LM117 is rated for operation from −55°C to +150°C, and the LM317 from 0°C to +125°C. The LM317T and LM317MP, rated for operation over a 0°C to +125°C range, are available in a TO-220 plastic package and a TO-202 package, respectively.

For applications requiring greater output current in excess of 3A and 5A, see LM150 series and LM138 series data sheets, respectively. For the negative complement, see LM137 series data sheet.

LM117 Series Packages and Power Capability

Device	Package	Rated Power Dissipation	Design Load Current
LM117	TO-3	20W	1.5A
LM317	TO-39	2W	0.5A
LM317T	TO-220	15W	1.5A
LM317M	TO-202	7.5W	0.5A

Features

- Adjustable output down to 1.2V
- Guaranteed 1.5A output current
- Line regulation typically 0.01%/V
- Load regulation typically 0.1%
- Current limit constant with temperature
- 100% electrical burn-in
- Eliminates the need to stock many voltages
- Standard 3-lead transistor package
- 80 dB ripple rejection
- Output is short-circuit protected

Typical Applications

1.2V–25V Adjustable Regulator

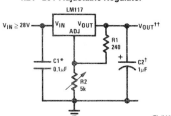

TL/H/9063–1

Full output current not available at high input-output voltages

*Needed if device is more than 6 inches from filter capacitors.

†Optional—improves transient response. Output capacitors in the range of 1 μF to 1000 μF of aluminum or tantalum electrolytic are commonly used to provide improved output impedance and rejection of transients.

$$\dagger\dagger V_{OUT} = 1.25V \left(1 + \frac{R2}{R1} \right) + I_{ADJ}(R2)$$

Digitally Selected Outputs

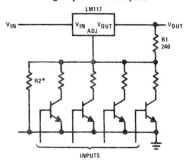

TL/H/9063–2

*Sets maximum V_{OUT}

Absolute Maximum Ratings

If Military/Aerospace specified devices are required, contact the National Semiconductor Sales Office/Distributors for availability and specifications. (Note 3)

Power Dissipation	Internally limited
Input—Output Voltage Differential	$+40V, -0.3V$
Operating Junction Temperature Range	
LM117	$-55°C$ to $+150°C$
LM317	$0°C$ to $+125°C$
Storage Temperature	$-65°C$ to $+150°C$
Lead Temperature (Soldering)	
Metal Package	$300°C$, 10 seconds
Plastic Package	$260°C$, 4 seconds
ESD rating	2k Volts

Preconditioning

Burn-In in Thermal Limit	100% All Devices

Electrical Characteristics (Note 1)

Parameter	Conditions	LM117			LM317			Units
		Min	Typ	Max	Min	Typ	Max	
Line Regulation	$T_j = 25°C$, $3V \leq (V_{IN} - V_{OUT}) \leq 40V$ (Note 2), $I_L = 10$ mA		0.01	0.02		0.01	0.04	%/V
Load Regulation	$T_J = 25°C$, 10 mA $\leq I_{OUT} \leq I_{MAX}$		0.1	0.3		0.1	0.5	%
Thermal Regulation	$T_J = 25°C$, 20 ms Pulse		0.03	0.07		0.04	0.07	%/W
Adjustment Pin Current			50	100		50.	100	μA
Adjustment Pin Current Change	10 mA $\leq I_L \leq I_{MAX}$ $3V \leq (V_{IN} - V_{OUT}) \leq 40V$		0.2	5		0.2	5	μA
Reference Voltage	$3V \leq (V_{IN} - V_{OUT}) \leq 40V$, (Note 3) 10 mA $\leq I_{OUT} \leq I_{MAX}$, $P \leq P_{MAX}$	1.20	1.25	1.30	1.20	1.25	1.30	V
Line Regulation	$3V \leq (V_{IN} - V_{OUT}) \leq 40V$, (Note 2)		0.02	0.05		0.02	0.07	%/V
Load Regulation	10 mA $\leq I_{OUT} \leq I_{MAX}$ (Note 2) $I_L = 10$ mA		0.3	1		0.3	1.5	%
Temperature Stability	$T_{MIN} \leq T_j \leq T_{MAX}$		1			1		%
Minimum Load Current	$(V_{IN} - V_{OUT}) = 40V$		3.5	5		3.5	10	mA
Current Limit	$(V_{IN} - V_{OUT}) \leq 15V$							
	K and T Package	1.5	2.2	3.4	1.5	2.2	3.4	A
	H and P Package	0.5	0.8	1.8	0.5		1.8	A
	$(V_{IN} - V_{OUT}) = 40V$, $T_j = +25°C$							
	K and T Package	0.30	0.4		0.15	0.4		A
	H and P Package	0.15	0.07		0.075	0.07		A
RMS Output Noise, % of V_{OUT}	$T_J = 25°C$, 10 Hz $\leq f \leq 10$ kHz		0.003			0.003		%
Ripple Rejection Ratio	$V_{OUT} = 10V$, $f = 120$ Hz		65			65		dB
	$C_{ADJ} = 10$ μF	66	80		66	80		dB
Long-Term Stability	$T_J = 125°C$		0.3	1		0.3	1	%
Thermal Resistance, Junction to Case	H Package		12	15		12	15	°C/W
	K Package		2.3	3		2.3	3	°C/W
	T Package					4		°C/W
	P Package					7		°C/W
Thermal Resistance, Junction to Ambient (No heat sink)	H Package		140			140		°C/W
	K Package		35			35		°C/W
	T Package					50		°C/W
	P Package					80		°C/W

Note 1: Unless otherwise specified, these specifications apply: $-55°C \leq T_j \leq +150°C$ for the LM117, and $0°C \leq T_J \leq +125°C$ for the LM317; $V_{IN} - V_{OUT} = 5V$; and $I_{OUT} = 0.1A$ for the TO-39 and TO-202 packages and $I_{OUT} = 0.5A$ for the TO-3 and TO-220 packages. Although power dissipation is internally limited, these specifications are applicable for power dissipations of 2W for the TO-39 and TO-202, and 20W for the TO-3 and TO-220. I_{MAX} is 1.5A for the TO-3 and TO-220 packages and 0.5A for the TO-39 and TO-202 packages.

Note 2: Regulation is measured at constant junction temperature, using pulse testing with a low duty cycle. Changes in output voltage due to heating effects are covered under the specification for thermal regulation.

Note 3: Refer to RETS117H drawing for LM117H or RETS117K drawing for LM117K military specifications.

Typical Performance Characteristics (K and T Packages)

Output Capacitor = 0 unless otherwise noted

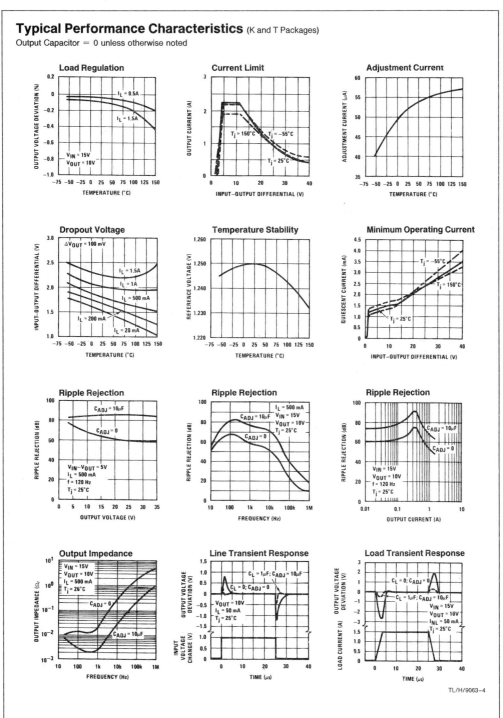

TL/H/9063-4

A33

Application Hints

In operation, the LM117 develops a nominal 1.25V reference voltage, V_{REF}, between the output and adjustment terminal. The reference voltage is impressed across program resistor R1 and, since the voltage is constant, a constant current I_1 then flows through the output set resistor R2, giving an output voltage of

$$V_{OUT} = V_{REF}\left(1 + \frac{R2}{R1}\right) + I_{ADJ}R2$$

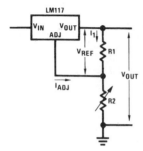

TL/H/9063–5

FIGURE 1

Since the 100 μA current from the adjustment terminal represents an error term, the LM117 was designed to minimize I_{ADJ} and make it very constant with line and load changes. To do this, all quiescent operating current is returned to the output establishing a minimum load current requirement. If there is insufficient load on the output, the output will rise.

External Capacitors

An input bypass capacitor is recommended. A 0.1 μF disc or 1 μF solid tantalum on the input is suitable input bypassing for almost all applications. The device is more sensitive to the absence of input bypassing when adjustment or output capacitors are used but the above values will eliminate the possibility of problems.

The adjustment terminal can be bypassed to ground on the LM117 to improve ripple rejection. This bypass capacitor prevents ripple from being amplified as the output voltage is increased. With a 10 μF bypass capacitor 80 dB ripple rejection is obtainable at any output level. Increases over 10 μF do not appreciably improve the ripple rejection at frequencies above 120 Hz. If the bypass capacitor is used, it is sometimes necessary to include protection diodes to prevent the capacitor from discharging through internal low current paths and damaging the device.

In general, the best type of capacitors to use is solid tantalum. Solid tantalum capacitors have low impedance even at high frequencies. Depending upon capacitor construction, it takes about 25 μF in aluminum electrolytic to equal 1 μF solid tantalum at high frequencies. Ceramic capacitors are also good at high frequencies; but some types have a large decrease in capacitance at frequencies around 0.5 MHz. For this reason, 0.01 μF disc may seem to work better than a 0.1 μF disc as a bypass.

Although the LM117 is stable with no output capacitors, like any feedback circuit, certain values of external capacitance can cause excessive ringing. This occurs with values be-

tween 500 pF and 5000 pF. A 1 μF solid tantalum (or 25 μF aluminum electrolytic) on the output swamps this effect and insures stability. Any increase of the load capacitance larger than 10 μF will merely improve the loop stability and output impedance.

Load Regulation

The LM117 is capable of providing extremely good load regulation but a few precautions are needed to obtain maximum performance. The current set resistor connected between the adjustment terminal and the output terminal (usually 240Ω) should be tied directly to the output (case) of the regulator rather than near the load. This eliminates line drops from appearing effectively in series with the reference and degrading regulation. For example, a 15V regulator with 0.05Ω resistance between the regulator and load will have a load regulation due to line resistance of 0.05$\Omega \times I_L$. If the set resistor is connected near the load the effective line resistance will be 0.05Ω (1 + R2/R1) or in this case, 11.5 times worse.

Figure 2 shows the effect of resistance between the regulator and 240Ω set resistor.

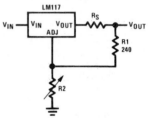

TL/H/9063–6

FIGURE 2. Regulator with Line Resistance in Output Lead

With the TO-3 package, it is easy to minimize the resistance from the case to the set resistor, by using two separate leads to the case. However, with the TO-5 package, care should be taken to minimize the wire length of the output lead. The ground of R2 can be returned near the ground of the load to provide remote ground sensing and improve load regulation.

Protection Diodes

When external capacitors are used with *any* IC regulator it is sometimes necessary to add protection diodes to prevent the capacitors from discharging through low current points into the regulator. Most 10 μF capacitors have low enough internal series resistance to deliver 20A spikes when shorted. Although the surge is short, there is enough energy to damage parts of the IC.

When an output capacitor is connected to a regulator and the input is shorted, the output capacitor will discharge into the output of the regulator. The discharge current depends on the value of the capacitor, the output voltage of the regulator, and the rate of decrease of V_{IN}. In the LM117, this discharge path is through a large junction that is able to sustain 15A surge with no problem. This is not true of other types of positive regulators. For output capacitors of 25 μF or less, there is no need to use diodes.

Application Hints (Continued)

The bypass capacitor on the adjustment terminal can discharge through a low current junction. Discharge occurs when *either* the input or output is shorted. Internal to the LM117 is a 50Ω resistor which limits the peak discharge current. No protection is needed for output voltages of 25V or less and 10 μF capacitance. *Figure 3* shows an LM117 with protection diodes included for use with outputs greater than 25V and high values of output capacitance.

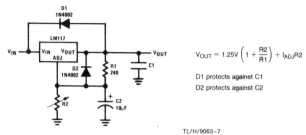

$$V_{OUT} = 1.25V \left(1 + \frac{R2}{R1}\right) + I_{ADJ}R2$$

D1 protects against C1
D2 protects against C2

TL/H/9063-7

FIGURE 3. Regulator with Protection Diodes

Schematic Diagram

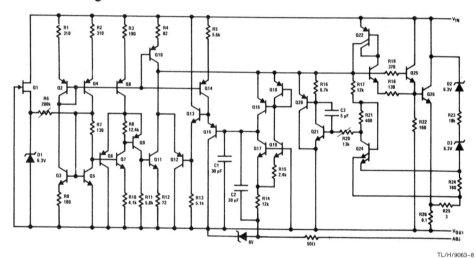

TL/H/9063-8

National
Semiconductor
Corporation

LM118/LM218/LM318
Operational Amplifiers

General Description

The LM118 series are precision high speed operational amplifiers designed for applications requiring wide bandwidth and high slew rate. They feature a factor of ten increase in speed over general purpose devices without sacrificing DC performance.

The LM118 series has internal unity gain frequency compensation. This considerably simplifies its application since no external components are necessary for operation. However, unlike most internally compensated amplifiers, external frequency compensation may be added for optimum performance. For inverting applications, feedforward compensation will boost the slew rate to over 150V/μs and almost double the bandwidth. Overcompensation can be used with the amplifier for greater stability when maximum bandwidth is not needed. Further, a single capacitor can be added to reduce the 0.1% settling time to under 1 μs.

The high speed and fast settling time of these op amps make them useful in A/D converters, oscillators, active filters, sample and hold circuits, or general purpose amplifiers. These devices are easy to apply and offer an order of magnitude better AC performance than industry standards such as the LM709.

The LM218 is identical to the LM118 except that the LM218 has its performance specified over a $-25°C$ to $+85°C$ temperature range. The LM318 is specified from 0°C to $+70°C$.

Features

- 15 MHz small signal bandwidth
- Guaranteed 50V/μs slew rate
- Maximum bias current of 250 nA
- Operates from supplies of $\pm5V$ to $\pm20V$
- Internal frequency compensation
- Input and output overload protected
- Pin compatible with general purpose op amps

Connection Diagrams

Metal Can Package*

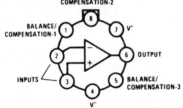

TL/H/7766-2

Top View

*Pin connections shown on schematic diagram and typical applications are for TO-5 package.

Order Number LM118H, LM218H or LM318H
See NS Package Number H08C

Dual-In-Line Package

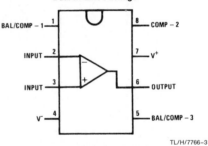

TL/H/7766-3

Top View
Order Number LM118J-8, LM318J-8,
LM318M or LM318N
See NS Package Number J08A, M08A or N08B

Absolute Maximum Ratings

If Military/Aerospace specified devices are required, contact the National Semiconductor Sales Office/ Distributors for availability and specifications. (Note 6)

Supply Voltage	±20V
Power Dissipation (Note 1)	500 mW
Differential Input Current (Note 2)	±10 mA
Input Voltage (Note 3)	±15V
Output Short-Circuit Duration	Indefinite

Operating Temperature Range	
LM118	−55°C to +125°C
LM218	−25°C to +85°C
LM318	0°C to +70°C
Storage Temperature Range	−65°C to +150°C
Lead Temperature (Soldering, 10 sec.)	
Hermetic Package	300°C
Plastic Package	260°C
Soldering Information	
Dual-In-Line Package	
Soldering (10 sec.)	260°C
Small Outline Package	
Vapor Phase (60 sec.)	215°C
Infrared (15 sec.)	220°C

See AN-450 "Surface Mounting Methods and Their Effect on Product Reliability" for other methods of soldering surface mount devices.

ESD rating to be determined.

Electrical Characteristics (Note 4)

Parameter	Conditions	LM118/LM218			LM318			Units
		Min	Typ	Max	Min	Typ	Max	
Input Offset Voltage	$T_A = 25°C$		2	4		4	10	mV
Input Offset Current	$T_A = 25°C$		6	50		30	200	nA
Input Bias Current	$T_A = 25°C$		120	250		150	500	nA
Input Resistance	$T_A = 25°C$	1	3		0.5	3		MΩ
Supply Current	$T_A = 25°C$		5	8		5	10	mA
Large Signal Voltage Gain	$T_A = 25°C$, $V_S = ±15V$ $V_{OUT} = ±10V$, $R_L \geq 2$ kΩ	50	200		25	200		V/mV
Slew Rate	$T_A = 25°C$, $V_S = ±15V$, $A_V = 1$ (Note 5)	50	70		50	70		V/µs
Small Signal Bandwidth	$T_A = 25°C$, $V_S = ±15V$		15			15		MHz
Input Offset Voltage				6			15	mV
Input Offset Current				100			300	nA
Input Bias Current				500			750	nA
Supply Current	$T_A = 125°C$		4.5	7				mA
Large Signal Voltage Gain	$V_S = ±15V$, $V_{OUT} = ±10V$ $R_L \geq 2$ kΩ	25			20			V/mV
Output Voltage Swing	$V_S = ±15V$, $R_L = 2$ kΩ	±12	±13		±12	±13		V
Input Voltage Range	$V_S = ±15V$	±11.5			±11.5			V
Common-Mode Rejection Ratio		80	100		70	100		dB
Supply Voltage Rejection Ratio		70	80		65	80		dB

Note 1: The maximum junction temperature of the LM118 is 150°C, the LM218 is 110°C, and the LM318 is 110°C. For operating at elevated temperatures, devices in the TO-5 package must be derated based on a thermal resistance of 150°C/W, junction to ambient, or 45°C/W, junction to case. The thermal resistance of the dual-in-line package is 100°C/W, junction to ambient.

Note 2: The inputs are shunted with back-to-back diodes for overvoltage protection. Therefore, excessive current will flow if a differential input voltage in excess of 1V is applied between the inputs unless some limiting resistance is used.

Note 3: For supply voltages less than ±15V, the absolute maximum input voltage is equal to the supply voltage.

Note 4: These specifications apply for ±5V ≤ V_S ≤ ±20V and −55°C ≤ T_A ≤ +125°C (LM118), −25°C ≤ T_A ≤ +85°C (LM218), and 0°C ≤ T_A ≤ +70°C (LM318). Also, power supplies must be bypassed with 0.1 µF disc capacitors.

Note 5: Slew rate is tested with $V_S = ±15V$. The LM118 is in a unity-gain non-inverting configuration. V_{IN} is stepped from −7.5V to +7.5V and vice versa. The slew rates between −5.0V and +5.0V and vice versa are tested and guaranteed to exceed 50V/µs.

Note 6: Refer to RETS118X for LM118H and LM118J-8 military specifications.

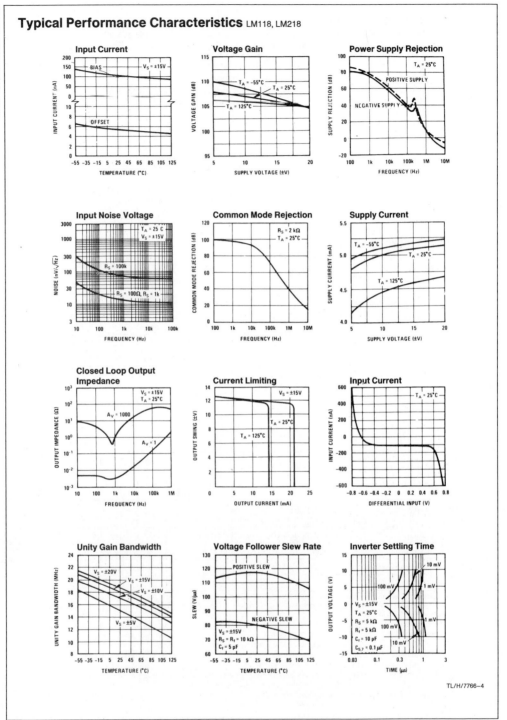

Typical Performance Characteristics LM118, LM218

TL/H/7766–4

Typical Performance Characteristics LM118, LM218 (Continued)

Typical Performance Characteristics LM318

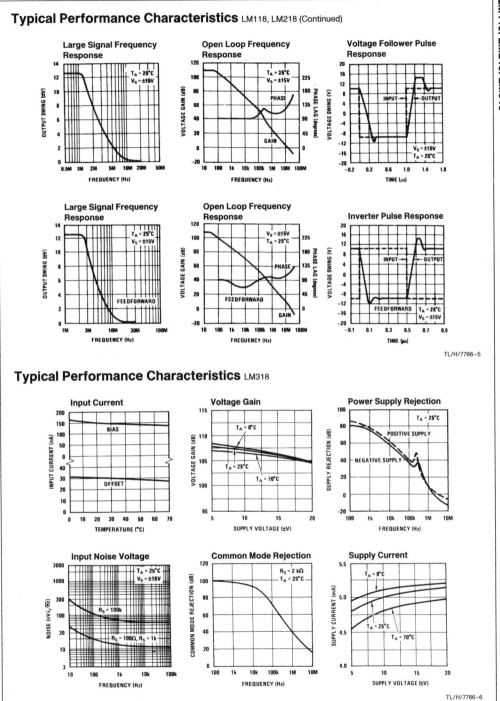

TL/H/7766–5

TL/H/7766–6

Typical Performance Characteristics LM318 (Continued)

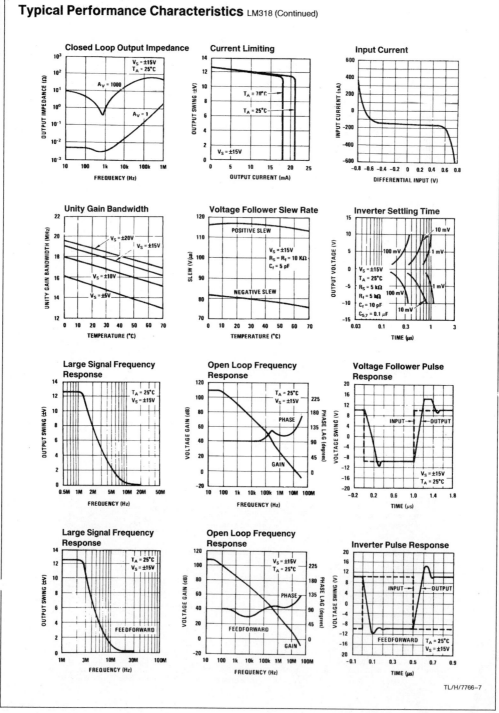

TL/H/7766–7

A40

LM125/LM325/LM325A, LM126/LM326 Voltage Regulators

General Description

These are dual polarity tracking regulators designed to provide balanced positive and negative output voltages at current up to 100 mA, the devices are set for ±15V and ±12V outputs respectively. Input voltages up to ±30V can be used and there is provision for adjustable current limiting. These devices are available in three package types to accommodate various power requirements and temperature ranges.

Features

- ±15V and ±12V tracking outputs
- Output current to 100 mA
- Output voltage balanced to within 1% (LM125, LM126, LM325A)
- Line and load regulation of 0.06%
- Internal thermal overload protection
- Standby current drain of 3 mA
- Externally adjustable current limit
- Internal current limit

Schematic and Connection Diagrams

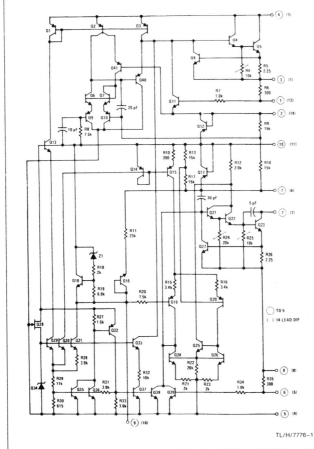

TL/H/7776–1

Dual-In-Line Package

Top View

Order Number LM325AN, LM325N or LM326N
See NS Package Number N14A

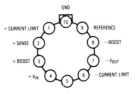

Metal Can Package

Case connected to −V$_{IN}$ TL/H/7776–3

Top View

Order Number LM125H, LM325H, LM126H or LM326H
See NS Package Number H10C

Absolute Maximum Ratings

If Military/Aerospace specified devices are required, contact the National Semiconductor Sales Office/Distributors for availability and specifications. (Note 5)

Input Voltage	±30V
Forced V_O^+ (Min) (Note 1)	−0.5V
Forced V_O^- (Max) (Note 1)	+0.5V
Power Dissipation (Note 2)	P_{MAX}
Output Short-Circuit Duration (Note 3)	Indefinite

Operating Conditions

Operating Free Temperature Range

LM125	−55°C to +125°C
LM325, LM325A	0°C to +70°C
Storage Temperature Range	−65°C to +150°C
Lead Temperature (Soldering, 10 sec.)	300°C

Electrical Characteristics LM125/LM325/LM325A (Note 2)

Parameter	Conditions	Min	Typ	Max	Units
Output Voltage	T_j = 25°C				
LM125/LM325A		14.8	15	15.2	V
LM325		14.5	15	15.5	V
Input-Output Differential		2.0			V
Line Regulation	V_{IN} = 18V to 30V, I_L = 20 mA, T_j = 25°C		2.0	10	mV
Line Regulation Over Temperature Range	V_{IN} = 18V to 30V, I_L = 20 mA,		2.0	20	mV
Load Regulation	I_L = 0 to 50 mA, V_{IN} = ±30V, T_j = 25°C				
V_O^+			3.0	10	mV
V_O^-			5.0	10	mV
Load Regulation Over Temperature Range	I_L = 0 to 50 mA, V_{IN} = ±30V				
V_O^+			4.0	20	mV
V_O^-			7.0	20	mV
Output Voltage Balance	T_j = 25°C				
LM125, LM325A				±150	mV
LM325				±300	mV
Output Voltage Over Temperature Range	$P \leq P_{MAX}$, $0 \leq I_O \leq 50$ mA, $18V \leq |V_{IN}| \leq 30$				
LM125, LM325A		14.65		15.35	V
LM325		14.27		15.73	V
Temperature Stability of V_O			±0.3		%
Short Circuit Current Limit	T_j = 25°C		260		mA
Output Noise Voltage	T_j = 25°C, BW = 100 − 10 kHz		150		μVrms
Positive Standby Current	T_j = 25°C		1.75	3.0	mA
Negative Standby Current	T_j = 25°C		3.1	5.0	mA
Long Term Stability			0.2		%/kHr
Thermal Resistance Junction to Case (Note 4)					
LM125H, LM325H			20		°C/W
Junction to Ambient	(Still Air)		215		°C/W
Junction to Ambient	(400 Lf/min Air Flow)		82		°C/W
Junction to Ambient	(Still Air)				
LM325AN, LM325N			90		°C/W

Note 1: That voltage to which the output may be forced without damage to the device.

Note 2: Unless otherwise specified these specifications apply for T_j = 55°C to +150°C on LM125, T_j = 0°C to +125°C on LM325A, T_j = 0°C to +125°C on LM325, V_{IN} = ±20V, I_L = 0 mA, I_{MAX} = 100 mA, P_{MAX} = 2.0W for the TO-5 H Package. I_{MAX} = 100 mA. I_{MAX} = 100 mA, P_{MAX} = 1.0W for the DIP N Package.

Note 3: If the junction temperature exceeds 150°C, the output short circuit duration is 60 seconds.

Note 4: Without a heat sink, the thermal resistance junction to ambient of the TO-5 Package is about 215°C/W. With a heat sink, the effective thermal resistance can only approach the junction to case values specified, depending on the efficiency of the sink.

Note 5: Refer to RETS125X drawing for military specification of LM125.

Absolute Maximum Ratings

If Military/Aerospace specified devices are required, contact the National Semiconductor Sales Office/Distributors for availability and specifications. (Note 5)

Input Voltage	± 30V
Forced V_O^+ (Min) (Note 1)	-0.5V
Forced V_O^- (Max) (Note 1)	$+0.5$V
Power Dissipation (Note 2)	Internally Limited
Output Short-Circuit Duration (Note 3)	Indefinite

Operating Conditions

Operating Free Temperature Range

LM126	$-55°C$ to $+125°C$
LM326	$0°C$ to $+70°C$
Storage Temperature Range	$-65°C$ to $+150°C$
Lead Temperature (Soldering, 10 sec.)	$300°C$

Electrical Characteristics LM126/LM326 (Note 2)

Parameter	Conditions	Min	Typ	Max	Units
Output Voltage LM126/LM326	$T_j = 25°C$	11.8 11.5	12	12.2 12.5	V V
Input-Output Differential		2.0			V
Line Regulation	$V_{IN} = 15V$ to $30V$ $I_L = 20$ mA, $T_j = 25°C$		2.0	10	mV
Line Regulation Over Temperature Range	$V_{IN} = 15V$ to $30V$, $I_L = 20$ mA		2.0	20	mV
Load Regulation V_O^+ V_O^-	$I_L = 0$ to 50 mA, $V_{IN} = \pm 30V$, $T_j = 25°C$		3.0 5.0	10 10	mV mV
Load Regulation Over Temperature Range V_O^+ V_O^-	$I_L = 0$ to 50 mA, $V_{IN} = \pm 30V$		4.0 7.0	20 20	mV mV
Output Voltage Balance LM126, LM326	$T_j = 25°C$			± 125 ± 250	mV mV
Output Voltage Over Temperature Range LM126 LM326	$P \le P_{MAX}, 0 \le I_O \le 50$ mA, $15V \le |V_{IN}| \le 30$	11.68 11.32		12.32 12.68	V V
Temperature Stability of V_O			± 0.3		%
Short Circuit Current Limit	$T_j = 25°C$		260		mA
Output Noise Voltage	$T_j = 25°C$, BW $= 100 - 10$ kHz		100		μVrms
Positive Standby Current	$T_j = 25°C$, $I_L = 0$		1.75	3.0	mA
Negative Standby Current	$T_j = 25°C$, $I_L = 0$		3.1	5.0	mA
Long Term Stability			0.2		%/kHr
Thermal Resistance Junction to Case (Note 4) LM126H, LM326H Junction to Ambient Junction to Ambient	 (Still Air) (400 Lf/min Air Flow)		20 215 82		°C/W °C/W °C/W
Junction to Ambient LM326N			150		°C/W

Note 1: That voltage to which the output may be forced without damage to the device.

Note 2: Unless otherwise specified these specifications apply for $T_j = 55°C$ to $+150°C$ on LM126, $T_j = 0°C$ to $+125°C$ on LM326, $V_{IN} = \pm 20V$, $I_L = 0$ mA, $I_{MAX} = 100$ mA, $P_{MAX} = 2.0W$ for the TO-5 H Package. $I_{MAX} = 100$ mA. $I_{MAX} = 100$ mA, $P_{MAX} = 1.0W$ for the DIP N Package.

Note 3: If the junction temperature exceeds 150°C, the output short circuit duration is 60 seconds.

Note 4: Without a heat sink, the thermal resistance junction to ambient of the TO-5 Package is about 215°C/W. With a heat sink, the effective thermal resistance can only approach the junction to case values specified, depending on the efficiency of the sink.

Note 5: Refer to RETS126X drawing for military specification of LM126.

Typical Performance Characteristics

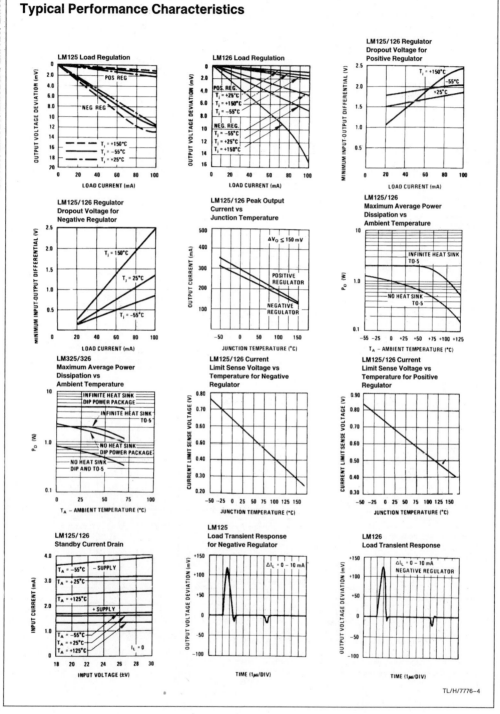

TL/H/7776–4

A44

LM386 Low Voltage Audio Power Amplifier

General Description

The LM386 is a power amplifier designed for use in low voltage consumer applications. The gain is internally set to 20 to keep external part count low, but the addition of an external resistor and capacitor between pins 1 and 8 will increase the gain to any value up to 200.

The inputs are ground referenced while the output is automatically biased to one half the supply voltage. The quiescent power drain is only 24 milliwatts when operating from a 6 volt supply, making the LM386 ideal for battery operation.

Features

- Battery operation
- Minimum external parts
- Wide supply voltage range — 4V–12V or 5V–18V
- Low quiescent current drain — 4 mA

- Voltage gains from 20 to 200
- Ground referenced input
- Self-centering output quiescent voltage
- Low distortion
- Eight pin dual-in-line package

Applications

- AM-FM radio amplifiers
- Portable tape player amplifiers
- Intercoms
- TV sound systems
- Line drivers
- Ultrasonic drivers
- Small servo drivers
- Power converters

Equivalent Schematic and Connection Diagrams

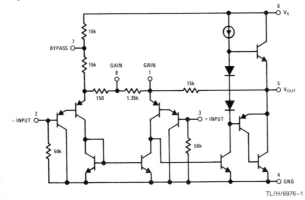

TL/H/6976–1

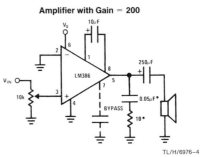

Dual-In-Line and Small Outline Packages

TL/H/6976–2

Top View

**Order Number LM386M-1,
LM386N-1, LM386N-3 or LM386N-4
See NS Package Number
M08A or N08E**

Typical Applications

**Amplifier with Gain = 20
Minimum Parts**

TL/H/6976–3

*Required for LM386N-4 only.

Amplifier with Gain = 200

TL/H/6976–4

*Required for LM386N-4 only.

A45

Absolute Maximum Ratings

If Military/Aerospace specified devices are required, contact the National Semiconductor Sales Office/Distributors for availability and specifications.

Supply Voltage (LM386N-1, -3, LM386M-1)	15V
Supply Voltage (LM386N-4)	22V
Package Dissipation (Note 1) (LM386N-4)	1.25W
Input Voltage	±0.4V
Storage Temperature	−65°C to +150°C
Operating Temperature	0°C to +70°C

Junction Temperature	+150°C
Soldering Information	
Dual-In-Line Package	
Soldering (10 sec)	+260°C
Small Outline Package	
Vapor Phase (60 sec)	+215°C
Infrared (15 sec)	+220°C

See AN-450 "Surface Mounting Methods and Their Effect on Product Reliability" for other methods of soldering surface mount devices.

Electrical Characteristics $T_A = 25°C$

Parameter	Conditions	Min	Typ	Max	Units
Operating Supply Voltage (V_S)					
LM386N-1, -3, LM386M-1		4		12	V
LM386N-4		5		18	V
Quiescent Current (I_Q)	$V_S = 6V$, $V_{IN} = 0$		4	8	mA
Output Power (P_{OUT})					
LM386N-1, LM386M-1	$V_S = 6V$, $R_L = 8\Omega$, THD = 10%	250	325		mW
LM386N-3	$V_S = 9V$, $R_L = 8\Omega$, THD = 10%	500	700		mW
LM386N-4	$V_S = 16V$, $R_L = 32\Omega$, THD = 10%	700	1000		mW
Voltage Gain (A_V)	$V_S = 6V$, f = 1 kHz		26		dB
	10 μF from Pin 1 to 8		46		dB
Bandwidth (BW)	$V_S = 6V$, Pins 1 and 8 Open		300		kHz
Total Harmonic Distortion (THD)	$V_S = 6V$, $R_L = 8\Omega$, $P_{OUT} = 125$ mW f = 1 kHz, Pins 1 and 8 Open		0.2		%
Power Supply Rejection Ratio (PSRR)	$V_S = 6V$, f = 1 kHz, $C_{BYPASS} = 10 \mu$F Pins 1 and 8 Open, Referred to Output		50		dB
Input Resistance (R_{IN})			50		kΩ
Input Bias Current (I_{BIAS})	$V_S = 6V$, Pins 2 and 3 Open		250		nA

Note 1: For operation in ambient temperatures above 25°C, the device must be derated based on a 150°C maximum junction temperature and 1) a thermal resistance of 80°C/W junction to ambient for the dual-in-line package and 2) a thermal resistance of 170°C/W for the small outline package.

Application Hints

GAIN CONTROL

To make the LM386 a more versatile amplifier, two pins (1 and 8) are provided for gain control. With pins 1 and 8 open the 1.35 kΩ resistor sets the gain at 20 (26 dB). If a capacitor is put from pin 1 to 8, bypassing the 1.35 kΩ resistor, the gain will go up to 200 (46 dB). If a resistor is placed in series with the capacitor, the gain can be set to any value from 20 to 200. Gain control can also be done by capacitively coupling a resistor (or FET) from pin 1 to ground.

Additional external components can be placed in parallel with the internal feedback resistors to tailor the gain and frequency response for individual applications. For example, we can compensate poor speaker bass response by frequency shaping the feedback path. This is done with a series RC from pin 1 to 5 (paralleling the internal 15 kΩ resistor). For 6 dB effective bass boost: R $\cong$ 15 kΩ, the lowest value for good stable operation is R = 10 kΩ if pin 8 is open. If pins 1 and 8 are bypassed then R as low as 2 kΩ can be used. This restriction is because the amplifier is only compensated for closed-loop gains greater than 9.

INPUT BIASING

The schematic shows that both inputs are biased to ground with a 50 kΩ resistor. The base current of the input transistors is about 250 nA, so the inputs are at about 12.5 mV when left open. If the dc source resistance driving the LM386 is higher than 250 kΩ it will contribute very little additional offset (about 2.5 mV at the input, 50 mV at the output). If the dc source resistance is less than 10 kΩ, then shorting the unused input to ground will keep the offset low (about 2.5 mV at the input, 50 mV at the output). For dc source resistances between these values we can eliminate excess offset by putting a resistor from the unused input to ground, equal in value to the dc source resistance. Of course all offset problems are eliminated if the input is capacitively coupled.

When using the LM386 with higher gains (bypassing the 1.35 kΩ resistor between pins 1 and 8) it is necessary to bypass the unused input, preventing degradation of gain and possible instabilities. This is done with a 0.1 μF capacitor or a short to ground depending on the dc source resistance on the driven input.

Typical Performance Characteristics

Quiescent Supply Current vs Supply Voltage

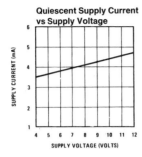

Power Supply Rejection Ratio (Referred to the Output) vs Frequency

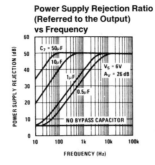

Peak-to-Peak Output Voltage Swing vs Supply Voltage

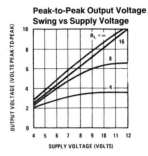

Voltage Gain vs Frequency

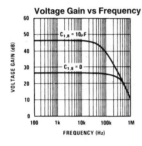

Distortion vs Frequency

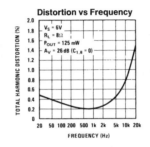

Distortion vs Output Power

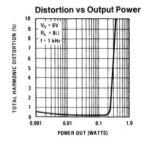

Device Dissipation vs Output Power—4Ω Load

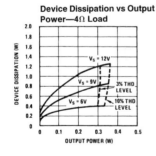

Device Dissipation vs Output Power—8Ω Load

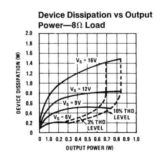

Device Dissipation vs Output Power—16Ω Load

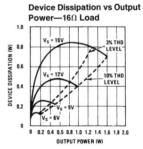

TL/H/6976–5

A47

**National
Semiconductor
Corporation**

LM2900/LM3900, LM3301, LM3401 Quad Amplifiers

General Description

The LM2900 series consists of four independent, dual input, internally compensated amplifiers which were designed specifically to operate off of a single power supply voltage and to provide a large output voltage swing. These amplifiers make use of a current mirror to achieve the non-inverting input function. Application areas include: ac amplifiers, RC active filters, low frequency triangle, squarewave and pulse waveform generation circuits, tachometers and low speed, high voltage digital logic gates.

Features

- Wide single supply voltage   4 V_{DC} to 32 V_{DC}
  Range or dual supplies ± 2 V_{DC} to ± 16 V_{DC}
- Supply current drain independent of supply voltage
- Low input biasing current   30 nA
- High open-loop gain   70 dB
- Wide bandwidth   2.5 MHz (unity gain)
- Large output voltage swing $(V^+ - 1)$ Vp-p
- Internally frequency compensated for unity gain
- Output short-circuit protection

Schematic and Connection Diagrams

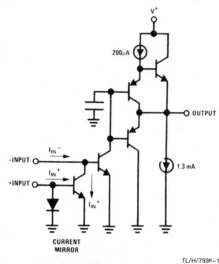

TL/H/7936–1

**Order Number LM3900M
See NS Package Number M14A**

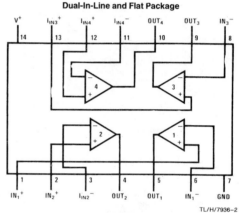

Dual-In-Line and Flat Package

TL/H/7936–2

Top View

**Order Number LM2900N, LM3900N, LM3301N or LM3401N
See NS Package Number N14A**

Absolute Maximum Ratings

If Military/Aerospace specified devices are required, contact the National Semiconductor Sales Office/Distributors for availability and specifications.

	LM2900/LM3900	LM3301	LM3401
Supply Voltage	32 V_{DC} ±16 V_{DC}	28 V_{DC} ±14 V_{DC}	18 V_{DC} ±9 V_{DC}
Power Dissipation (T_A = 25°C) (Note 1) Molded DIP	1220 mW 1080 mW	1080 mW	1080 mW
Input Currents, I_{IN}^+ or I_{IN}^-	20 mA_{DC}	20 mA_{DC}	20 mA_{DC}
Output Short-Circuit Duration—One Amplifier T_A = 25°C (See Application Hints)	Continuous	Continuous	Continuous
Operating Temperature Range LM2900 LM3900	 −40°C to +85°C 0°C to +70°C	−40°C to +85°C	0°C to +75°C
Storage Temperature Range	−65°C to +150°C	−65°C to +150°C	−65°C to +150°C
Lead Temperature (Soldering, 10 sec.)	260°C	260°C	260°C
Soldering Information Dual-In-Line Package Soldering (10 sec.) Small Outline Package Vapor Phase (60 sec.) Infrared (15 sec.)	 260°C 215°C 220°C	 260°C 215°C 220°C	 260°C 215°C 220°C

See AN-450 "Surface Mounting Methods and Their Effect on Product Reliability" for other methods of soldering surface mount devices.

ESD rating to be determined.

Electrical Characteristics T_A = 25°C, V^+ = 15 V_{DC}, unless otherwise stated

Parameter		Conditions	LM2900			LM3900			LM3301			LM3401			Units
			Min	Typ	Max	Min	Typ	Max	Min	Typ	Max	Min	Typ	Max	
Open Loop	Voltage Gain	Over Temp.										0.8			V/mV
	Voltage Gain	ΔV_O = 10 V_{DC} Inverting Input	1.2	2.8		1.2	2.8		1.2	2.8		1.2	2.8		
	Input Resistance			1			1			1		0.1	1		MΩ
	Output Resistance			8			8			9			8		kΩ
Unity Gain Bandwidth		Inverting Input		2.5			2.5			2.5			2.5		MHz
Input Bias Current		Inverting Input, V^+ = 5 V_{DC} Inverting Input		30	200		30	200		30	300		30	300 500	nA
Slew Rate		Positive Output Swing Negative Output Swing		0.5 20			0.5 20			0.5 20			0.5 20		V/μs
Supply Current		R_L = ∞ On All Amplifiers		6.2	10		6.2	10		6.2	10		6.2	10	mA_{DC}
Output Voltage Swing	V_{OUT} High	R_L = 2k, V^+ = 15.0 V_{DC}	I_{IN}^- = 0, I_{IN}^+ = 0	13.5			13.5			13.5			13.5			
	V_{OUT} Low		I_{IN}^- = 10 μA, I_{IN}^+ = 0		0.09	0.2		0.09	0.2		0.09	0.2		0.09	0.2	V_{DC}
	V_{OUT} High	V^+ = Absolute Maximum Ratings	I_{IN}^- = 0, I_{IN}^+ = 0 R_L = ∞,	29.5			29.5			26.0			16.0			
Output Current Capability	Source		6	18		6	10		5	18		5	10		
	Sink	(Note 2)	0.5	1.3		0.5	1.3		0.5	1.3		0.5	1.3		mA_{DC}
	I_{SINK}	V_{OL} = 1V, I_{IN}^- = 5 μA		5			5			5			5		

Electrical Characteristics (Note 6), V$^+$ = 15 V$_{DC}$, unless otherwise stated (Continued)

Parameter	Conditions	LM2900			LM3900			LM3301			LM3401			Units
		Min	Typ	Max	Min	Typ	Max	Min	Typ	Max	Min	Typ	Max	
Power Supply Rejection	T$_A$ = 25°C, f = 100 Hz		70			70			70			70		dB
Mirror Gain	@ 20 μA (Note 3)	0.90	1.0	1.1	0.90	1.0	1.1	0.90	1	1.10	0.90	1	1.10	μA/μA
	@ 200 μA (Note 3)	0.90	1.0	1.1	0.90	1.0	1.1	0.90	1	1.10	0.90	1	1.10	
ΔMirror Gain	@ 20 μA to 200 μA (Note 3)		2	5		2	5		2	5		2	5	%
Mirror Current	(Note 4)		10	500		10	500		10	500		10	500	μA$_{DC}$
Negative Input Current	T$_A$ = 25°C (Note 5)		1.0			1.0			1.0			1.0		mA$_{DC}$
Input Bias Current	Inverting Input		300			300								nA

Note 1: For operating at high temperatures, the device must be derated based on a 125°C maximum junction temperature and a thermal resistance of 92°C/W which applies for the device soldered in a printed circuit board, operating in a still air ambient.

Note 2: The output current sink capability can be increased for large signal conditions by overdriving the inverting input. This is shown in the section on Typical Characteristics.

Note 3: This spec indicates the current gain of the current mirror which is used as the non-inverting input.

Note 4: Input V$_{BE}$ match between the non-inverting and the inverting inputs occurs for a mirror current (non-inverting input current) of approximately 10 μA. This is therefore a typical design center for many of the application circuits.

Note 5: Clamp transistors are included on the IC to prevent the input voltages from swinging below ground more than approximately −0.3 V$_{DC}$. The negative input currents which may result from large signal overdrive with capacitance input coupling need to be externally limited to values of approximately 1 mA. Negative input currents in excess of 4 mA will cause the output voltage to drop to a low voltage. This maximum current applies to any one of the input terminals. If more than one of the input terminals are simultaneously driven negative smaller maximum currents are allowed. Common-mode current biasing can be used to prevent negative input voltages; see for example, the "Differentiator Circuit" in the applications section.

Note 6: These specs apply for −40°C ≤ T$_A$ ≤ +85°C, unless otherwise stated.

Application Hints

When driving either input from a low-impedance source, a limiting resistor should be placed in series with the input lead to limit the peak input current. Currents as large as 20 mA will not damage the device, but the current mirror on the non-inverting input will saturate and cause a loss of mirror gain at mA current levels—especially at high operating temperatures.

Precautions should be taken to insure that the power supply for the integrated circuit never becomes reversed in polarity or that the unit is not inadvertently installed backwards in a test socket as an unlimited current surge through the resulting forward diode could cause fusing of the internal conductors and result in a destroyed unit.

Output short circuits either to ground or to the positive power supply should be of short time duration. Units can be destroyed, not as a result of the short circuit current causing metal fusing, but rather due to the large increase in IC chip dissipation which will cause eventual failure due to excessive junction temperatures. For example, when operating from a well-regulated +5 V$_{DC}$ power supply at T$_A$ = 25°C with a 100 kΩ shunt-feedback resistor (from the output to the inverting input) a short directly to the power supply will not cause catastrophic failure but the current magnitude will be approximately 50 mA and the junction temperature will be above T$_J$ max. Larger feedback resistors will reduce the current, 11 MΩ provides approximately 30 mA, an open circuit provides 1.3 mA, and a direct connection from the output to the non-inverting input will result in catastrophic failure when the output is shorted to V$^+$ as this then places the base-emitter junction of the input transistor directly across the power supply. Short-circuits to ground will have magnitudes of approximately 30 mA and will not cause catastrophic failure at T$_A$ = 25°C.

Unintentional signal coupling from the output to the non-inverting input can cause oscillations. This is likely only in breadboard hook-ups with long component leads and can be prevented by a more careful lead dress or by locating the non-inverting input biasing resistor close to the IC. A quick check of this condition is to bypass the non-inverting input to ground with a capacitor. High impedance biasing resistors used in the non-inverting input circuit make this input lead highly susceptible to unintentional AC signal pickup.

Operation of this amplifier can be best understood by noticing that input currents are differenced at the inverting-input terminal and this difference current then flows through the external feedback resistor to produce the output voltage. Common-mode current biasing is generally useful to allow operating with signal levels near ground or even negative as this maintains the inputs biased at +V$_{BE}$. Internal clamp transistors (see note 5) catch negative input voltages at approximately −0.3 V$_{DC}$ but the magnitude of current flow has to be limited by the external input network. For operation at high temperature, this limit should be approximately 100 μA.

This new "Norton" current-differencing amplifier can be used in most of the applications of a standard IC op amp. Performance as a DC amplifier using only a single supply is not as precise as a standard IC op amp operating with split supplies but is adequate in many less critical applications. New functions are made possible with this amplifier which are useful in single power supply systems. For example, biasing can be designed separately from the AC gain as was shown in the "inverting amplifier," the "difference integrator" allows controlling the charging and the discharging of the integrating capacitor with positive voltages, and the "frequency doubling tachometer" provides a simple circuit which reduces the ripple voltage on a tachometer output DC voltage.

Typical Performance Characteristics

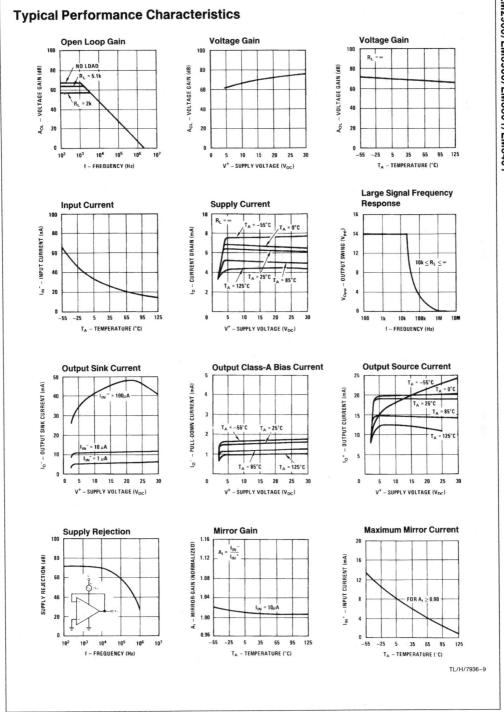

TL/H/7936–9

**National
Semiconductor
Corporation**

LM723/LM723C Voltage Regulator

General Description

The LM723/LM723C is a voltage regulator designed primarily for series regulator applications. By itself, it will supply output currents up to 150 mA; but external transistors can be added to provide any desired load current. The circuit features extremely low standby current drain, and provision is made for either linear or foldback current limiting.

The LM723/LM723C is also useful in a wide range of other applications such as a shunt regulator, a current regulator or a temperature controller.

The LM723C is identical to the LM723 except that the LM723C has its performance guaranteed over a 0°C to +70°C temperature range, instead of −55°C to +125°C.

Features

■ 150 mA output current without external pass transistor
■ Output currents in excess of 10A possible by adding external transistors
■ Input voltage 40V max
■ Output voltage adjustable from 2V to 37V
■ Can be used as either a linear or a switching regulator

Connection Diagrams

Dual-In-Line Package

NC	1	14	NC
CURRENT LIMIT	2	13	FREQUENCY COMPENSATIONS
CURRENT SENSE	3	12	V^+
INVERTING INPUT	4	11	V_C
NON−INVERTING INPUT	5	10	V_{OUT}
V_{REF}	6	9	V_Z
V^-	7	8	NC

TL/H/8563–2

Top View

**Order Number LM723J, LM723CJ,
LM723CM or LM723CN
See NS Package J14A, M14A or N14A**

Metal Can Package

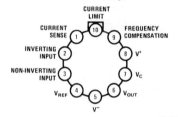

Note: Pin 5 connected to case.
Top View

**Order Number LM723H or LM723CH
See NS Package H10C**

Equivalent Circuit*

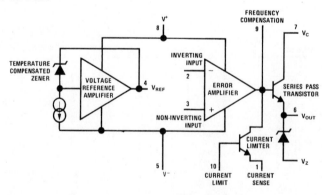

*Pin numbers refer to metal can package.

TL/H/8563–4

Absolute Maximum Ratings

If Military/Aerospace specified devices are required, contact the National Semiconductor Sales Office/ Distributors for availability and specifications. (Note 9)

Pulse Voltage from V^+ to V^- (50 ms)	50V
Continuous Voltage from V^+ to V^-	40V
Input-Output Voltage Differential	40V
Maximum Amplifier Input Voltage (Either Input)	8.5V
Maximum Amplifier Input Voltage (Differential)	5V
Current from V_Z	25 mA
Current from V_{REF}	15 mA

Internal Power Dissipation Metal Can (Note 1)		800 mW
Cavity DIP (Note 1)		900 mW
Molded DIP (Note 1)		660 mW
Operating Temperature Range LM723		$-55°C$ to $+150°C$
LM723C		$0°C$ to $+70°C$
Storage Temperature Range Metal Can		$-65°C$ to $+150°C$
Molded DIP		$-55°C$ to $+150°C$
Lead Temperature (Soldering, 4 sec. max.)		
Hermetic Package		300°C
Plastic Package		260°C

Electrical Characteristics (Note 2)

Parameter	Conditions	LM723 Min	LM723 Typ	LM723 Max	LM723C Min	LM723C Typ	LM723C Max	Units
Line Regulation	$V_{IN} = 12V$ to $V_{IN} = 15V$		0.01	0.1		0.01	0.1	% V_{OUT}
	$-55°C \leq T_A \leq +125°C$			0.3				% V_{OUT}
	$0°C \leq T_A \leq +70°C$						0.3	% V_{OUT}
	$V_{IN} = 12V$ to $V_{IN} = 40V$		0.02	0.2		0.1	0.5	% V_{OUT}
Load Regulation	$I_L = 1$ mA to $I_L = 50$ mA		0.03	0.15		0.03	0.2	% V_{OUT}
	$-55°C \leq T_A \leq +125°C$			0.6				% V_{OUT}
	$0°C \leq T_A \leq +70°C$						0.6	% V_{OUT}
Ripple Rejection	$f = 50$ Hz to 10 kHz, $C_{REF} = 0$		74			74		dB
	$f = 50$ Hz to 10 kHz, $C_{REF} = 5 \mu F$		86			86		dB
Average Temperature Coeffic-	$-55°C \leq T_A \leq +125°C$		0.002	0.015				%/°C
ient of Output Voltage (Note 8)	$0°C \leq T_A \leq +70°C$					0.003	0.015	%/°C
Short Circuit Current Limit	$R_{SC} = 10\Omega$, $V_{OUT} = 0$		65			65		mA
Reference Voltage		6.95	7.15	7.35	6.80	7.15	7.50	V
Output Noise Voltage	BW = 100 Hz to 10 kHz, $C_{REF} = 0$		86			86		μVrms
	BW = 100 Hz to 10 kHz, $C_{REF} = 5 \mu F$		2.5			2.5		μVrms
Long Term Stability			0.05			0.05		%/1000 hrs
Standby Current Drain	$I_L = 0$, $V_{IN} = 30V$		1.7	3.5		1.7	4.0	mA
Input Voltage Range		9.5		40	9.5		40	V
Output Voltage Range		2.0		37	2.0		37	V
Input-Output Voltage Differential		3.0		38	3.0		38	V
θ_{JA}	Molded DIP		105			105		°C/W
θ_{JA}	Cavity DIP		150			150		°C/W
θ_{JA}	TO-5 Board Mount in Still Air		225			225		°C/W
θ_{JA}	TO-5 Board Mount in 400 LF/Min Air Flow		90			90		°C/W
θ_{JA}	SO					125		°C/W
θ_{JC}			25			25		°C/W

Note 1: See derating curves for maximum power rating above 25°C.

Note 2: Unless otherwise specified, $T_A = 25°C$, $V_{IN} = V^+ = V_C = 12V$, $V^- = 0$, $V_{OUT} = 5V$, $I_L = 1$ mA, $R_{SC} = 0$, $C_1 = 100$ pF, $C_{REF} = 0$ and divider impedance as seen by error amplifier ≤ 10 kΩ connected as shown in *Figure 1*. Line and load regulation specifications are given for the condition of constant chip temperature. Temperature drifts must be taken into account separately for high dissipation conditions.

Note 3: L_1 is 40 turns of No. 20 enameled copper wire wound on Ferroxcube P36/22-3B7 pot core or equivalent with 0.009 in. air gap.

Note 4: Figures in parentheses may be used if R1/R2 divider is placed on opposite input of error amp.

Note 5: Replace R1/R2 in figures with divider shown in *Figure 13*.

Note 6: V^+ must be connected to a $+3V$ or greater supply.

Note 7: For metal can applications where V_Z is required, an external 6.2V zener diode should be connected in series with V_{OUT}.

Note 8: Guaranteed by correlation to other tests.

Note 9: Refer to RETS723X military specifications for the LM723.

Typical Performance Characteristics

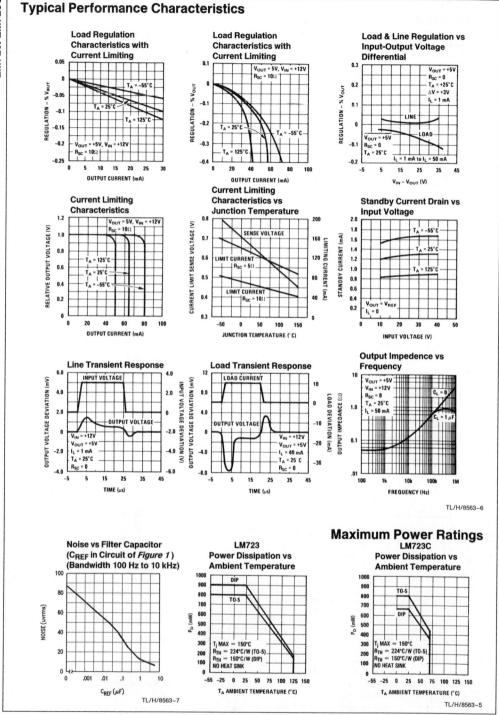

TL/H/8563–6

TL/H/8563–7

Maximum Power Ratings

TL/H/8563–5

TABLE I. Resistor Values (kΩ) for Standard Output Voltage

Positive Output Voltage	Applicable Figures	Fixed Output ±5%		Output Adjustable ±10% (Note 5)			Negative Output Voltage	Applicable Figures	Fixed Output ±5%		5% Output Adjustable ±10%		
	(Note 4)	R1	R2	R1	P1	R2			R1	R2	R1	P1	R2
+3.0	1, 5, 6, 9, 12 (4)	4.12	3.01	1.8	0.5	1.2	+100	7	3.57	102	2.2	10	91
+3.6	1, 5, 6, 9, 12 (4)	3.57	3.65	1.5	0.5	1.5	+250	7	3.57	255	2.2	10	240
+5.0	1, 5, 6, 9, 12 (4)	2.15	4.99	0.75	0.5	2.2	−6 (Note 6)	3, (10)	3.57	2.43	1.2	0.5	0.75
+6.0	1, 5, 6, 9, 12 (4)	1.15	6.04	0.5	0.5	2.7	−9	3, 10	3.48	5.36	1.2	0.5	2.0
+9.0	2, 4, (5, 6, 9, 12)	1.87	7.15	0.75	1.0	2.7	−12	3, 10	3.57	8.45	1.2	0.5	3.3
+12	2, 4, (5, 6, 9, 12)	4.87	7.15	2.0	1.0	3.0	−15	3, 10	3.65	11.5	1.2	0.5	4.3
+15	2, 4, (5, 6, 9, 12)	7.87	7.15	3.3	1.0	3.0	−28	3, 10	3.57	24.3	1.2	0.5	10
+28	2, 4, (5, 6, 9, 12)	21.0	7.15	5.6	1.0	2.0	−45	8	3.57	41.2	2.2	10	33
+45	7	3.57	48.7	2.2	10	39	−100	8	3.57	97.6	2.2	10	91
+75	7	3.57	78.7	2.2	10	68	−250	8	3.57	249	2.2	10	240

TABLE II. Formulae for Intermediate Output Voltages

Outputs from +2 to +7 volts (Figures 1, 5, 6, 9, 12, [4]) $V_{OUT} = \left(V_{REF} \times \dfrac{R2}{R1 + R2} \right)$	Outputs from +4 to +250 volts (Figure 7) $V_{OUT} = \left(\dfrac{V_{REF}}{2} \times \dfrac{R2 - R1}{R1} \right); R3 = R4$	Current Limiting $I_{LIMIT} = \dfrac{V_{SENSE}}{R_{SC}}$
Outputs from +7 to +37 volts (Figures 2, 4, [5, 6, 9, 12]) $V_{OUT} = \left(V_{REF} \times \dfrac{R1 + R2}{R2} \right)$	Outputs from −6 to −250 volts (Figures 3, 8, 10) $V_{OUT} = \left(\dfrac{V_{REF}}{2} \times \dfrac{R1 + R2}{R1} \right); R3 = R4$	Foldback Current Limiting $I_{KNEE} = \left(\dfrac{V_{OUT}\, R3}{R_{SC}\, R4} + \dfrac{V_{SENSE}\,(R3 + R4)}{R_{SC}\, R4} \right)$ $I_{SHORT\ CKT} = \left(\dfrac{V_{SENSE}}{R_{SC}} \times \dfrac{R3 + R4}{R4} \right)$

Typical Applications

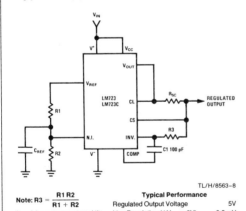

TL/H/8563–8

Note: $R3 = \dfrac{R1\ R2}{R1 + R2}$
for minimum temperature drift.

Typical Performance

Regulated Output Voltage	5V
Line Regulation (ΔV_IN = 3V)	0.5 mV
Load Regulation (ΔI_L = 50 mA)	1.5 mV

FIGURE 1. Basic Low Voltage Regulator
(V_{OUT} = 2 to 7 Volts)

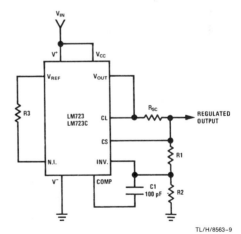

TL/H/8563–9

Note: $R3 = \dfrac{R1\ R2}{R1 + R2}$
for minimum temperature drift.
R3 may be eliminated for minimum component count.

Typical Performance

Regulated Output Voltage	15V
Line Regulation (ΔV_IN = 3V)	1.5 mV
Load Regulation (ΔI_L = 50 mA)	4.5 mV

FIGURE 2. Basic High Voltage Regulator
(V_{OUT} = 7 to 37 Volts)

Typical Applications (Continued)

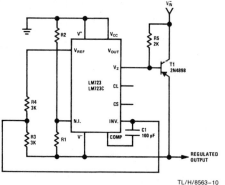

TL/H/8563–10

Typical Performance
Regulated Output Voltage   −15V
Line Regulation ($\Delta V_{IN} = 3V$)   1 mV
Load Regulation ($\Delta I_L = 100$ mA) 2 mV

FIGURE 3. Negative Voltage Regulator

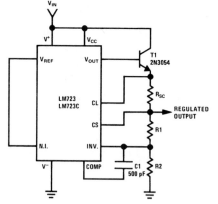

TL/H/8563–11

Typical Performance
Regulated Output Voltage   +15V
Line Regulation ($\Delta V_{IN} = 3V$)   1.5 mV
Load Regulation ($\Delta I_L = 1A$)   15 mV

FIGURE 4. Positive Voltage Regulator
(External NPN Pass Transistor)

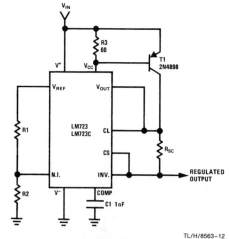

TL/H/8563–12

Typical Performance
Regulated Output Voltage   +5V
Line Regulation ($\Delta V_{IN} = 3V$)   0.5 mV
Load Regulation ($\Delta I_L = 1A$)   5 mV

FIGURE 5. Positive Voltage Regulator
(External PNP Pass Transistor)

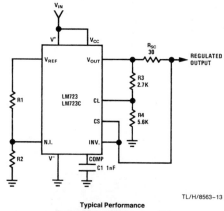

TL/H/8563–13

Typical Performance
Regulated Output Voltage   +5V
Line Regulation ($\Delta V_{IN} = 3V$)   0.5 mV
Load Regulation ($\Delta I_L = 10$ mA)   1 mV
Short Circuit Current   20 mA

FIGURE 6. Foldback Current Limiting

Typical Applications (Continued)

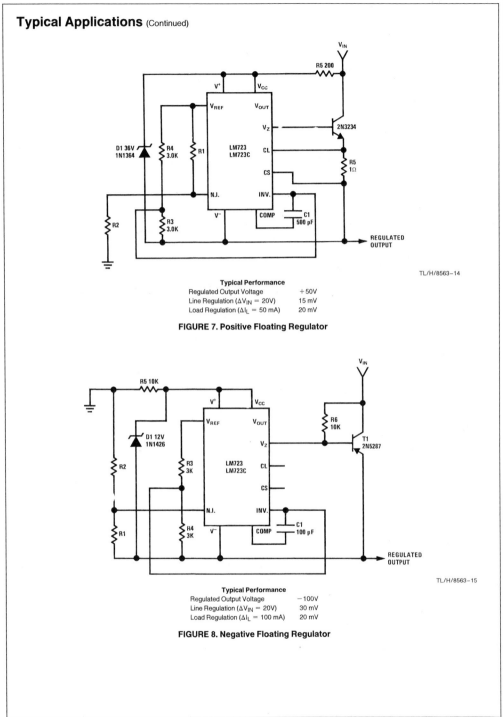

TL/H/8563–14

Typical Performance

Regulated Output Voltage	+50V
Line Regulation ($\Delta V_{IN} = 20V$)	15 mV
Load Regulation ($\Delta I_L = 50$ mA)	20 mV

FIGURE 7. Positive Floating Regulator

TL/H/8563–15

Typical Performance

Regulated Output Voltage	−100V
Line Regulation ($\Delta V_{IN} = 20V$)	30 mV
Load Regulation ($\Delta I_L = 100$ mA)	20 mV

FIGURE 8. Negative Floating Regulator

Typical Applications (Continued)

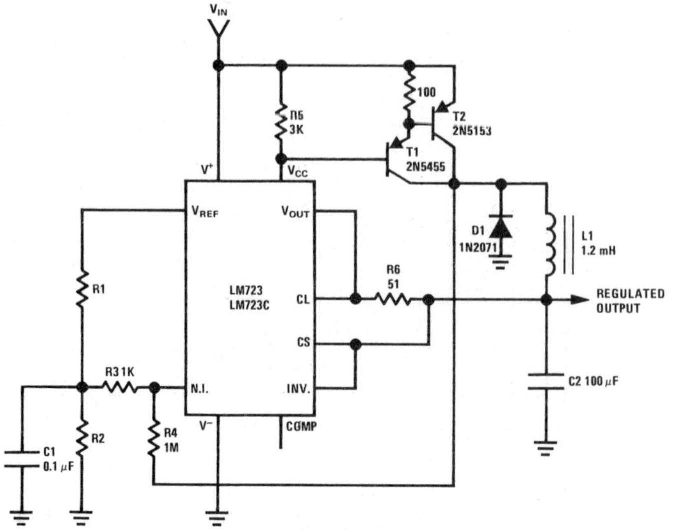

TL/H/8563–16

Typical Performance

Regulated Output Voltage	+5V
Line Regulation ($\Delta V_{IN} = 30V$)	10 mV
Load Regulation ($\Delta I_L = 2A$)	80 mV

FIGURE 9. Positive Switching Regulator

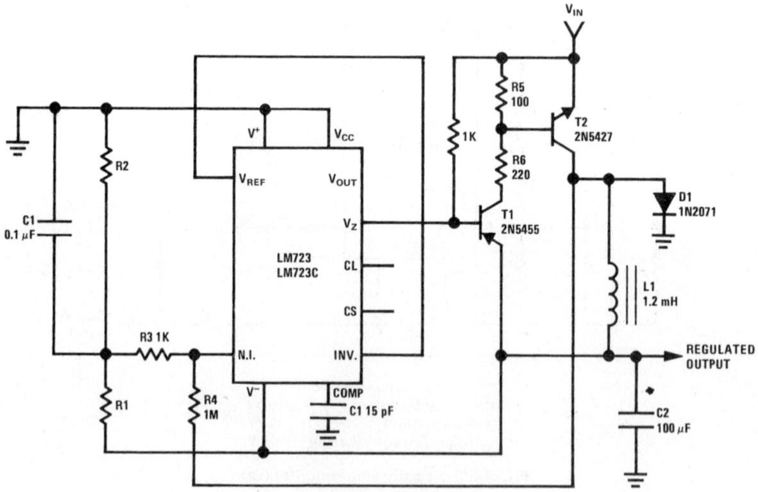

TL/H/8563–17

Typical Performance

Regulated Output Voltage	−15V
Line Regulation ($\Delta V_{IN} = 20V$)	8 mV
Load Regulation ($\Delta I_L = 2A$)	6 mV

FIGURE 10. Negative Switching Regulator

Typical Applications (Continued)

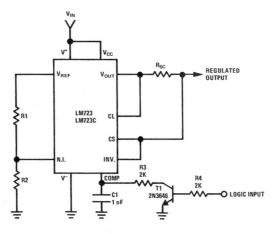

TL/H/8563–18

Note: Current limit transistor may be used for shutdown if current limiting is not required.

Typical Performance

Regulated Output Voltage	+5V
Line Regulation (ΔV_{IN} = 3V)	0.5 mV
Load Regulation (ΔI_L = 50 mA)	1.5 mV

FIGURE 11. Remote Shutdown Regulator with Current Limiting

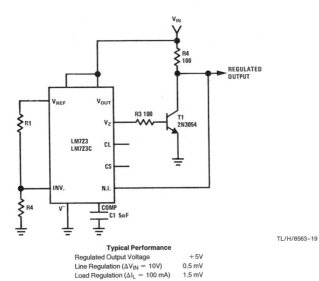

TL/H/8563–19

Typical Performance

Regulated Output Voltage	+5V
Line Regulation (ΔV_{IN} = 10V)	0.5 mV
Load Regulation (ΔI_L = 100 mA)	1.5 mV

FIGURE 12. Shunt Regulator

A59

Typical Applications (Continued)

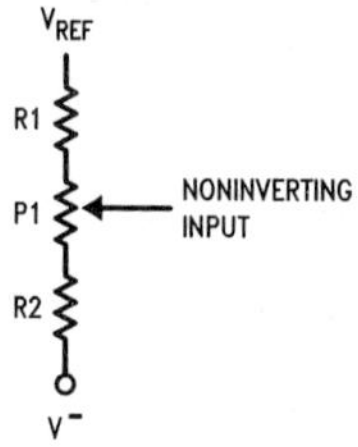

TL/H/8563–20

FIGURE 13. Output Voltage Adjust
(See Note 5)

Schematic Diagram

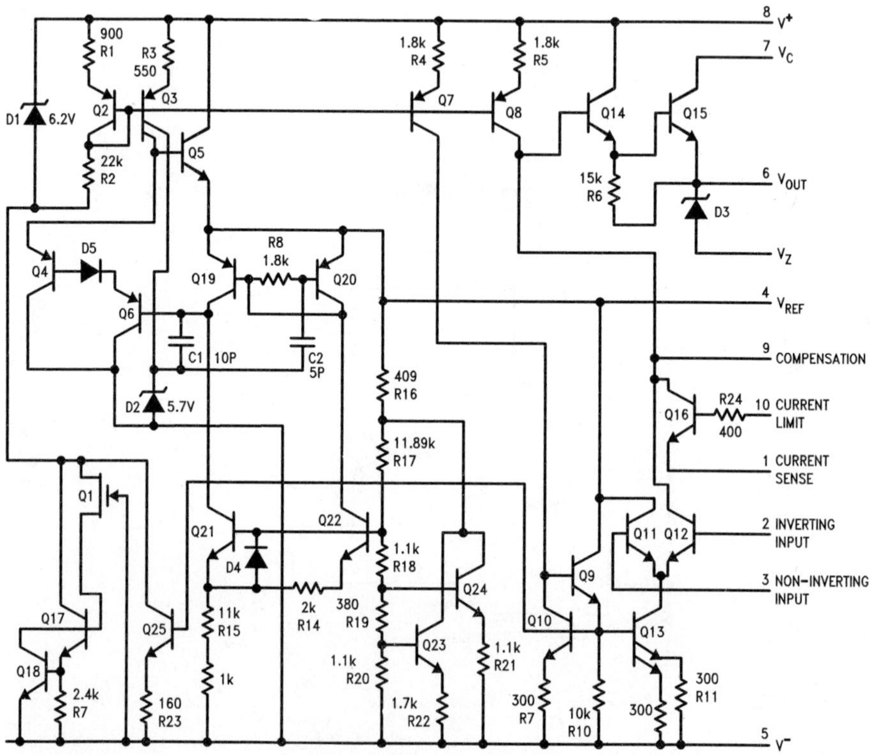

TL/H/8563–1

National Semiconductor Corporation

LM78XX Series Voltage Regulators

General Description

The LM78XX series of three terminal regulators is available with several fixed output voltages making them useful in a wide range of applications. One of these is local on card regulation, eliminating the distribution problems associated with single point regulation. The voltages available allow these regulators to be used in logic systems, instrumentation, HiFi, and other solid state electronic equipment. Although designed primarily as fixed voltage regulators these devices can be used with external components to obtain adjustable voltages and currents.

The LM78XX series is available in an aluminum TO-3 package which will allow over 1.0A load current if adequate heat sinking is provided. Current limiting is included to limit the peak output current to a safe value. Safe area protection for the output transistor is provided to limit internal power dissipation. If internal power dissipation becomes too high for the heat sinking provided, the thermal shutdown circuit takes over preventing the IC from overheating.

Considerable effort was expanded to make the LM78XX series of regulators easy to use and mininize the number of external components. It is not necessary to bypass the output, although this does improve transient response. Input bypassing is needed only if the regulator is located far from the filter capacitor of the power supply.

For output voltage other than 5V, 12V and 15V the LM117 series provides an output voltage range from 1.2V to 57V.

Features

- Output current in excess of 1A
- Internal thermal overload protection
- No external components required
- Output transistor safe area protection
- Internal short circuit current limit
- Available in the aluminum TO-3 package

Voltage Range

LM7805C	5V
LM7812C	12V
LM7815C	15V

Schematic and Connection Diagrams

TL/H/7746-1

Metal Can Package TO-3 (K) Aluminum

OUTPUT GND
INPUT

TL/H/7746-2

Bottom View

Order Number LM7805CK, LM7812CK or LM7815CK See NS Package Number KC02A

Plastic Package TO-220 (T)

GND → OUTPUT GND INPUT

TL/H/7746-3

Top View

Order Number LM7805CT, LM7812CT or LM7815CT See NS Package Number T03B

Absolute Maximum Ratings

If Military/Aerospace specified devices are required, contact the National Semiconductor Sales Office/ Distributors for availability and specifications.

Input Voltage (V_O = 5V, 12V and 15V)	35V
Internal Power Dissipation (Note 1)	Internally Limited
Operating Temperature Range (T_A)	0°C to +70°C

Maximum Junction Temperature	
(K Package)	150°C
(T Package)	150°C
Storage Temperature Range	−65°C to +150°C
Lead Temperature (Soldering, 10 sec.)	
TO-3 Package K	300°C
TO-220 Package T	230°C

Electrical Characteristics LM78XXC (Note 2) 0°C ≤ Tj ≤ 125°C unless otherwise noted.

	Output Voltage		5V			12V			15V			
	Input Voltage (unless otherwise noted)		10V			19V			23V			Units
Symbol	Parameter	Conditions	Min	Typ	Max	Min	Typ	Max	Min	Typ	Max	
V_O	Output Voltage	Tj = 25°C, 5 mA ≤ I_O ≤ 1A	4.8	5	5.2	11.5	12	12.5	14.4	15	15.6	V
		P_D ≤ 15W, 5 mA ≤ I_O ≤ 1A	4.75		5.25	11.4		12.6	14.25		15.75	V
		V_{MIN} ≤ V_{IN} ≤ V_{MAX}	(7.5 ≤ V_{IN} ≤ 20)			(14.5 ≤ V_{IN} ≤ 27)			(17.5 ≤ V_{IN} ≤ 30)			V
ΔV_O	Line Regulation	I_O = 500 mA, Tj = 25°C		3	50		4	120		4	150	mV
		ΔV_{IN}	(7 ≤ V_{IN} ≤ 25)			(14.5 ≤ V_{IN} ≤ 30)			(17.5 ≤ V_{IN} ≤ 30)			V
		0°C ≤ Tj ≤ +125°C			50			120			150	mV
		ΔV_{IN}	(8 ≤ V_{IN} ≤ 20)			(15 ≤ V_{IN} ≤ 27)			(18.5 ≤ V_{IN} ≤ 30)			V
		I_O ≤ 1A, Tj = 25°C			50			120			150	mV
		ΔV_{IN}	(7.5 ≤ V_{IN} ≤ 20)			(14.6 ≤ V_{IN} ≤ 27)			(17.7 ≤ V_{IN} ≤ 30)			V
		0°C ≤ Tj ≤ +125°C			25			60			75	mV
		ΔV_{IN}	(8 ≤ V_{IN} ≤ 12)			(16 ≤ V_{IN} ≤ 22)			(20 ≤ V_{IN} ≤ 26)			V
ΔV_O	Load Regulation	Tj = 25°C, 5 mA ≤ I_O ≤ 1.5A		10	50		12	120		12	150	mV
		250 mA ≤ I_O ≤ 750 mA			25			60			75	mV
		5 mA ≤ I_O ≤ 1A, 0°C ≤ Tj ≤ +125°C			50			120			150	mV
I_Q	Quiescent Current	I_O ≤ 1A, Tj = 25°C			8			8			8	mA
		0°C ≤ Tj ≤ +125°C			8.5			8.5			8.5	mA
ΔI_Q	Quiescent Current Change	5 mA ≤ I_O ≤ 1A			0.5			0.5			0.5	mA
		Tj = 25°C, I_O ≤ 1A			1.0			1.0			1.0	mA
		V_{MIN} ≤ V_{IN} ≤ V_{MAX}	(7.5 ≤ V_{IN} ≤ 20)			(14.8 ≤ V_{IN} ≤ 27)			(17.9 ≤ V_{IN} ≤ 30)			V
		I_O ≤ 500 mA, 0°C ≤ Tj ≤ +125°C			1.0			1.0			1.0	mA
		V_{MIN} ≤ V_{IN} ≤ V_{MAX}	(7 ≤ V_{IN} ≤ 25)			(14.5 ≤ V_{IN} ≤ 30)			(17.5 ≤ V_{IN} ≤ 30)			V
V_N	Output Noise Voltage	T_A = 25°C, 10 Hz ≤ f ≤ 100 kHz		40			75			90		μV
$\frac{\Delta V_{IN}}{\Delta V_{OUT}}$	Ripple Rejection	f = 120 Hz { I_O ≤ 1A, Tj = 25°C or	62	80		55	72		54	70		dB
		I_O ≤ 500 mA	62			55			54			dB
		0°C ≤ Tj ≤ +125°C										
		V_{MIN} ≤ V_{IN} ≤ V_{MAX}	(8 ≤ V_{IN} ≤ 18)			(15 ≤ V_{IN} ≤ 25)			(18.5 ≤ V_{IN} ≤ 28.5)			V
R_O	Dropout Voltage	Tj = 25°C, I_{OUT} = 1A		2.0			2.0			2.0		V
	Output Resistance	f = 1 kHz		8			18			19		mΩ
	Short-Circuit Current	Tj = 25°C		2.1			1.5			1.2		A
	Peak Output Current	Tj = 25°C		2.4			2.4			2.4		A
	Average TC of V_{OUT}	0°C ≤ Tj ≤ +125°C, I_O = 5 mA		0.6			1.5			1.8		mV/°C
V_{IN}	Input Voltage Required to Maintain Line Regulation	Tj = 25°C, I_O ≤ 1A		7.5			14.6			17.7		V

Note 1: Thermal resistance of the TO-3 package (K, KC) is typically 4°C/W junction to case and 35°C/W case to ambient. Thermal resistance of the TO-220 package (T) is typically 4°C/W junction to case and 50°C/W case to ambient.

Note 2: All characteristics are measured with capacitor across the input of 0.22 μF, and a capacitor across the output of 0.1 μF. All characteristics except noise voltage and ripple rejection ratio are measured using pulse techniques (t_w ≤ 10 ms, duty cycle ≤ 5%). Output voltage changes due to changes in internal temperature must be taken into account separately.

Typical Performance Characteristics

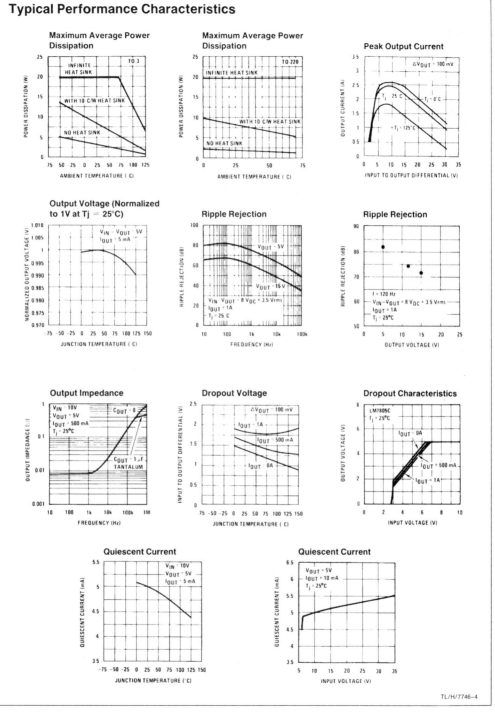

TL/H/7746-4

National
Semiconductor
Corporation

MF4 4th Order Switched Capacitor Butterworth Lowpass Filter

General Description

The MF4 is a versatile, easy to use, precision 4th order Butterworth low-pass filter. Switched-capacitor techniques eliminate external component requirements and allow a clock-tunable cutoff frequency. The ratio of the clock frequency to the low-pass cutoff frequency is internally set to 50 to 1 (MF4-50) or 100 to 1 (MF4-100). A Schmitt trigger clock input stage allows two clocking options, either self-clocking (via an external resistor and capacitor) for stand-alone applications, or for tighter cutoff frequency control an external TTL or CMOS logic compatible clock can be applied. The maximally flat passband frequency response together with a DC gain of 1 V/V allows cascading MF4 sections together for higher order filtering.

Features

- Low Cost
- Easy to use
- 8-pin mini-DIP or 14-pin wide-body S.O.
- No external components
- 5V to 14V supply voltage
- Cutoff frequency range of 0.1 Hz to 20 kHz
- Cutoff frequency accuracy of $\pm$ 0.3% typical
- Cutoff frequency set by external clock
- Separate TTL and CMOS/Schmitt-trigger clock inputs

Block and Connection Diagrams

Dual-In-Line Package

CLK IN	1	8	FILTER IN
CLK R	2	7	V$^+$
L. Sh	3	6	AGND
V$^-$	4	5	FILTER OUT

TL/H/5064–2

**Order Number MF4CN-50
or MF4CN-100
See NS Package Number N08E**

FILTER OUT 5

FILTER IN 8 — 4TH ORDER BUTTERWORTH LOWPASS FILTER

AGND 6 — ϕ_1 ϕ_2 — NON-OVERLAPPING CLOCK GENERATOR — V$^+$ V$^-$

CLK IN 1 — TRI-STATE® BUFFER — LEVEL SHIFT — 7 V$^+$ / 4 V$^-$

CLK R 2 / L. Sh 3

TL/H/5064–1

**Small-Outline
Wide-Body Package**

CLK IN	1	14	FILTER IN
NC	2	13	NC
CLK R	3	12	V$^+$
NC	4	11	NC
L. SH	5	10	AGND
NC	6	9	NC
V$^-$	7	8	FILTER OUT

TL/H/5064–25

**Top View
Order Number MF4CWM-50
or MF4CWM-100
See NS Package Number M14B**

Absolute Maximum Ratings (Notes 1, 2)

If Military/Aerospace specified devices are required, contact the National Semiconductor Sales Office/Distributors for availability and specifications.

Supply Voltage $(V^+ - V^-)$	14V
Voltage At Any Pin	$V^+ + 0.2V$
	$V^- - 0.2V$
Input Current at Any Pin (Note 14)	5 mA
Package Input Current (Note 14)	20 mA
Power Dissipation (Note 15)	500 mW
Storage Temperature	150°C
ESD Susceptibility (Note 13)	800 V

Soldering Information:
- N Package: 10 sec.   260°C
- SO Package: Vapor Phase (60 sec.)   215°C
-                    Infrared (15 sec.)   220°C

See AN-450 "Surface Mounting Methods and Their Effect on Product Reliability" for other methods of soldering surface mount devices.

Operating Ratings (Note 2)

Temperature Range	$T_{min} \leq T_A \leq T_{max}$	
MF4CN-50, MF4CN-100	$0°C \leq T_A \leq 70°C$	
MF4CWM-50, MF4CWM-100	$0°C \leq T_A \leq 70°C$	
Supply Voltage $(V^+ - V^-)$		5V to 14V

Filter Electrical Characteristics
The following specifications apply for $f_{CLK} \leq 250$ kHz (see Note 5) unless otherwise specified. **Boldface limits apply for T_{MIN} to T_{MAX};** all other limits $T_A = T_J = 25°C$.

Parameter		Conditions	MF4-50			MF4-100			Unit
			Typical (Note 10)	Tested Limit (Note 11)	Design Limit (Note 12)	Typical (Note 10)	Tested Limit (Note 11)	Design Limit (Note 12)	
$V^+ = +5V$, $V^- = -5V$									
f_c, Cutoff Frequency Range (Note 3)	Min				0.1			0.1	Hz
	Max				20k			10k	
Supply Current		$f_{clk} = 250$ kHz	2.5	3.5	**3.5**	2.5	3.5	**3.5**	mA
Maximum Clock Feedthrough (Peak-to-Peak)	Filter Output	$V_{in} = 0V$	25			25			mV
H_0, DC Gain		$R_{source} \leq 2$ kΩ	0.0	±0.15	±**0.15**	0.0	±0.15	±**0.15**	dB
f_{clk}/f_c, Clock to Cutoff Frequency Ratio			49.96 ±0.3%	49.96 ±0.8%	**49.96** ±**0.6%**	99.09 ±0.3%	99.09 ±1.0%	**99.09** ±**0.6%**	
f_{clk}/f_c Temperature Coefficient			±15			±30			ppm/°C
Stopband Attenuation (Min)		at $2 f_c$	−25.0	−24.0	**−24.0**	−25.0	−24.0	**−24.0**	dB
DC Offset Voltage			−200			−400			mV
Minimum Output Swing		$R_L = 10$ kΩ	+4.0	+3.5	**+3.5**	+4.0	+3.5	**+3.5**	V
			−4.5	−4.0	**−4.0**	−4.5	−4.0	**−4.0**	V
Output Short Circuit Current (Note 8)	Source		50			50			mA
	Sink		1.5			1.5			mA
Dynamic Range (Note 4)			80			82			dB
Additional Magnitude Response Test Points (Note 6) $f_{clk} = 250$ kHz		f = 6000 Hz		−7.57 ±0.27	**−7.57** ±**0.27**				dB
		f = 4500 Hz		−1.44 ±0.12	**−1.44** ±**0.12**				
		f = 3000 Hz					−7.21 ±0.2	**−7.21** ±**0.2**	dB
		f = 2250 Hz					−1.39 ±0.1	**−1.39** ±**0.1**	

Filter Electrical Characteristics

The following specifications apply for $f_{CLK} \leq 250$ kHz (see Note 5) unless otherwise specified. **Boldface limits apply for T_{MIN} to T_{MAX};** all other limits $T_A = T_J = 25°C$. (Continued)

Parameter		Conditions	MF4-50			MF4-100			Unit
			Typical (Note 10)	Tested Limit (Note 11)	Design Limit (Note 12)	Typical (Note 10)	Tested Limit (Note 11)	Design Limit (Note 12)	
$V^+ = +2.5V$, $V^- = -2.5V$									
f_c Cutoff Frequency Range (Note 3)	min max				0.1 10k			0.1 5k	Hz
Supply Current		$f_{clk} = 250$ kHz	1.5	2.25	**2.25**	1.5	2.25	**2.25**	mA
Maximum Clock Feedthrough (Peak-to-Peak)	Filter Output	$V_{in} = 0V$	15			15			mV
H_o, DC Gain		$R_{source} \leq 2$ kΩ	0.0	±0.15	**±0.15**	0.0	±0.15	**±0.15**	dB
f_{clk}/f_c, Clock to Cutoff Frequency Ratio			50.07 ±0.3%	50.07 ±1.0%	**50.07 ±0.6%**	99.16 ±0.3%	99.16 ±1.0%	**99.16 ±0.6%**	
f_{CLK}/f_C Temperature Coefficient			±25			±60			ppm/°C
Stopband Attenuation (Min)		at 2 f_c	−25.0	−24.0	**−24.0**	−25.0	−24.0	**−24.0**	dB
DC Offset Voltage			−150			−300			mV
Minimum Output Swing		$R_L = 10$ kΩ	+1.5 −2.2	+1.0 −1.7	**+1.0 −1.7**	+1.5 −2.2	+1.0 −1.7	**+1.0 −1.7**	V V
Output Short Circuit Current (Note 8)	Source Sink		28 0.5			28 0.5			mA mA
Dynamic Range (Note 4)			78			78			dB
Additional Magnitude Response Test Points (Note 6) ($f_c = 5$ kHz) Magnitude at		$f_{clk} = 250$ kHz							
		$f = 6000$ Hz		−7.57 ±0.27	**−7.57 ±0.27**				dB
		$f = 4500$ Hz		−1.46 ±0.12	**−1.46 ±0.12**				dB
($f_c = 2.5$ kHz) Magnitude		$f = 3000$ Hz					−7.21 ±0.2	**−7.21 ±0.2**	dB
		$f = 2250$ Hz					−1.39 ±0.1	**−1.39 ±0.1**	

Logic Input-Output Characteristics

The following specifications apply for $V^- = 0V$ (see Note 7) unless otherwise specified. **Boldface limits apply for T_{MIN} to T_{MAX};** all other limits $T_A = T_J = 25°C$.

Parameter		Conditions	Typical (Note 10)	Tested Limit (Note 11)	Design Limit (Note 12)	Unit
SCHMITT TRIGGER						
V_{T+}, Positive Going Threshold Voltage	Min Max	$V^+ = 10V$	7.0	6.1	**6.1 8.9**	V
	Min Max	$V^+ = 5V$	3.5	3.1 4.4	**3.1 4.4**	V

Logic Input-Output Characteristics

The following specifications apply for $V^- = 0V$ (see Note 7) unless otherwise specified. **Boldface limits apply for T_{MIN} to T_{MAX};** all other limits $T_A = t_J = 25°C$. (Continued)

Parameter		Conditions		Typical (Note 10)	Tested Limit (Note 11)	Design Limit (Note 12)	Unit
SCHMITT TRIGGER (Continued)							
V_{T-}, Negative Going Threshold Voltage	Min	$V^+ = 10V$		3.0	1.3	**1.3**	V
	Max				3.8	**3.8**	
	Min	$V^+ = 5V$		1.5	0.6	**0.6**	V
	Max				1.9	**1.9**	
Hysteresis ($V_{T+} - V_{T-}$)	Min	$V^+ = 10V$		4.0	2.3	**2.3**	V
	Max				7.6	**7.6**	
	Min	$V^+ = 5V$		2.0	1.2	**1.2**	V
	Max				3.8	**3.8**	
Minimum Logical "1" Output Voltage (pin 2)		$I_0 = -10 \mu A$	$V^+ = 10V$		9.0	**9.0**	V
			$V^+ = 5V$		4.5	**4.5**	V
Maximum Logical "0" Output Voltage (pin 2)		$I_0 = 10 \mu A$	$V^+ = 10V$		1.0	**1.0**	V
			$V^+ = 5V$		0.5	**0.5**	V
Minimum Output Source Current (pin 2)		CLK R Shorted to Ground	$V^+ = 10V$	6.0	3.0	**3.0**	mA
			$V^+ = 5V$	1.5	0.75	**0.75**	mA
Maximum Output Sink Current (pin 2)		CLK R Shorted to V^+	$V^+ = 10V$	5.0	2.5	**2.5**	mA
			$V^+ = 5V$	1.3	0.65	**0.65**	mA
TTL CLOCK INPUT, CLK R PIN (Note 9)							
Maximum V_{IL}, Logical "0" Input Voltage				0.8			V
Minimum V_{IH}, Logical "1" Input Voltage				2.0			V
Maximum Leakage Current at CLK R Pin		L. Sh Pin at Mid-Supply		2.0			μA

Note 1: Absolute Maximum Ratings indicate limits beyond which damage to the device may occur. AC and DC electrical specifications do not apply when operating the device beyond its specified operating conditions.

Note 2: All voltages are with respect to GND.

Note 3: The cutoff frequency of the filter is defined as the frequency where the magnitude response is 3.01 dB less than the DC gain of the filter.

Note 4: For $\pm 5V$ supplies the dynamic range is referenced to 2.82 Vrms (4V peak) where the wideband noise over a 20 kHz bandwidth is typically 280 μVrms for the MF4-50 and 230 μVrms for the MF4-100. For $\pm 2.5V$ supplies the dynamic range is referenced to 1.06 Vrms (1.5V peak) where the wideband noise over a 20 kHz bandwidth is typically 130 μVrms for both the MF4-50 and the MF4-100.

Note 5: The specifications for the MF4 have been given for a clock frequency (f_{CLK}) of 250 kHz or less. Above this clock frequency the cutoff frequency begins to deviate from the specified error band of $\pm 0.6\%$ but the filter still maintains its magnitude characteristics. See Application Hints.

Note 6: Besides checking the cutoff frequency (f_c) and the stopband attenuation at 2 f_c, two additional frequencies are used to check the magnitude response of the filter. The magnitudes are referenced to a DC gain of 0.0 dB.

Note 7: For simplicity all the logic levels have been referenced to $V^- = 0V$ (except for the TTL input logic levels). The logic levels will scale accordingly for $\pm 5V$ and $\pm 2.5V$ supplies.

Note 8: The short circuit source current is measured by forcing the output that is being tested to its maximum positive voltage swing and then shorting that output to the negative supply. The short circuit sink current is measured by forcing the output that is being tested to its maximum negative voltage and then shorting that output to the positive supply. These are worst case conditions.

Note 9: The MF4 is operating with symmetrical split supplies and L. Sh is tied to ground.

Note 10: Typicals are at 25°C and represent most likely parametric norm.

Note 11: Guaranteed to National's Average Outgoing Quality Level (AOQL).

Note 12: Guaranteed, but not 100% production tested. These limits are not used to determine outgoing quality levels.

Note 13: Human body model; 100 pF discharged through a 1.5 kΩ resistor.

Note 14: When the input voltage (V_{IN}) at any pin exceeds the power supply rails ($V_{IN} < V^-$ or $V_{IN} > V^+$) the absolute value of current at that pin should be limited to 5 mA or less. The 20 mA package input current limits the number of pins that can exceed the power supply boundaries with a 5 mA current limit to four.

Note 15: Thermal Resistance

θ_{JA} (Junction to Ambient) N Package . . . . 105°C/W.

θ_{JA} M Package . . . . 95°C/W.

Typical Performance Characteristics

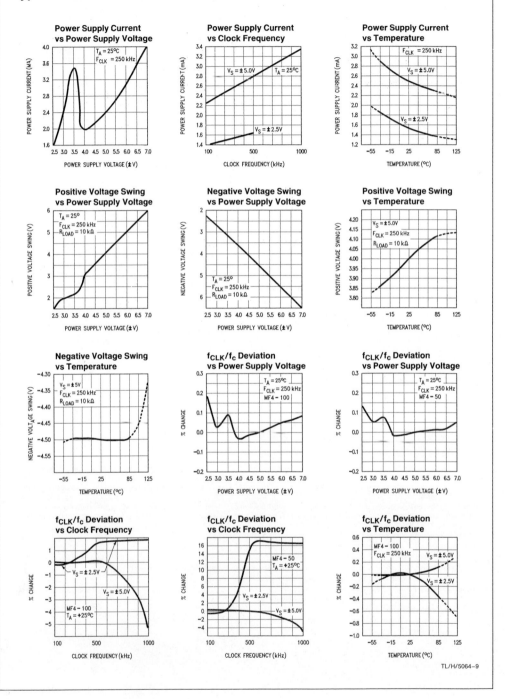

TL/H/5064–9

Typical Performance Characteristics (Continued)

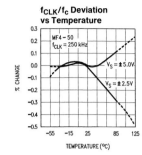

f_{CLK}/f_c Deviation vs Temperature

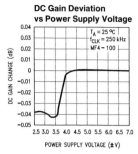

DC Gain Deviation vs Power Supply Voltage

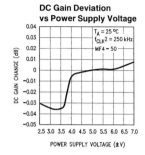

DC Gain Deviation vs Power Supply Voltage

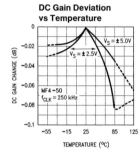

DC Gain Deviation vs Temperature

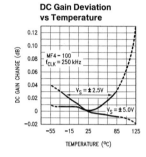

DC Gain Deviation vs Temperature

TL/H/5064–10

Pin Descriptions

(Numbers in ( ) are for 14-pin package.)

Pin #	Pin Name	Function
1 (1)	CLK IN	A CMOS Schmitt-trigger input to be used with an external CMOS logic level clock. Also used for self clocking Schmitt-trigger oscillator (see section 1.1).
2 (3)	CLK R	A TTL logic level clock input when in split supply operation (± 2.5V to ± 7V) with L. Sh tied to system ground. This pin becomes a low impedance output when L. Sh is tied to V^-. Also used in conjunction with the CLK IN pin for a self clocking Schmitt-trigger oscillator (see section 1.1). The TTL input signal must not exceed the supply voltages by more than 0.2V.
3 (5)	L. Sh	Level shift pin; selects the logic threshold levels for the clock. When tied to V^- it enables an internal tri-state buffer stage between the Schmitt trigger and the internal clock level shift stage thus enabling the CLK IN Schmitt-trigger input and making the CLK R pin a low impedance output. When the voltage level at this input exceeds 25% $(V^+ - V^-) + V^-$ the internal tri-state buffer is disabled allowing the CLK R pin to become the clock input for the internal clock level-shift stage. The CLK R threshold level is now 2V above the voltage on the L. Sh pin. The CLK R pin will be compatible with TTL logic levels when the MF4 is operated on split supplies with the L. Sh pin connected to system ground.
5 (8)	FILTER OUT	The output of the low-pass filter. It will typically sink 0.9 mA and source 3 mA and swing to within 1V of each supply rail.
6 (10)	AGND	The analog ground pin. This pin sets the DC bias level for the filter section and must be tied to the system ground for split supply operation or to mid-supply for single supply operation (see section 1.2). When tied to mid-supply this pin should be well bypassed.
7, 4 (7, 12)	V^+, V^-	The positive and negative supply pins. The total power supply range is 5V to 14V. Decoupling these pins with 0.1 μF capacitors is highly recommended.
8 (14)	FILTER IN	The input to the low-pass filter. To minimize gain errors the source impedance that drives this input should be less than 2K (see section 3). For single supply operation the input signal must be biased to mid-supply or AC coupled through a capacitor.

1.0 MF4 Application Hints

The MF4 is a non-inverting unity gain low-pass fourth-order Butterworth switched-capacitor filter. The switched-capacitor topology makes the cutoff frequency (where the gain drops 3.01 dB below the DC gain) a direct ratio (100:1 or 50:1) of the clock frequency supplied to the filter. Internal integrator time constants set the filter's cutoff frequency. The resistive element of these integrators is actually a capacitor which is "switched" at the clock frequency (for a detailed discussion see Input Impedance Section). Varying the clock frequency changes the value of this resistive element and thus the time constant of the integrators. The clock-to-cutoff-frequency ratio ($f_{CLK}f_c$) is set by the ratio of the input and feedback capacitors in the integrators. The higher the clock-to-cutoff-frequency ratio the closer this approximation is to the theoretical Butterworth response. The MF4 is available in f_{CLK}/f_c ratios of 50:1 (MF4-50) or 100:1 (MF4-100).

1.1 CLOCK INPUTS

The MF4 has a Schmitt-trigger inverting buffer which can be used to construct a simple R/C oscillator. Pin 3 is connected to V^- which makes Pin 2 a low impedance output. The oscillator's frequency is nominally

$$f_{CLK} = \frac{1}{RC \ln\left[\left(\frac{V_{CC} - V_{t-}}{V_{CC} - V_{t+}}\right)\left(\frac{V_{t+}}{V_{t-}}\right)\right]} \quad (1)$$

which, is typically

$$f_{CLK} \cong \frac{1}{1.69\,RC} \quad (1a)$$

for $V_{CC} = 10$V.

Note that f_{CLK} is dependent on the buffer's threshold levels as well as the resistor/capacitor tolerance (see *Figure 1*). Schmitt-trigger threshold voltage levels can change significantly causing the R/C oscillator's frequency to vary greatly from part to part.

Where accurate cutoff frequency is required, an external clock can be used to drive the CLK R input of the MF4. This input is TTL logic level compatible and also presents a very light load to the external clock source ($\sim 2\ \mu$A). With split supplies and the level shift (L. Sh) tied to system ground, the logic level is about 2V. (See the Pin Description for L. Sh).

1.2 POWER SUPPLY

The MF4 can be powered from a single supply or split supplies. The split supply mode shown in *Figure 2* is the most flexible and easiest to implement. Supply voltages of ± 5V to ± 7V enable the use of TTL or CMOS clock logic levels. *Figure 3* shows AGND resistor-biased to $V^+/2$ for single supply operation. In this mode only CMOS clock logic levels can be used, and input signals should be capacitor-coupled or biased near mid-supply.

1.3 INPUT IMPEDANCE

The MF4 low-pass filter input (FILTER IN) is not a high impedance buffer input. This input is a switched-capacitor resistor equivalent, and its effective impedance is inversely proportional to the clock frequency. The equivalent circuit of the filter's input can be seen in *Figure 4*. The input capacitor charges to V_{in} during the first half of the clock period; during the second half the charge is transferred to the feedback capacitor. The total transfer of charge in one clock cycle is therefore $Q = C_{in}V_{in}$, and since current is defined as the flow of charge per unit time, the average input current becomes

$$I_{in} = Q/T$$

1.0 MF4 Application Hints (Continued)

(where T equals one clock period) or

$$I_{in} = \frac{C_{in}V_{in}}{T} = C_{in}V_{in}f_{CLK}$$

The equivalent input resistor (R_{in}) then can be expressed as

$$R_{in} = \frac{V_{in}}{I_{in}} = \frac{1}{C_{in}f_{CLK}}$$

The input capacitor is 2 pF for the MF4-50 and 1 pF for the MF4-100, so for the MF4-100

$$R_{in} = \frac{1 \times 10^{12}}{f_{CLK}} = \frac{1 \times 10^{12}}{f_c \times 100} = \frac{1 \times 10^{10}}{f_c}$$

and

$$R_{in} = \frac{5 \times 10^{11}}{f_{CLK}} = \frac{5 \times 10^{11}}{f_c \times 50} = \frac{1 \times 10^{10}}{f_c}$$

for the MF4-50. The above equation shows that for a given cutoff frequency (f_c), the input resistance of the MF4-50 is the same as that of the MF4-100. The higher the clock-to-cutoff-frequency ratio, the greater equivalent input resistance for a given clock frequency.

This input resistance will form a voltage divider with the source impedance (R_{source}). Since R_{in} is inversely proportional to the cutoff frequency, operation at higher cutoff frequencies will be more likely to load the input signal which would appear as an overall decrease in gain to the output of the filter. Since the filter's ideal gain is unity, the overall gain is given by:

$$A_v = \frac{R_{in}}{R_{in} + R_{source}}$$

If the MF4-50 or the MF-100 were set up for a cutoff frequency of 10 kHz the input impedance would be:

$$R_{in} = \frac{1 \times 10^{10}}{10 \text{ kHz}} = 1 \text{ M}\Omega$$

In this example with a source impedance of 10K the overall gain, if the MF4 had an ideal gain of 1 or 0 dB, would be:

$$A_v = \frac{1 \text{ M}\Omega}{10 \text{ k}\Omega + 1 \text{ M}\Omega} = 0.99009 \text{ or } -0.086 \text{ dB}$$

Since the maximum overall gain error for the MF4 is ± 0.15 dB with $R_s \leq 2$ kΩ the actual gain error for this case would be $+0.06$ dB to -0.24 dB.

1.4 CUTOFF FREQUENCY RANGE

The filter's cutoff frequency (f_c) has a lower limit due to leakage currents through the internal switches draining the charge stored on the capacitors. At lower clock frequencies these leakage currents can cause millivolts of error, for example:

$$f_{CLK} = 100 \text{ Hz}, I_{leakage} = 1 \text{ pA}, C = 1 \text{ pF}$$

$$V = \frac{1 \text{ pA}}{1 \text{ pF } (100 \text{ Hz})} = 10 \text{ mV}$$

The propagation delay in the logic and the settling time required to acquire a new voltage level on the capacitors limit the filter's accuracy at high clock frequencies. The amplitude characteristic on ± 5V supplies will typically stay flat until f_{CLK} exceeds 750 kHz and then peak at about 0.5 dB at the corner frequency with a 1 MHz clock. As supply voltage drops to ± 2.5V, a shift in the f_{CLK}/f_c ratio occurs

which will become noticeable when the clock frequency exceeds 250 kHz. The response of the MF4 is still a good approximation of the ideal Butterworth low-pass characteristic shown in *Figure 5*.

2.0 Designing With The MF4

Given any low-pass filter specification, two equations will come in handy in trying to determine whether the MF4 will do the job. The first equation determines the order of the low-pass filter required to meet a given response specification:

$$n = \frac{\log [(10^{0.1 A_{min}} - 1)/(10^{0.1 A_{max}} - 1)]}{2 \log (f_s/f_b)} \quad (2)$$

where n is the order of the filter, A_{min} is the minimum stopband attenuation (in dB) desired at frequency f_s, and A_{max} is the passband ripple or attenuation (in dB) at cutoff frequency f_b. If the result of this equation is greater than 4, more than a single MF4 is required.

The attenuation at any frequency can be found by the following equation:

$$\text{Attn} (f) = 10 \log [1 + (10^{0.1 A_{max}} - 1) (f/f_b)^{2n}] \text{ dB} \quad (3)$$

where n = 4 for the MF4.

2.1 A LOW-PASS DESIGN EXAMPLE

Suppose the amplitude response specification in *Figure 6* is given. Can the MF4 be used? The order of the Butterworth approximation will have to be determined using (1):

$$A_{min} = 18 \text{ dB}, A_{max} = 1.0 \text{ dB}, f_s = 2 \text{ kHz, and } f_b = 1 \text{ kHz}$$

$$n = \frac{\log [(10^{1.8} - 1)/(10^{0.1} - 1)]}{2 \log(2)} = 3.95$$

Since n can only take on integer values, n = 4. Therefore the MF4 can be used. In general, if n is 4 or less a single MF4 stage can be utilized.

Likewise, the attenuation at f_s can be found using (3) with the above values and n = 4:

$$\text{Attn} (2 \text{ kHz}) = 10 \log [1 + 10^{0.1} - 1) (2 \text{ kHz/1 kHz})^8] = 18.28 \text{ dB}$$

This result also meets the design specification given in *Figure 6* again verifying that a single MF4 section will be adequate.

Since the MF4's cutoff frequency (f_c), which corresponds to a gain attenuation of -3.01 dB, was not specified in this example, it needs to be calculated. Solving equation 3 where f = f_c as follows:

$$f_c = f_b \left[\frac{(10^{0.1(3.01 \text{ dB})} - 1)}{(10^{0.1 A_{max}} - 1)} \right]^{1/(2n)}$$

$$= 1 \text{ kHz} \left[\frac{10^{0.301} - 1}{10^{0.1} - 1} \right]^{1/8}$$

$$= 1.184 \text{ kHz}$$

where $f_c = f_{CLK}/50$ or $f_{CLK}/100$. To implement this example for the MF4-50 the clock frequency will have to be set to $f_{CLK} = 50(1.184 \text{ kHz}) = 59.2$ kHz, or for the MF4-100, $f_{CLK} = 100 (1.184 \text{ kHz}) = 118.4$ kHz.

2.2 CASCADING MF4s

When a steeper stopband attenuation rate is required, two MF4s can be cascaded (*Figure 7*) yielding an 8th order

A71

2.0 Designing With The MF4 (Continued)

slope of 48 dB per octave. Because the MF4 is a Butterworth filter and therefore has no ripple in its passband when MF4s are cascaded, the resulting filter also has no ripple in its passband. Likewise the DC and passband gains will remain at 1V/V. The resulting response is shown in *Figure 9*.

In determining whether the cascaded MF4s will yield a filter that will meet a particular amplitude response specification, as above, equations 3 and 4 can be used, shown below.

$$n = \frac{\log[(10^{0.05A_{min}} - 1)/(10^{0.05A_{max}} - 1)]}{2 \log (f_s/f_c)} \quad (2)$$

$$\text{Attn (f)} = 10 \log [1 + (10^{0.05A_{max}} - 1) (f/f_c)^2] \text{ dB} \quad (3)$$

where n = 4 (the order of each filter).

Equation 2 will determine whether the order of the filter is adequate (n ≤ 4) while equation 3 can determine the actual stopband attenuation and cutoff frequency (f_c) necessary to obtain the desired frequency response. The design procedure would be identical to the one shown in section 2.0.

2.3 CHANGING CLOCK FREQUENCY INSTANTANEOUSLY

The MF4 will respond favorably to an instantaneous change in clock frequency. If the control signal in *Figure 9* is low the MF4-50 has a 100 kHz clock making f_c = 2 kHz; when this signal goes high the clock frequency changes to 50 kHz yielding f_c = 1 kHz. As the Figure illustrates, the output signal changes quickly and smoothly in response to a sudden change in clock frequency.

The step response of the MF4 in *Figure 10* is dependent on f_c. The MF4 responds as a classical fourth-order Butterworth low-pass filter.

2.4 ALIASING CONSIDERATIONS

Aliasing effects have to be considered when input signal frequencies exceed half the sampling rate. For the MF4 this equals half the clock frequency (f_{CLK}). When the input signal contains a component at a frequency higher than half the clock frequency $f_{CLK}/2$, as in *Figure 11a*, that component will be "reflected" about $f_{CLK}/2$ into the frequency range below $f_{CLK}/2$, as in *Figure 11b*. If this component is within the passband of the filter and of large enough amplitude it can cause problems. Therefore, if frequency components in the input signal exceed $f_{CLK}2$ they must be attenuated before being applied to the MF4 input. The necessary amount of attenuation will vary depending on system requirements. In critical applications the signal components above $f_{CLK}/2$ will have to be attenuated at least to the filter's residual noise level.

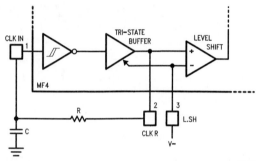

$$f = \frac{1}{FC \ln \left[\left(\dfrac{V_{CC} - V_{t-}}{V_{CC} - V_{t+}} \right) \left(\dfrac{V_{t+}}{V_{t-}} \right) \right]}$$

$$f \approx \frac{1}{1.69 \ RC}$$

$$(V_{CC} = 10V)$$

TL/H/5064–11

FIGURE 1. Schmitt Trigger R/C Oscillator

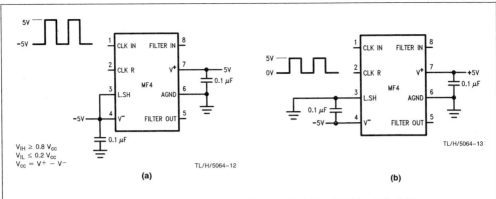

$V_{IH} \geq 0.8\ V_{CC}$
$V_{IL} \leq 0.2\ V_{CC}$
$V_{CC} = V^+ - V^-$

TL/H/5064-12

(a)

TL/H/5064-13

(b)

FIGURE 2. Split Supply Operation with CMOS Level Clock (a) and TTL Level Clock (b)

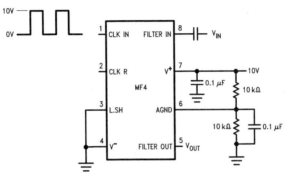

TL/H/5064-14

FIGURE 3. Single Supply Operation. ANGD Resistor Biased to $V^+/2$

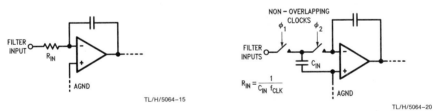

TL/H/5064-15

a) Equivalent Circuit for MF4 Filter Input

$R_{IN} = \dfrac{1}{C_{IN}\ f_{CLK}}$

TL/H/5064-20

b) Actual Circuit for MF4 Filter Input

FIGURE 4. MF4 Filter Input

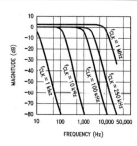

**FIGURE 5a. MF4-100 Amplitude
Response with ±5V Supplies**

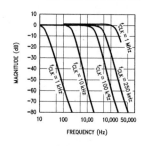

**FIGURE 5b. MF4-50 Amplitude
Response with ±5V Supplies**

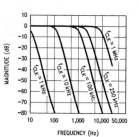

**FIGURE 5c. MF4-100 Amplitude
Response with ±2.5V Supplies**

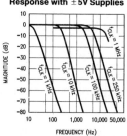

**FIGURE 5d. MF4-50 Amplitude
Response with ±2.5V Supplies**

TL/H/5064–21

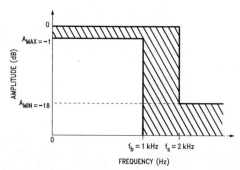

TL/H/5064–22

**FIGURE 6. Design Example Magnitude Response Specification where the Response of
the Filter Design must fall within the shaded area of the specification**

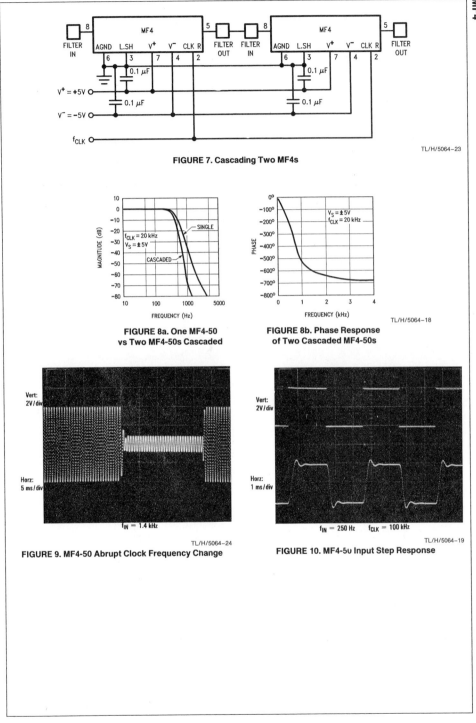

TL/H/5064–23

FIGURE 7. Cascading Two MF4s

FIGURE 8a. One MF4-50
vs Two MF4-50s Cascaded

FIGURE 8b. Phase Response
of Two Cascaded MF4-50s

TL/H/5064–18

TL/H/5064–24

FIGURE 9. MF4-50 Abrupt Clock Frequency Change

TL/H/5064–19

FIGURE 10. MF4-5υ Input Step Response

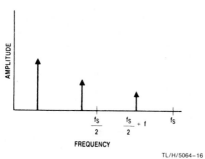

FREQUENCY

TL/H/5064–16

(a) input signal spectrum

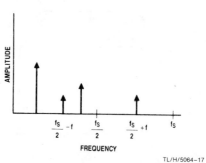

FREQUENCY

TL/H/5064–17

(b) Output signal spectrum. Note that the input signal at $f_c/2 + f$ causes an output signal to appear at $f_c/2 - f$.

FIGURE 11. The phenomenon of aliasing in sampled-data systems. An input signal whose frequency is greater than one-half the sampling frequency will cause an output to appear at a frequency lower than one-half the sampling frequency. In the MF4, $f_s = f_{CLK}$.

LIFE SUPPORT POLICY

National Semiconductor Corporation	National Semiconductor GmbH	National Semiconductor Japan Ltd.	National Semiconductor Hong Kong Ltd.	National Semicondutores Do Brasil Ltda.	National Semiconductor (Australia) PTY, Ltd.
2900 Semiconductor Drive	Industriestrasse 10	Sanseido Bldg. 5F	Suite 513, 5th Floor	Av. Brig. Faria Lima, 1383	1st Floor, 441 St. Kilda Rd.
P.O. Box 58090	D-8080 Furstenfeldbruck	4-15 Nishi Shinjuku	Chinachem Golden Plaza,	6.0 Andor-Conj. 62	Melbourne, 3004
Santa Clara, CA 95052-8090	West Germany	Shinjuku-Ku,	77 Mody Road, Tsimshatsui East,	01451 Sao Paulo, SP, Brasil	Victory, Australia
Tel: (408) 721-5000	Tel: (0-81-41) 103-0	Tokyo 160, Japan	Kowloon, Hong Kong	Tel: (55/11) 212-5066	Tel: (03) 267-5000
TWX: (910) 339-9240	Telex: 527-649	Tel: 3-299-7001	Tel: 3-7231290	Fax: (55/11) 211-1181 NSBR BR	Fax: 61-3-2677458
	Fax: (08141) 103554	FAX: 3-299-7000	Telex: 52996 NSSEA HX		
			Fax: 3-3112536		

Signetics Company

μA741

NE5534

The specification sheets for the μA741 General Purpose Amplifier and the NE5534 Dual and Single Low Noise Op Amp are reprinted courtesy of Philips Components-Signetics from *Linear Data Manual Volume 2: Industrial*. For current product information, contact

Signetics Company
811 E. Arques Avenue
P. O. Box 3409
Sunnyvale CA 94088-3409

Signetics

µA741/µA741C/SA741C
General Purpose Operational Amplifier

Product Specification

Linear Products

DESCRIPTION

The µA741 is a high performance operational amplifier with high open-loop gain, internal compensation, high common mode range and exceptional temperature stability. The µA741 is short-circuit-protected and allows for nulling of offset voltage.

FEATURES

- **Internal frequency compensation**
- **Short circuit protection**
- **Excellent temperature stability**
- **High input voltage range**

PIN CONFIGURATION

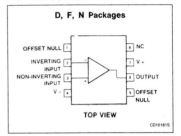

D, F, N Packages

TOP VIEW

ORDERING INFORMATION

DESCRIPTION	TEMPERATURE RANGE	ORDER CODE
8-Pin Plastic DIP	−55°C to +125°C	µA741N
8-Pin Plastic DIP	0 to +70°C	µA741CN
8-Pin Plastic DIP	−40°C to +85°C	SA741CN
8-Pin Cerdip	−55°C to +125°C	µA741F
8-Pin Cerdip	0 to +70°C	µA741CF
8-Pin SO	0 to +70°C	µA741CD

EQUIVALENT SCHEMATIC

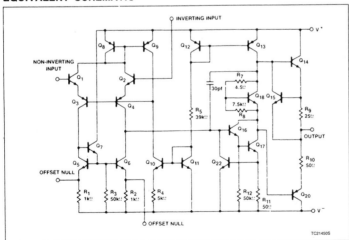

A79

General Purpose Operational Amplifier   µA741/µA741C/SA741C

ABSOLUTE MAXIMUM RATINGS

SYMBOL	PARAMETER	RATING	UNIT
V_S	Supply voltage µA741C µA741	±18 ±22	V V
P_D	Internal power dissipation D package N package F package	780 1170 800	mW mW mW
V_{IN}	Differential input voltage	±30	V
V_{IN}	Input voltage[1]	±15	V
I_{SC}	Output short-circuit duration	Continuous	
T_A	Operating temperature range µA741C SA741C µA741	0 to +70 −40 to +85 −55 to +125	°C °C °C
T_{STG}	Storage temperature range	−65 to +150	°C
T_{SOLD}	Lead soldering temperature (10sec max)	300	°C

NOTE:
1. For supply voltages less than ±15V, the absolute maximum input voltage is equal to the supply voltage.

DC ELECTRICAL CHARACTERISTICS (µA741, µA741C) T_A = 25°C, V_S = ±15V, unless otherwise specified.

SYMBOL	PARAMETER	TEST CONDITIONS	µA741			µA741C			UNIT
			Min	Typ	Max	Min	Typ	Max	
V_{OS} $\Delta V_{OS}/\Delta T$	Offset voltage	R_S = 10kΩ R_S = 10kΩ, over temp.		1.0 1.0 10	5.0 6.0		2.0 10	6.0 7.5	mV mV µV/°C
I_{OS} $\Delta I_{OS}/\Delta T$	Offset current	 Over temp. T_A = +125°C T_A = −55°C		20 7.0 20 200	200 200 500		20 300 200	200	nA nA nA nA pA/°C
I_{BIAS} $\Delta I_B/\Delta T$	Input bias current	 Over temp. T_A = +125°C T_A = −55°C		80 30 300 1	500 500 1500		80 800 1	500	nA nA nA nA nA/°C
V_{OUT}	Output voltage swing	R_L = 10kΩ R_L = 2kΩ, over temp.	±12 ±10	±14 ±13		±12 ±10	±14 ±13		V V
A_{VOL}	Large-signal voltage gain	R_L = 2kΩ, V_O = ±10V R_L = 2kΩ, V_O = ±10V, over temp.	50 25	200		20 15	200		V/mV V/mV
	Offset voltage adjustment range			±30			±30		mV
PSRR	Supply voltage rejection ratio	R_S ⩽ 10kΩ R_S ⩽ 10kΩ, over temp.		 10	150		10	150	µV/V µV/V
CMRR	Common-mode rejection ratio	 Over temp.	70	90		70	90	dB dB
I_{CC}	Supply current	 T_A = +125°C T_A = −55°C		1.4 1.5 2.0	2.8 2.5 3.3		1.4	2.8	mA mA mA

General Purpose Operational Amplifier   μA741/μA741C/SA741C

DC ELECTRICAL CHARACTERISTICS (Continued) (μA741, μA741C) T_A = 25°C, V_S = ±15V, unless otherwise specified.

SYMBOL	PARAMETER	TEST CONDITIONS	μA741			μA741C			UNIT
			Min	Typ	Max	Min	Typ	Max	
V_{IN} R_{IN}	Input voltage range Input resistance	(μA741, over temp.)	±12 0.3	±13 2.0		±12 0.3	±13 2.0		V MΩ
P_D	Power consumption	T_A = +125°C T_A = −55°C		50 45 45	85 75 100		50	85	mW mW mW
R_{OUT}	Output resistance			75			75		Ω
I_{SC}	Output short-circuit current		10	25	60	10	25	60	mA

DC ELECTRICAL CHARACTERISTICS (SA741C) T_A = 25°C, V_S = ±15V, unless otherwise specified.

SYMBOL	PARAMETER	TEST CONDITIONS	SA741C			UNIT
			Min	Typ	Max	
V_{OS} $\Delta V_{OS}/\Delta T$	Offset voltage	R_S = 10kΩ R_S = 10kΩ, over temp.		2.0 10	6.0 7.5	mV mV μV/°C
I_{OS} $\Delta I_{OS}/\Delta T$	Offset current	Over temp.		20 200	200 500	nA nA pA/°C
I_{BIAS} $\Delta I_B/\Delta T$	Input bias current	Over temp.		80 1	500 1500	nA nA nA/°C
V_{OUT}	Output voltage swing	R_L = 10kΩ R_L = 2kΩ, over temp.	±12 ±10	±14 ±13		V V
A_{VOL}	Large-signal voltage gain	R_L = 2kΩ, V_O = ±10V R_L = 2kΩ, V_O = ±10V, over temp.	20 15	200		V/mV V/mV
	Offset voltage adjustment range			±30		mV
PSRR	Supply voltage rejection ratio	R_S ⩽ 10kΩ		10	150	μV/V
CMRR	Common mode rejection ration		70	90		dB
V_{IN}	Input voltage range	Over temp.	±12	±13		V
R_{IN}	Input resistance		0.3	2.0		MΩ
P_d	Power consumption			50	85	mW
R_{OUT}	Output resistance			75		Ω
I_{SC}	Output short-circuit current			25		mA

AC ELECTRICAL CHARACTERISTICS T_A = 25°C, V_S = ±15V, unless otherwise specified.

SYMBOL	PARAMETER	TEST CONDITIONS	μA741, μA741C			UNIT
			Min	Typ	Max	
R_{IN}	Parallel input resistance	Open-loop, f = 20Hz	0.3			MΩ
C_{IN}	Parallel input capacitance	Open-loop, f = 20Hz		1.4		pF
	Unity gain crossover frequency	Open-loop		1.0		MHz
t_R SR	Transient response unity gain Rise time Overshoot Slew rate	V_{IN} = 20mV, R_L = 2kΩ, C_L ⩽ 100pF C ⩽ 100pF, R_L ⩾ 2kΩ, V_{IN} = ±10V		0.3 5.0 0.5		μs % V/μs

A81

General Purpose Operational Amplifier

µA741/µA741C/SA741C

TYPICAL PERFORMANCE CHARACTERISTICS

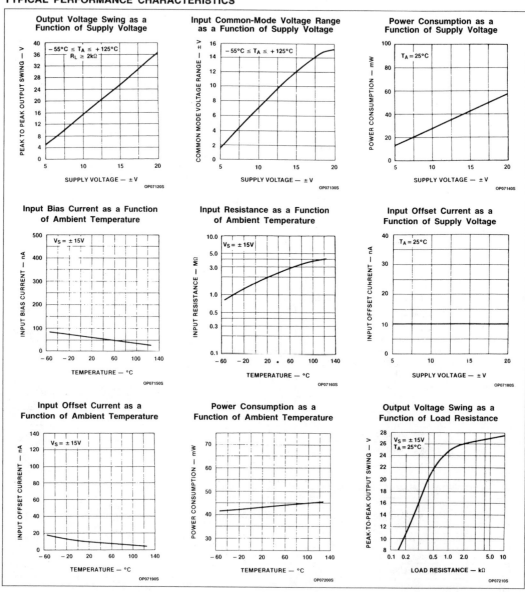

General Purpose Operational Amplifier

μA741/μA741C/SA741C

TYPICAL PERFORMANCE CHARACTERISTICS (Continued)

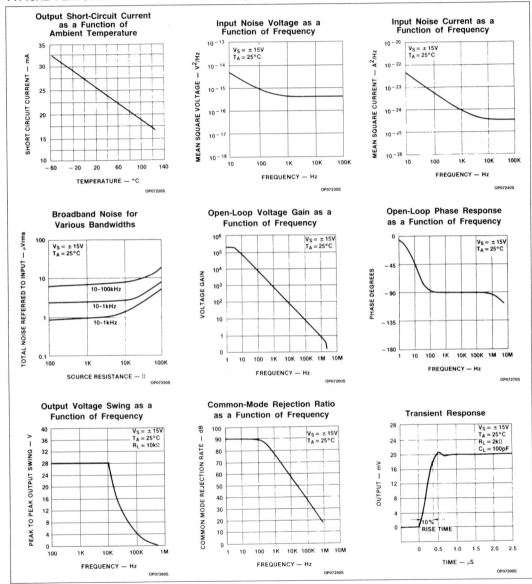

Signetics

Linear Products

NE5533/5533A
NE/SA/SE5534/5534A
Dual and Single Low Noise Op Amp

Product Specification

DESCRIPTION

The 5533/5534 are dual and single high-performance low noise operational amplifiers. Compared to other operational amplifiers, such as TL083, they show better noise performance, improved output drive capability and considerably higher small-signal and power bandwidths.

This makes the devices especially suitable for application in high quality and professional audio equipment, in instrumentation and control circuits and telephone channel amplifiers. The op amps are internally compensated for gain equal to, or higher than, three. The frequency response can be optimized with an external compensation capacitor for various applications (unity gain amplifier, capacitive load, slew rate, low overshoot, etc.) If very low noise is of prime importance, it is recommended that the 5533A/5534A version be used which has guaranteed noise specifications.

FEATURES

- Small-signal bandwidth: 10MHz
- Output drive capability: 600Ω, 10V$_{RMS}$ at V$_S$ = ± 18V
- Input noise voltage: 4nV/$\sqrt{Hz}$
- DC voltage gain: 100000
- AC voltage gain: 6000 at 10kHz
- Power bandwith: 200kHz
- Slew rate: 13V/μs
- Large supply voltage range: ± 3 to ± 20V
- 5534 MIL-STD processing available

APPLICATIONS

- Audio equipment
- Instrumentation and control circuits
- Telephone channel amplifiers
- Medical equipment

PIN CONFIGURATIONS

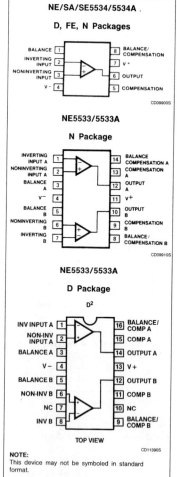

ORDERING INFORMATION

DESCRIPTION	TEMPERATURE RANGE	ORDER CODE
14-Pin Plastic DIP	0 to +70°C	NE5533N
16-Pin Plastic SO package	0 to +70°C	NE5533AD
14-Pin Plastic DIP	0 to +70°C	NE5533AN
16-Pin Plastic SO package	0 to +70°C	NE5533D
8-Pin Plastic SO package	0 to +70°C	NE5534D
8-Pin Hermetic Cerdip	0 to +70°C	NE5534FE
8-Pin Plastic DIP	0 to +70°C	NE5534N
8-Pin Plastic SO package	0 to +70°C	NE5534AD
8-Pin Hermetic Cerdip	0 to +70°C	NE5534AFE
8-Pin Plastic DIP	0 to +70°C	NE5534AN
8-Pin Plastic DIP	−40°C to +85°C	SA5534N
8-Pin Plastic SO package	−40°C to +85°C	SA5534AD
8-Pin Plastic DIP	−40°C to +85°C	SA5534AN
8-Pin Hermetic Cerdip	−55°C to +125°C	SE5534AFE
8-Pin Plastic DIP	−55°C to +125°C	SE5534N
8-Pin Hermetic Cerdip	−55°C to +125°C	SE5534AFE
8-Pin Plastic DIP	−55°C to +125°C	SE5534AN

Dual and Single Low
Noise Op Amp

EQUIVALENT SCHEMATIC

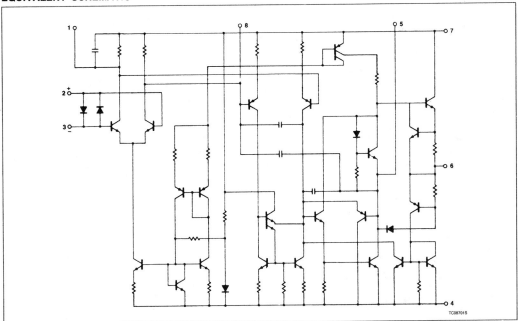

TC08701S

Dual and Single Low
Noise Op Amp

NE5533/5533A
NE/SA/SE5534/5534A

ABSOLUTE MAXIMUM RATINGS

SYMBOL	PARAMETER	RATING	UNIT
V_S	Supply voltage	± 22	V
V_{IN}	Input voltage	± V supply	V
V_{DIFF}	Differential input voltage[1]	± 0.5	V
T_A	Operating temperature range SE SA NE	 −55 to +125 −40 to +85 0 to +70	 °C °C °C
T_{STG}	Storage temperature range	−65 to +150	°C
T_J	Junction temperature	150	°C
P_D	Power dissipation at 25°C[2] 5533D 5533N 5534D 5534FE 5534N	 1350 1500 750 800 1150	 mW mW mW mW mW
	Output short-circuit duration[3]	Indefinite	
T_{SOLD}	Lead soldering temperature (10sec max)	300	°C

NOTES:
1. Diodes protect the inputs against over voltage. Therefore, unless current-limiting resistors are used, large currents will flow if the differential input voltage exceeds 0.6V. Maximum current should be limited to ± 10mA.
2. For operation at elevated temperature, derate packages based on the following junction-to-ambient thermal resistance:
     8-pin ceramic DIP 150°C/W
     8-pin plastic DIP 105°C/W
     8-pin plastic SO 160°C/W
     14-pin plastic DIP 80°C/W
     16-pin plastic SO 90°C/W
3. Output may be shorted to ground at $V_S = ± 15V$, $T_A = 25°C$. Temperature and/or supply voltages must be limited to ensure dissipation rating is not exceeded.

Dual and Single Low Noise Op Amp

NE5533/5533A
NE/SA/SE5534/5534A

DC ELECTRICAL CHARACTERISTICS $T_A = 25°C$, $V_S = \pm 15V$, unless otherwise specified. [1, 2, 3]

SYMBOL	PARAMETER	TEST CONDITIONS	SE5534/5534A			NE5533/5533A NE/SA5534/5534A			UNIT
			Min	Typ	Max	Min	Typ	Max	
V_{OS}	Offset voltage	Over temperature		0.5 3	2		0.5	4 5	mV mV
$\Delta V_{OS}/\Delta T$				5			5		$\mu V/°C$
I_{OS}	Offset current	Over temperature		10	200 500		20	300 400	nA nA
$\Delta I_{OS}/\Delta T$				200			200		pA/°C
I_B	Input current	Over temperature		400	800 1500		500	1500 2000	nA nA
$\Delta I_B/\Delta T$				5			5		nA/°C
I_{CC}	Supply current per op amp	Over temperature		4	6.5 9		4	8 10	mA mA
V_{CM} CMRR PSRR	Common mode input range Common mode rejection ratio Power supply rejection ratio		± 12 80	± 13 100 10	50	± 12 70	± 13 100 10	100	V dB $\mu V/V$
A_{VOL}	Large-signal voltage gain	$R_L \geqslant 600\Omega$, $V_O = \pm 10V$ Over temperature	50 25	100		25 15	100		V/mV V/mV
V_{OUT}	Output swing	$R_L \geqslant 600\Omega$ Over temperature $R_L \geqslant 600\Omega$, $V_S = \pm 18V$ $R_L \geqslant 2k\Omega$ Over temperature	± 12 ± 10 ± 15 ± 13 ± 12	± 13 ± 12 ± 16 ± 13.5 ± 12.5		± 12 ± 10 ± 15 ± 13 ± 12	± 13 ± 12 ± 16 ± 13.5 ± 12.5		V V V V V
R_{IN}	Input resistance		50	100		30	100		$k\Omega$
I_{SC}	Output short circuit current			38			38		mA

NOTES:
1. For NE5533/5533A/5534/5534A, $T_{MIN} = 0°C$, $T_{MAX} = 70°C$.
2. For SE5534/5534A, $T_{MIN} = -55°C$, $T_{MAX} = +125°C$.
3. For SA5534/5534A, $T_{MIN} = -40°C$, $T_{MAX} = +125°C$.

Dual and Single Low
Noise Op Amp

NE5533/5533A
NE/SA/SE5534/5534A

AC ELECTRICAL CHARACTERISTICS $T_A = 25°C$, $V_S = \pm 15V$, unless otherwise specified

SYMBOL	PARAMETER	TEST CONDITIONS	SE5534/5534A			NE5533/5533A NESA5534/5534A			UNIT
			Min	Typ	Max	Min	Typ	Max	
R_{OUT}	Output resistance	$A_V = 30dB$ closed-loop $f = 10kHz$, $R_L = 600\Omega$, $C_C = 22pF$		0.3			0.3		Ω
	Transient response	Voltage-follower, $V_{IN} = 50mV$ $R_L = 600\Omega$, $C_C = 22pF$, $C_L = 100pF$							
t_R	Rise time			20			20		ns
	Overshoot			20			20		%
	Transient response	$V_{IN} = 50mV$, $R_L = 600\Omega$ $C_C = 47pF$, $C_L = 500pF$							
t_R	Rise time			50			50		ns
	Overshoot			35			35		%
A_V	Gain	$f = 10kHz$, $C_C = 0$ $f = 10kHz$, $C_C = 22pF$		6 2.2			6 2.2		V/mV V/mV
GBW	Gain bandwidth product	$C_C = 22pF$, $C_L = 100pF$		10			10		MHz
SR	Slew rate	$C_C = 0$ $C_C = 22pF$		13 6			13 6		V/μs V/μs
	Power bandwidth	$V_{OUT} = \pm 10V$, $C_C = 0$ $V_{OUT} = \pm 10V$, $C_C = 22pF$ $V_{OUT} = \pm 14V$, $R_L = 600\Omega$ $C_C = 22pF$, $V_{CC} = \pm 18V$		200 95 70			200 95 70		kHz kHz kHz

ELECTRICAL CHARACTERISTICS $T_A = 25°C$, $V_S = 15V$, unless otherwise specified.

SYMBOL	PARAMETER	TEST CONDITIONS	5533/5534			5533A/5534A			UNIT
			Min	Typ	Max	Min	Typ	Max	
V_{NOISE}	Input noise voltage	$f_O = 30Hz$ $f_O = 1kHz$		7 4			5.5 3.5	7 4.5	nV/$\sqrt{Hz}$ nV/$\sqrt{Hz}$
I_{NOISE}	Input noise current	$f_O = 30Hz$ $f_O = 1kHz$		2.5 0.6			1.5 0.4		pA/$\sqrt{Hz}$ pA/$\sqrt{Hz}$
	Broadband noise figure	$f = 10Hz - 20kHz$, $R_S = 5k\Omega$					0.9		dB
	Channel separation	$f = 1kHz$, $R_S = 5k\Omega$		110			110		dB

Dual and Single Low
Noise Op Amp

NE5533/5533A
NE/SA/SE5534/5534A

TYPICAL PERFORMANCE CHARACTERISTICS

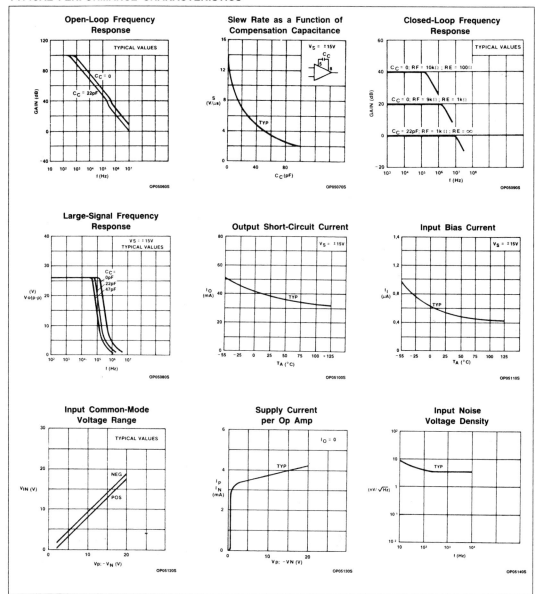

Dual and Single Low
Noise Op Amp

NE5533/5533A
NE/SA/SE5534/5534A

TYPICAL PERFORMANCE CHARACTERISTICS (Continued)

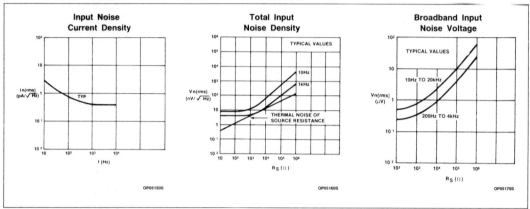

TEST LOAD CIRCUITS

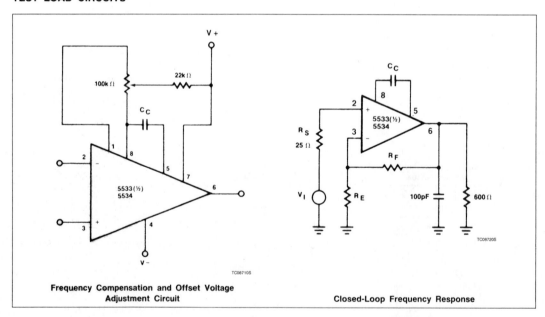

Frequency Compensation and Offset Voltage Adjustment Circuit

Closed-Loop Frequency Response

Appendix B:
Answers to Selected Problems

Chapter 1

1. a. 10 dB   b. 19 dB   c. 27 dB
    d. 0 dB   e. -7 dB   f. -15.2 dB

3. 33 dB

5. $G = 501$, $p_{out} = 12.53$ W

7. a. 1.06   b. 1   c. 199.5   d. 3.43   e. .398   f. .188

9. $A_v = 8.57$, $A'_v = 18.66$ dB

11. a. 0 dBW   b. 13.6 dBW   c. 8.13 dBW
    d. -7 dBW   e. -26.4 dBW   f. 30.8 dBW
    g. -43.5 dBW   h. -65.2 dBW   i. -172.5 dBW

13. a. 150 dBf   b. 163.6 dBf   c. 158.1 dBf
    d. 143 dBf   e. 123.6 dBf   f. 180.8 dBf
    g. 106.5 dBf   h. 84.8 dBf   i. -22.5 dBf

15. $G' = 28$ dB, $G = 631$

17. For $P_{in} = -4$ dBm, outputs are 6 dBm, 0 dBm, and 15 dBm.
For $P_{in} = -34$ dBW, outputs are -24 dBW, -30 dBW, and -15 dBW.

19. a. 200 mW

21. Final output $= 11.2$ V (21 dBv); stage 1 output $= 1.58$ V (4 dBv); stage 2 output $= 2.82$ V (9 dBv)

23. a. 15 V

25. At 50 kHz: $-.022$ dB, $-4.09°$
At 700 kHz: -3 dB, $-45°$
At 10 kHz: -23.1 dB, $-86°$
$T_r = .5$ μS

27. Amplitude does not change; phases are $-188.5°$, $-225°$, $-258.7°$.

29.

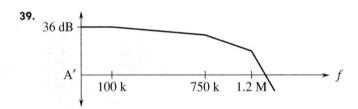

31. 4 Hz: 78.7°, 17.9 dB; 20 Hz: 45°, 29 dB; 100 Hz: 11.3°, 31.8 dB

33.

35. 20 kHz: 35.87 dB, 51.7°; 100 kHz: 40 dB, $-5.1°$; 800 kHz: 30.5 dB, $-70.5°$

37.

39.

41. 60 dB/decade above 1.2 MHz

43. $r'_e = 42.3\ \Omega$, $A_v = 65$

45. .48 sin $2\pi 1000t$

47. 1.656 sin $2\pi 2000t$

49. .775 V

51. 360 W

53. Output $= -71.5$ dBV

55. Greater than 30 Hz

Chapter 2   **1.** a. v_{out} is unknown.   b. $v_{out} = -V_{sat}$ c. $v_{out} = +V_{sat}$

3. a. $v_{out} = -V_{sat}$ b. $v_{out} = +V_{sat}$ c. $v_{out} = -V_{sat}$

d. $v_{out} = -V_{sat}$

5.

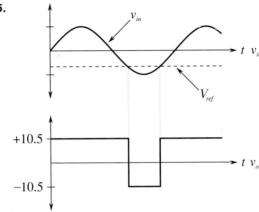

7. 55°C

9. 4.967 V

11. 125 Ω

Chapter 3   **1.** $S = 100$ at low frequencies, $S = 10$ at 1 kHz.

3. $A_{cl} = 19.96$; Approximate $A_{cl} = 20$; at 1 kHz, $A_{cl} = 19.6$

5. $S = 1000$, $f_2 = 25$ kHz

7. Margins are: 60° to spare, 12 dB to spare

9. $A_{cl} = 50$

11. $S = 2000$

13. Negative feedback will not help the S/N.

15. Closed-loop $f_1 = 5 \times 10^{-3}$ Hz

17. gm = .5 mS

19. THD = .0233%

21. $Z_{out} = 166.7$ MΩ

23. $S = 200.96$, $Z_{in} = 199$ Ω, $Z_{out} = 4.98$ Ω

25. Transresistance = 100 Ω, $\beta = .01$

27. a. Reduced high frequency gain

b. Increased high frequency gain

c. Output appears clipped (like an overdriven amplifier)

Chapter 4
1. $A_v = 7.67$, $Z_{in} \approx 470$ k
3. $A_v = 10$, $Z_{in} \approx \infty$ Ω
5. $v'_{out} = -9.7$ dBV
7. $v_{out} = -660$ mV
9. $R_i = 20$ k, $R_f = 50$ kΩ
11. $i_{out} = 100$ μA (1/3 full scale)
13. $R = 333.3$ Ω
15. $i_{out(max)} = 2.29$ mA, $i_{in(max)} = 208$ μA
17. $R'_i = 3.33$ k, $R'_f = 16.67$ k

19.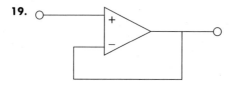

21.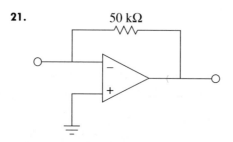

23.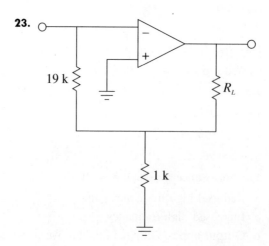

25. $i_{out} = 200\ \mu\text{A}$

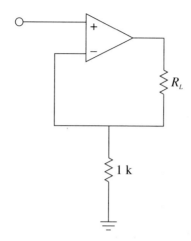

27. The figure below is one possible answer:

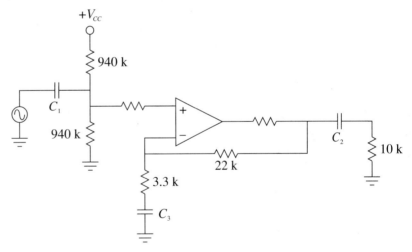

29.

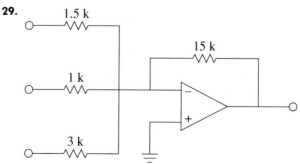

31. For C_1 dominant (C_2 and C_3 increased by $10\times$): $C_1 = 3.18\ \mu F$, $C_2 = 7.23\ \mu F$, $C_3 = 1.45\ \mu F$

33. $A_{max} = 9.15$, $A_{min} = 6.45$

35. The figure below is one possible answer:

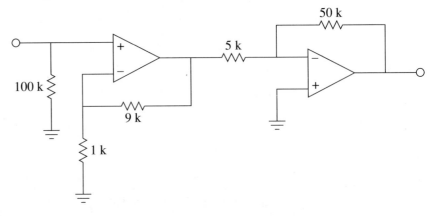

37. 1.88 V to 2.12 V

39. The figure below is one possible answer:

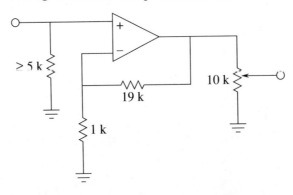

41. Loss $\approx .0119$ (ignoring loading and phase angle)

43. $A_v = 11$, $Z_{in} \approx \infty\ \Omega$

Chapter 5   **1.** $A_{noise} = 6$, $f_2 = 667$ kHz

3. $A_{noise} = 6$, $f_{unity} \geq 1.5$ MHz

5. $f_{max} = 207$ kHz

7. Slew rate $= 1.26$ V/μS

9. Slew rate = 6.79 V/μS

11. Slew rate = 1.25 V/μS

13. $V_{out\text{-}offset} \approx \pm 4$ mV

15. $V_{out\text{-}ripple} = .5$ μV

17. $A_{noise} = 21$, $v_{out\text{-}noise} = 16.8$ μV, input referred noise = .84 μV

19. S/N = 80.6 dB

21. 1.25 V peak triangle at 100 kHz

23. $V_{drift} \approx \pm 7.49$ mV, $R_{off} = 952$ Ω

25. $C = 2.4$ μF

27.

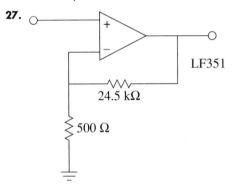

LF351

24.5 kΩ

500 Ω

29.

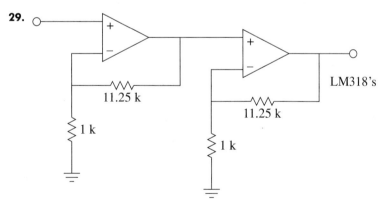

LM318's

11.25 k

1 k

11.25 k

1 k

Note: A 3-stage LF351 design is also possible.

31. $e_n = 5.68$ μV RMS

33. Total offset = ± 112.5 mV

35. Stage 1: 741, $f_2 = 1$ MHz; stage 2: 351, $f_2 = 667$ kHz; stage 3: 318, $f_2 = 750$ kHz

Chapter 6

1. Desired = 200 mV, hum = 4.5 μV

3. v_{out} = 1.69 V

5. I_{set} = 14.4 μA

7. I_{set} = 7.2 μA, A_v = 10.4, f_2 = 57.7 kHz, f_{max} = 6.37 kHz

9. $C \geq 1 \mu$F

11. P_D = .8 W

13. I_{abc} = 218 μA

15. C = 212 nF

17.

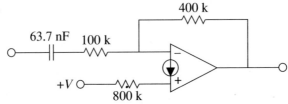

19.

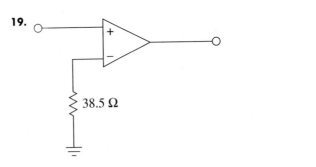

21.

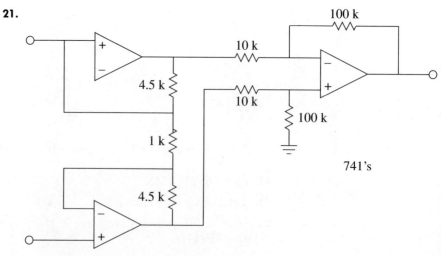

23. As outlined in Figure 6.15, but add the following between pins 1 and 8:

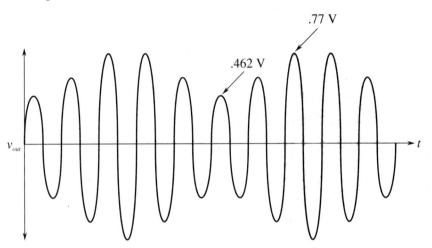

1080 Ω

25. I_{abc} = .15 mA, gm = 3 mS, net gain = .924.
The result is an amplitude-modulated 300 kHz wave, max peak = .77 V, min peak = .462 V.

.77 V

.462 V

v_{out}

t

Chapter 7   **1.**

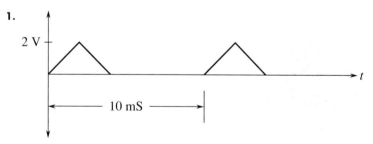

2 V

10 mS

t

3.

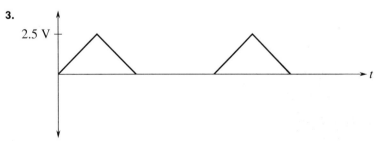

2.5 V

t

5. 1 V DC

7.

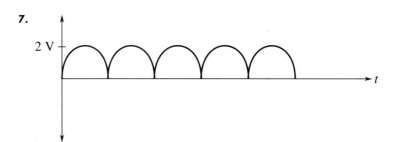

9.

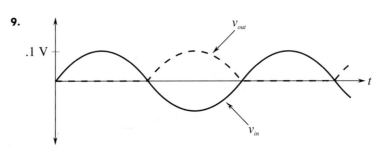

11.

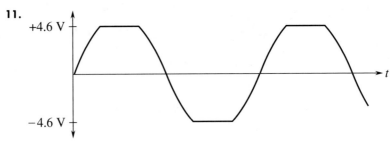

13.

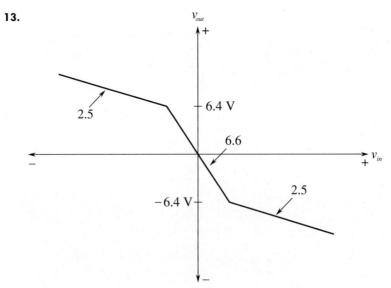

15. 8.97 V peak square wave

17.

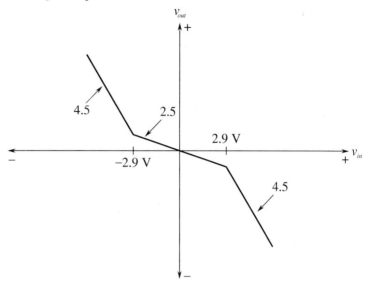

19.

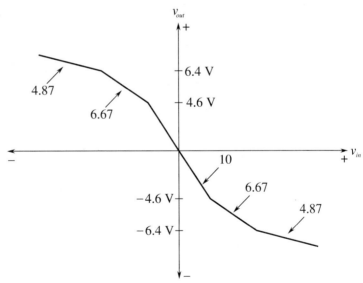

21. ± 10.4 V

23. $v_{out} = +5$ V

25. $v_{out} = -72.9$ mV

27. $.2 \sin 2\pi 1000t$, $-.4 \sin 2\pi 1000t$, $1 \sin 2\pi 1000t$

29. $-12.5 \sin 2\pi 2000t$

31. $R_L = 500$ kΩ

33. $R_f = 60$ k, $R_i = 6$ k, $R_a = 6$ k

35. $R_1 = 20$ k, $R_2 = R_6 = 5$ k, $R_3 = R_7 = 34.1$ k, $R_4 = R_8 = 2.87$ k,
$R_5 = R_9 = 13.7$ k

37. $R_i = 22.1$ k

39. $R_1 = R_2 = 1.35$ kΩ

43. Use Figures 7.41 and 7.42 as a guide.
$R_f = 10$ k, $R_1 = 10$ k, $R_2 = R_6 = 90$ k, $R_3 = R_7 = 785$ k,
$R_4 = R_8 = 72$ k, $R_5 = R_9 = 279$ k

45. Threshold $= \pm 2.88$ V

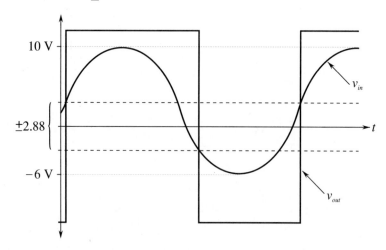

Chapter 8   **1.** $P_D = 9$ W

3. See figure below.

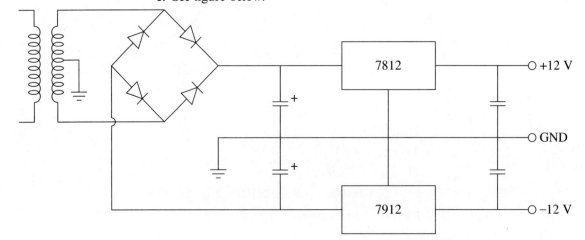

5. $P_D = 3.86$ W

7. $R_3 = 10$ k, $R_2 = 35.5$ k, $I_D = 2$ mA, $R_1 = 5.85$ k, $\beta \geq 25$

9. $R_2 = 2064\ \Omega$

11. Follow Figure 8.11b. $R_2 = 720\ \Omega$ (5 V), 2064 Ω (12 V), 2640 Ω (15 V)

13. $R_{SC} = 6.5\ \Omega$ (Use the general form of the "Basic High Voltage Regulator" in Figure 8.14)
$R_2 = 10$ k, $R_3 = 2.06$ k, $R_1 = 2.59$ k

15. $R_{pos} = 9.86\ \Omega$, $R_{neg} = 8.43\ \Omega$ (Refer to Figure 8.19)

17. $R_2 = 10$ k, $R_1 = 40$ k, $R_3 = .15\ \Omega$, $C_1 = 1$ nF, $C_3 = 20$ pF,
$L_1 = .278$ mH, $C_2 \geq 26.6\ \mu$F

19.

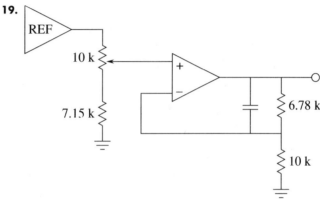

21. $P_D = 8.33$ W

Chapter 9   **1.** $f_o = 2.4$ kHz

3. $f_{o(max)} = 4081$ Hz, $f_{o(min)} = 371$ Hz

5. $f_o = 229$ Hz

7. $f_o = 9917$ Hz

9. 6583 Hz, 13.25 kHz

11. $f_o = 27.9$ kHz

13. 66.9 kHz, 14.8 kHZ

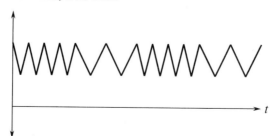

15. $V_{threshold} = 9$ V, $f_o = 10.3$ kHz

17. 13.6 Hz to 25 Hz, 136 Hz to 250 Hz, 1.36 kHz to 2.5 kHZ

19. $R_3 = R_4 = 8.47$ kΩ

21. $R_3 = R_4 = 81.2$ Ω, $P_1 = P_2 = 730.8$ Ω

23. $R_a = 1.5$ k, $R_{b_1} = 2.2$ k, $R_{b_2} = 1.8$ k (All else unchanged.)

25. $C_1 = 4.59$ nF, $C_2 = .459$ nF, $C_3 = 45.9$ pF

27. $R = 3938$ Ω

29. $C = 5.56$ nF

31. $R_2/R_3 = .37{:}1$

33. $R = 10$ kΩ, $C = 523$ pF

35. $C_1 = 1.15$ nF, $f_l = \pm 86.7$ kHz, $C_2 = 2.04$ nF

37. Increase all capacitors by a factor of 71.7.

39. Feed the sine wave into a comparator that has a variable reference.

41.

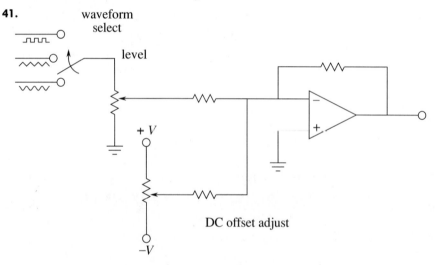

Chapter 10

1. 4.17 V peak-to-peak triangle at 1 kHz

3. $v_{out} = .663 \sin 2\pi 10{,}000t$

5. $A = -20$

7. 6.06 mV peak-to-peak triangle at 50 kHz

9. v_{out} eventually reaches -1 V.

11. 80 mV peak-to-peak square wave at 100 Hz

13. $v_{out} = 2.26 \sin 2\pi 60t$

15. $v_{out} = -.04t$

17.

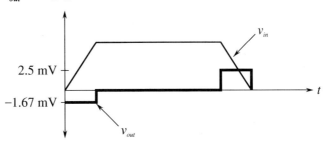

19. 150 mV peak square wave at 2.5 kHz

21. $R = 1\ \text{k}\Omega$

23. $C_f = 31.8\ \text{nF}$

25. $R_i = 50\ \Omega$, $C = 31.8\ \text{nF}$, $R_f = 3774\ \Omega$, $C_f = 422\ \text{pF}$

27. $v_{out}(t) = 6.37 \sum\limits_{n=1}^{\infty} 1/(2n-1)^2 \cos 2\pi500(2n-1)t$

29. $v_{out}(t) = -R_f/L \int v_{in}(t)\, dt$

31. $v_{out}(t) = .314 \sum\limits_{n=1}^{\infty} 1/(2n-1) \sin 2\pi500(2n-1)t$

Chapter 11

1. 500 Hz: ≈ 0; 1 kHz: 3 dB; 2 kHz: 13 dB; 4 kHz: 24 dB

3. 200 Hz: 10 dB; 500 Hz; 3 dB; 2 kHz: $\approx .2$ dB

5. Butterworth: ≈ 0 dB
Bessel: $\approx .75$ dB
1 dB Cheby: ≈ 1 dB

7. Order 2   15 dB
        3   20 dB
        4   24 dB
        5   27 dB
        6   30 dB

9. Order 2   1.75 dB
        3   2.5 dB
        4   ≈ 0 dB
        5   2.5 dB
        6   2 dB

11. Butterworth: ≈ 0 dB; Bessel: 0 dB; 1 dB Cheby: 3 dB

13. Butterworth: fourth order; 1 dB Cheby: third order; 3 dB Cheby: third order

15. BW = 1.5 kHz, f_0 = 8.72 kHz, Q = 5.81

17.

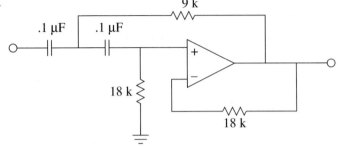

19.

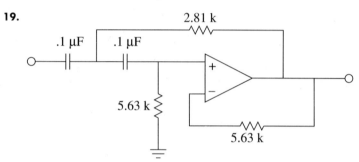

21.

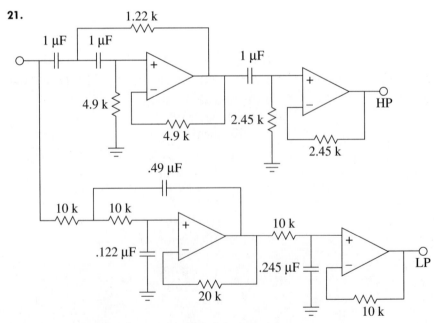

23.

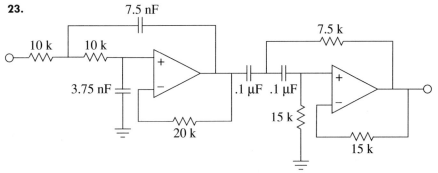

25.

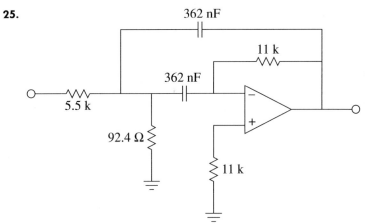

27. See figure below.

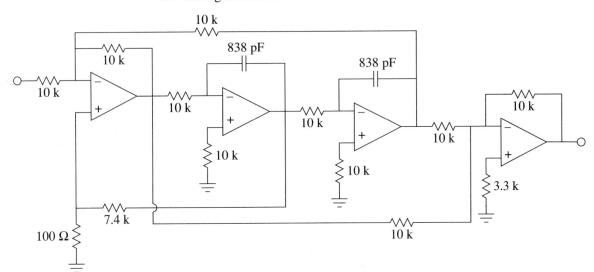

29.

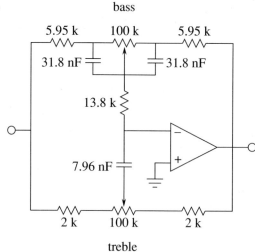

bass

5.95 k   100 k   5.95 k

31.8 nF   31.8 nF

13.8 k

7.96 nF

2 k   100 k   2 k

treble

31. Using the internal clock, $C = 11.8$ nF, $R = 100\ \Omega$ in series with a 5 kΩ potentiometer.

33.

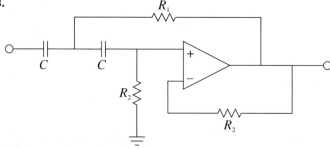

Cascade three stages:

Stage 1:	$C = .345\ \mu F$	$R_1 = 6.57$ k	$R_2 = 15.22$ k
Stage 2:	$C = .729\ \mu F$	$R_1 = 2.28$ k	$R_2 = 44$ k
Stage 3:	$C = .972\ \mu F$	$R_1 = 625\ \Omega$	$R_2 = 160$ k

—————   Chapter 12   **1.** 1024

3. 60.2 dB, 84.3 dB (for 16,384)

5. 20 μS

7. 11.1 μS

9. 20 kilobytes per second

11. 403.2 megabytes for stereo

13. $f_N = 20.8$ kHz

15. 4.09 V

17. a. .4 μS   b. 3.2 μS   c. 102.4 μS

19. a.

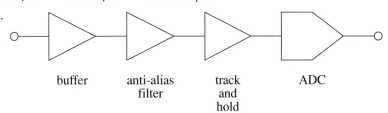

buffer   anti-alias   track   ADC
                     filter   and
                                          hold

   b. 9 bits   c. $f_s = 960$ kHz   d. $f_c = 400$ kHz
   e. subranging   f. CS5412 converter

21. In BASIC, if A( ) is the array and N is its size,

```
MAX = 0
FOR X = 1 TO N
    TEMP = ABS(A(X))
    IF TEMP > MAX THEN MAX = TEMP
NEXT X
```

Index